Elementary probability

Elementary probability

DAVID STIRZAKER

Mathematical Institute, University of Oxford

CAMBRIDGE
UNIVERSITY PRESS

Published by the Press Syndicate of the University of Cambridge
The Pitt Building, Trumpington Street, Cambridge CB2 1RP
40 West 20th Street, New York, NY 10011–4211, USA
10 Stamford Road, Oakleigh, Melbourne 3166, Australia

First published 1994
Reprinted 1995

Printed in Great Britain at the University Press, Cambridge

A catalogue record for this book is available from the British Library

Library of Congress cataloguing in publication data

Stirzaker, David.
Elementary probability / David Stirzaker.
p. cm.
Includes index.
ISBN 0-521-42028-8 (hbk). – ISBN 0-521-42183-7 (pbk)
1. Probabilities. I. Title.
A273.S7534 1994
519.2–dc20 93-12917 CIP

ISBN 0 521 42028 8 hardback
ISBN 0 521 42183 7 paperback

KT

Contents

Preface

This book provides an introduction to elementary probability and some of its simple applications. In particular, a principal purpose of the book is to help the student to solve problems. Probability is now being taught to an ever wider audience, not all of whom can be assumed to have a high level of problem-solving skills and mathematical background. It is also characteristic of probability that, even at an elementary level, few problems are entirely routine. Successful problem-solving requires flexibility and imagination on the part of the student. Commonly these skills are developed by observation of examples and practice at exercises, both of which this text aims to supply.

With these targets in mind, in each chapter of the book, the theoretical exposition is accompanied by a large number of examples, and is followed by worked examples incorporating a cluster of exercises. The examples and exercises have been chosen, firstly to illustrate the subject, secondly to help the student solve the kind of problems typical of examinations, and thirdly for their entertainment value. (For besides its practical importance, probability is without doubt one of the most entertaining branches of mathematics). Each chapter concludes with problems; solutions to many of these appear in an appendix, together with the solutions to the exercises.

The book has three main parts. The first introduces the basic ideas of probability, conditional probability, and independence. It is assumed that the reader has some knowledge of elementary set theory. (We adopt the now conventional formal definition of probability. This is not because of high principles, but merely because the alternative intuitive approach seems to lead more students into errors.) The second part introduces discrete random variables, mass functions, and expectation. It is assumed that the reader can do simple things with functions and series. The third part considers continuous random variables, and for this a knowledge of the simpler techniques of calculus is desirable.

In addition there are chapters on combinatorial methods in probability, the use of probability (and other) generating functions, and the basic theory of Markov chains in discrete and continuous time. These sections can be omitted at a first reading if so desired.

In general the material is presented in a conventional order, which roughly corresponds to increasing levels of knowledge and dexterity on the part of the reader. Those who start with a sufficient level of basic skills have more freedom to choose the order in which they read the book. For example, you might wish to read chapters 4 and 7 together, (and then 5 and 8 together), regarding discrete and continuous random variables as two varieties of the same species (which they are). Also, much of chapter 9 could be read immediately after chapter 5, if you wish.

The ends of examples, proofs, and definitions are indicated by the symbols ●, ■, and ▲, respectively.

Finally you should note that the book contains a random number of errors. I entreat readers to inform me of all those they find.

Oxford
September 1993

1
Probability

1.1 Introduction

The idea of probability underlies statements such as: 'Rain is likely today', 'This egg is probably infected with salmonella', and so on. These uses of probability can be expressed mathematically in various ways. Here, we start with the concept of an *experiment*, which is any procedure that has a well-defined set of outcomes; the set of outcomes is called the *sample space*. Subsets of the sample space are called *events*, and statements like those above typically assess how likely an event is to occur. The collection of events of interest is the *event space*, and *probability* is a function defined on each event in the event space. By convention a probability can be any number between zero and one, inclusive.

Of course the brief summary above calls for much explanation and elaboration. In the next few sections of this chapter we shall provide a few simple rules (or axioms) defining the properties of events and their probabilities. This choice of rules is guided by our experience of real events and their likelihoods, but our experience and intuition cannot *prove* that these rules are true or say what probability 'really' is. What we can say is that, starting with these rules, we can derive a theory that provides an elegant and accurate description of many random phenomena, ranging from the behaviour of queues in supermarkets to the behaviour of nuclear reactors.

1.2 Events

Suppose we are considering some experiment such as tossing a coin. To say that the experiment is well-defined means that we can list all the possible outcomes; in the case of a tossed coin the list reads: {head, tail}. For a general (unspecified) experiment any particular outcome is denoted by ω; the collection of all outcomes is called the *sample space* and is denoted by Ω.

Any specified collection of outcomes in Ω is called an *event*. Upper case letters such as A, B, C, are used to denote events; these may have suffices or other adornments such as A_i, $\bar{B}$, C^*, and so on. If the outcome of the experiment is ω, and $\omega \in A$, then A is said to *occur*. The set of outcomes not in A is called the complement of A, and is denoted by A^c.

In particular, the event which contains all possible outcomes is the *certain event*, and is denoted by Ω. Also, the event containing no outcomes is the *impossible event* and is denoted by ϕ. Obviously $\phi = \Omega^c$.

Example: Coins If a coin is tossed once, then $\Omega = \{\text{head, tail}\}$. In line with the notation above, we usually write

$$\Omega = \{H, T\}.$$

The event that the coin shows a head should strictly be denoted by $\{H\}$, but in common with most other writers we omit the braces in this case, and denote a head by H. Obviously $H^c = T$, and $T^c = H$.

Likewise, if a coin is tossed twice, then

$$\Omega = \{HH, HT, TH, TT\},$$

and so on. This experiment is performed even more often in probability textbooks than it is in real life. ●

Since events are sets we use the usual notation for combining them, thus:

$A \cap B$ denotes outcomes in both A and B, their *intersection*;
$A \cup B$ denotes outcomes in either A or B or both, their *union*;
$A \triangle B$ denotes outcomes in either A or B, but not both, their *symmetric difference*;
$A \backslash B$ denotes outcomes in A which are not in B, their *difference*;
$\bigcup_{j=1}^{\infty} A_j$ denotes outcomes which are in at least one of the countable collection $(A_j; j \geqslant 1)$, their *countable union*;
[Countable sets are in one–one correspondence with a subset of the positive integers.]
$A \subseteq B$ denotes that every outcome in A is also in B, this is *inclusion*;
$A = \{\omega_1, \omega_2, \omega_3, \dots, \omega_n\}$ denotes that the event A consists of the outcomes $\omega_1, \dots, \omega_n$;
$A \times B$ denotes the product of A and B, that is the set of all ordered pairs (ω_a, ω_b) where $\omega_a \in A$ and $\omega_b \in B$.

These methods of combining events give rise to many equivalent ways of denoting an event. Some of the more useful identities for any events A and B are:

(1) $$A \triangle B = (A \cap B^c) \cup (A^c \cap B)$$

(2) $$A = (A \cap B) \cup (A \cap B^c)$$

(3) $$A \backslash B = A \cap B^c$$

(4) $$A^c = \Omega \backslash A$$

(5) $$A \cap A^c = \phi$$

(6) $$A \cup A^c = \Omega.$$

These are all easily verified by checking that every element of the left hand side is included in the right hand side, and vice versa. You should do this.

Such relationships are often very conveniently represented by simple diagrams.

We illustrate this by providing some basic examples in Figures 1.1 and 1.2. Similar relationships hold between combinations of three or more events and some of these are given in the problems at the end of this chapter.

When $A \cap B = \phi$ we say that A and B are disjoint (or mutually exclusive).

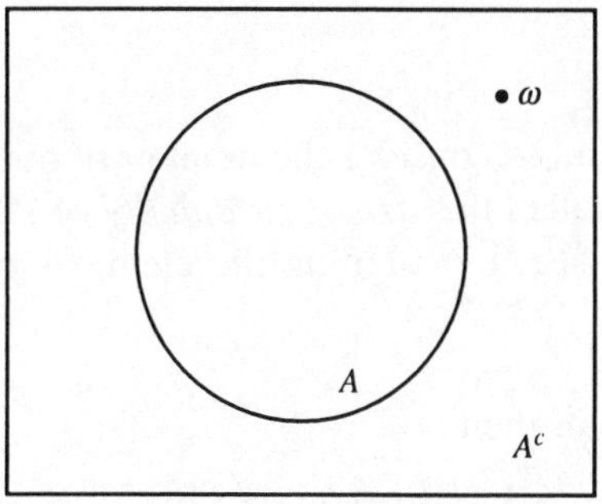

Figure 1.1 The interior of the rectangle represents the sample space Ω, and the interior of the circle represents an event A. The point ω represents an outcome in the event A^c. The diagram clearly illustrates the identities $A^c \cup A = \Omega$ and $\Omega \backslash A = A^c$.

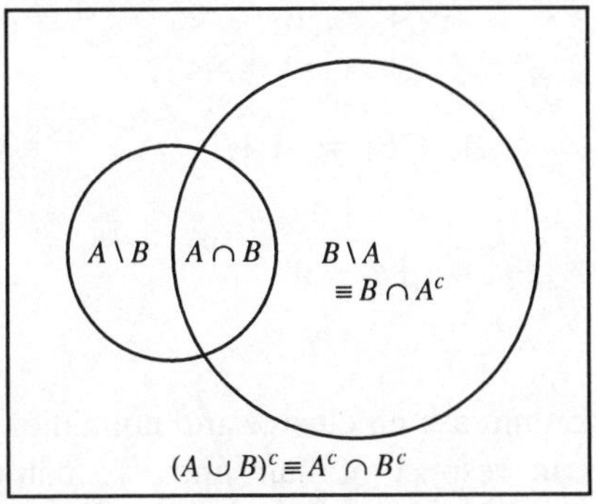

Figure 1.2 The interior of the smaller circle represents the event A; the interior of the larger circle represents the event B. The diagram illustrates a number of simple relationships; for example, the region common to both circles is $A \cap B \equiv (A^c \cup B^c)^c$. For another example, observe that $A \triangle B = A^c \triangle B^c$.

(7) **Example** A die* is rolled. The outcome is one of the integers from 1 to 6. We may denote these by $\{\omega_1, \omega_2, \omega_3, \omega_4, \omega_5, \omega_6\}$, or more directly by $\{1, 2, 3, 4, 5, 6\}$, as we choose. Define:

A the event that the outcome is even,
B the event that the outcome is odd,
C the event that the outcome is prime,
D the event that the outcome is perfect (a perfect number is the sum of its prime factors).

*Unless otherwise specified, a die is always a cube bearing the numbers 1, 2, 3, 4, 5 and 6 on its faces, one to each face.

Then the above notation compactly expresses obvious statements about these events. For example:

$$A \cap B = \phi \qquad A \cup B = \Omega$$
$$A \cap D = \{\omega_6\} \qquad C\backslash A = B\backslash\{\omega_1\}$$

and so on. ●

It is natural and often useful to consider the number of outcomes in an event A. This is denoted by $|A|$, and is called the *size* or *cardinality* of A.

It is straightforward to see, by counting the elements on each side, that size has the following properties.

If A and B are disjoint then

(8) $$|A \cup B| = |A| + |B|,$$

and more generally, for any A and B

(9) $$|A \cup B| + |A \cap B| = |A| + |B|.$$

If $A \subseteq B$ then

(10) $$|A| \leqslant |B|.$$

For the product $A \times B$,

(11) $$|A \times B| = |A||B|.$$

Finally

(12) $$|\phi| = 0.$$

(13) Example The Shelmikedmu are an elusive and nomadic tribe whose members are unusually heterogeneous in respect of hair and eye colour, and skull shape. A persistent anthropologist establishes the following facts:

(i) 75% have dark hair, the rest have fair hair;
(ii) 80% have brown eyes, the rest have blue eyes;
(iii) no narrow-headed person has fair hair and blue eyes;
(iv) the proportion of blue-eyed broad-headed tribespeople is the same as the proportion of blue-eyed narrow-headed tribespeople;
(v) those who are blue-eyed and broad-headed are fair-haired or dark-haired in equal proportion;
(vi) half the tribe is dark-haired and broad-headed;
(vii) the proportion who are brown-eyed, fair-haired and broad-headed is equal to the proportion who are brown eyed, dark-haired and narrow-headed.

The anthropologist also finds n, the proportion of the tribe who are narrow-headed, but unfortunately this information is lost in a clash with a crocodile on the difficult journey home. Is another research grant and field trip required to find n? Fortunately not, if the anthropologist uses set theory. Let

B be the set of those with blue eyes
C the set of those with narrow heads
D the set of those with dark hair.

Then the division of the tribe into its heterogeneous sets can be represented by Figure 1.3. This type of representation of sets and their relationships is known as a *Venn diagram*. The proportion of the population in each set is denoted by the lower case letter in each compartment, so

$$a = |B^c \cap C^c \cap D^c|/|\Omega|,$$
$$b = |B \cap C^c \cap D^c|/|\Omega|,$$

and so on. The required proportion having narrow heads is

$$n = |C|/|\Omega| = e + f + g + h$$

and of course $a + b + c + d + e + f + g + h = 1$. The information in (i)–(vii), which survived the crocodile, yields the following relationships:

(i) $$c + d + e + f = 0.75$$
(ii) $$a + d + e + h = 0.8$$
(iii) $$g = 0$$
(iv) $$f + g = b + c$$
(v) $$b = c$$
(vi) $$c + d = 0.5$$
(vii) $$a = e$$

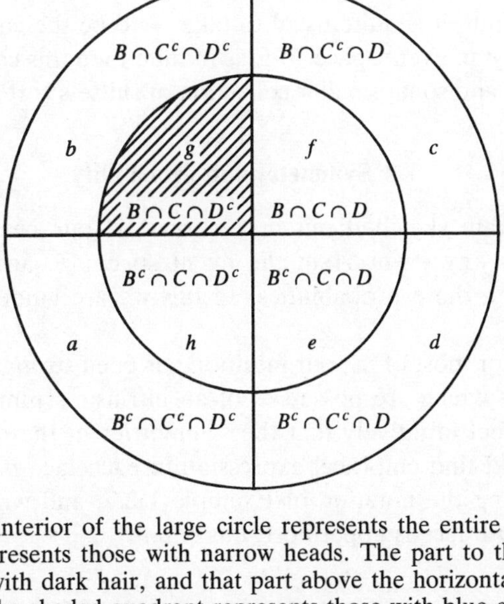

Figure 1.3 Here the interior of the large circle represents the entire tribe, and the interior of the small circle represents those with narrow heads. The part to the right of the vertical line represents those with dark hair, and that part above the horizontal line represents those with blue eyes. Thus the shaded quadrant represents those with blue eyes, narrow heads and fair hair, and as it happens this set is empty by (iii). That is to say $B \cap C \cap D^c = \phi$, and so $g = 0$.

The anthropologist (who has a pretty competent knowledge of algebra) solves this set of equations to find that

$$n = e + f + g + h = e + f + h = 0.15 + 0.1 + 0.05 = 0.3$$

Thus three tenths of the tribe are narrow-headed. ●

This section concludes with a technical note (which you may omit on a first reading). We have noted that events are subsets of Ω. A natural question is, which subsets of Ω are entitled to be called events?

It seems obvious that if A and B are events then all of $A \cup B$, A^c, $A \cap B$, and so on should be entitled to be events also. This is a bit vague, to be precise, we say that a subset A of Ω can be an event if it belongs to a collection $\mathcal{F}$ of subsets of Ω obeying the following three rules:

(14) $$\Omega \in \mathcal{F};$$

(15) $$\text{if } A \in \mathcal{F} \text{ then } A^c \in \mathcal{F};$$

(16) $$\text{if } A_j \in \mathcal{F} \text{ for } j \geq 1, \text{ then } \bigcup_{j=1}^{\infty} A_j \in \mathcal{F}.$$

The collection $\mathcal{F}$ is called an event space, or a σ-field.

Notice that using (1)–(6) shows that if A and B are in $\mathcal{F}$ then so are $A \backslash B$, $A \triangle B$ and $A \cap B$.

(17) Example (7) revisited It is easy for you to check that $\{\phi, A, B, \Omega\}$ is an event space, and $\{\phi, A \cup C, B \backslash C, \Omega\}$ is an event space. However, $\{\phi, A, \Omega\}$ and $\{\phi, A, B, D, \Omega\}$ are *not* event spaces. ●

In general, if Ω is finite it is quite usual to take $\mathcal{F}$ to be the collection of all subsets of Ω, which is clearly an event space. If Ω is infinite then this collection is sometimes too big to be useful, and some smaller collection of subsets is required.

1.3 Symmetry and probability

Following the program sketched out in Section 1.1, our next task is to assign a probability $\mathbf{P}(A)$ to any event A in the event space $\mathcal{F}$, and then to define the relationships between these probabilities. In this we are guided by experience and intuition.

Since childhood, for most of us, our intuition has been strongly guided by a special class of experiments which are possessed of an intrinsic symmetry. When rolling a well-made die, we feel intuitively that the symmetries of the die (under reflections and rotations) should find empirical expression in each face being equally likely to end uppermost. Using the notation of Example 1.2.7, and writing $\mathbf{P}(\{\omega_j\})$ for the probability that the jth face is uppermost, this entails

$$\mathbf{P}(\{\omega_i\}) = \mathbf{P}(\{\omega_j\}),$$

$$\mathbf{P}(\{\omega_1, \omega_2\}) = \mathbf{P}(\{\omega_3, \omega_4\}),$$

and so on.

Recalling the convention that probabilities lie between zero and one, it is tempting and natural, given an event A, to set

(1) $$\mathbf{P}(A) = \frac{|A|}{|\Omega|},$$

so that for any $A \subseteq \Omega$, we do indeed have

(2) $$0 \leqslant \mathbf{P}(A) \leqslant 1.$$

(3) Example (1.2.7) revisited With the probability of any event now defined by (1), it is elementary to find the probability of any of the events that may occur when we roll a die. The probability that it shows an even number is

$$\mathbf{P}(A) = \mathbf{P}(\{\omega_2, \omega_4, \omega_6\}) = \frac{|A|}{|\Omega|} = \frac{3}{6} = \frac{1}{2}.$$

Likewise, and equally trivially, we find that

$$\mathbf{P}(\text{odd}) = \mathbf{P}(B) = \frac{1}{2}$$

$$\mathbf{P}(\text{prime}) = \mathbf{P}(C) = \mathbf{P}(\{\omega_2, \omega_3, \omega_5\}) = \frac{1}{2}$$

$$\mathbf{P}(\text{perfect}) = \mathbf{P}(D) = \mathbf{P}(\{\omega_6\}) = \frac{1}{6}.$$

These values of the probabilities are not inconsistent with our ideas about how the symmetries of this die should express themselves when it is rolled. ●

Recalling the properties of the size $|A|$ of an event A, listed in (1.2.8), (1.2.9), (1.2.10) and (1.2.12), we see that the function $\mathbf{P}(.)$ defined by (1) has the following properties.

For any events A and B

(4) $$\mathbf{P}(A \cup B) + \mathbf{P}(A \cap B) = \mathbf{P}(A) + \mathbf{P}(B).$$

In particular, if A and B are disjoint then

$$\mathbf{P}(A \cup B) = \mathbf{P}(A) + \mathbf{P}(B).$$

More generally, if $A_1, A_2, \ldots, A_n$ form a collection of disjoint events (which is to say that $A_i \cap A_j = \phi$ when $i \neq j$), then

(5) $$\mathbf{P}\left(\bigcup_{i=1}^{n} A_i\right) = \sum_{i=1}^{n} \mathbf{P}(A_i).$$

This property (5) is known as *finite additivity*.

If $A \subseteq B$, then

(6) $$\mathbf{P}(A) \leqslant \mathbf{P}(B)$$

and finally

(7) $$\mathbf{P}(\phi) = 0.$$

Once again, these statements are quite consistent with our intuition about likelihoods.

Historically the theory of probability has its roots firmly based in just this kind of speculation about games of chance employing cards, dice and lotteries.

(8) **Example** Three dice are rolled and the numbers on the upper faces are added together. The outcomes 9 and 10 can each be obtained in six distinct ways, thus:

$$10 = 1 + 3 + 6 = 1 + 4 + 5 = 2 + 2 + 6 = 2 + 3 + 5 = 2 + 4 + 4 = 3 + 3 + 4$$

$$9 = 1 + 2 + 6 = 1 + 3 + 5 = 1 + 4 + 4 = 2 + 2 + 5 = 2 + 3 + 4 = 3 + 3 + 3.$$

Some time before 1642, Galileo was asked to explain why, despite this, the outcome 10 is more likely than the outcome 9, as shown by repeated experiment. He observed that the sample space Ω has in fact $6^3 = 216$ outcomes, being all possible triples of numbers from 1 to 6. Of these, 27 sum to 10, and 25 sum to 9, so $\mathbf{P}(10) = \frac{27}{216}$ and $\mathbf{P}(9) = \frac{25}{216}$.

This provides an explanation for the preponderance of 10 over 9. ●

It is just this kind of agreement between theory and experiment that justifies our rather arbitrary adoption of rules (4)–(7) above. We shall see many more examples of this.

1.4 Properties of probability

Most experiments do not possess strong symmetry, nor is it often plausible on other grounds that all outcomes should be equally likely. We are nevertheless still faced with the problem of defining a probability function $\mathbf{P}(.): \mathcal{F} \to [0, 1]$ so that $\mathbf{P}(A)$ tells us how likely the event A is to occur.

Here our intuition doesn't tell us much. However, it seems unthinkable that our function, when we produce it, should not be relevant when we roll a die. It is thus natural and convenient, and as you will discover very rewarding, to make the following definition, guided by the discussion of Section 1.3.

Definition The function $\mathbf{P}(.): \mathcal{F} \to [0, 1]$ is a *probability function* if

(1) $$\mathbf{P}(A) \geqslant 0 \text{ for all } A \in \mathcal{F},$$

(2) $$\mathbf{P}(\Omega) = 1$$

and

(3) $$\mathbf{P}\left(\bigcup_{j=1}^{\infty} A_j\right) = \sum_{j=1}^{\infty} \mathbf{P}(A_j)$$

whenever $A_1, A_2, \ldots$ are disjoint events (which is to say that $A_i \cap A_j = \phi$ whenever $i \neq j$). ▲

In passing, we note that (3) is known as the property of *countable additivity*. Obviously it implies *finite additivity* so that, in particular, if $A \cap B = \phi$, then $\mathbf{P}(A \cup B) = \mathbf{P}(A) + \mathbf{P}(B)$.

From these three rules we can derive many important and useful relationships, for example:

(4) $$\mathbf{P}(\phi) = 0,$$

(5) $$\mathbf{P}(A^c) = 1 - \mathbf{P}(A),$$

(6) $$\mathbf{P}(A \backslash B) = \mathbf{P}(A) - \mathbf{P}(A \cap B),$$

(7) $$\mathbf{P}(A \cup B) = \mathbf{P}(A) + \mathbf{P}(B) - \mathbf{P}(A \cap B),$$

(8) $$\mathbf{P}\left(\bigcup_1^n A_i\right) = \sum_1^n \mathbf{P}(A_i) - \sum_{i<j} \mathbf{P}(A_i \cap A_j) + \sum_{i<j<k} \mathbf{P}(A_i \cap A_j \cap A_k) + \ldots + (-)^{n+1}\mathbf{P}(A_1 \cap A_2 \cap \ldots \cap A_n).$$

The following examples begin to demonstrate the importance and utility of (1)–(8).

(9) Example Let us prove (4), (5) and (7) above. First, by (2) and (3), for any $A \in \mathcal{F}$,

$$1 = \mathbf{P}(\Omega) = \mathbf{P}(A \cup A^c) = \mathbf{P}(A) + \mathbf{P}(A^c)$$

which proves (5). Now setting $A = \Omega$ establishes (4). Finally, using (3) repeatedly we obtain

(10) $$\mathbf{P}(B) = \mathbf{P}(B \cap A) + \mathbf{P}(B \cap A^c),$$

and

$$\mathbf{P}(A \cup B) = \mathbf{P}(A \cup (B \cap A^c)) = \mathbf{P}(A) + \mathbf{P}(B \cap A^c) = \mathbf{P}(A) + \mathbf{P}(B) - \mathbf{P}(B \cap A) \quad \text{by (1)},$$

which proves (7). ●

You should now prove (6) as an elementary exercise; the proof of (8) is part of Problem 12.

(11) Example: Inequalities for P(.) (i) If $A \subseteq B$ then $B \cap A = A$, so from (10)

$$\mathbf{P}(B) = \mathbf{P}(A) + \mathbf{P}(B \cap A^c) \geqslant \mathbf{P}(A) \quad \text{by (1)}.$$

(ii) For any A, B we have *Boole's inequalities*

(12)
$$\begin{aligned}
\mathbf{P}(A) + \mathbf{P}(B) &\geqslant \mathbf{P}(A \cup B) && \text{by (7)} \\
&\geqslant \max\{\mathbf{P}(A), \mathbf{P}(B)\} && \text{by part (i)} \\
&\geqslant \mathbf{P}(A \cap B) && \text{by part (i) again} \\
&\geqslant \mathbf{P}(A) + \mathbf{P}(B) - 1 && \text{by (7) again.}
\end{aligned}$$

●

(13) Example: Lottery An urn contains 1000 lottery tickets numbered from 1 to 1000. One is selected at random. A fairground performer offers to pay \$3 to anyone who has already paid him \$2, if the number on the ticket is divisible by 2, 3 or 5. Would you pay him your \$2 before the draw? (If the ticket number is not divisible by 2, 3 or 5 you lose your \$2.)

Solution Let D_k be the event that the number drawn is divisible by k. Then

$$\mathbf{P}(D_2) = \frac{500}{1000} = \frac{1}{2}$$

and so on. Also

$$\mathbf{P}(D_2 \cap D_3) = \mathbf{P}(D_6) = \frac{166}{1000}$$

and so forth. Using (8) with $n = 3$, and making several similar calculations we have

$$\begin{aligned}\mathbf{P}(D_2 \cup D_3 \cup D_5) &= \mathbf{P}(D_2) + \mathbf{P}(D_3) + \mathbf{P}(D_5) + \mathbf{P}(D_2 \cap D_3 \cap D_5) \\ &\quad -\mathbf{P}(D_2 \cap D_3) - \mathbf{P}(D_3 \cap D_5) - \mathbf{P}(D_2 \cap D_5) \\ &= 10^{-3}(500 + 333 + 200 + 33 - 166 - 66 - 100) = \frac{367}{500}.\end{aligned}$$

The odds on winning are thus better than 2:1, and you should accept his very generous offer. ●

1.5 Sequences of events

This section is important, but may be omitted at a first reading.

Very often we will be confronted by an infinite sequence of events $(A_n; n \geqslant 1)$ such that $A = \lim_{n\to\infty} A_n$ exists. In particular, if $A_n \subseteq A_{n+1}$ for all n then

(1) $$\lim_{n\to\infty} A_n = \bigcup_{j=1}^{\infty} A_j = A,$$

and A is an event by (1.2.16). It is of interest, and also often useful, to know $\mathbf{P}(A)$. The following theorem is therefore important as well as attractive.

(2) Theorem If $A_n \subseteq A_{n+1} \in \mathcal{F}$ for all $n \geqslant 1$, and $A = \lim_{n\to\infty} A_n$, then

$$\mathbf{P}(A) = \lim_{n\to\infty} \mathbf{P}(A_n).$$

Proof Since $A_n \subseteq A_{n+1}$, we have $(A_{j+1}\backslash A_j) \cap (A_{k+1}\backslash A_k) = \phi$, for $k \neq j$. Also, setting $A_0 = \phi$

$$\bigcup_{1}^{n}(A_j\backslash A_{j-1}) = A_n.$$

Furthermore,

(3) $$\mathbf{P}(A_{n+1}\backslash A_n) = \mathbf{P}(A_{n+1}) - \mathbf{P}(A_n).$$

Hence, since $A_0 = \phi$

$$\begin{aligned}\mathbf{P}(A) = \mathbf{P}(\lim_{n\to\infty} A_n) &= \mathbf{P}\Big(\bigcup_{j=0}^{\infty} A_{j+1}\backslash A_j\Big) \\ &= \sum_{j=0}^{\infty} \mathbf{P}(A_{j+1}\backslash A_j) && \text{by (1.4.3)} \\ &= \lim_{n\to\infty} \mathbf{P}(A_n) && \text{by (3).} \quad ■\end{aligned}$$

From this result it is a simple matter to deduce that if $A_n \supseteq A_{n+1}$ for all n then

(4) $$\lim_{n\to\infty} \mathbf{P}(A_n) = \mathbf{P}(A).$$

With a bit more work one can show more generally that if $\lim_{n\to\infty} A_n = A$, then (4) is still true. Because of this, the function $\mathbf{P}(.)$ is said to be a *continuous set function*.

1.6 Remarks

Simple problems in probability typically require the calculation of the probability $\mathbf{P}(E)$ of some event E, or at least the calculation of bounds for $\mathbf{P}(E)$. The underlying experiment may be implicit or explicit; in any case the first step is always to choose the sample space. (Naturally, we choose the one which makes finding $\mathbf{P}(E)$ easiest.)

Then you may set about finding $\mathbf{P}(E)$ by using the rules and results of Section 1.4, and the usual methods for manipulating sets. Useful aids at this simple level include Venn diagrams, and identities such as de Morgan's laws, namely:

(1) $$\left(\bigcup_{i=1}^{n} A_i\right)^c = \bigcap_{i=1}^{n} A_i^c$$

and

(2) $$\left(\bigcap_{i=1}^{n} A_i\right)^c = \bigcup_{i=1}^{n} A_i^c.$$

These are readily established directly, or by induction as follows. First, draw a Venn diagram to see that $(A \cup B)^c = A^c \cap B^c$.

Now if $\bigcap_{j=1}^{n} A_j = B_n$ then

$$\left(\bigcap_{j=1}^{n+1} A_j\right)^c = (A_{n+1} \cup B_n)^c = A_{n+1}^c \cap B_n^c.$$

Hence (1) follows by induction on n. The result (2) can also be proved directly, or by induction, or by using (1). You can do this, and you should do it now.

We end this section with a note for the more demanding reader. It has been claimed above that probability theory is of wide applicability, yet most of the examples in this chapter deal with the behaviour of coins, dice, lotteries and urns. For most of us, weeks or even months can pass without any involvement with dice or urns. (The author has never even seen an urn, let alone removed a red ball from one.) The point is that such simple problems present their probabilistic features to the reader unencumbered by strained assumptions about implausible models of reality. The penalty for simplicity is that popular problems may become hackneyed through overuse in textbooks. We take this risk, but reassure the reader that more realistic problems feature largely in later chapters. In addition, when considering some die or urn, students may be pleased to know that they are treading in the footsteps of many very eminent mathematicians as they perform these calculations. Euler or Laplace may have pondered over exactly the same difficulty as you, though not for so long perhaps.

WORKED EXAMPLES AND EXERCISES

1.7 Example: Dice

You roll two dice. What is the probability of the events:

(a) they show the same?
(b) their sum is seven or eleven?
(c) they have no common factor greater than unity?

Solution First we must choose the sample space. A natural representation is as ordered pairs of numbers (i, j), where each number refers to the face shown by one of the dice. We require the dice to be distinguishable (one red and one green say), so that $1 \le i \le 6$ and $1 \le j \le 6$. Because of the symmetry of a perfect die, we assume that these 36 outcomes are equally likely.

(a) Of these 36 outcomes, just 6 are of the form (i, i), so using (1.3.1) the required probability is

$$\frac{|A|}{|\Omega|} = \frac{6}{36} = \frac{1}{6}.$$

(b) There are 6 outcomes of the form $(i, 7 - i)$, whose sum is 7, so the probability that the sum is 7 is $\frac{6}{36} = \frac{1}{6}$.

There are 2 outcomes whose sum is 11, namely (5, 6) and (6, 5), so the probability that the sum is 11 is $\frac{2}{36} = \frac{1}{18}$.

Hence, using (1.4.3), the required probability is $\frac{1}{6} + \frac{1}{18} = \frac{2}{9}$.

(c) It is routine to list the outcomes which do have a common factor greater than unity. They are thirteen in number, namely:

$$\{(i, i); i \ge 2\}, (2, 4), (4, 2), (2, 6), (6, 2), (3, 6), (6, 3), (4, 6), (6, 4).$$

This is the complementary event, so by (1.4.5), the required probability is

$$1 - \frac{13}{36} = \frac{23}{36}.$$

Doing it this way gives a slightly quicker enumeration than the direct approach.

(1) **Exercise** What is the probability that the sum of the numbers is 2, 3, or 12?

(2) **Exercise** What is the probability that:
(a) the sum is odd?
(b) the difference is odd?
(c) the product is odd?

(3) **Exercise** What is the probability that one number divides the other?

(4) **Exercise** What is the probability that the first die shows a smaller number than the second?

(5) **Exercise** What is the probability that different numbers are shown and the smaller of the two numbers is r, $1 \le r \le 6$?

Remark It was important to distinguish the two dice. Had we not done so, the sample space would have been $\{(i, j); 1 \le i \le j \le 6\}$, and these 21 outcomes are not equally likely, either intuitively or empirically. Note that the dice need not be different colours in fact, it is enough for us to be able to suppose that they are.

1.8 Example: Urn

An urn contains n heliotrope and n tangerine balls. Two balls are removed from the urn together, at random.

(a) What is the sample space?
(b) What is the probability of drawing two balls of different colours?
(c) Find the probability p_n that the balls are the same colour, and evaluate $\lim_{n\to\infty} p_n$.

Solution (a) As balls of the same colour are otherwise indistinguishable, and they are drawn together, a natural sample space is $\Omega = \{HH, HT, TT\}$.

(b) The outcomes in Ω exhibit no symmetry. Taking our cue from the previous example, we choose to distinguish the balls by numbering them from 1 to $2n$, and also suppose that they are drawn successively. Then the sample space is the collection of ordered pairs of the form (i, j) where $1 \leqslant i, j \leqslant 2n$, and $i \neq j$, because we cannot pick the same ball twice. These $2n(2n-1)$ outcomes are equally likely, by symmetry. In n^2 of them we draw heliotrope followed by tangerine, and in n^2 we draw tangerine followed by heliotrope. Hence

$$\mathbf{P}(HT) = \frac{n^2 + n^2}{2n(2n-1)} = \frac{n}{2n-1}.$$

(c) By (1.4.5)

$$p_n = 1 - \mathbf{P}(HT) = \frac{n-1}{2n-1}$$
$$\to \frac{1}{2} \quad \text{as } n \to \infty.$$

(1) Exercise Find $\mathbf{P}(HH)$ when the sample space is taken to be all unordered pairs of distinguishable balls.

(2) Exercise What is the probability that (a) the first ball is tangerine? (b) the second ball is tangerine?

(3) Exercise Half the balls are removed and placed in a box. One of those remaining in the urn is removed. What is the probability that it is tangerine?

(4) Exercise A fair die with n sides is rolled. If the rth face is shown, r balls are removed from the urn and placed in a bag. What is the probability that a ball removed at random from the bag is tangerine?

1.9 Example: Cups and saucers

A teaset has four cups and saucers with two cups and saucers in each of two different colours. If the cups are placed at random on the saucers, what is the probability that no cup is on a saucer of the same colour?

***Solution* I** Call the colours azure and blue; let A be the event that an azure cup is on an azure saucer, and B the event that a blue cup is on a blue saucer. Since there are only two places for blue cups not to be on blue saucers, we see that A occurs if and only if B occurs, so $A = B$ and

$$\mathbf{P}((A \cup B)^c) = \mathbf{P}(A^c) = 1 - \mathbf{P}(A).$$

Now $\mathbf{P}(A) = \mathbf{P}(A_1 \cup A_2)$, where A_1 and A_2 denote the events that the first and second azure cups respectively are on azure saucers. There are 24 equally probable ways of putting the four cups on the four saucers. In 12 of them A_1 occurs; in 12 of them A_2 occurs; and in 4 of them $A_1 \cap A_2$ occurs, by enumeration.

Hence, by (1.4.7),

$$\mathbf{P}(A) = \frac{12}{24} + \frac{12}{24} - \frac{4}{24} = \frac{5}{6},$$

and the required probability is $\frac{1}{6}$.

***Solution* II** Alternatively, instead of considering all the ways of placing cups on saucers, we may consider only the distinct ways of arranging the cups by colour with the saucers fixed. There are only six of these, namely:

$$aabb;\ abba;\ abab;\ baab;\ baba;\ bbaa;$$

and by symmetry they are equally likely. By inspection, in only one of these arrangements is no cup on a saucer of the same colour, so the required probability is $\frac{1}{6}$.

Remark In this example, considering a *smaller* sample space makes the problem easier. This is in contrast to our solutions to Examples 1.7 and 1.8 where we used larger sample spaces to simplify things.

(1) **Exercise** What is the probability that exactly
(a) one cup is on a saucer of the same colour?
(b) two cups are on saucers of the same colour?

(2) **Exercise** What is the probability that no cup is on a saucer of the same colour if the set comprises four cups and saucers in four distinct colours?

1.10 Example: Sixes

Three players, Achilles, Briseis and Chryseis, take it in turns to roll a die in the order ABC, ABC, A ... Each player drops out of the game immediately upon throwing a six.

(a) What is the sample space for this experiment?
(b) Suppose the game stops when two players have rolled a six. What is the sample space for this experiment?
(c) What is the probability that Achilles is the second player to roll a six?
(d) Let D_n be the event that the third player to roll a six does so on the nth roll. Describe the event E given by

$$E = \left(\bigcup_{n=1}^{\infty} D_n\right)^c.$$

Solution (a) Let U be the collection of all sequences $x_1, \ldots, x_n$ for all $n \geqslant 1$, such that

$$x_j \in \{1, 2, 3, 4, 5\} \quad \text{for } 1 \leqslant j < n,$$
$$x_n = 6.$$

Then each player's rolls generate such a sequence, and the sample space consists of all selections of the triple (u_1, u_2, u_3) where $u_i \in U$ for $1 \leqslant i \leqslant 3$. We may denote this by $U \times U \times U$, or even U^3 if we wish.

(b) Let V be the collection of all sequences $x_1, x_2, \ldots, x_n$, for $n \geqslant 1$, such that

$$x_j \in \{1, 2, 3, 4, 5\} \quad \text{for } 1 \leqslant j \leqslant n.$$

Then the sample space consists of two selections u_1, u_2 from U, corresponding to the players who roll sixes, and one selection v from V corresponding to the player who does not. The length of v equals the longer of u_1 and u_2 if this turn in the round comes before the second player to get a six, or it is one less than the longer of u_1 and u_2 if this turn in the round is later than the second player to get a six.

(c) Despite the answers to (a) and (b) we use a different sample space to answer this question. Suppose that the player who is first to roll a six continues to roll the die when the turn comes round, these rolls being ignored by the others. This does not affect the respective chances of the other two to be the next player (of these two) to roll a six. We therefore let Ω be the sample space consisting of all sequences of length $3r + 1$, for $r \geqslant 0$, using the integers 1, 2, 3, 4, 5 or 6. This represents $3r + 1$ rolls of the die, and by the assumed symmetry the 6^{3r+1} possible outcomes are all equally likely for each r.

Suppose Achilles is the second player to roll a six on the $3r + 1$th roll. Then his $r + 1$ rolls include no six except his last roll; this can occur in 5^r ways. If Briseis was first to roll a six, then her r rolls include at least one six; this may be accomplished in $6^r - 5^r$ ways. In this case Chryseis rolled no six in r attempts; this can be done in 5^r ways. Hence Achilles is second to Briseis in $5^r.5^r.(6^r - 5^r)$ outcomes. Likewise he is second to Chryseis in $5^r.5^r.(6^r - 5^r)$ outcomes. Hence the probability that he is second to roll a six on the $3r + 1$th roll is

$$p_r = \frac{2(6^r - 5^r)5^{2r}}{6^{3r+1}}, \quad \text{for } r \geqslant 1.$$

By (1.4.3) therefore, the total probability that Achilles is second to roll a six is the sum of these, namely

$$\sum_{r=1}^{\infty} p_r = \frac{300}{1001}.$$

(d) The event $\bigcup_{n=1}^{\infty} D_n$ is the event that the game stops at the nth roll for some $n \geqslant 1$. Therefore E is the event that they never stop.

(1) Exercise For each of the three players, find the probability that he or she is the first to roll a six.

(2) Exercise Show that $\mathbf{P}(E) = 0$.

(3) Exercise Find the probability that the Achilles rolls a six before Briseis rolls a six. [**Hint**: use a *smaller* sample space.]

(4) Exercise Show that the probability that Achilles is last to throw a six is $\frac{305}{1001}$. Are you surprised that he is more likely to be last than to be second?

Remark The interesting thing about the solution to (c) is that the sample space Ω includes outcomes that are not in the original experiment, whose sample space is described in (a). The point is that the event in question has the same probability in the original experiment and in the modified experiment, but the required probability is obtained rather more easily in the second case because the outcomes are equally likely. This idea of augmenting the sample space was first used by Fermat in the seventeenth century.

In fact we shall find easier methods for evaluating this probability in Chapter 2, using new concepts.

1.11 Example: Family planning

A woman planning her family considers the following schemes, on the assumption that boys and girls are equally likely at each delivery:

(a) have three children;
(b) bear children until the first girl is born or until three are born, whichever is sooner, and then stop;
(c) bear children until there is one of each sex or until there are three, whichever is sooner and then stop.

Let B_i denote the event that i boys are born, and let C denote the event that more girls are born than boys. Find $\mathbf{P}(B_1)$ and $\mathbf{P}(C)$ in each of the cases (a) and (b).

Solution (a) If we do not consider order there are four distinct possible families: BBB; GGG; GGB; BBG, but these are not equally likely. With order included, there are eight possible families in this larger sample space:

(1) $$\{BBB;\ BBG;\ BGB;\ GBB;\ GGB;\ GBG;\ BGG;\ GGG\} = \Omega$$

and by symmetry they are equally likely. Now, by (1.3.1), $\mathbf{P}(B_1) = \frac{3}{8}$ and $\mathbf{P}(C) = \frac{1}{2}$. The fact that $\mathbf{P}(C) = \frac{1}{2}$ is also clear by symmetry.

Now consider (b). There are four possible families: $F_1 = G$, $F_2 = BG$, $F_3 = BBG$, and $F_4 = BBB$.

Once again these outcomes are not equally likely, but as we have now done several times we can use a different sample space. One way is to use the sample space in (1), remembering that if we do this then some of the later births are fictitious. The advantage is that outcomes are equally likely by symmetry. With this choice F_2 corresponds to $\{BGG \cup BGB\}$ and so $\mathbf{P}(B_1) = \mathbf{P}(F_2) = \frac{1}{4}$. Likewise $F_1 = \{GGG \cup GGB \cup GBG \cup GBB\}$ and so $\mathbf{P}(C) = \frac{1}{2}$.

(2) **Exercise** Find $\mathbf{P}(B_1)$ and $\mathbf{P}(C)$ in case (c).
(3) **Exercise** Find $\mathbf{P}(B_2)$ and $\mathbf{P}(B_3)$ in all three cases.
(4) **Exercise** Let E be the event that the completed family contains equal numbers of boys and girls. Find $\mathbf{P}(E)$ in all three cases.

1.12 Example: Craps

You roll two fair dice. If the sum of the numbers shown is 7 or 11 you win, if it is 2, 3 or 12 you lose. If it is any other number j, you continue to roll two dice until the sum is j or 7, whichever is sooner. If it is 7 you lose, if it is j you win. What is the probability p that you win?

Solution Suppose that you roll the dice n times. That experiment is equivalent to rolling $2n$ fair dice, with the sample space Ω_{2n} being all possible sequences of length $2n$, of the numbers 1, 2, 3, 4, 5, 6, for any $n \geqslant 1$. By symmetry these 6^{2n} outcomes are equally likely, and whether you win or you lose at or before the nth roll of the pair of dice is determined by looking at the sum of successive pairs of numbers in these outcomes.

The sample space for the roll of a pair of dice (Ω_2), has 36 equally likely outcomes. Let n_j denote the number of outcomes in which the sum of the numbers shown is j, $2 \leqslant j \leqslant 12$. Now let A_k be the event that you win by rolling a pair with sum k, and consider the eleven distinct cases:

(a) $$\mathbf{P}(A_2) = \mathbf{P}(A_3) = \mathbf{P}(A_{12}) = 0,$$
because you always lose with these.

(b) For A_7 to occur you must get 7 on the first roll. Since $n_7 = 6$

$$\mathbf{P}(A_7) = \frac{n_7}{|\Omega_2|} = \frac{6}{36} = \frac{1}{6}.$$

(c) Likewise

$$\mathbf{P}(A_{11}) = \frac{n_{11}}{36} = \frac{2}{36} = \frac{1}{18}, \quad \text{since } n_{11} = 2.$$

(d) For A_4 to occur you must get 4 on the first roll and on the nth roll, for some $n \geqslant 2$, with no 4 or 7 in the intervening $n-2$ rolls. You can do this in $n_4^2(36 - n_4 - n_7)^{n-2}$ ways, and therefore

$$\mathbf{P}(A_4) = \sum_{n=2}^{\infty} \frac{n_4^2(36 - n_4 - n_7)^{n-2}}{6^{2n}} \qquad \text{by (1.4.3)}$$

$$= \frac{n_4^2}{36(n_4 + n_7)} = \frac{1}{36} \qquad \text{because } n_4 = 3.$$

(e) Likewise

$$\mathbf{P}(A_5) = \frac{n_5^2}{36(n_5 + n_7)}$$

$$= \frac{2}{45} \qquad \text{because } n_5 = 4$$

$$= \mathbf{P}(A_9) \qquad \text{because } n_9 = 4.$$

Finally

$$\mathbf{P}(A_6) = \mathbf{P}(A_8)$$

$$= \frac{25}{396} \qquad \text{because } n_6 = n_8 = 5$$

and

$$\mathbf{P}(A_{10}) = \mathbf{P}(A_4) \qquad \text{because } n_{10} = n_4 = 3.$$

Therefore the probability that you win is

$$\mathbf{P}(A_7) + \mathbf{P}(A_{11}) + 2\mathbf{P}(A_4) + 2\mathbf{P}(A_5) + 2\mathbf{P}(A_6) = \frac{1}{6} + \frac{1}{18} + \frac{1}{18} + \frac{4}{45} + \frac{25}{198}$$
$$\simeq 0.493.$$

(1) Exercise What is the probability that you win on or before the second roll?

(2) Exercise What is the probability that you win on or before the third roll?

(3) Exercise What is the probability that you win if, on the first roll,

(a) the first die shows 2?

(b) the first die shows 6?

(4) Exercise If you could fix the number to be shown by one die of the two on the first roll, what number would you choose?

1.13 Example: Murphy's law

A fair coin is tossed repeatedly. Let s denote any fixed sequence of heads and tails of length r. Show that with probability one the sequence s will eventually appear in r consecutive tosses of the coin.

[The usual statement of Murphy's law says that anything that can go wrong, will go wrong].

Solution If a fair coin is tossed r times, there are 2^r distinct equally likely outcomes and one of them is s. We consider a fair die with 2^r faces; each face corresponds to one of the 2^r outcomes of tossing the coin r times and one of them is face s. Now roll the die repeatedly.

Let A_k be the event that face s appears for the first time on the kth roll. There are 2^{rk} distinct outcomes of k rolls, and by symmetry they are equally likely. In $(2^r - 1)^{k-1}$ of them A_k occurs, so by (1.3.1)

$$\mathbf{P}(A_k) = \frac{(2^r - 1)^{k-1}}{2^{rk}}.$$

Since $A_k \cap A_j = \phi$ for $k \neq j$ we have by (1.4.3) that

(1)
$$\mathbf{P}\left(\bigcup_1^m A_k\right) = \sum_1^m \mathbf{P}(A_k) = 1 - \left(\frac{2^r - 1}{2^r}\right)^m$$

which is the probability that face s appears at all in m rolls.

Now consider n tosses of the coin, and let $m = [\frac{n}{r}]$ (where $[x]$ is the integer part of x). The n tosses can thus be divided up into m sequences of length r with a remainder $n - mr$. Let B_n be the event that none of these m sequences is s, and let C_n be the event that the sequence s does not occur anywhere in the n tosses. Then

$$C_n \subseteq B_n = \left(\bigcup_1^m A_k\right)^c,$$

since rolling the die m times and tossing the coin mr times yield the same sample space of equally likely outcomes. Hence by Example 1.4.11(i) and (1.4.5) and (1),

$$\mathbf{P}(C_n) \leq \mathbf{P}(B_n) = \left(\frac{2^r - 1}{2^r}\right)^m \to 0$$

as $n \to \infty$. Now the event that s eventually occurs is $\lim_{n\to\infty} (C_n^c)$, so by (1.4.5) and (1.5.4)

$$\mathbf{P}(\lim_{n\to\infty} C_n^c) = \lim_{n\to\infty} \mathbf{P}(C_n^c) = 1 - \lim_{n\to\infty} \mathbf{P}(C_n) = 1.$$

(2) Exercise If the coin is tossed n times, show that the probability that it shows heads on an odd number of tosses (and tails on the rest) is $\frac{1}{2}$.

(3) Exercise If the coin is tossed an unbounded number of times, show that the probability that a head is first shown on an odd numbered toss is $\frac{2}{3}$.

(4) Exercise If Malone tosses his coin m times, and Watt tosses his coin n times, show that the probability that they get the same number of heads each is equal to the probability that Beckett gets m heads in $m + n$ tosses of his coin.

PROBLEMS

N.B. Unless otherwise stated, coins are fair, dice are regular cubes and pack of cards are well-shuffled with four suits of thirteen cards.

1 You are given a conventional pack of cards. What is the probability that the top card is an ace?

2 You count a pack of cards (face down) and find it defective (having only 49 cards!). What is the probability that the top card is an ace?

3 A class contains seven boys and eight girls.
(a) If two are selected at random to leave the room, what is the probability that they are of different sexes?
(b) On two separate occasions a child is selected at random to leave the room. What is the probability that the two choices result in children of different sexes?

4 An urn contains one hundred balls numbered from 1 to 100. Four are removed at random without being replaced. Find the probability that the number on the last ball is smaller than the number on the first ball.

5 Let $\mathcal{F}$ be an event space. Show that the total number of events in $\mathcal{F}$ cannot be exactly six. What integers can be the number of events in a finite event space?

6 In order to start playing a game of chance with a die, it is necessary first to throw a six.
(a) What is the probability that you throw your first six at your third attempt?
(b) What is the probability that you require more than three attempts?
(c) What is the most likely number of attempts up to your first throwing a six?
(d) After how many throws would your probability of having thrown a six be at least 0.95?

7 Let A, B and C be events. Write down expressions for the events
(a) at least two of A, B and C occur;
(b) exactly two of A, B and C occur;
(c) at most two of A, B and C occur;
(d) exactly one of A, B and C occurs.

8 A die is loaded in such a way that the probability that a 6 is thrown is five times that of any other number, each of them being equally probable.

(a) By what factor is the probability of a total of 24 from 4 throws greater than that for an unloaded die?

(b) Show that for the loaded die, the probability of obtaining a total of 6 from 4 throws is two and half times that of obtaining 5, and compare the probability of obtaining 23 with that of obtaining 24 from 4 throws.

9 A fair coin is tossed four times. What is the probability of
(a) at least three heads?
(b) exactly three heads?
(c) a run of three or more consecutive heads?
(d) a run of exactly three consecutive heads?

10 Find the probability that in 24 throws of two dice, double six fails to appear.

11 Two dice are rolled and their scores are denoted by S_1 and S_2. What is the probability that the quadratic $x^2 + xS_1 + S_2 = 0$ has real roots?

12 (a) If $\mathbf{P}(A)$ is the probability that an event A occurs, prove that

$$\mathbf{P}\left(\bigcup_{i=1}^{n} A_i\right) = \sum_{i=1}^{n} \mathbf{P}(A_i) - \sum_{i<j\leq n} \mathbf{P}(A_i \cap A_j) + \sum_{i<j<k\leq n} \mathbf{P}(A_i \cap A_j \cap A_k) + \dots$$
$$+ (-1)^{n+1}\mathbf{P}(A_1 \cap A_2 \cap \dots \cap A_n),$$

where$A_1, A_2, \dots, A_n$ are events.

(b) A teaset consists of six cups and saucers with two cups and saucers in each of three different colours. The cups are placed randomly on the saucers. What is the probability that no cup is on a saucer of the same colour?

13 An urn contains three tickets, numbered 1, 2 and 3, and they are drawn successively without replacement. What is the probability that there will be at least one value of r ($r = 1, 2, 3$) such that on the rth drawing a ticket numbered r will be drawn?

14 Four red balls and two blue balls are placed at random into two urns so that each urn contains three balls. What is the probability of getting a blue ball if
(a) you select a ball at random from the first urn?
(b) you select an urn at random and then select a ball from it at random?
(c) you discard two balls from the second urn and select the last ball?

15 Four fair dice are rolled and the four numbers shown are multiplied together. What is the probability that this product
(a) is divisible by 5?
(b) has last digit 5?

16 Suppose that n fair dice are rolled, and let M_n be the product of the numbers shown.
(a) Show that the probability that the last digit of M_n is 5 is a non-increasing function of n.
(b) Show that the probability that M_n is divisible by 5 is a non-decreasing function of n.
(c) Find the limits of the probabilities in (a) and (b) and interpret this.

17 The consecutive integers 1, 2, ..., n are inscribed on n balls in an urn. Let D_r be the event that the number on a ball drawn at random is divisible by r.
(a) What are $\mathbf{P}(D_3)$, $\mathbf{P}(D_4)$, $\mathbf{P}(D_3 \cup D_4)$ and $\mathbf{P}(D_3 \cap D_4)$?
(b) Find the limits of these probabilities as $n \to \infty$.
(c) What would your answers be if the n consecutive numbers began at a number $a \neq 1$?

18 Show that if A and B are events then

$$\begin{aligned}\mathbf{P}(A \cap B) - \mathbf{P}(A)\mathbf{P}(B) &= \mathbf{P}(A)\mathbf{P}(B^c) - \mathbf{P}(A \cap B^c)\\ &= \mathbf{P}(A^c)\mathbf{P}(B) - \mathbf{P}(A^c \cap B)\\ &= \mathbf{P}((A \cup B)^c) - \mathbf{P}(A^c)\mathbf{P}(B^c)\end{aligned}$$

19 Show that
(a) $\min\{1, \mathbf{P}(A) + \mathbf{P}(B)\} \geq \mathbf{P}(A \cup B) \geq \max\{\mathbf{P}(A), \mathbf{P}(B)\}$.
(b) $\min\{\mathbf{P}(A), \mathbf{P}(B)\} \geq \mathbf{P}(A \cap B) \geq \max\{0, \mathbf{P}(A) + \mathbf{P}(B) - 1\}$.
(c) $\mathbf{P}\left(\bigcap_1^n A_i\right) \geq \sum_{i=1}^n \mathbf{P}(A_i) - (n-1)$.

20 The function $d(x, y)$ is defined on the event space by $d(A, B) = \mathbf{P}(A \triangle B)$.
(a) Show that for any events A, B and C,

$$d(A, B) + d(B, C) - d(A, C) = 2(\mathbf{P}(A \cap B^c \cap C) + \mathbf{P}(A^c \cap B \cap C^c)).$$

(b) When is $d(A, B)$ zero?
(c) Let $A_1, A_2, \ldots$ be a monotone sequence of events such that $A_i \subseteq A_j$ for $i \leq j$. Show that for $i \leq j \leq k$,

$$d(A_i, A_k) = d(A_i, A_j) + d(A_j, A_k).$$

21 An urn contains $x \geq 2$ xanthic balls and $y \geq 1$ yellow balls. Two balls are drawn at random without replacement; let p be the probability that both are xanthic.
(a) If $p = \frac{1}{2}$, find the smallest possible value of x in the two cases when y is odd or even.
(b) If $p = \frac{1}{8}$, find the smallest possible value of x.
(c) If $p = r^{-2}$ where r is an integer, show that $r \geq 6$, and find values of x and y which yield $p = \frac{1}{36}$.

22 When are the following true?
(a) $A \cup (B \cap C) = (A \cup B) \cap (A \cup C)$
(b) $A \cap (B \cap C) = (A \cap B) \cap C$
(c) $A \cup (B \cup C) = A \backslash (B \backslash C)$
(d) $(A \backslash B) \backslash C = A \backslash (B \backslash C)$
(e) $A \triangle (B \triangle C) = (A \triangle B) \triangle C$
(f) $A \backslash (B \cap C) = (A \backslash B) \cup (A \backslash C)$
(g) $A \backslash (B \cup C) = (A \backslash B) \cap (A \backslash C)$.

23 **Birthdays** If m students born in 1985 are attending a lecture, show that the probability that at least two of them share a birthday is

$$p = 1 - \frac{(365)!}{(365-m)!(365)^m}.$$

Show that if $m \geq 23$ then $p > \frac{1}{2}$. What difference would it make if they were born in 1988?

24 Let $(A_n; n > 1)$ be a collection of events. Show that the event that infinitely many of the A_n occur is given by $\bigcap_{n \geq 1} \bigcup_{m=n}^{\infty} A_m$.

25 **Boole's inequality** Show that

$$\mathbf{P}\left(\bigcup_1^n A_i\right) \leq \sum_{i=1}^n \mathbf{P}(A_i).$$

2
Conditional probability and independence

2.1 Conditional probability

Suppose you have a well-shuffled conventional pack of cards. Obviously (by symmetry) the probability $\mathbf{P}(T)$ of the event T that the top card is an ace is

$$\mathbf{P}(T) = \frac{4}{52} = \frac{1}{13}.$$

However, suppose you notice that the bottom card is the ace of spades S_A. What now is the probability that the top card is an ace? There are 51 possibilities, and three of them are aces, so by symmetry again the required probability is $\frac{3}{51}$. To distinguish this from the original probability, we denote it by $\mathbf{P}(T|S_A)$, and call it the conditional probability of T given that the bottom card is the ace of spades.

Similarly, had you observed that the bottom card was the king of spades S_K, you would conclude that the probability that the top card is an ace is

$$\mathbf{P}(T|S_K) = \frac{4}{51}.$$

Here is a less trivial example.

Example: Poker [**Note:** *in this example the symbol* $\binom{n}{r}$ *denotes the number of ways of choosing r cards from n cards. If you are unfamiliar with this notation, omit this example at a first reading.*]

Suppose you are playing poker. As the hand is dealt, you calculate the chance of being dealt a royal flush R, assuming that all hands of five cards are equally likely. (A *royal flush* comprises 10, J, Q, K, A in a single suit.) Just as you get the answer

$$\mathbf{P}(R) = 4\binom{52}{5}^{-1} = \frac{1}{649\,740},$$

the dealer deals your last card face up. It is the ace of spades, S_A. If you accept the card, what now is your chance of picking up a royal flush?

Intuitively it seems unlikely still to be $\mathbf{P}(R)$ above, as the conditions for getting one have changed. Now you need your first four cards to be the ten to king of spades precisely. (Also, had your last card been the two of spades, S_2, your chance of a royal flush would be definitely zero.) As above, to distinguish this new probability we call it the conditional probability of R given S_A, and denote it by $\mathbf{P}(R|S_A)$.

Is it larger or smaller than $\mathbf{P}(R)$? At least you do have an ace, which is a start, so it might be greater. But you cannot now get a flush in any suit but spades, so it might be smaller. To resolve the uncertainty you assume that any set of four cards from the remaining 51 cards is equally likely to complete your hand, and calculate that

$$\mathbf{P}(R|S_A) = \binom{51}{4}^{-1} = \frac{13}{5}\mathbf{P}(R).$$

Your chances of a royal flush have more than doubled. ●

Let us investigate these ideas in a more general setting. As usual we are given a sample space, an event space $\mathcal{F}$, and a probability function $\mathbf{P}(.)$. We suppose that some event $B \in \mathcal{F}$ definitely occurs, and denote the conditional probability of any event A, given B, by $\mathbf{P}(A|B)$. As we did for $\mathbf{P}(.)$, we observe that $\mathbf{P}(.|B)$ is a function defined on $\mathcal{F}$, which takes values in $[0, 1]$. But what function is it?

Clearly $\mathbf{P}(A)$ and $\mathbf{P}(A|B)$ are not equal in general, because, even when $\mathbf{P}(B^c) \neq 0$, we always have

$$\mathbf{P}(B^c|B) = 0.$$

Secondly, we note that given the occurrence of B, the event A can occur if and only if $A \cap B$ occurs. This makes it natural to require that

$$\mathbf{P}(A|B) \propto \mathbf{P}(A \cap B).$$

Finally, and trivially,

$$\mathbf{P}(B|B) = 1.$$

After a moment's thought about these three observations, it appears that an attractive candidate to play the role of $\mathbf{P}(A|B)$ is $\mathbf{P}(A \cap B)/\mathbf{P}(B)$. We make these intuitive reflections formal as follows.

Definition Let A and B be events with $\mathbf{P}(B) > 0$. Given that B occurs, the *conditional probability* that A occurs is denoted by $\mathbf{P}(A|B)$, and defined by

(1)
$$\mathbf{P}(A|B) = \frac{\mathbf{P}(A \cap B)}{\mathbf{P}(B)}.$$
▲

When $\mathbf{P}(B) = 0$, the conditional probability $\mathbf{P}(A|B)$ is not defined by (1). However, in order to avoid an endless stream of tiresome reservations about special cases, it is convenient to adopt the convention that, even when $\mathbf{P}(B) = 0$, we may still write $\mathbf{P}(A \cap B) = \mathbf{P}(A|B)\mathbf{P}(B)$, both sides having the value zero. Thus, whether $\mathbf{P}(B) > 0$ or not, it is true that

$$\mathbf{P}(A \cap B) = \mathbf{P}(A|B)\mathbf{P}(B).$$

Likewise

$$\mathbf{P}(A \cap B^c) = \mathbf{P}(A|B^c)\mathbf{P}(B^c)$$

and hence for any events A and B we have proved the following:

(2) Theorem
$$\begin{aligned}\mathbf{P}(A) &= \mathbf{P}(A \cap B) + \mathbf{P}(A \cap B^c) \\ &= \mathbf{P}(A|B)\mathbf{P}(B) + \mathbf{P}(A|B^c)\mathbf{P}(B^c).\end{aligned}$$
■

The reader will come to realize the crucial importance of (1) and (2) as he or she discovers more about probability. We begin with a trivial example.

Example: Poker revisited Let us check that Definition 1 is consistent with our informal discussion earlier in this section. By (1)

$$\mathbf{P}(R|S_A) = \mathbf{P}(R \cap S_A)/\mathbf{P}(S_A) = \frac{1}{\binom{52}{5}} \Bigg/ \frac{\binom{51}{4}}{\binom{52}{5}} = \binom{51}{4}^{-1},$$

as required. ●

Here is a more complicated example.

Example: Lemons An industrial conglomerate manufactures a certain type of car in three towns called Farad, Gilbert and Henry. Of 1000 made in Farad 20% are defective; of 2000 made in Gilbert 10% are defective, and of 3000 made in Henry 5% are defective. You buy a car from a distant dealer. Let D be the event that it is defective, F the event that it was made in Farad and so on. Find: (a) $\mathbf{P}(F|H^c)$; (b) $\mathbf{P}(D|H^c)$; (c) $\mathbf{P}(D)$; (d) $\mathbf{P}(F|D)$. Assume that you are equally likely to have bought any one of the 6000 cars produced.

Solution (a)
$$\mathbf{P}(F|H^c) = \frac{\mathbf{P}(F \cap H^c)}{\mathbf{P}(H^c)} \quad \text{by (1)},$$
$$= \frac{\mathbf{P}(F)}{\mathbf{P}(H^c)} \quad \text{since } F \subseteq H^c,$$
$$= \frac{1000}{6000}\Bigg/\frac{3000}{6000} = \frac{1}{3}.$$

(b)
$$\mathbf{P}(D|H^c) = \frac{\mathbf{P}(D \cap H^c)}{\mathbf{P}(H^c)} \quad \text{by (1)}$$
$$= \frac{\mathbf{P}(D \cap (F \cup G))}{\mathbf{P}(H^c)} \quad \text{because } H^c = F \cup G,$$
$$= \frac{\mathbf{P}(D \cap F) + \mathbf{P}(D \cap G)}{\mathbf{P}(H^c)} \quad \text{since } F \cap G = \phi$$
$$= \frac{\mathbf{P}(D|F)\mathbf{P}(F) + \mathbf{P}(D|G)\mathbf{P}(G)}{\mathbf{P}(H^c)} \quad \text{by (1)}$$
$$= \frac{\frac{1}{5}\cdot\frac{1}{6} + \frac{1}{10}\cdot\frac{1}{3}}{\frac{1}{2}} \quad \text{on using the data in the question,}$$
$$= \frac{2}{15}.$$

(c) $$\mathbf{P}(D) = \mathbf{P}(D|H)\mathbf{P}(H) + \mathbf{P}(D|H^c)\mathbf{P}(H^c) \quad \text{by (2)}$$
$$= \frac{1}{20}\cdot\frac{1}{2} + \frac{2}{15}\cdot\frac{1}{2} \quad \text{on using the data and (b)}$$
$$= \frac{11}{120}.$$

(d) $$\mathbf{P}(F|D) = \frac{\mathbf{P}(F \cap D)}{\mathbf{P}(D)} \quad \text{by (1)}$$
$$= \frac{\mathbf{P}(D|F)\mathbf{P}(F)}{\mathbf{P}(D)} \quad \text{by (1)}$$
$$= \frac{1}{5}\cdot\frac{1}{6}\Big/\frac{11}{120} \quad \text{on using the data and (c)}$$
$$= \frac{4}{11}.$$ ●

We often have occasion to use the following elementary generalization of Theorem 2.

(3) **Theorem** We have $$\mathbf{P}(A) = \sum_i \mathbf{P}(A|B_i)\mathbf{P}(B_i)$$

whenever $A \subseteq \bigcup_i B_i$ and $B_i \cap B_j = \phi$ for $i \neq j$.

Proof This is immediate from (1.4.3) and (1). ■

For example, with the notation of (3), we may write

$$\mathbf{P}(B_j|A) = \mathbf{P}(B_j \cap A)/\mathbf{P}(A) = \frac{\mathbf{P}(A|B_j)\mathbf{P}(B_j)}{\mathbf{P}(A)},$$

and expanding the denominator using (3), we have proved the following celebrated result:

Bayes's theorem If $A \subseteq \bigcup_1^n B_i$, and $B_i \cap B_j = \phi$ for $i \neq j$, then

(4) $$\mathbf{P}(B_j|A) = \frac{\mathbf{P}(A|B_j)\mathbf{P}(B_j)}{\sum_1^n \mathbf{P}(A|B_i)\mathbf{P}(B_i)}; \quad \mathbf{P}(A) > 0.$$ ■

The following is a typical example of how (4) is applied in practice.

Example: False positives You have a blood test for some rare disease which occurs by chance in 1 in every 100 000 people. The test is fairly reliable; if you have the disease it will correctly say so with probability 0.95; if you do not have the disease, the test will wrongly say you do with probability 0.005. If the test says you do have the disease, what is the probability that this is a correct diagnosis?

Solution Let D be the event that you have the disease, and T the event that the test says you do. Then we require $\mathbf{P}(D|T)$, which is given by

$$\mathbf{P}(D|T) = \frac{\mathbf{P}(T|D)\mathbf{P}(D)}{\mathbf{P}(T|D)\mathbf{P}(D) + \mathbf{P}(T|D^c)\mathbf{P}(D^c)} \quad \text{by (4)}$$
$$= \frac{(0.95)(0.00001)}{(0.95)(0.00001) + (0.99999)(0.005)} \simeq 0.002.$$

Despite appearing to be a pretty good test, for a disease as rare as this the test is almost useless. ●

It is important to note that conditional probability is a probability function in the sense defined in Section 1.4. Thus $\mathbf{P}(\Omega|B) = 1$, and, if $A_i \cap A_j = \phi$ for $i \neq j$, we have

(5) $$\mathbf{P}\left(\bigcup_i A_i | B\right) = \sum_i \mathbf{P}(A_i|B).$$

From these we may deduce a variety of useful identities (as we did in Section 1.4); for example:

(6) $$\mathbf{P}(A \cap B \cap C) = \mathbf{P}(A|B \cap C)\mathbf{P}(B|C)\mathbf{P}(C),$$

(7) $$\mathbf{P}\left(\bigcap_1^n A_i\right) = \mathbf{P}\left(A_1 | \bigcap_2^n A_i\right)\mathbf{P}\left(A_2 | \bigcap_3^n A_i\right) \ldots \mathbf{P}(A_n)$$

(8) $$\mathbf{P}(A|B) = 1 - \mathbf{P}(A^c|B),$$

(9) $$\mathbf{P}(A \cup B|C) = \mathbf{P}(A|C) + \mathbf{P}(B|C) - \mathbf{P}(A \cap B|C),$$

and so on.

(10) Example Let us prove (5), (6), (7), (8) and (9). First

$$\mathbf{P}\left(\bigcup_i A_i | B\right) = \mathbf{P}\left(\left(\bigcup_i A_i\right) \cap B\right) \Big/ \mathbf{P}(B) \quad \text{by (1).}$$
$$= \mathbf{P}\left(\bigcup_i (A_i \cap B)\right) \Big/ \mathbf{P}(B)$$
$$= \sum_i \mathbf{P}(A_i \cap B)/\mathbf{P}(B) \quad \text{by (1.4.3), because the } A_i \text{ are disjoint,}$$
$$= \sum_i \mathbf{P}(A_i|B) \quad \text{by (1) again,}$$

and we have proved (5). Second, by repeated use of (1),

$$\mathbf{P}(A|B \cap C)\mathbf{P}(B|C)\mathbf{P}(C) = \frac{\mathbf{P}(A \cap B \cap C)}{\mathbf{P}(B \cap C)} \cdot \frac{\mathbf{P}(B \cap C)}{\mathbf{P}(C)} . \mathbf{P}(C) = \mathbf{P}(A \cap B \cap C),$$

if the denominator is not zero. If the denominator is zero then (6) still holds by convention, both sides taking the value zero.

The relation (7) follows by induction using (6); and (8) and (9) are trivial consequences of (5). ●

(11) Example: Repellent and attractive events The event A is said to be *attracted* to B if $\mathbf{P}(A|B) > \mathbf{P}(A)$.

If $\mathbf{P}(A|B) < \mathbf{P}(A)$, then A is *repelled* by B, and A is *indifferent* to B if

(12) $$\mathbf{P}(A|B) = \mathbf{P}(A).$$

(a) Show that if B attracts A, then A attracts B, and B^c repels A.
(b) A flimsy slip of paper is in one of n bulging box files. The event that it is in the jth box file is B_j, where $\mathbf{P}(B_j) = b_j > 0$. The event that a cursory search of the jth box file fails to discover the slip is F_j, where $\mathbf{P}(F_j|B_j) = \phi_j < 1$. Show that B_j and F_j are mutually repellent, but F_j attracts B_i, for $i \neq j$.

Solution (a) Since B attracts A, by (1), $\mathbf{P}(A \cap B) > \mathbf{P}(A)\mathbf{P}(B)$, whence, on dividing by $\mathbf{P}(A)$, we have $\mathbf{P}(B|A) > \mathbf{P}(B)$. Furthermore, by Theorem 2,

$$\begin{aligned}\mathbf{P}(A|B^c)\mathbf{P}(B^c) &= \mathbf{P}(A) - \mathbf{P}(A|B)\mathbf{P}(B) < \mathbf{P}(A)(1 - \mathbf{P}(B)), \quad \text{because } B \text{ attracts } A,\\ &= \mathbf{P}(A)\mathbf{P}(B^c).\end{aligned}$$

So B^c repels A (on dividing through by $\mathbf{P}(B^c) \neq 0$).

(b) By Bayes' theorem (4),

$$\mathbf{P}(B_j|F_j) = \frac{\mathbf{P}(F_j|B_j)\mathbf{P}(B_j)}{\sum_{i=1}^{n}\mathbf{P}(F_j|B_i)\mathbf{P}(B_i)} = \frac{\phi_j b_j}{1 - b_j + \phi_j b_j}$$

since, obviously, for $i \neq j$, $\mathbf{P}(F_j|B_i) = 1$. Hence

$$\mathbf{P}(B_j) - \mathbf{P}(B_j|F_j) = \frac{b_j(1 - b_j)(1 - \phi_j)}{1 - b_j + \phi_j b_j} > 0.$$

Therefore B_j is repelled by F_j. Also, for $i \neq j$,

$$\mathbf{P}(B_i|F_j) - \mathbf{P}(B_i) = \frac{b_i}{1 - b_j + \phi_j b_j} - b_i = \frac{b_i b_j(1 - \phi_j)}{1 - b_j + \phi_j b_j} > 0$$

so F_j attracts B_i, for $i \neq j$. ●

Notice that this agrees with our intuition. We feel quite strongly that if we look in a file for a slip and fail to find it, then it is more likely (than before the search) to be elsewhere. (Try to think about the consequences if the opposite were true.) This conclusion of Example 11 was not incorporated in our axioms, but follows from them. It therefore lends a small but valuable boost to their credibility.

Finally we consider sequences of conditional probabilities. Since conditional probability is a probability function (see (5)), we expect it to be continuous in the sense of Section 1.5. Thus if (as $n \to \infty$) $A_n \to A$ and $B_n \to B$ then, by Theorem 1.5.2 we have

$$\lim_{n\to\infty} \mathbf{P}(A_n|B) = \mathbf{P}(A|B)$$

and

(13) $$\lim_{n\to\infty} \mathbf{P}(A|B_n) = \mathbf{P}(A|B).$$

2.2 Independence

It may happen that the conditional probability $\mathbf{P}(A|B)$ is the same as the unconditional probability $\mathbf{P}(A)$, so that

$$\mathbf{P}(A) = \mathbf{P}(A|B) = \frac{\mathbf{P}(A \cap B)}{\mathbf{P}(B)}.$$

This idea leads to the following:

(1) **Definition** (a) Events A and B are *independent* when

$$\mathbf{P}(A \cap B) = \mathbf{P}(A)\mathbf{P}(B).$$

(b) A collection of events $(A_i;\ i \geqslant 1)$ is independent when

$$\mathbf{P}\left(\bigcap_{i\in F} A_i\right) = \prod_{i\in F} \mathbf{P}(A_i)$$

for any finite set F of indices.

(c) Events A and B are *conditionally* independent, given C, when

$$\mathbf{P}(A \cap B|C) = \mathbf{P}(A|C)\mathbf{P}(B|C).$$

This does not imply independence unless $C = \Omega$.

(d) A collection of events $(A_i;\ i \geqslant 1)$ is *pairwise* independent if

$$\mathbf{P}(A_i \cap A_j) = \mathbf{P}(A_i)\mathbf{P}(A_j) \quad \text{for } i \neq j.$$

This does not imply independence in general. ▲

It is easy to see that independence is equivalent to the idea of indifference defined in (2.1.12), but the term 'indifference' is not in general use. It is usually, but not always, clear when two events are independent, as the next two examples illustrate.

(2) **Example: Sport** Prior to a game of football you toss a coin for the kick-off. Let C be the event that you win the toss, and let M be the event that you win the match.

(a) Show that the outcome of the match is independent of whether or not you win the toss if and only if, for some p and p', with $0 < p,\ p' < 1$,

$$\mathbf{P}(C \cap M) = pp',$$
$$\mathbf{P}(C \cap M^c) = p(1 - p'),$$
$$\mathbf{P}(C^c \cap M) = (1 - p)p',$$

and

$$\mathbf{P}(C^c \cap M^c) = (1 - p)(1 - p').$$

(b) Let B be the event that you win both or lose both, so $B = \{(C \cap M) \cup (C^c \cap M^c)\}$. Suppose that C and M are indeed independent. Show that C and B are independent if and only if $p' = \frac{1}{2}$.

Solution (a) If C and M are independent, and $\mathbf{P}(C) = p$ and $\mathbf{P}(M) = p'$, then by definition $\mathbf{P}(C \cap M) = pp'$ and so on.

Conversely, for the given probabilities

$$\mathbf{P}(C) = \mathbf{P}(C \cap M) + \mathbf{P}(C \cap M^c) = pp' + p(1 - p') = p$$

and similarly we have $\mathbf{P}(M) = p'$. Hence

$$\mathbf{P}(C)\mathbf{P}(M) = pp' = \mathbf{P}(C \cap M).$$

This, together with three similar identities (exercises for you), demonstrates the independence.

(b) Trivially $\mathbf{P}(C \cap B) = \mathbf{P}(C \cap M)$. Hence C and B are independent if

$$pp' = \mathbf{P}(C \cap M) = \mathbf{P}(C)\mathbf{P}(B) = p(pp' + (1 - p)(1 - p')).$$

That is if $(1 - p)(1 - 2p') = 0$. Since $p \neq 1$, it follows that $p' = \frac{1}{2}$. The converse is trivial. ●

(3) **Example: Flowers** A plant gets two independent genes for flower colour, one from each parent plant. If the genes are identical, then the flowers are uniformly of that colour; if they are different, then the flowers are striped in those two colours. The genes for the colours pink, crimson and red occur in the population in the proportions $p{:}q{:}r$, where $p + q + r = 1$. A given plant's parents are selected at random; let A be the event that its flowers are at least partly pink, and let B be the event that its flowers are striped.

(a) Find $\mathbf{P}(A)$ and $\mathbf{P}(B)$.
(b) Show that A and B are independent if $p = \frac{2}{3}$ and $r = q = \frac{1}{6}$.
(c) Are these the only values of p, q and r such that A and B are independent?

Solution (a) With an obvious notation (P for pink, C for crimson and R for red), we have

$$\begin{aligned}\mathbf{P}(PP) &= \mathbf{P}(P)\mathbf{P}(P), \quad \text{by parents' independence,}\\ &= p^2,\end{aligned}$$

since P occurs with probability p. Likewise

$$\mathbf{P}(PR) = \mathbf{P}(R)\mathbf{P}(P) = rp = \mathbf{P}(RP).$$

Hence

$$\begin{aligned}\mathbf{P}(A) &= \mathbf{P}(PP \cup PR \cup RP \cup PC \cup CP)\\ &= p^2 + 2pr + 2pq \quad \text{by (1.4.3),}\\ &= 1 - (1 - p)^2,\end{aligned}$$

since $p + q + r = 1$. (Can you see how to get this last expression directly?) Similarly

$$\mathbf{P}(B) = \mathbf{P}(PC \cup PR \cup RC) = 2(pq + qr + rp).$$

(b) The events A and B are independent if and only if:

$$\mathbf{P}(A)\mathbf{P}(B) = \mathbf{P}(A \cap B) = \mathbf{P}(PC \cup PR) = 2(pq + pr).$$

From part (a), this is equivalent to

(4) $$(1-(1-p)^2)(pq+qr+pr)=p(q+r),$$

and this is satisfied by the given values of p, q and r.

(c) No. Rearranging (4), we see that A and B are independent for any values of q and r lying on the curve $rq = 2rq(q+r) + r^3 + q^3$, in the r–q plane. You may care to amuse yourself by showing that this is a loop from the origin. Outside the loop A and B are attractive; inside the loop A and B are repellent. ●

(5) **Example 1.12 revisited: Craps** Let us reconsider this game using conditional probability and independence. Recall that A_k is the event that you win by rolling a pair with sum k. Let S_k be the event that any given roll yields sum k. Now, for example, A_4 occurs only if S_4 occurs at the first roll and S_4 occurs before S_7 in later rolls. However, all the rolls after the first until the first occurrence of S_4 or S_7 are irrelevant, and rolls are independent. Hence

$$\mathbf{P}(A_4) = \mathbf{P}(S_4)\mathbf{P}(S_4|S_4 \cup S_7) = \frac{(\mathbf{P}(S_4))^2}{\mathbf{P}(S_4 \cup S_7)} = \frac{\left(\frac{3}{36}\right)^2}{\frac{3}{36}+\frac{6}{36}} = \frac{1}{36}.$$

Now performing a similar calculation for A_5, A_6, A_8, A_9 and A_{10} yields the solution to Example 1.12. ●

(6) **Example** Suppose A and B are independent, and B and C are independent.

(a) Are A and C independent in general?
(b) Is B independent of $A \cup C$?
(c) Is B independent of $A \cap C$?

Solution (a) No. Take $A \equiv C$. (And then do Problem 4(a).)

(b) No. Consider Example 2(b), with $p = p' = \frac{1}{2}$, and let $A = M$. Then B is independent of M and C, but

$$\mathbf{P}(B \cap (C \cup M)) = \mathbf{P}(C \cap M) = \frac{1}{4}$$

and

$$\mathbf{P}(B)\mathbf{P}(C \cup M) = (pp' + (1-p)(1-p'))(1-(1-p')(1-p)) = \frac{3}{8} \neq \frac{1}{4}.$$

(c) No. With the same notation as in (b), following Example 2(b) again, we have

$$\mathbf{P}(B \cap C \cap M) = \mathbf{P}(C \cap M) = \frac{1}{4},$$

and

$$\mathbf{P}(B)\mathbf{P}(C \cap M) = (pp' + (1-p')(1-p))pp' = \frac{1}{8} \neq \frac{1}{4}.$$ ●

2.3 Recurrence and difference equations

Many problems in probability have a structure in which the repetition of some procedure is essential. At a trivial level, one may repeatedly roll dice, catch fish, have children, and so on; more important problems involve the same idea. In Chapter 1 it was necessary to suppose that all the probabilities on the sample space were given, or that all outcomes were equally likely. Conditional probability provides a more natural way of defining such problems; conditional on the procedure having reached some stage, it supplies the probabilities of the consequent events.

By emphasizing this recurrent aspect of some experiment, conditional probability enables us to tackle problems by deriving recurrence relations. These often turn out to be simple difference equations. Many of the worked examples illustrate these ideas; the following results are useful.

(1) Theorem Let $(a_k;\, k \geqslant 0)$ be a sequence of real numbers.

(i) If the sequence $(u_k;\, k \geqslant 0)$ satisfies

$$u_{k+1} - a_k u_k = 0$$

then

$$u_k = u_0 \prod_0^{k-1} a_j.$$

(ii) If $(u_k;\, k \geqslant 0)$ satisfies

$$u_{k+1} - u_k = c\alpha^k,$$

where α and c are constants, then

$$u_k - u_0 = c\frac{\alpha^k - 1}{\alpha - 1}.$$

(iii) If $(u_k;\, k \geqslant 0)$ satisfies

$$u_{k+1} - u_k = c,$$

for some constant c, then

$$u_k - u_0 = kc.$$

(iv) If for some constants a, α, c_1 and c_2, we have

$$u_{k+1} - au_k = c_1 + c_2\alpha^k,\ a \neq \alpha,$$

then

$$u_k = u_0 a^k + \frac{c_1(1 - a^k)}{1 - \alpha} + \frac{c_2(\alpha^k - a^k)}{\alpha - a}.$$

Proof All of these results are verified simply by substitution. ∎

(2) Example: Fly A room has four walls, a floor and a ceiling. A fly moves between these surfaces. If it leaves the floor or ceiling then it is equally likely to alight on any one of the four walls or the surface it has just left. If it leaves a wall, then it is equally likely to alight on any of the other three walls, or the floor, or the ceiling. Initially it is on the ceiling.

Let F_k denote the event that it is on the floor after k moves. What is $f_k = \mathbf{P}(F_k)$?

Solution Let C_k denote the event that it is on the ceiling after k moves, and N_k, E_k, W_k, S_k denote the corresponding event for the four walls. Set $c_k = \mathbf{P}(C_k)$, and so on. Then by Theorem 2.1.3,

(3)
$$\begin{aligned}\mathbf{P}(F_k) &= \mathbf{P}(F_k|F_{k-1})\mathbf{P}(F_{k-1}) + \mathbf{P}(F_k|C_{k-1})\mathbf{P}(C_{k-1}) + \mathbf{P}(F_k|N_{k-1})\mathbf{P}(N_{k-1})\\ &\quad + \mathbf{P}(F_k|E_{k-1})\mathbf{P}(E_{k-1}) + \mathbf{P}(F_k|W_{k-1})\mathbf{P}(W_{k-1}) + \mathbf{P}(F_k|S_{k-1})\mathbf{P}(S_{k-1})\\ &= \frac{1}{5}f_{k-1} + 0 + \frac{4}{5}w_{k-1},\end{aligned}$$

where we have used the fact that, by symmetry, $n_k = e_k = w_k = s_k$. Likewise

$$w_k = \frac{3}{5}w_{k-1} + \frac{1}{5}f_{k-1} + \frac{1}{5}c_{k-1} = \frac{3}{5}w_{k-1} + \frac{1}{5}(1 - 4w_{k-1}),$$

on using the fact that

(4)
$$f_k + c_k + 4w_k = 1,$$

which follows from the observation that the fly has to be somewhere in the room. Hence

(5)
$$w_k = -\frac{1}{5}w_{k-1} + \frac{1}{5}.$$

Because the fly starts on the ceiling, $w_0 = 0$, and so by Theorem 1(iv)

(6)
$$w_k = \frac{1}{6}\left(1 - \left(-\frac{1}{5}\right)^k\right).$$

Substituting into (3) gives

$$f_k = \frac{1}{5}f_{k-1} + \frac{2}{15}\left(1 - \left(-\frac{1}{5}\right)^{k-1}\right).$$

Hence, for some constant A,

$$f_k = A\left(\frac{1}{5}\right)^k + \frac{1}{3}\left(-\frac{1}{5}\right)^k + \frac{1}{6}.$$

The arbitrary constant A is determined by recalling that the fly starts on the ceiling. Thus $f_0 = 0$, and

(7)
$$f_k = -\frac{1}{2}\left(\frac{1}{5}\right)^k + \frac{1}{3}\left(-\frac{1}{5}\right)^k + \frac{1}{6}.$$

Alternatively, we may substitute (5) into (3) to get directly:

(8)
$$f_k = \frac{1}{25}f_{k-2} + \frac{4}{25}.$$

It is a simple matter to check that, for any constants A and B,

$$f_k = A\left(\frac{1}{5}\right)^k + B\left(-\frac{1}{5}\right)^k + \frac{1}{6}$$

satisfies (8). Since $f_0 = f_1 = 0$, the solution (7) is recovered immediately. Notice that as $k \to \infty$, (6) and (7) yield $f_k \to \frac{1}{6}$ and $w_k \to \frac{1}{6}$. It follows from (4) that $c_k \to \frac{1}{6}$. In the long run the fly is equally likely to be on any surface. ●

2.4 Remarks

Independence and conditioning greatly add to your armoury of weapons for attacking problems. If an event requires the occurrence of a number of independent events, then calculations are simplified by using Definition 2.2.1.

Where independence fails, it may be possible to find a family of disjoint events B_i whose union includes the event A of interest, and for which $\mathbf{P}(A|B_i)$ is easily calculated. The required probability is then calculated using Theorem 2.1.3.

Such events B_i can also be found to yield $\mathbf{P}(A)$ as the solution of some recurrence relation, as exemplified in Section 2.3.

Finally, you are warned to avoid the painful student error which asserts that A and B are independent when $A \cap B = \phi$. This is wrong, except possibly when one of $\mathbf{P}(A)$ or $\mathbf{P}(B)$ is zero, which is not a case of any great interest in general.

WORKED EXAMPLES AND EXERCISES

2.5 Example: Sudden death

Two golfers (Atropos and Belladonna) play a series of holes. Atropos wins each hole with probability p, Belladonna wins each hole with probability q, and holes are halved with probability r. Holes are independent, and the game stops on the first occasion when a hole is not halved. What is the probability that Atropos wins?

We give two methods of solution.

***Solution* I** Let A_n be the event that Atropos wins the match at the nth hole; H_k the event that the kth hole is halved, and W_k the event that Atropos wins the kth hole. Then

(1) $$\mathbf{P}(A_n) = \mathbf{P}(H_1 \cap H_2 \cap \ldots \cap H_{n-1} \cap W_n) = \prod_1^{n-1} \mathbf{P}(H_k)\mathbf{P}(W_n) \quad \text{by independence}$$
$$= r^{n-1}p.$$

Hence the probability that Atropos wins the match is

(2) $$\mathbf{P}\left(\bigcup_1^\infty A_n\right) = \sum_1^\infty \mathbf{P}(A_n)$$

by (1.4.3) because $A_k \cap A_j = \phi$ for $k \neq j$. Now

$$\sum_1^\infty \mathbf{P}(A_n) = \sum_{n=1}^\infty r^{n-1}p = \frac{p}{1-r} = \frac{p}{p+q}.$$

***Solution* II** Let V be the event that Atropos wins the match. Then by Theorem 2.1.3

$$\mathbf{P}(V) = \mathbf{P}(V|W_1)\mathbf{P}(W_1) + \mathbf{P}(V|H_1)\mathbf{P}(H_1) + \mathbf{P}(V|W_1^c \cap H_1^c)\mathbf{P}(W_1^c \cap H_1^c). \tag{3}$$

Now

$$\mathbf{P}(V|W_1) = 1$$

and

$$\mathbf{P}(V|H_1^c \cap W_1^c) = 0.$$

Also, because holes are independent

$$\mathbf{P}(V|H_1) = \mathbf{P}(V).$$

Hence, substituting into (3),

$$\mathbf{P}(V) = p + \mathbf{P}(V)r,$$

so $\mathbf{P}(V) = p/(1 - r)$, in agreement with Solution I.

Remark The first solution harks back to Chapter 1, by dividing up the sample space into disjoint events and using (1.4.3). The second solution exploits the power of conditional probability, by conditioning on the outcome of the first hole. You will use this second idea frequently in tackling problems in probability.

(4) **Exercise** Show that the probability u_n that Atropos wins at or before the nth hole is $p(1 - r^n)/(1 - r)$.

(5) **Exercise** Given that Atropos wins at or before the nth hole, show that:

(a) the probability that the first hole was halved is $r(1 - r^{n-1})/(1 - r^n)$,

(b) the probability that the first hole was won is $(1 - r)/(1 - r^n)$.

(6) **Exercise** Given that Atropos wins, what is the probability that she has won before the third hole?

(7) **Exercise** What is the probability that Atropos wins, given that exactly n holes have been played when the match is won? Use this to solve the example by a third method.

2.6 Example: Polya's urn

An urn contains b blue balls and c cyan balls. A ball is drawn at random, its colour is noted, and it is returned to the urn together with d further balls of the same colour. This procedure is repeated indefinitely. What is the probability that:

(a) the second ball drawn is cyan?
(b) the first ball drawn is cyan given that the second ball drawn is cyan?

Solution Let C_n denote the event that the nth drawn ball is cyan. Then

(a) $$\mathbf{P}(C_2) = \mathbf{P}(C_2|C_1)\mathbf{P}(C_1) + \mathbf{P}(C_2|C_1^c)\mathbf{P}(C_1^c).$$

Now given C_1, the urn contains $c + d$ cyan balls on the second drawing, so

$$\mathbf{P}(C_2|C_1) = \frac{c + d}{b + c + d}.$$

Likewise, given C_1^c the urn contains c cyan balls on the second drawing, so

$$\mathbf{P}(C_2|C_1^c) = \frac{c}{b+c+d}.$$

Hence

$$\mathbf{P}(C_2) = \frac{c+d}{b+c+d}\cdot\frac{c}{b+c} + \frac{c}{b+c+d}\cdot\frac{b}{b+c} = \frac{c}{b+c} = \mathbf{P}(C_1).$$

(b) $\mathbf{P}(C_1|C_2) = \mathbf{P}(C_1 \cap C_2)/\mathbf{P}(C_2) = \mathbf{P}(C_2|C_1)\mathbf{P}(C_1)/\mathbf{P}(C_2) = \dfrac{c+d}{b+c+d}.$

using the results of (a).

(1) Exercise Show that $\mathbf{P}(C_n) = \mathbf{P}(C_1)$ for all $n \geqslant 1$.

(2) Exercise Find the probability that the first drawn ball is cyan given that the nth drawn ball is cyan.

(3) Exercise Find the probability that the first drawn ball is cyan given that the following n drawn balls are all cyan. What is the limit of this probability as $n \to \infty$?

(4) Exercise Show that for any j, k, $\mathbf{P}(C_k|C_j) = \mathbf{P}(C_j|C_k)$

(5) Exercise Show that in $m + n$ drawings, the probability that m cyan balls are followed by n blue balls is the same as the probability that n blue balls are followed by m cyan balls. Generalize this result.

2.7 Example: Complacency

In a factory, if the most recent accident occurred exactly k days before today, then the probability that an accident occurs today is p_k; there is no accident with probability $1 - p_k$. During the n successive days immediately after an accident, what is the probability that

(a) there are no accidents?
(b) there is exactly one accident?

Solution (a) Let A_n be the event that the n days following an accident are free of accidents, $n \geqslant 1$. We are given that

(1) $$\mathbf{P}(A_n|A_{n-1}) = 1 - p_n, \quad n \geqslant 2$$

and $\mathbf{P}(A_1) = 1 - p_1$. The crucial observation is that

(2) $$\mathbf{P}(A_n) = \mathbf{P}(A_n|A_{n-1})\mathbf{P}(A_{n-1}) = (1 - p_n)\mathbf{P}(A_{n-1}).$$

This is almost completely obvious, but we labour the point by giving two detailed verifications.

I Notice that $A_n \subseteq A_{n-1}$. Hence

$$\mathbf{P}(A_n) = \mathbf{P}(A_n \cap A_{n-1}) = \mathbf{P}(A_n|A_{n-1})\mathbf{P}(A_{n-1}) \quad \text{by (2.1.1).}$$

II Alternatively, by Theorem 2.1.2

$$\mathbf{P}(A_n) = \mathbf{P}(A_n|A_{n-1})\mathbf{P}(A_{n-1}) + \mathbf{P}(A_n|A_{n-1}^c)\mathbf{P}(A_{n-1}^c) = \mathbf{P}(A_n|A_{n-1})\mathbf{P}(A_{n-1})$$

because $\mathbf{P}(A_n|A_{n-1}^c) = 0$.

Returning to (1), we iterate this relation to get

$$\mathbf{P}(A_n) = \prod_1^n (1 - p_j) = \pi_n \quad \text{(say)}.$$

(b) Let B_k be the event that the first accident in the n day period occurs on the kth day. Then

$$\begin{aligned}\mathbf{P}(B_k) &= \mathbf{P}(A_k^c \cap A_{k-1}) = \mathbf{P}(A_k^c | A_{k-1})\mathbf{P}(A_{k-1}) \qquad \text{by (2.1.1)}\\ &= p_k \prod_1^{k-1} (1 - p_j) = p_k \pi_{k-1}\end{aligned}$$

Now, given an accident on the kth day, the event that there are no accidents in the succeeding $n - k$ days has the same probability as A_{n-k}. Hence the probability of exactly one accident is

$$\sum_{k=1}^n \mathbf{P}(B_k)\mathbf{P}(A_{n-k}) = \sum_{k=1}^n \pi_{k-1} p_k \pi_{n-k} = s_n \quad \text{(say)}.$$

(3) **Exercise** Show that if p_n is non-decreasing in n (and $p_n > 0$), then an accident is certain to occur sometime.

(4) **Exercise** Evaluate π_n and s_n when $p_n = p$.

(5) **Exercise** What is the probability that in the n days following an accident

(a) there is at least one accident?

(b) there are exactly two accidents?

(c) Evaluate these probabilities when $p_n = p$.

(6) **Exercise** Show that if c_i is a collection of numbers satisfying $0 \leq c_i \leq 1$, $i \geq 1$, then

$$c_1 + \sum_{i=2}^{\infty} c_i \prod_{j=1}^{i-1} (1 - c_j) + \prod_{i=1}^{\infty} (1 - c_i) = 1.$$

(7) **Exercise** What condition on $(p_n;\ n \geq 1)$ would allow a non-zero chance of no sceond accident?

2.8 Example: Dogfight

Algy, Biggles and the Commodore are fighting an air battle. In the first round each fires one shot in alphabetical order, and each may fire at any unhit opponent. Anyone hit drops out of the battle immediately. Survivors play successive rounds with the same rules as the first round until one winner remains.

On any shot aimed at an opponent, Algy hits with probability α, the Commodore hits with probability γ, and Biggles never misses. Show that if shots are independent and

$$\gamma > \alpha > 1 - \frac{\gamma}{1 - \gamma}$$

then Algy should fire his first shot into the ground.

Solution Suppose that Algy were to fire at Biggles and hit him. Algy would then be the Commodore's target, and the battle would continue with shots alternating between these two until a hit is scored. Let C_A be the event that the Commodore wins this two-man battle, and consider the following three events:

$A_1 \equiv$ the Commodore scores a hit with his first shot

$A_2 \equiv$ the Commodore misses and Algy scores a hit with his first returned shot

$A_3 \equiv$ the Commodore and Algy both miss their first shots.

If A_3 occurs then the next round begins under the same conditions, hence

$$\mathbf{P}(C_A|A_3) = \mathbf{P}(C_A).$$

Also $\mathbf{P}(C_A|A_1) = 1$, and $\mathbf{P}(C_A|A_2) = 0$. Since $A_i \cap A_j = \phi$ for $i \neq j$, and $\bigcup_1^3 A_i = \Omega$, we may use Theorem 2.1.3 to give

$$\mathbf{P}(C_A) = \gamma + 0 + (1-\gamma)(1-\alpha)\mathbf{P}(C_A)$$

yielding

$$\mathbf{P}(C_A) = \frac{\gamma}{1-(1-\gamma)(1-\alpha)}.$$

On the other hand, if Algy misses, then Biggles will certainly fire at the Commodore, since $\alpha < \gamma$, and hit him. Then Algy can win only if his second round shot at Biggles hits, for otherwise Biggles surely hits him with his second round shot. Thus in this case Algy wins with probability α.

Hence, missing Biggles gives Algy a better chance if

$$1 - \mathbf{P}(C_A) = \frac{\alpha(1-\gamma)}{1-(1-\gamma)(1-\alpha)} > \alpha$$

that is if $\alpha > 1 - \gamma/(1-\gamma)$.

(1) Exercise If Algy does fire his first shot at Biggles, what is the probability that he wins the battle?

(2) Exercise Algy is not a competent probabilist, and decides whether or not to shoot at Biggles by tossing a coin (heads he does, tails he doesn't). Given that the battle ends with the fourth shot, what is the probability that Algy aimed to miss?

(3) Exercise Suppose that Biggles is not infallible, in fact the probability that any shot of his hits is β. If $\alpha = 0.5$; $\beta = 0.875$ and $\gamma = 0.75$, where should Algy aim his first shot?

2.9 Example: Smears

In a population of women a proportion p have abnormal cells on the cervix. The Pap. test entails taking a sample of cells from the surface of the cervix and examining the sample to detect any abnormality.

(i) In a case where abnormal cells are present, the sample will fail to include any with probability μ.

(ii) In a sample including abnormal cells, examination fails to observe them with probability ν.

(iii) In a sample free of abnormal cells, normal cells are wrongly classified as abnormal with probability π.

All sampling and identification errors are independent.

If a randomly selected woman has such a test:

(a) what is the probability that the result is wrong?
(b) if an abnormality is reported, what is the probability that in fact no abnormal cells are present?

Solution (a) Let E denote the event that the result is in error, A denote the event that abnormal cells are present, and S denote the event that the sample fails to include abnormal cells when they are present. Using Theorem 2.1.2 we have

$$\mathbf{P}(E) = \mathbf{P}(E|A)\mathbf{P}(A) + \mathbf{P}(E|A^c)\mathbf{P}(A^c) = \mathbf{P}(E|A)p + \mathbf{P}(E|A^c)(1-p).$$

By (iii), $\mathbf{P}(E|A^c) = \pi$. Also by (2.1.5)

$$\mathbf{P}(E|A) = \mathbf{P}(E \cap S|A) + \mathbf{P}(E \cap S^c|A).$$

By (iii) and (i)

$$\mathbf{P}(E \cap S|A) = (1-\pi)\mu$$

and by (ii) and (i)

$$\mathbf{P}(E \cap S^c|A) = \nu(1-\mu).$$

Hence

$$\mathbf{P}(E) = p((1-\pi)\mu + \nu(1-\mu)) + (1-p)\pi.$$

(b) Let D denote the event that an abnormality is reported. By (2.1.1)

$$\mathbf{P}(A^c|D) = \frac{\mathbf{P}(A^c \cap D)}{\mathbf{P}(D)}.$$

Now by Theorem 2.1.2

$$\begin{aligned}\mathbf{P}(D) &= \mathbf{P}(D|A)\mathbf{P}(A) + \mathbf{P}(D|A^c)\mathbf{P}(A^c)\\ &= \mathbf{P}(D \cap S|A)\mathbf{P}(A) + \mathbf{P}(D \cap S^c|A)\mathbf{P}(A) + \pi(1-p) \quad \text{by (2.1.5)}\\ &= \pi\mu p + (1-\nu)(1-\mu)p + \pi(1-p)\end{aligned}$$

and

$$\mathbf{P}(A^c \cap D) = \pi(1-p).$$

Hence

$$\mathbf{P}(A^c|D) = \frac{\pi(1-p)}{\pi\mu p + \pi(1-p) + (1-\nu)(1-\mu)p}.$$

Notice that this is an example of Bayes' Theorem (2.1.4).

(1) **Exercise** Evaluate $\mathbf{P}(E)$ and $\mathbf{P}(A^c|D)$ when
(a) $p = 10\%$ and $\mu = \nu = \pi = 10^{-1}$, and when
(b) $p = 50\%$ and $\mu = 10^{-1}$ and $\nu = \pi = 10^{-2}$.

(2) **Exercise** What is the probability that the result is wrong if no abnormality is reported? Evaluate this in the above two cases and compare $\mathbf{P}(A^c|D)$ and $\mathbf{P}(A|D^c)$.

(3) **Exercise** Whatever the result of the test, it is recorded wrongly in the letter to the patient with the probability ρ independently of other errors. Let L be the event that the letter is wrong, and let M be the event that the letter reports abnormalities to the patient. Find $\mathbf{P}(L)$, $\mathbf{P}(A^c|M)$ and $\mathbf{P}(A|M^c)$.

2.10 Example: Gambler's ruin

You enter a casino with $\$k$, and on each spin of a roulette wheel you bet \$1 at evens on the event R that the result is red. The wheel is not fair, so $\mathbf{P}(R) = p < \frac{1}{2}$. If you lose all $\$k$ you must leave, and if you ever possess $\$K \geqslant \$k$, you choose to leave immediately. What is the probability that you leave with nothing? (Assume spins of the wheel are independent.)

Solution Let p_k be the probability that you leave with nothing. If the first spin results in red you gain \$1, and are in the same position as if you had just entered with $\$k + 1$. Thus conditional on R, your chance of leaving with nothing is p_{k+1}. Similarly if the first spin results in black (or zero) you have $\$k - 1$ and your chance of leaving with nothing is p_{k-1}. Hence

(1) $$p_k = pp_{k+1} + (1-p)p_{k-1}, \quad 0 < k < K.$$

If $k = 0$ then you certainly leave with nothing, and if $k = K$ you leave before betting. Hence $p_0 = 1$, and $p_K = 0$.
Writing (1) as

$$p_{k+1} - p_k = \frac{1-p}{p}(p_k - p_{k-1}) \quad \text{for } p > 0,$$

gives (on using $p_0 = 1$)

$$p_{k+1} - p_k = \left(\frac{1-p}{p}\right)^k (p_1 - 1), \quad \text{by Theorem 2.3.1(a).}$$

Hence

$$p_k = 1 + (p_1 - 1)\frac{\left(\frac{1-p}{p}\right)^k - 1}{\frac{1-p}{p} - 1}, \quad \text{by Theorem 2.3.1(b).}$$

Since $p_K = 0$, setting $k = K$ in this gives

$$0 = 1 + (p_1 - 1)\frac{\left(\frac{1-p}{p}\right)^K - 1}{\frac{1-p}{p} - 1},$$

and now eliminating p_1 gives finally

(2) $$p_k = \frac{\left(\frac{1-p}{p}\right)^K - \left(\frac{1-p}{p}\right)^k}{\left(\frac{1-p}{p}\right)^K - 1}.$$

(3) Exercise Show that as $K \to \infty$ in (2), $p_k \to 1$.

(4) Exercise Find p_k when $p = \frac{1}{2}$.

(5) Exercise Show that with probability one you do not remain in the casino forever.

(6) Exercise Given that you leave with nothing, what is the probability that you never possessed more than your initial $\$k$?

(7) Exercise Let $n(k, K, r)$ be the number of sequences of red and black which result in your leaving the casino with $\$K$ on the rth spin of the wheel. Show that the probability of this event is $n(k, K, r)p^{(r+K-k)/2}(1-p)^{(r-K+k)/2}$.

(8) Exercise Let $K = 2k$. Show that the probability that you leave the casino on the rth spin given that you leave with nothing is the same as the probability that you leave the casino on the rth spin given that you leave with $\$K$.

(9) Exercise Show that doubling the stakes increases your chance of leaving with $\$K$. What does this tell you about most gamblers?

2.11 Example: Accidents and insurance

In any given year the probability that a given male driver has a mishap entailing a claim from his insurance company is μ, independently of other years. The equivalent probability in female drivers is λ.

Assume there are equal numbers of male and female drivers insured with the Acme Assurance Association, which selects one of them at random.

(a) What is the probability that the selected driver makes a claim this year?
(b) What is the probability that the selected driver makes a claim in two consecutive years?
(c) If the insurance company picks a claimant at random, what is the probability that this claimant makes another claim in the following year?

Solution (a) Let A_1 and A_2 be the events that a randomly chosen driver makes a claim in each of the first and second years. Then conditioning on the sex of the driver (M or F) yields

$$\mathbf{P}(A_1) = \mathbf{P}(A_1|M)\mathbf{P}(M) + \mathbf{P}(A_1|F)\mathbf{P}(F) = \frac{1}{2}(\mu + \lambda)$$

because $\mathbf{P}(F) = \mathbf{P}(M) = \frac{1}{2}$.

(b) Likewise

$$\mathbf{P}(A_1 \cap A_2) = \mathbf{P}(A_1 \cap A_2|M)\mathbf{P}(M) + \mathbf{P}(A_1 \cap A_2|F)\mathbf{P}(F) = \frac{1}{2}(\mu^2 + \lambda^2).$$

(c) By definition

$$\mathbf{P}(A_2|A_1) = \mathbf{P}(A_2 \cap A_1)/\mathbf{P}(A_1) = \frac{\mu^2 + \lambda^2}{\mu + \lambda}$$

(1) Exercise Note that A_1 and A_2 are conditionally independent given the sex of the driver. Are they ever independent?

(2) Exercise Show that $\mathbf{P}(A_2|A_1) \geqslant \mathbf{P}(A_2)$. When does equality hold?

(3) Exercise

(a) Find the probability that a driver makes a claim in a third year given that the driver has claimed in each of the two preceding years.
(b) Find the probability that a driver claims in year n, given that this driver has claimed in all of the preceding n years.
(c) Find the limit in (b) as $n \to \infty$.

(4) **Exercise** Find the probability that a claimant is
(a) male
(b) female.

(5) **Exercise** Find the probability that a driver claiming in n successive years is male.

(6) **Exercise** Now rework the example and exercises without assuming that equal numbers of male and female drivers are insured with the AAA.

Remark The fact that a claimant is more likely to have a second accident, even though accidents were assumed independent for a given driver, is an example of what is sometimes called a sampling paradox (though it is not a paradox). It is the reason why insurance companies offer no-claims discounts (or at least, one of the reasons). It is the case in practice that $\mu > \lambda$.

2.12 Example: Protocols

Part A: Boys and girls

Consider the following question: 'Tom has exactly one sibling. What is the probability that it is a girl?'

(a) Do you think this question has a well-defined answer?
(b) If so, write down your answer, and then consider the following arguments:
   (i) There are three family possibilities; two girls, two boys, or one of each. Two girls is impossible, which leaves equal chances that the sibling is a boy or a girl. The answer is $\frac{1}{2}$.
   (ii) Families with a child called Tom arise in four equally likely ways: TB, BT, TG, GT. So Tom has a brother as often as he has a sister. The answer is $\frac{1}{2}$.
   (iii) There are four cases: BB, BG, GB, GG. The last is impossible, and in two of the remaining three cases the sibling is a girl. The answer is $\frac{2}{3}$.
   (iv) Assuming that the sex of siblings is independent, the other sibling is equally likely to be a girl or a boy. The answer is $\frac{1}{2}$.

Are any of these correct? Is yours correct? (You may assume that any given birth gives rise to one girl or one boy with equal probability.)

Solution (a) The question is ill-posed; there is no correct answer because the sample space is not defined. This is the same as saying that the underlying experiment (selecting Tom) is not described.

(b) We may consider some well-posed questions.

I A woman has two children which are independently equally likely to be a boy or a girl. One of them at least (Tom) is male. Now the sample space has four equally likely outcomes:

(1) $$\Omega = \{BG, GB, BB, GG\};$$

the event of interest is $A = \{BG, GB\}$, and we are given that B occurs, where $B = \{BG, GB, BB\}$. Hence

$$\mathbf{P}(A|B) = \frac{\mathbf{P}(A \cap B)}{\mathbf{P}(B)} = \frac{\mathbf{P}(A)}{\mathbf{P}(B)} = \frac{2}{3}.$$

II A woman has two children which are independently equally likely to be a boy or a girl. Her first son is called Tom with probability $p_1 < 1$. If she has two sons, and the oldest is not Tom, the second son is called Tom with probability p_2. The sample space is

(2) $$\Omega = \{BB, BT, TB, TG, BG, GT, GB, GG\}$$

where, for example, $\mathbf{P}(BT) = \frac{1}{4}(1 - p_1)p_2$. Then the required probability is, using Theorem 2.1.4,

$$\frac{\mathbf{P}(GT) + \mathbf{P}(TG)}{\mathbf{P}(TB) + \mathbf{P}(BT) + \mathbf{P}(TG) + \mathbf{P}(GT)} = \frac{p_1 + p_2}{2p_1 + 2p_2 - p_1 p_2}.$$

Notice that this is equal to $\frac{1}{2}$ if either $p_1 = 0$ or $p_2 = 0$, but not both. It is also $\frac{1}{2}$ if $p_1 = 1$ and we define p_2 to be zero. In any case

$$\frac{1}{2} \leqslant \frac{p_1 + p_2}{2p_1 + 2p_2 - p_1 p_2} \leqslant \frac{2}{3}.$$

Notice that we have assumed that families and names are independent, that is that women are not more (or less) likely to have boys because they want to call them Tom (or not), and that having a girl does not change the chance that a boy is called Tom.

III A boy is selected at random from a number of boys who have one sibling. This sample space has four equally likely outcomes

$$\Omega = \{B^*B, BB^*, B^*G, GB^*\}$$

where the star denotes the boy (Tom) who was picked at random. (The experiment amounts to picking one of the B-symbols in (1) with equal chance of picking any.) Hence the event of interest is $A = \{B^*G, GB^*\}$, and the required probability is $\frac{1}{2}$.

IV A chance acquaintance is introduced as Tom who has just one sibling. What is the chance that it is a sister?
The sample space is the set of your chance acquaintances. This is too vague to allow further progress.

Remark The arguments of (b) (i), (ii), (iii) and (iv) all appeared in letters to *The Guardian*, in June 1989. An answer can only be defined when the exact procedure (also known as a *protocol*) for selecting Tom is decided. If, for example, you meet Tom at a club for identical twins, the problem is different again.

Notice also that parts of this example are getting rather distant from the type of experiment used to justify our axioms. You may well see no particular reason to suppose that our theory of probability is relevant in, say, Case IV, or even in Case II.

(3) Exercise In the framework of Case II consider the following two procedures:
(a) select one of her two children at random
(b) select one of her sons (if any) at random.
In each case find the probability that the child is the elder given that his name is Tom.

(4) Exercise In the framework of Case II, can it be the case the TG, TB, BT and GT are equally likely outcomes?

(5) Exercise Suppose a woman has three children and each is independently equally likely to be male or female. Show that the event 'they are either all girls or all boys' is independent of the event 'at least two children are boys'.

Part B: Goats and cars

Suppose yourself to be participating in the following bizarre contest. You have a choice of three doors. Behind one door is a costly automobile, behind the other two doors are cheap goats. You choose the first door, whereupon the master of ceremonies opens the third door to reveal a goat; he then offers you the opportunity to change your choice of door. Can you calculate the probability that the car lies behind the second door? (You are given the object behind the door you open.)

Solution No, you cannot. To see this, let C_i be the event that the car lies behind the ith door, and let G be the event that a goat is revealed to you behind the third door. You require $\mathbf{P}(C_2|G)$ which we can write as

(6) $$\mathbf{P}(C_2 \cap G)/\mathbf{P}(G) = \frac{\mathbf{P}(G|C_2)\mathbf{P}(C_2)}{\mathbf{P}(G|C_1)\mathbf{P}(C_1) + \mathbf{P}(G|C_2)\mathbf{P}(C_2)} = \frac{\mathbf{P}(G|C_2)}{\mathbf{P}(G|C_1) + \mathbf{P}(G|C_2)},$$

on the reasonable assumption that the car is equally likely to be behind any door, so that $\mathbf{P}(C_2) = \mathbf{P}(C_1)$. Now observe that all three terms in the denominator and numerator of (6) depend on the decisions of the master of ceremonies. His rules for making his decision once again form a *protocol*. If you do not know his protocol for the contest you cannot calculate $\mathbf{P}(C_2|G)$.

Remark This problem was recently presented in *Parade* magazine (1990, 1991, distributed in the USA) and generated an extensive correspondence in that and several other periodicals. Almost all participants assumed (wrongly) that the problem as stated has one solution, and chiefly disputed as to whether the answer should be $\frac{1}{2}$ or $\frac{2}{3}$.

(7) Exercise Show that if you have paid the master of ceremonies enough to ensure that you win, then $\mathbf{P}(C_2|G) = 1$.

(8) Exercise Show that if the master of ceremonies has decided that: (i) whatever you choose he will show you a goat, and (ii) if he has a choice of two goats he will pick one at random, then $\mathbf{P}(C_2|G) = \frac{2}{3}$.

(9) Exercise Show that if the master of ceremonies has decided that: (i) whatever you choose he will open a different door, and (ii) he will pick it at random, then $\mathbf{P}(C_2|G) = \frac{1}{2}$.

(10) Exercise Show that if the master of ceremonies has decided that: (i) if a goat is behind the first door he will open it for you, and (ii) if a car lies behind the first door then he will open another door, then $\mathbf{P}(C_2|G) = 0$.

Remark There are many famous problems equivalent to these two, to all of which the correct answer is, *there is no unique answer*; (for example, the 'Prisoners' paradox', and 'Red Ace'). There seems to be no way of preventing the futile, acrimonious and incorrect discussions accompanying their regular appearance in the popular press. The so-called 'Doomsday Argument' provides a similar example of this error. (See *The Mathematical Intelligencer*, 1992 and 1993.)

2.13 Example: Eddington's controversy

Four men each tell the truth independently with probability $\frac{1}{3}$. D makes a statement which C reports to B, and B then reports C's statement to A. If A asserts that B denies that C claims that D is a liar, what is the probability that D spoke the truth?

Solution Let S_A be the event that A makes the given statement. Further let A denote the event that A tells the truth, and A^c denote the event that A lies, and so on. Then obviously

$$D \cap C \cap B \cap A \subseteq S_A,$$

for if they all tell the truth then A makes the given statement. Also, by the independence

$$\mathbf{P}(D \cap C \cap B \cap A) = \frac{1}{81}.$$

Likewise, by following through the chain of assertions, we see that $D \cap C^c \cap B^c \cap A$, $D \cap C^c \cap B \cap A^c$ and $D \cap C \cap B^c \cap A^c$ are all included in S_A, each having probability $\frac{4}{81}$.

When D lies,

$$D^c \cap C^c \cap B^c \cap A^c \subseteq S_A$$

for A also makes the given statement if they are all liars. Here

$$\mathbf{P}(D^c \cap C^c \cap B^c \cap A^c) = \frac{16}{81},$$

and likewise $D^c \cap C^c \cap B \cap A$, $D^c \cap C \cap B^c \cap A$ and $D^c \cap C \cap B \cap A^c$ are all included in S_A, each having probability $\frac{4}{81}$. These mutually exclusive outcomes exhaust the possibilities, so by conditional probability

$$\mathbf{P}(D|S_A) = \frac{\mathbf{P}(D \cap S_A)}{\mathbf{P}(D \cap S_A) + \mathbf{P}(D^c \cap S_A)} = \frac{\frac{13}{81}}{\frac{13}{81} + \frac{28}{81}} = \frac{13}{41}.$$

(1) **Exercise** What is the probability that C did claim that D is a liar, given S_A?

(2) **Exercise** What is the probability that both C and D lied, given S_A?

(3) **Exercise** Prove the result of the example more laboriously by using conditional probability rather than by listing outcomes.

(4) **Exercise** Eddington himself gave the answer to this problem as $\frac{25}{71}$. This is the controversy! Can you reconstruct the argument that led him to this answer?

Remark This example is similar to Example 2.12, in that it is entertaining but of no real practical value. Our theory of probability does not pretend to include this type of problem, and nothing can be said about the credibility of real reports by these methods. Despite this, the first attempt to do so was made in the seventeenth century, and such attempts have been repeated sporadically ever since.

PROBLEMS

1 The probability that an archer hits the target when it is windy is 0.4, when it is not windy her probability of hitting the target is 0.7. On any shot, the probability of a gust of wind is 0.3. Find the probability that:
(a) on a given shot there is a gust of wind and she hits the target.
(b) she hits the target with her first shot.
(c) she hits the target exactly once in two shots.
(d) there was no gust of wind on an occasion when she missed.

2 Let A, B be two events with $\mathbf{P}(B) > 0$. Show that
(a) if $B \subset A$, then $\mathbf{P}(A|B) = 1$,
(b) if $A \subset B$, then $\mathbf{P}(A|B) = \mathbf{P}(A)/\mathbf{P}(B)$.

3 Three biased coins C_1, C_2, C_3 lie on a table. Their respective probabilities of falling heads when tossed are $\frac{1}{3}$, $\frac{2}{3}$ and 1. A coin is picked at random, tossed, and observed to fall heads. Calculate the probability that it is C_k for each $k = 1, 2, 3$.

Given that a coin has been tossed once and observed to fall heads, calculate the probability that a second throw of the same coin will also produce heads.

The experiment is begun again with the same three coins. This time the coin selected is tossed twice and observed to fall heads both times. Calculate the probability that it is C_k for each $k = 1, 2, 3$.

Given that a coin has been tossed twice and observed to fall heads both times, calculate the probability that a third throw of the same coin will also produce heads.

4 (a) An event E is independent of itself. Show that it has probability either 0 or 1.
(b) Events A and B are disjoint. Can you say whether they are dependent or independent?
(c) Prove that if events A and B are independent then so are the events A^c and B, and the events A^c and B^c.

5 Candidates are allowed at most three attempts at a given test. Given $j - 1$ previous failures, the probability that a candidate fails at his jth attempt is p_j. If $p_1 = 0.6$, $p_2 = 0.4$ and $p_3 = 0.75$, find the probability that a candidate:
(a) passes at the second attempt;
(b) passes at the third attempt;
(c) passes given that he failed at the first attempt;
(d) passes at the second attempt given that he passes.

6 Dick throws a die once. If the upper face shows j, he then throws it a further $j - 1$ times and adds all j scores shown. If this sum is 3, what is the probability that he only threw the die
(a) once altogether?
(b) twice altogether?

7 A man has five coins in his pocket. Two are double-headed, one is double-tailed and two are normal. They can be distinguished only by looking at them.

(a) The man shuts his eyes, chooses a coin at random and tosses it. What is the probability that the lower face of the coin is a head?

(b) He opens his eyes and sees that the upper face is a head. What is the probability that the lower face is a head?
(c) He shuts his eyes again, picks up the coin and tosses it again. What is the probability that the lower face is a head?
(d) He opens his eyes and sees that the upper face is a head. What is the probability that the lower face is a head?

8 An urn contains four dice, one red, one green and two blue.
(a) One is selected at random; what is the probability that it is blue?
(b) The first is not replaced, and a second die is removed. What is the chance that it is: (i) blue? or (ii) red?
(c) The two dice are thrown. What is the probability that they show the same numbers and are the same colour?
(d) Now the two remaining in the urn are tossed. What is the probability that they show the same number and are the same colour, given that the first two did not show the same number and colour?

9 A twelve-sided die A has 9 green faces and 3 white faces, whereas another twelve-sided die B has 3 green faces and 9 white faces. A fair coin is tossed once. It it falls heads, a series of throws is made with die A alone; if it falls tails then only the die B is used.
(a) Show that the probability that green turns up at the first throw is $\frac{1}{2}$.
(b) If green turns up at the first throw, what is the probability that die A is being used?
(c) Given that green turns up at the first two throws, what is the probability that green turns up at the third throw?

10 Suppose that any child is male with probability p, or female with probability $1-p$, independently of other children. In a family with four children, let A be the event that there is at most one girl, and B the event that there are children of both sexes. Show that there is a value of p, with $0 < p < \frac{1}{2}$, such that A and B are independent.

11 Suppose that parents are equally likely to have (in total) one, two or three offspring. A girl is selected at random; what is the probability that the family includes no older girl? [Assume that children are independent and equally likely to be male or female.]

12 Two roads join Ayton to Beaton, and two further roads join Beaton to the City. Ayton is directly connected to the City by a railway. All four roads and the railway are each independently blocked by snow with probability p. I am at Ayton.
(a) Find the probability that I can drive to the City.
(b) Find the probability that I can travel to the City.
(c) Given that I can travel to the City, what is the probability that the railway is blocked?

13 An urn contains b blue and r red balls, which may be withdrawn at random according to one of the following three schemes.
(a) The balls are removed at random one at a time until all those remaining are of the same colour.
(b) The balls are removed until a ball differs in colour from its predecessor. This first different ball is replaced in the urn; this process is then continued until the remaining balls are all of the same colour.
(c) The balls are removed one by one and inspected. The first is discarded. Each succeeding ball that is the same colour as its predecessor is replaced, the others are discarded, until the remaining balls are all of the same colour.
In each case find the probability that the remaining balls are all red.

14 Let $A_1, A_2, \ldots, A_n$ be independent events. Sow that the probability that none of the events $A_1, \ldots, A_n$ occur is less than $\exp(-\sum_1^n \mathbf{P}(A_i))$.

15 Let A and B be independent events. Show that

$$\max\{\mathbf{P}((A \cup B)^c), \mathbf{P}(A \cap B), \mathbf{P}(A \triangle B)\} \geqslant \frac{4}{9}.$$

16 A coin is tossed repeatedly; on each toss a head is shown with probability p, or a tail with probability $1 - p$. All tosses are mutually independent. Let E denote the event that the first run of r successive heads occurs earlier than the first run of s successive tails. Let A denote the outcome of the first toss. Show that

$$\mathbf{P}(E|A = \text{head}) = p^{r-1} + (1 - p^{r-1})\mathbf{P}(E|A = \text{tail}).$$

Find a similar expression for $\mathbf{P}(E|A = \text{tail})$ and hence find $\mathbf{P}(E)$.

17 After marking the papers of a certain student the examiners are unable to decide whether he really understands the subject or is just bluffing. They reckon that the probability that he is a bluffer is p, $0 < p < 1$, and the probability that he understands is $q = (1 - p)$. They therefore give him a *viva voce* consisting of n independent questions, each of which has a probability u of being answered by someone who understands the subject. Unfortunately there is also a probability b, $0 < b < 1$, that the answer can be guessed by someone who does not understand. Show that the probability that the student understands given that he manages to answer k questions correctly is given by r, where

$$r = \frac{qu^k(1-u)^{n-k}}{qu^k(1-u)^{n-k} + pb^k(1-b)^{n-k}}.$$

Show that if the student gets every single question right and $u > b$ then as n increases the probability that the student really understands tends to 1. How many questions must the student get right in order to convince the examiners that it is more likely that he understands the subject than that he is bluffing?

18 A team of three students Amy, Bella and Carol answer questions in a quiz. A question is answered by Amy or Bella or Carol with probability $\frac{1}{2}$ or $\frac{1}{3}$ or $\frac{1}{6}$ respectively. The probability of Amy or Bella or Carol answering a question correctly is $\frac{4}{5}$ or $\frac{3}{5}$ or $\frac{3}{5}$ respectively. What is the probability that the team answers a question correctly? Find the probability that Carol answered the question given that the team answered incorrectly.

The team starts the contest with one point, and gains (loses) one point for each correct (incorrect) answer. The contest ends when the team's score reaches zero points or ten points. Find the probability that the team will win the contest by scoring ten points, and show that this is approximately $\frac{4}{7}$.

19 A and B play a sequence of games, in each of which A has a probability p of winning and B has a probability q $(= 1 - p)$ of winning. The sequence is won by the first player to achieve a lead of two games. By considering what may happen in the first two games, or otherwise, show that the probability that A wins the sequence is $p^2/(1 - 2pq)$.

If the rules are changed so that the sequence is won by the player who first wins two consecutive games, show that the probability that A wins the sequence becomes $p^2(1 + q)/(1 - pq)$.

Which set of rules gives the weaker player the better chance of winning the sequence?

20 You toss a coin. If it shows a tail, you roll one die and your score is the number it shows. If the coin shows a head, you toss five more coins and your score is the total number of heads shown (including the first coin). If you tell me only that your score is two, what is the probability that you rolled a die?

21 Three fair dice labelled A, B and C are rolled on to a sheet of paper. If a pair show the same number a straight line is drawn joining them. Show that the event that the line AB is drawn is independent of the event that BC is drawn. What is the probability that a complete triangle is drawn? [The dice are not colinear.]

22 You roll a fair die n times. What is the probability that
(a) You have rolled an odd number of sixes?
(b) You have not rolled a six on two successive rolls?
(c) You rolled a one before you rolled a six, given that you have rolled at least one of each?

23 Irena throws at a target. After each throw she moves further away so that the probability of a hit is two-thirds of the probability of a hit on the previous throw. The probability of a hit on the first throw is $\frac{1}{4}$. Find the probability of a hit on the nth throw. Deduce that the probability of never hitting the target is greater than $\frac{1}{4}$.

24 A fair coin is tossed three times. What is the probability that it lands 'heads' at least once?

In a coin-tossing game, a player tosses five fair coins. If he is content with the result, he stops. If not, he picks up one or more of the coins and tosses them a second time. If he is still dissatisfied, he may for one last time pick up and throw again one or more of the coins. Show that if the player's aim is to finish with five heads showing, and if he uses the best strategy, then the probability that he will succeed is $(\frac{7}{8})^5$.

A second player plays the same game but aims to finish with *either* all heads *or* all tails showing. What is the probability of his succeeding?

25 Alf and Bert play a game which each wins with probability $\frac{1}{2}$. The winner then plays Charlie whose probability of winning is always θ. The three continue in turn, the winner of each game always playing the next game against the third player, until the tournament is won by the first player to win two successive games. Let p_A, p_B, p_C be the probabilities that Alf, Bert and Charlie respectively win the tournament. Show that $p_C = 2\theta^2/(2 - \theta + \theta^2)$. Find p_A and p_B, and find the value of θ for which p_A, p_B, p_C are all equal. (Games are independent.)

If Alf wins the tournament, what is the probability that he also won the first game?

26 Box A contains 3 red balls and 2 white balls; box B contains 2 red balls and 2 white balls. A fair die is thrown. If the upper face of the die shows 1 or 2, a ball is drawn at random from box A and put in box B and then a ball is drawn at random from box B. If the upper face of the die shows 3, 4, 5 or 6, a ball is drawn at random from box B and put in box A and then a ball is drawn at random from box A.

What are the probabilities
(a) that the second ball drawn is white?
(b) that both balls drawn are red?
(c) that the upper face of the red die showed 3, given that one ball drawn is white and the other red?

27 A fair six-sided die, with faces numbered from 1 to 6, is thrown repeatedly on to a flat surface until it first lands with the 6 face uppermost. Find the probability that this requires:
(a) n throws, and
(b) an even number of throws.
(c) Show that the probability that the 5 face appears at least once before the first 6 is $\frac{1}{2}$, and find the probability that all of the faces 1 to 5 appear before the first 6.

28 Suppose that n water-lily leaves are placed so that the base of each leaf lies on a circle. A frog is initially on leaf L_1; she hops clockwise to the adjacent leaf L_2 with probability p, or anticlockwise to leaf L_n with probability q. Succeeding hops are independent, and go to the nearest leaf clockwise with probability p or the nearest leaf anticlockwise with probability q.

Find the probability that:
(a) the frog returns to L_1 before visiting all n leaves, and
(b) the first hop on to L_1 has the same orientation as the first hop off L_1.
(c) What is the probability that the first hop on to L_1 is clockwise?

29 Anselm and Bill toss a fair coin repeatedly. Initially Anselm has m marks, where $1 \leqslant m \leqslant n-1$. If the coin shows a head then Anselm gains a mark from Bill, otherwise he forfeits a mark to Bill. Whenever Anselm has n marks he must immediately give one to Bill.

Let p_m^k be the probability that Anselm has n marks on k occasions before the first moment at which he has no marks. Write down a difference equation for p_m^k, with appropriate boundary conditions, and deduce that for $k \geqslant 1$,

$$p_m^k = \frac{m}{n^2}\left(1 - \frac{1}{n}\right)^{k-1}.$$

Explain how you could have shown this without solving the equation for p_m^k. Show that Anselm is certain to lose all his marks eventually.

30 A and B each have \$60. They play a sequence of independent games at each of which A wins \$$x$ from B with probability p, or loses \$$x$ to B with probability q, where $p + q = 1$. The stake x is determined by rolling a fair die once, and setting x as the number shown by the die; $1 \leqslant x \leqslant 6$.

(a) What is the probability that A wins all his opponent's fortune before losing all his own?

(b) If A could choose the stake to be an integer x such that $1 \leqslant x \leqslant 6$, and $p < q$, what value should he choose for x?

31 A document is equally likely to be in any of three boxfiles. A search of the ith box will discover the document (if it is indeed there) with probability p_i.

What is the probability that the document is in the first box:

(a) given that I have searched the first box once and not found it?

(b) given that I have searched the first box twice and not found it?

(c) given that I have searched all three boxes once and not found it?

Assume searches are independent.

32 A network forming the edges of a cube is constructed using twelve wires, each one metre long. An ant is placed on one corner and walks around the network, leaving a trail of scent as it does so. It never turns around in the middle of an edge, and when it reaches a corner:

(i) if it has previously walked along both the other edges, it returns along the edge on which it has just come;

(ii) if it has previously walked along just one of the other edges, it continues along the edge along which it has not previously walked;

(iii) otherwise it chooses one of the other edges arbitrarily.

Show that the probability that the ant passes through the corner opposite where it started after walking along just three edges is $\frac{1}{2}$, but that it is possible that it never reaches the opposite corner. In the latter case, determine the probability of this occurring. What is the greatest distance that the ant has to walk before an outside observer (who knows the rules) will know whether or not the ant will ever reach the corner opposite where it started?

Show that the rules may be modified to guarantee that the ant (whose only sense is smell) will be able to reach the corner opposite the corner where it started by walking not more than a certain maximum distance which should be determined. The ant can count.

33 **Pooling** You have $(n!)^2$ jars of fluid, one of which is contaminated. Any jar is equally likely to be the contaminated one, and you need to identify it. Luckily you have an infallible test.

(a) If you test the jars one at a time, find the probability that you require t tests to identify the contaminated jar.

(b) Alternatively you may arrange the jars in j groups of size k, where $jk = (n!)^2$. A sample from each of the jars in a group is pooled in one jar, and this pooled sample is tested. On finding the contaminated pooled sample, each jar of this group is tested separately. Find the probability that you require t tests to find the contaminated jar.

(c) What is the best choice for j and k?

34 **Simpson's paradox** Two drugs are being tested. Of 200 patients given drug A, 60 are cured; and of 1100 given drug B, 170 are cured. If we assume a homogeneous group of patients, find the probabilities of successful treatment with A or B. Now closer investigation reveals that the 200 patients given drug A were in fact 100 men, of whom 50 were cured, and 100 women of whom 10 were cured. Further, of the 1100 given drug B, 100 were men of whom 60 were cured, and 1000 were women of whom 110 were cured. Calculate the probability of cure for men and women receiving each drug; note that B now seems better than A. [Results of this kind indicate how much care is needed in the design of experiments.]

35 In Problem 34, given that a randomly chosen patient is cured, find:

(a) the probability that the patient is male;

(b) the probability that the patient is female.

36 **Prisoners' paradox** Three prisoners are informed by their warder that one of them is to be released, and the other two shipped to Devil's Island, but the warder cannot inform any prisoner of that prisoner's fate. Prisoner A thus knows his chance of release to be $\frac{1}{3}$. He asks the warder to name some one of the other two who is destined for Devil's Island, and the warder names B. Can A now calculate the conditional probability of his release?

37 Let A and B be events. Show that $\mathbf{P}(A \cap B | A \cup B) \leqslant \mathbf{P}(A \cap B | A)$. When does equality hold?

38 Explain the following 'paradox' posed by Lewis Carroll. We are provided with a supply of balls which are independently equally likely to be black or white.

Proposition If an urn contains two such balls, then one is black and the other white.

Proof Initially $\mathbf{P}(BB) = \mathbf{P}(BW) = \mathbf{P}(WB) = \mathbf{P}(WW) = \frac{1}{4}$. Add a black ball, so that now

$$\mathbf{P}(BBB) = \mathbf{P}(BBW) = \mathbf{P}(BWB) = \mathbf{P}(BWW) = \tfrac{1}{4}.$$

Now pick a ball at random. By (2.1.3)

$$\mathbf{P}(\text{black ball drawn}) = 1.\tfrac{1}{4} + \tfrac{2}{3}\cdot\tfrac{1}{4} + \tfrac{2}{3}\cdot\tfrac{1}{4} + \tfrac{1}{3}\cdot\tfrac{1}{4} = \tfrac{2}{3}.$$

But if I pick a ball at random from three, with probability $\frac{2}{3}$ of drawing a black ball, then two are black and one is white. Hence before adding the black ball the urn contained one white ball and one black ball. *QED*

39 Let $M_1, M_2, \ldots, M_n$ be a sequence of men such that M_j reports to M_{j+1} on a statement made by M_{j-1}. Let R_n be the event that M_n reports that M_{n-1} reports that ... that M_2 reports that M_1 is a liar. If each reporter lies independently with probability p, find p_n, the probability that M_1 told the truth given R_n. Show that as $n \to \infty$, $p_n \to 1 - p$.

40 Suppose that for events S, A and B,

$$(*) \qquad \mathbf{P}(S|A) \geqslant \mathbf{P}(S)$$

$$\mathbf{P}(A|S \cap B) \geqslant \mathbf{P}(A|S)$$

and

$$\mathbf{P}(A|S^c) \geqslant \mathbf{P}(A|S^c \cap B).$$

(a) Show that, except in trivial cases, $\mathbf{P}(S|A \cap B) \geqslant \mathbf{P}(S|B)$.

(b) Show that $\mathbf{P}(S|A) \geqslant \mathbf{P}(A)$

(c) Show that if $(*)$ is replaced by $\mathbf{P}(S|B) \geqslant \mathbf{P}(S)$, then $\mathbf{P}(S|A \cap B) \geqslant \mathbf{P}(S|A)$.

41 You have to play Alekhine, Botvinnik and Capablanca once each. You win each game with respective probabilities p_a, p_b and p_c, where $p_a > p_b > p_c$. You win the tournament if you win two consecutive games, otherwise you lose, but you can choose in which order to play the three games.

Show that to maximize your chance of winning you should play Alekhine second.

42 Show that the events A and B are mutually attractive if and only if $\mathbf{P}(B|A) > \mathbf{P}(B|A^c)$.

43 **Weather** Days can be sunny or cloudy. The weather tomorrow is the same as the weather today with probability p, or it is different with probability q, where $p + q = 1$. If it is sunny today, show that the probability s_n that it will be sunny n days hence satisfies

$$s_n = (p - q)s_{n-1} + q; \qquad n \geqslant 1$$

where $s_0 = 1$. Deduce that

$$s_n = \tfrac{1}{2}(1 + (p - q)^n); \qquad n \geqslant 0.$$

44 **Flats, sharps and craps** Dice can be crooked (or weighted) in various ways. One way is to shorten the distance between one pair of opposite faces, thus making them more likely; these are called *flats* (also known as *broads* or *doctors*).

Another way is to taper all four sides of the die, creating a truncated pyramid, or to insert a weight in the base. The top face becomes more likely and the base less likely; these are called *sharps*.

You have three pairs of dice:

(i) a fair pair for which $p_1 = p_2 = p_3 = p_4 = p_5 = p_6 = \frac{1}{6}$,
(ii) a pair of 1–6 flats for which $p_1 = p_6 = \frac{1}{4}$ and $p_2 = p_3 = p_4 = p_5 = \frac{1}{8}$, and
(iii) a pair of 5–2 sharps for which $p_5 = \frac{1}{4}$, $p_2 = \frac{1}{12}$ and $p_1 = p_3 = p_4 = p_6 = \frac{1}{6}$.

With which pair would you prefer to play craps? (See Example 1.12 for the rules.)

3
Counting

This chapter deals with a special subject, and may be omitted on a first reading. Its contents are important and useful, but are not a prerequisite for most of the following chapters.

3.1 First principles

We have seen that many interesting problems in probability can be solved by counting the number of outcomes in an event. Such counting often turns out to be useful in more general contexts also. This chapter sets out some simple methods of dealing with the commonest counting problems

The basic principles are pleasingly easy, and are perfectly illustrated in the following examples.

(1) Principle If I have m garden forks and n fish forks, then I have $m + n$ forks altogether.

(2) Principle If I have m different knives and n different forks, then there are mn distinct ways of taking a knife and fork.

These principles can be rephrased in general terms involving objects, operations or symbols and their properties, but the idea is already obvious. The important points are that in (1), the two sets in question are disjoint; that is a fork cannot be both a garden fork and a fish fork. In (2), my choice of knife in no way alters my freedom to choose any fork (and vice versa).

Real problems involve, for example, catching different varieties of fish, drawing various balls from a number of urns, and dealing hands at numerous types of card game. In the standard terminology for such problems we say that a number n (say) of *objects* or *things* are to be *divided* or *distributed* into r *classes* or *groups*.

The number of ways in which this distribution can take place depends on whether

(i) the objects can be distinguished or not;
(ii) the classes can be distinguished or not;
(iii) the order of objects in a class is relevant or not;

(iv) the order of classes is relevant or not;
(v) the objects can be used more than once or not at all;
(vi) empty classes are allowed or not.

We shall generally consider only the cases having applications in probability problems. Other aspects are explored in books devoted to combinatorial theory.

(3) Example (a) Six dice are rolled. What is the probability that they all show different faces?
(b) What is the probability that five dice show different faces when rolled?

Solution (a) Let A be the event that they all show different faces. Then because of the assumed symmetry of the dice

$$\mathbf{P}(A) = \frac{|A|}{|\Omega|}.$$

Now the upper face of each die may be freely chosen in six different ways, so by Principle 2

$$|\Omega| = 6^6 = 46656.$$

However, for outcomes in A the upper faces are required to be different. Thus when the upper face of one die is freely chosen in 6 ways, the upper face of the next can be freely chosen in 5 ways (different from the first choice). The next may be freely chosen in 4 ways, and so on. Hence, by Principle 2, $|A| = 6!$ and

$$\mathbf{P}(A) = \frac{6!}{6^6} = \frac{5}{324}.$$

(b) Let $\hat{A}$ be the event that the 5 dice show different faces. By the same argument as above $|\Omega| = 6^5$.

For outcomes in $\hat{A}$ we may first make a free choice of which different five faces are to be shown; this is the same as choosing one face not to be shown, which we can do in 6 ways. Then the first face is freely chosen in 5 ways, the second in 4 ways and so on. Hence, $|\hat{A}| = 6!$ and

$$\mathbf{P}(\hat{A}) = \frac{6!}{6^5} = \frac{5}{54}. \qquad \bullet$$

3.2 Permutations: ordered selection

Suppose that a number of objects are placed randomly in a row; playing cards or lottery numbers provide trite examples (but important if you hold a ticket). The number of ways in which this arrangement may occur depends on how many objects there are, whether they are all distinct or not, whether they may be repeated, and so on. Such arrangements are called *permutations*.

(1) Theorem Given n distinct symbols, the number of distinct permutations (without repetition) of length $r \leqslant n$ is

$$n(n-1)\ldots(n-r+1) = \frac{n!}{(n-r)!}.$$

(2) Theorem Given n distinct symbols which may be repeated any number of times, the number of permutations of length r is n^r.

Proof Theorems (1) and (2) are easily proved by induction. You do it. ■

(3) Theorem Given $n = \sum_{i=1}^{r} n_i$ symbols of r distinct types, where n_i are of type i and are otherwise indistinguishable, the number of permutations (without repetition) of all n symbols is

$$M_n(n_1, \ldots, n_r) = \frac{n!}{\prod_{i=1}^{r} n_i!}.$$

Proof Suppose that the symbols of each type are numbered so that they are all distinguishable. Then in each originally unnumbered permutation, the symbols of type 1 can be permuted in $n_1!$ ways, the symbol of type 2 in $n_2!$ ways, and so on. Thus the total number of permutations is $M_n(n_1, \ldots, n_r)n_1!n_2! \ldots n_r!$. However, we already know from (1) that the number of permutations of n objects is $n!$. Hence

$$M_n(n_1, \ldots, n_r)n_1! \ldots n_r! = n!$$

which proves (3). ■

The number M_n is known as a *multinomial coefficient*. A particularly important case that arises frequently is when $r = 2$. This is a *binomial coefficient*, and it has its own special notation:

(4)
$$\begin{aligned} M_n(k, n-k) &= \binom{n}{k} && \text{in most books,} \\ &= {}^nC_k && \text{in some older books,} \\ &= \frac{n!}{k!(n-k)!}. \end{aligned}$$

(5) Example You are playing bridge, and when you pick up your hand you notice that the suits are already grouped, that is the clubs are all adjacent to each other, the hearts likewise, and so on. Given that your hand contains 4 spades, 4 hearts, 3 diamonds and 2 clubs, what is the probability $\mathbf{P}(G)$ of this event G?

Solution There are 13! permutations of your hand, which we assume are equally likely by symmetry. Now there are 4! permutations in which the spades are adjacent in any given position, 4! where the hearts are adjacent, and so on. Furthermore, there are 4! permutations of the order in which the respective suits may be placed in their adjacent blocks. Hence, the number of permutations in which G occurs is

4!4!4!3!2! and

(6) $$\mathbf{P}(G) = \frac{(4!)^3 3!2!}{13!} = \frac{4!}{M_{13}(4,4,3,2)}.$$

Alternatively, you may observe from (3) that there are $M_{13}(4,4,3,2)$ permutations of your hand where cards of the same suit are regarded as indistinguishable. For each order of suits, only one of these is in G. Since there are 4! permutations of the suit order, we immediately recover (6) again. ●

Finally we remark that the definition of the symbol $\binom{n}{r}$ is sometimes extended to cases other than that when n and r are integers with $0 \leqslant r \leqslant n$.

Definition For real x and non-negative integer r

$$\binom{x}{r} = \frac{x(x-1)\dots(x-r+1)}{r!}.$$ ▲

This definition can provide more compact expressions occasionally, for example we have

$$\binom{-x}{r} = (-)^r \binom{x+r-1}{r}.$$

3.3 Combinations: unordered selection

In a bridge hand, or an election, the order in which you get your cards, or the politician his votes, is irrelevant. In problems of this type we do not arrange, we choose; a choice of objects or symbols is also called a *combination*.

(1) **Theorem** The number of ways of choosing a set of r symbols from a set of n distinct symbols without repetition is

$$\frac{n!}{r!(n-r)!} = \binom{n}{r}.$$

Proof This is just a special case of Theorem 3.2.3. ■

(2) **Theorem** The number of ways of dividing n distinct objects into r distinct groups of sizes $n_1, n_2, \dots, n_r$, where $\sum_{i=1}^r n_i = n$, is $n!/\prod_{i=1}^r n_i!$.

Proof This is also a simple corollary of Theorem 3.2.3. ■

(3) **Theorem** Out of $n = \sum_1^r n_i$ symbols (where the ith distinct set contains n_i indistinguishable symbols), we can select $\prod_{i=1}^r (n_i + 1) - 1$ combinations.

Proof Note that we can select any number of symbols from zero to n_i, from each of the r sets, but we cannot take zero from all of them. ■

(4) **Theorem** Given a set of n distinct symbols which may be repeated any number of times, the number of ways of choosing a set of size r is $\binom{n+r-1}{r}$.

Proof A proof of Theorem (4) may be found in Example 3.11. ■

(5) **Example: Ark** The wyvern is an endangered species in the wild. You wish to form a captive breeding colony, and you estimate that a viable colony should initially contain r males and r females. You therefore trap a sequence of animals, each of which is independently male with probability p, or female with probability $q = 1 - p$, where $p \neq q$ (the females are more wary). Find the probability p_n that it is necessary to capture n animals in order to create your viable colony of r males and r females.

Solution Let A_n be the event that you first possess r of each sex with the nth capture, and let M be the event that the nth animal is male. Then of the previous $n - 1$ captured animals, $r - 1$ are male and $n - r$ are female. For any fixed order of these sexes, the probability of being captured in that order is $p^r q^{n-r}$. The number of ways of ordering r males and $n - r$ females, with a male last, is just the same as the number of ways of choosing $r - 1$ of the first $n - 1$ captures to be male. By (1) this is $\binom{n-1}{r-1}$. Hence, for $n \geqslant 2r$,

$$\mathbf{P}(A_n \cap M) = \binom{n-1}{r-1} p^r q^{n-r}.$$

Likewise when the last animal is female

$$\mathbf{P}(A_n \cap M^c) = \binom{n-1}{r-1} q^r p^{n-r},$$

and so

$$p_n = \binom{n-1}{r-1} q^r p^r (p^{n-2r} + q^{n-2r}), \quad n \geqslant 2r. \quad \bullet$$

3.4 Inclusion–exclusion

If a group of N men contains $N(b_1)$ who are bald, $N(b_2)$ who are bearded and $N(b_1, b_2)$ who are bald and bearded, how many altogether are bald, bearded or both? The answer is

(1) $$N_1 = N(b_1) + N(b_2) - N(b_1, b_2),$$

because anyone who is both is counted once in all three terms on the right hand side, and so contributes just one to the total N_1 in (1).

More generally if a group of N objects may each have up to r distinct properties $b_1, \ldots, b_r$, then the number possessing at least one is

(2) $$N_1 = \sum_{b_i} N(b_i) - \sum_{b_i < b_j} N(b_i, b_j) + \ldots + (-)^{r-1} N(b_1, \ldots, b_r).$$

This is proved either by induction or by noting that any object having exactly k of the properties is counted $\binom{k}{t}$ times in the tth term, and it is the case that

$$\sum_{t=1}^{k}(-)^{t+1}\binom{k}{t} = 1.$$

(See Problem 3.28(a).) Hence this object contributes just one to the total N_1, as required. Notice that N_0, the number of objects possessing none of the r properties, is just

(3) $$N_0 = N - N_1.$$

(4) **Example: Derangements** A permutation of the first n integers is called a *derangement* if no integer is in its natural position. Thus $(3, 2, 1)$ is not a derangement of $(1, 2, 3)$, but $(2, 3, 1)$ is a derangement. Suppose one of the $n!$ permutations of $(1, \ldots, n)$ is picked at random. Find p_n, the probability that it is a derangement, and show that as $n \to \infty$, $p_n \to e^{-1}$.

Solution Let b_k be the property that the integer k is in its natural position (the kth place). Then the number of derangements is $n! - N_1$, where N_1 is given by (2). Now $N(b_k)$ is the number of permutations of $(1, \ldots, n)$ with k fixed, namely $(n-1)!$. Likewise $N(b_i, b_j)$ is the number of permutations of $(1, 2, \ldots, n)$ with i and j fixed, namely $(n-2)!$, and so on. Hence

$$N_1 = \sum_{i=1}^{n}(n-1)! - \sum_{1\leq i<j\leq n}(n-2)! + \ldots + (-)^{n-1}$$

$$= n(n-1)! - \binom{n}{2}(n-2)! + \ldots + (-)^{n-1}.$$

Now since all permutations were supposed equally likely, the required probability is

$$p_n = \frac{1}{n!}(n! - N_1) = 1 - \frac{1}{1!} + \frac{1}{2!} - \frac{1}{3!} + \ldots + (-)^n\frac{1}{n!} \to e^{-1}$$

as $n \to \infty$. ●

3.5 Recurrence relations

The answer a to a counting problem usually depends on some given parameters, such as the original number n of objects, or the number r which are red (say), or the number s selected. We can make this explicit by writing, for example, $a = a(n, r, s)$. It is often possible to find relationships between two or more of $a(n, r, s)$, and then the problem is reduced to solving a recurrence relation. Some of these can indeed be solved.

(1) **Theorem** If $(a_n;\ n \geq 0)$ satisfies

$$a_n = c(n)a_{n-1}, \quad n \geq 1,$$

then

$$a_n = \prod_{k=1}^{n} c(k) a_0.$$

Proof Trivial, by induction. ■

(2) **Theorem** If $(a_n;\, n \geqslant 0)$ satisfies

$$a_{n+2} + 2b a_{n+1} + c a_n = 0, \quad n \geqslant 0,$$

and $x = \alpha$ and $x = \beta$ are two distinct solutions of $x^2 + 2bx + c = 0$, then

$$a_n = \frac{1}{\alpha - \beta}((a_1 - \beta a_0)\alpha^n - (a_1 - \alpha a_0)\beta^n).$$

Proof Straightforward by induction. ■

Higher order linear difference equations can be solved similarly; we omit the details. The recurrence can be in more than one variable, as the following example shows.

(3) **Example** Let $a(n, k)$ be the number of ways of choosing k objects from n objects. Show that

$$a(n, k) = \binom{n}{k}.$$

Solution Suppose we add an additional $(n + 1)$th object, and choose k of these $n + 1$. This may be done in $a(n + 1, k)$ ways. But also we may consider whether the additional object was among those chosen or not. The number of choices when it is *not* is $a(n, k)$. If it *is*, then it is necessary to choose $k - 1$ further objects from the remaining n, which we may do in $a(n, k - 1)$ ways. Then $a(n, k)$ satisfies the difference equation

(4) $$a(n + 1, k) = a(n, k) + a(n, k - 1), \quad 0 \leqslant k \leqslant n,$$

with solution by inspection $a(n, k) = \binom{n}{k}$. ●

The array generated by $a(n, k);\, 0 \leqslant k \leqslant n$, as n increases, is called *Pascal's triangle*; it has many curious and interesting properties, which we have no time to explore.

3.6 Generating functions

Even quite straightforward counting problems can lead to laborious and lengthy calculations. These are often greatly simplified by using generating functions (introduced by de Moivre and Euler in the early eighteenth century). Later examples will show the utility of generating functions; in this section we give a fairly bald list of basic definitions and properties, for ease of reference.

$$\begin{array}{ccccccccc} & & & & \binom{0}{0} & & & & \\ & & & \binom{1}{0} & & \binom{1}{1} & & & \\ & & \binom{2}{0} & & \binom{2}{1} & & \binom{2}{2} & & \\ & \binom{3}{0} & & \binom{3}{1} & & \binom{3}{2} & & \binom{3}{3} & \\ \binom{4}{0} & & \binom{4}{1} & & \binom{4}{2} & & \binom{4}{3} & & \binom{4}{4} \\ \dots & \dots & \dots & \dots & \dots & \dots & \dots & \dots & \dots \end{array} \equiv \begin{array}{ccccccccccc} & & & & & 1 & & & & & \\ & & & & 1 & & 1 & & & & \\ & & & 1 & & 2 & & 1 & & & \\ & & 1 & & 3 & & 3 & & 1 & & \\ & 1 & & 4 & & 6 & & 4 & & 1 & \\ 1 & & 5 & & 10 & & 10 & & 5 & & 1 \\ \dots & \dots & \dots & \dots & \dots & \dots & \dots & \dots & \dots & \dots & \dots \end{array}$$

Figure 3.1 Pascal's triangle written as an array of binomial coefficients and as the array of their values. Observe that by (4), each term is obtained by adding its two neighbours in the row above (except for $\binom{0}{0}$ which is defined to be 1).

(1) Definition Given a collection of numbers $(a_i;\ i \geq 0)$ the function

$$g_a(x) = \sum_{i=0}^{\infty} a_i x^i$$

is called the *generating function* of (a_i). (It is of course necessary that $\sum a_i x^i$ converges somewhere if g_a is defined as a function of x. If we regard g_a as an element of a ring of polynomials, such convergence is not necessary.) ▲

(2) Definition Given $(a_i;\ i \geq 0)$ the function

$$h_a(x) = \sum_{i=0}^{\infty} \frac{a_i x^i}{i!}$$

is the *exponential generating function* of (a_i). ▲

(3) Definition Given a collection of functions $(f_n(y);\ n \geq 0)$, the function

$$g(x, y) = \sum_{n=0}^{\infty} x^n f_n(y)$$

is a *bivariate generating function*. ▲

The following crucial result is a corollary of Taylor's Theorem. We omit the proof.

(4) Theorem (Uniqueness.) If for some x_0 and x_1 we have

$$g_a(x) = g_b(x) < \infty \quad \text{for } x_0 < x < x_1$$

then $a_i = b_i$ for all i.

Generating functions help to tackle difference equations; the following result is typical.

(5) **Theorem** If $(a_n;\ n \geqslant 0)$ satisfies a recurrence relation

$$a_{n+2} + 2ba_{n+1} + ca_n = d_n; \quad n \geqslant 0$$

then

$$g_a(x) = \frac{x^2 g_d(x) + a_0 + a_1 x + 2ba_0 x}{1 + 2bx + cx^2},$$

with a corresponding result for higher order equations.

Proof To prove (5) multiply each side of the recurrence by x^{n+2} and sum over n. ■

The next four theorems are proved by rearranging the summation on the right hand side in each case. You should do at least one as an exercise.

(6) **Theorem (Convolution)** If (a_n) can be written as a *convolution* of the sequences (b_n) and (c_n), so

$$a_n = \sum_{i=0}^{n} c_i b_{n-i}, \quad n \geqslant 0,$$

then

$$g_a(x) = g_c(x) g_b(x).$$

(7) **Theorem (Tails)** If

$$b_n = \sum_{i=1}^{\infty} a_{n+i}, \quad n \geqslant 0,$$

then

$$g_b(x) = \frac{g_a(1) - g_a(x)}{1 - x}.$$

(8) **Theorem** If

$$c_n = \sum_{i=0}^{n} a_i,$$

then

$$g_c(x) = \frac{g_a(x)}{1 - x}.$$

(9) **Theorem (Exponential function)** Let the function $e(x)$ be defined by

$$e(x) = \sum_{k=0}^{\infty} \frac{x^k}{k!},$$

then

$$e(x + y) = e(x)e(y) = e^{x+y} = e^x e^y.$$

Finally we have the celebrated

(10) Binomial theorems For integral n

(11)
$$(1 + x)^n = \sum_{k=0}^{n} \binom{n}{k} x^k, \quad n \geq 0,$$

and

(12)
$$(1 - x)^{-n} = \sum_{k=0}^{\infty} \binom{n + k - 1}{k} x^k \quad n \geq 0, |x| < 1.$$

(13) Example Let us prove the binomial theorems.

Proof of (11) Considering the product $(1 + x)^n = (1 + x)(1 + x) \dots (1 + x)$, we see that a term x^k is obtained by taking x from any choice of k of the brackets, and taking 1 from the rest. Since there are $\binom{n}{k}$ ways of choosing k brackets, the term x^k occurs $\binom{n}{k}$ times in the expansion. Since this is true for any k, the result follows.

Alternatively you can prove (11) by induction on n, using the easily verified identity

$$\binom{n}{k} = \binom{n - 1}{k - 1} + \binom{n - 1}{k}.$$

Proof of (12) Multiplication verifies that $1 - x^{m+1} = (1 - x)(1 + x + \dots + x^m)$. Hence, for $|x| < 1$, we let $m \to \infty$ to find that (12) is true for $n = 1$,

$$\frac{1}{1 - x} = \sum_{0}^{\infty} x^k.$$

Now we can prove (12) for arbitrary n in a number of ways; one possibility is by induction on n. Alternatively we observe that a term x^k is obtained in the product $(1 + x + x^2 + \dots)^n$ if we choose a set of k x's from the n brackets, where we may take any number of x's from each bracket. But by Theorem 3.3.4 this may be done in just $\binom{n+k-1}{k}$ ways. The negative binomial theorem (12) follows. ●

Here is a classic example.

(14) Example: The coupon collector's problem Each packet of an injurious product is equally likely to contain any one of n different types of coupon. If you buy r packets, what is the probability $p(n, r)$ that you obtain at least one of each type of coupon?

Solution This famous problem can be approached in many different ways, as we shall see later, but the exponential generating function offers a particularly elegant answer.

First recall from Theorem 3.2.3 that the probability of getting $r = \sum_1^n t_i$ coupons of n distinct types, where t_i are of type i, is

$$n^{-r}M_r(t_1, \ldots, t_n) = \frac{r!}{n^r t_1! \ldots t_n!}.$$

Then $p(n, r)$ is the sum of all such expressions in which $t_i \geq 1$ for all i (so we have at least one coupon of each type).

But now expanding $(\sum_{t=1}^{\infty}(s/n)^t/t!)^n$ by the multinomial theorem shows that the coefficient of s^r in this expansion is just $p(n, r)/r!$. Hence

$$\left(\exp\left(\frac{s}{n}\right) - 1\right)^n = \sum_{r=n}^{\infty} s^r p(n, r)/r!,$$

and we have obtained the exponential generating function of the $p(n, r)$. ●

Suppose that you are a more demanding collector who requires two complete sets of coupons, let $p_2(n, r)$ be the probability that r packets yield two complete sets of n coupons. Then exactly similar arguments show that

$$\left(\exp\left(\frac{s}{n}\right) - 1 - \frac{s}{n}\right)^n = \sum_{r=2n}^{\infty} \frac{s^r}{r!} p_2(n, r),$$

and so on for more sets.

In conclusion it is worth remarking that multivariate generating functions are often useful, though they will not appear much at this early stage. We give one example.

(15) **Multinomial theorem** Recall the multinomial coefficients defined in Theorem 3.2.3,

$$M_n(n_1, \ldots, n_r) = \frac{n!}{\prod_{i=1}^{r} n_i!}; \quad \sum_{i=1}^{r} n_i = n.$$

We have

$$(x_1 + x_2 + \ldots + x_r)^n = \sum M_n(n_1, \ldots, n_r) x_1^{n_1} x_2^{n_2} \ldots x_r^{n_r},$$

where the sum is over all $(n_1, \ldots, n_r)$ such that $\sum n_i = n$.

Proof This is immediate from the definition of M_n as the number of ways of permuting n symbols of which n_i are of type i. In this case x_i is of type i, of course. ■

Corollary Setting $x_1 = x_2 = \ldots = x_r = 1$ gives $\sum M_n(n_1, \ldots, n_r) = r^n$.

3.7 Techniques

When evaluating any probability by counting, it is essential first to be clear what the sample space is, and exactly which outcomes are in the event of interest. Neglect of this obvious but essential step has led many a student into lengthy but nugatory calculations.

Second, it is even more important than usual to be flexible and imaginative in your approach. As the following examples will show, a simple reinterpretation or reformulation can turn a hard problem into a trivial one. The main mechanical methods are:

(i) using the theorems giving the numbers of ordered and unordered selections, recalling in particular that the number of ordered selections of r objects is the number of unordered selections multiplied by $r!$

(ii) use of the inclusion–exclusion principle.

(iii) setting up recurrence relations.

(iv) use of generating functions.

Finally we remark that the implicit assumption of this chapter can be turned on its head. That is, we have developed counting techniques to solve probability problems, but the solution of probability problems can just as well be used to prove combinatorial identities.

(1) **Example** Prove the following remarkable identity:

(2)
$$\sum_{k=0}^{n} \binom{n+k}{k} 2^{-(n+k)} = 1.$$

Hint: consider an ant walking on a square lattice.

Solution An ant walks on the square lattice of points with non-negative integer coordinates (i, j), $i \geq 0$, $j \geq 0$. It starts at $(0, 0)$. If it is at (x, y) it proceeds next either to $(x + 1, y)$ or to $(x, y + 1)$ with equal probability $\frac{1}{2}$. Therefore, at some transition (certainly less than $2n + 1$ transitions) it leaves the square

$$(0 \leq x \leq n, 0 \leq y \leq n).$$

It does so either with an x-step from (n, y) to $(n + 1, y)$ where $0 \leq y \leq n$ or with a y-step from (x, n) to $(x, n + 1)$, where $0 \leq x \leq n$. Let these $2n + 2$ events be S_y $(0 \leq y \leq n)$ and S_x $(0 \leq x \leq n)$ respectively. Then by symmetry

(3)
$$\sum_{x=0}^{n} \mathbf{P}(S_x) = \sum_{y=0}^{n} \mathbf{P}(S_y) = \frac{1}{2}.$$

However S_y occurs if the ant has taken exactly y vertical steps before its $(n + 1)$th horizontal step. There are $\binom{n+y+1}{y}$ choices for the y vertical steps and each route to $(n + 1, y)$ has probability $2^{-(n+y+1)}$. Hence

(4)
$$\mathbf{P}(S_y) = \binom{n+y+1}{y} 2^{-(n+y+1)}.$$

Substituting (4) into (3) yields (2). ●

WORKED EXAMPLES AND EXERCISES

3.8 Example: Railway trains

Suppose you work in a marshalling yard. Given n similar wagons and r similar locomotives, how many distinct ways are there of making up r trains if empty trains are allowed, and all the wagons are used, and either:

(a) the wagons and locomotives bear distinct numbers?

or

(b) the wagons are anonymous but the locomotives are numbered?

or

(c) wagons and locomotives are anonymous?

What are the corresponding numbers if each train must contain at least m wagons?

Solution (a) First consider how many ways the trains may be sent out of the yard. This is just the number of permutations of $n + r$ objects with the one constraint that the first object has to be a locomotive! The number is $r(n + r - 1)!$. As regards just making up trains, we do not require an ordering of the locomotives, so we divide by $r!$ to get the answer

$$\frac{r(n+r-1)!}{r!} = \frac{(n+r-1)!}{(r-1)!}.$$

(b) If the wagons are not numbered then we cannot distinguish the $n!$ permutations of wagons, so the answer is

$$\frac{(n+r-1)!}{n!(r-1)!} = \binom{n+r-1}{r-1},$$

which could alternatively have been obtained by observing that we can make up the trains by choosing r places to intersperse the locomotives with the wagons, given that one locomotive must be at the front.

(c) If neither locomotives nor wagons bear numbers, then the number of ways of making up r trains is the number of distinct partitions of n into at most r non-negative integers, denoted by $p_r(n)$.

For example, the number 5 has altogether seven integral partitions

$$5 = 4+1 = 3+2 = 3+1+1 = 2+2+1 = 2+1+1+1 = 1+1+1+1+1$$

so $p_\infty(5) = 7$; of these, four are of three or fewer integers, so $p_3(5) = 4$. There are no simple expressions for $p_r(n)$, but we use the following result, which we state without proof.

Theorem The number of partitions of n with at most k parts is the same as the number of partitions of n in which no part is greater than k.

Now if we define the generating function

$$g_r(x) = \sum_{n=1}^{\infty} p_r(n)x^n$$

we can see that

$$1 + g_r(x) = \frac{1}{1-x}\frac{1}{1-x^2}\cdots\frac{1}{1-x^r}$$

by first expanding the right hand side (by the negative binomial theorem 3.6.12), and then observing that the term x^n arises just as often as n can be written as the sum of positive integers, no one of which is greater than r.

(b′) If each train must contain at least m wagons then we require $mr \geqslant n$. In this case, first suppose the wagons do not have numbers. We attach m to each train in essentially one way, leaving $n - mr$ to be distributed in any way. Using (b) we get the answer $\binom{n-rm+r-1}{r-1}$ ways.

(a′) If the wagons have numbers then any of the $n!$ permutations is distinct, giving $n!\binom{n-rm+r-1}{r-1}$ ways.

(c′) If neither the locomotives nor wagons have numbers, and each train must contain m wagons at least, then we require the number of partitions of m into at most r integers, all of which are not less than m. This is the same as the number of partitions of $n - mr$ into at most r integers, that is $p_r(n - mr)$, which is the coefficient of x^{n-mr} in $g_r(x)$.

Remark The results about partitions of n are not used elsewhere. They are included as an example of the power of generating functions, which are used extensively throughout probability and related subjects.

(1) **Exercise** How many ways of making up the r trains are there if you do not have to use all the wagons?

(2) **Exercise** In how many ways can five oranges be distributed among seven boys? How many ways are there on the more realistic assumption that boys are indistinguishable?

(3) **Exercise** The hard disk on your PC stores 10 mb. In how many ways can you divide it up so that no section has less than 3 mb? [Use only integral multiples of 1 mb.]

3.9 Example: Genoese lottery

Suppose that n cards in an urn each bear one of n consecutive integers; all the cards have a different number. Five cards are randomly selected from the urn without replacement. What is the probability that their numbers can be arranged as a run of three consecutive numbers and a non-adjacent run of two consecutive numbers? (For example: 34578 or 23789.)

Solution There are $\binom{n}{5}$ ways of choosing a set of five numbers, and these are assumed to be equally likely. We must count the number of sets providing the two required runs. If the run of three is first, and it starts at k, then the run of two may start at $n - k - 4$ places ($1 \leqslant k \leqslant n - 5$). Hence the runs may occur, in this order, in

$$\sum_{k=1}^{n-5} n - k - 4 = \sum_{j=1}^{n-5} j = \frac{1}{2}(n-5)(n-4)$$

ways.

The run of two is first in another $\frac{1}{2}(n-5)(n-4)$ cases, so the required probability is $(n-5)(n-4)/\binom{n}{5}$.

Remark This lottery is called Genoese because it was introduced by a member of the senate of Genoa named Benedetto Gentile in 1623. It was used to raise money in 1757 by the treasury of Louis XV of France on the advice of Casanova, who had a licence to sell tickets at 6% commission. The treasury also consulted d'Alembert; this was shortly after he had asserted that the chance of HH or TT in two spins of a fair coin is $\frac{2}{3}$.

When Frederick II of Prussia used this lottery to raise money for his treasury, he asked Euler to calculate the odds again.

(1) **Exercise** As above, n cards bear n consecutive integers. Find the probability that:
(a) if three cards are drawn, their numbers can form a run of length three.
(b) if three cards are drawn, their numbers include a run of length exactly two.
(c) if five cards are drawn, their numbers include two runs of length exactly two.

3.10 Example: Ringing birds

A wood contains n birds, none of which is ringed. Each day one bird is caught, ringed (if it does not already have a ring), and released. Each bird is equally likely to be caught on each day. This procedure is repeated on r successive days, $r \geqslant n$. Show that the probability of ringing all the birds is $p(r, n)$ where

$$p(r, n) = \sum_{j=0}^{n} (-)^j \binom{n}{j}\left(1 - \frac{j}{n}\right)^r$$

***Solution* I** The total number of outcomes (birds being distinguishable), is n^r.

Let $N(r, n)$ be the number of outcomes in which every bird is caught at least once. Since birds are distinct, some bird may be called the first bird. Let $N_k(r, n)$ be the number of outcomes in which the first bird is caught exactly k times, and all the birds are caught at least once. There are $\binom{r}{k}$ ways of choosing the k days on which the first bird is caught, and $N(r-k, n-1)$ ways in which the remaining $n-1$ birds may be caught on the other $r-k$ occasions.

Hence by Principle 3.1.2

(1) $$N_k(r, n) = \binom{r}{k} N(r-k, n-1),$$

and by Principle 3.1.1

(2) $$N(r, n) = \sum_{k=1}^{r} N_k(r, n).$$

Of course

(3) $$N(r, 1) = 1.$$

Substituting (1) into (2) yields a rather complicated recurrence relation. It can be solved by using the exponential generating function:

(4) $$G_n(s) = \sum_{r=1}^{\infty} \frac{s^r}{r!} N(r, n); \quad n > 1.$$

By (3) we have

(5) $$G_1(s) = e^s - 1.$$

Now multiply (2) by $s^r/r!$ and sum over r (using (1)), to give

$$\begin{aligned}
G_n(s) &= \sum_{r=1}^{\infty} \sum_{k=1}^{r} \frac{s^{r-k}s^k}{(r-k)!k!} N(r-k, n-1) \\
&= \sum_{k=1}^{\infty} \frac{s^k}{k!} G_{n-1}(s) && \text{by the convolution theorem 3.6.6,} \\
&= (e^s - 1)G_{n-1}(s) \\
&= (e^s - 1)^n && \text{on iterating and using (5),} \\
&= e^{ns} \sum_{j=0}^{n} (-e^{-s})^j \binom{n}{j} && \text{by the binomial theorem} \\
&= \sum_{r=0}^{\infty} \sum_{j=0}^{n} \frac{(n-j)^r}{r!} (-1)^j \binom{n}{j} s^r && \text{on expanding } e^{(n-j)s}.
\end{aligned}$$

Hence

$$N(r, n) = \sum_{j=0}^{n} (-1)^j \binom{n}{j} (n-j)^r,$$

using (4) and Theorem 3.6.4. Dividing by n^r yields the required result.

Solution II We may alternatively use the principle of inclusion and exclusion. In the terminology of Section 3.4 'objects' are outcomes and an 'object with the kth property' is an outcome in which k birds remain unringed. Now the number of ways of selecting k birds to remain unringed is $\binom{n}{k}$, and the number of outcomes with k given birds unringed is $(n-k)^r$. Hence, by (3.4.1) and (3.4.3),

$$N(r, n) = n^r - \binom{n}{1}(n-1)^r + \ldots + (-1)^{n-1}\binom{n}{n-1}(n-(n-1))^r$$

which yields the required result.

(6) **Exercise** Do the example again using (1.4.8).

(7) **Exercise** What is the probability that no bird is caught more than twice?

(8) **Exercise** What is the probability that every bird is caught at least twice?

(9) **Exercise** As $r \to \infty$, show that the probability of ringing all the birds converges to one.

3.11 Example: Lottery

In each draw of a lottery, an integer is picked independently at random from the first n integers $1, 2, \ldots, n$. What is the probability that in a sample of r successive draws the numbers are drawn in a non-decreasing sequence?

***Solution* I** There are n^r possible sequences in all. We may take any non-decreasing sample sequence of r numbers, together with the first n integers, and arrange these $n + r$ integers as a non-decreasing sequence. Now place a bar between adjacent different numbers in this sequence, and place a star between adjacent equal numbers. The number to the right of each star is a member of the original sample sequence; there are r stars and $n + r - 1$ places to choose to put them. Furthermore each such choice corresponds to just one possible sample sequence. By Theorem 3.3.1 there are therefore $\binom{n+r-1}{r}$ non-decreasing sample sequences, so the required probability is $\binom{n+r-1}{r}n^{-r}$.

***Solution* II** Take any sample sequence $\mathbf{s} = \{s_1, s_2, \ldots, s_r\}$ and add $j - 1$ to s_j, to get a unique new sequence $\mathbf{t} = \{s_1, s_2 + 1, s_3 + 2, \ldots, s_r + r - 1\}$. The sequence $\mathbf{t}$ is a selection of r numbers without replacement from $\{1, 2, 3, \ldots, n + r - 1\}$, and subtracting $j - 1$ from t_j yields a unique sequence $\mathbf{s}$, which is selected with replacement from $\{1, 2, \ldots, n\}$.

Since there are $\binom{n+r-1}{r}$ ways of choosing the sequence $\mathbf{t}$, this is also the number of ways of choosing $\mathbf{s}$.

Remark Observe that this solution includes a proof of Theorem 3.3.4.

(1) Exercise What is the probability that the r numbers are drawn in a strictly increasing sequence?

(2) Exercise Show that the probability that no two drawn numbers are consecutive (i.e. differ by unity) is $r!\binom{n-r+1}{r}n^{-r}$.

(3) Exercise Integers are picked at random from $\{1, 2, \ldots, n\}$ until every integer k, $1 \leq k \leq n$, has been drawn at least once. What is the probability that m draws are required to achieve this?

3.12 Example: The ménages problem

Suppose that n married couples are seated randomly around a circular table so that men and women alternate.

(a) Find the number of such seatings in which, for a given set of k couples, the husband and wife are sitting in adjacent seats (some other couples may be adjacent also).

(b) Hence deduce the probability that no couple is seated next to each other.

Solution There are two ways of choosing in which seats the women will sit, and $n!$ ways in which they can be arranged in them. The men can be seated in the remaining seats (alternating with the women) in $n!$ ways. Thus by Principle 3.1.2 there are $2(n!)^2$ possible seating arrangements, which are equally likely by hypothesis.

(a) To count the number of arrangements in which a given set of k couples are adjacent, it is first necessary to count the number of ways of choosing k pairs of adjacent seats for them to occupy. First consider $2n - k$ chairs in a row. Now choose k of these (which we may do in $\binom{2n-k}{k}$ ways), and place an extra chair by each of the

k chosen chairs. This provides a choice of k distinct non-overlapping pairs of adjacent chairs in a row of $2n$. Conversely, for any choice of k disjoint pairs of adjacent chairs from $2n$, we may discard a chair from each pair to give a choice of k from $2n - k$ chairs. This one–one correspondence shows that the number of ways of choosing k disjoint pairs of chairs from $2n$ is just $\binom{2n-k}{k}$.

Now number the seats round the circular table from 1 to $2n$. By Principle 3.1.1, the number of choices of k disjoint pairs of adjacent seats is the sum of the choices in which $(1, 2n)$ is such a pair (which number $\binom{2n-2-(k-1)}{k-1}$ by the above result), and the choices in which $(1, 2n)$ is not a pair (which number $\binom{2n-k}{k}$).

Hence the disjoint pairs of seats can be chosen in

$$M_k = \binom{2n-k-1}{k-1} + \binom{2n-k}{k} = \left(\frac{2n}{2n-k}\right)\binom{2n-k}{k}$$

ways.

The k couples to occupy these seats can be chosen in $\binom{n}{k}$ ways, the women's seats chosen in two ways, the k chosen couples arranged in $k!$ ways, and the remaining men and women arranged in $((n-k)!)^2$ ways. Hence, using Principle 3.1.2, the number of seatings for which the k couples are in adjacent seats is

(1) $$S_k = \frac{2n}{2n-k} \cdot \binom{2n-k}{k} \cdot 2k!((n-k)!)^2 = \frac{4n(2n-k-1)!}{(2n-2k)!}((n-k)!)^2.$$

(b) Recalling the principle of inclusion and exclusion, we may interpret 'an object with the kth property' as 'a seating with the kth couple adjacent'. Hence by using (3.4.3) and (3.4.2) the probability that no couple is adjacent is

(2) $$\sum_{k=0}^{n}(-)^k\binom{n}{k}\frac{S_k}{2(n!)^2} = \sum_{0}^{n}(-)^k\frac{2n}{2n-k}\binom{2n-k}{k}\frac{(n-k)!}{n!}.$$

(3) Exercise Why does it not matter whether we assume the table has a principal seat (head) or not?

(4) Exercise Suppose that n pairs of twins are seated randomly at a round table. What is the probability that no pair of twins sit next to each other? What is the limit of this probability as $n \to \infty$?

(5) Exercise What is the limit of (2) as $n \to \infty$?

Remark The problem was first discussed by E Lucas in 1891 (*ménage* is French for *household*). This method of solution is due to K Bogart and P Doyle, *American Mathematical Monthly*, 1986.

3.13 Example: Identity

Show that

(1) $$\sum_{j=0}^{r}\binom{j+k-1}{k-1} = \binom{r+k}{k}.$$

Solution I Suppose we place r balls randomly in m urns. This is essentially the same as arranging r balls and $m-1$ stars in a row, where the balls between successive stars are placed in successive urns. The positions for the stars may be chosen in $\binom{m+r-1}{r-1}$ ways. Now, in how many of these arrangements do we find b balls in the first urn? The answer is, in just as many ways as the other $r-b$ balls can be put in the other $m-1$ urns, that is in $\binom{m+r-b-2}{m-2}$ ways. Hence by Principle 3.1.1

$$\sum_{b=0}^{r}\binom{m+r-b-2}{m-2}=\binom{m+r-1}{m-1}.$$

Setting $m-1=k$, $r-b=j$ gives the required result (1).

***Solution* II** Multiply each side of (1) by x^r and sum from $r=0$. By Theorem 3.6.8 we have

$$\frac{1}{1-x}\sum_{r=0}^{\infty}x^r\binom{r+k-1}{k-1}=\sum_{r=0}^{\infty}\binom{r+k}{k}x^r,$$

and by (3.6.10) both sides equal $(1-x)^{-(k+1)}$. The identity (1) follows by Theorem 3.6.4.

***Solution* III** Obviously

$$\frac{1}{(1-x)^k}=\frac{1}{(1-x)^{k-1}}\frac{1}{1-x}.$$

Equating the coefficient of x^r on each side of this identity yields (1).

Remark Up to a point these three methods are really the same, but Solution III does illustrate how useful generating functions can be if you happen to hit upon an appropriate identity when you need it.

(2) Exercise Show that for any number x and integer j, $\binom{x}{j-1}+\binom{x}{j}=\binom{x+1}{j}$, and hence prove (1) by a fourth method.

(3) Exercise Show that $\sum_{j=0}^{n}(-1)^j\binom{x}{j}=(-1)^n\binom{x-1}{n}$.

(4) Exercise Show that $\sum_{j=0}^{n}\binom{x-j}{r}=\binom{x+1}{r+1}-\binom{x-n}{r+1}$.

3.14 Example: Runs

A fair coin is tossed repeatedly. A run of heads is all the heads shown between one tail and the next; if the first toss is a head there is an opening run up to the first tail, likewise there may be a closing run after the last tail. A zero run is no run.

(a) If the coin is tossed n times, show that the most likely number of runs (of heads, including the opening and closing runs) is $[\frac{n}{4}]+1$ when n is large.

(b) Also, prove the identity

$$\sum_{i=0}^{m}\binom{i+k-1}{k-1}\binom{m+k-i}{k}=\binom{2k+m}{m}.$$

Solution (a) There are 2^n possible outcomes in all. We can choose any outcome with k head runs in the following way. Visualize the n coins in a row. They provide

$n+1$ intervening spaces (including that before the first and that after the last coin). Now place $2k$ stars in the spaces, and let the coins between the $(2r+1)$th and $(2r+2)$th stars be heads ($r = 0, 1, \ldots, k-1$); the other coins are tails. There are $\binom{n+1}{2k}$ ways to do this, so

$$f_k = \mathbf{P}(k \text{ head runs}) = 2^{-n}\binom{n+1}{2k}.$$

Now considering f_k/f_{k+1} we find

$$\frac{f_k}{f_{k+1}} \gtrless 1 \quad \text{according as} \quad k \gtrless \frac{n^2+n-2}{4n}.$$

Therefore the most likely number of runs is the integer next after $\frac{n^n+n-2}{4n}$, which is $[\frac{n}{4}]+1$ for large n.

(b) Suppose that the n tosses result in a heads and b tails. First we divide the a heads into k non-empty groups which form the runs. Imagining the heads in a row, there are $a-1$ places to put $k-1$ dividing lines, so the number of ways of getting the k runs of heads is $\binom{a-1}{k-1}$. The head runs alternate with runs of tails, so the b tails are divided into $k+1$ groups of which the first and last may be empty (providing an opening and/or closing run). If we add an auxiliary tail to each of these, the same argument as used for head runs shows that the number of ways of arranging the tails is

$$\binom{b+2-1}{k+1-1} = \binom{b+1}{k}.$$

Hence the number of ways of getting k head runs in n tosses showing a heads is

$$r_k = \binom{a-1}{k-1}\binom{b+1}{k}.$$

The total number of ways of getting k head runs is therefore

$$\begin{aligned}\sum_{a=k}^{n-k+1} r_k &= \sum_{i=0}^{n-2k+1}\binom{i+k-1}{k-1}\binom{n-i-k+1}{k} \\ &= \sum_{i=0}^{m}\binom{i+k-1}{k-1}\binom{m+k-i}{k}; \quad m = n-2k+1, \\ &= \binom{n+1}{2k} \quad \text{by (a)} \\ &= \binom{2k+m}{m}\end{aligned}$$

as required.

Note: in the exercises, n tosses yield a heads and $n-a$ tails.

(1) Exercise What is the probability that the first run is a head run of length k?

(2) Exercise What is the probability that the last run is a head run of length k?

(3) Exercise What is the probability that the second run is a head run of length k?

(4) Exercise What is the probability that the first run is of length k?

3.15 Example: Fish

A lake contains b bream and c chub. Any fish of each species is equally likely to take a hook.

(a) If you catch n fish, and don't throw any back, what is the probability that you have caught x bream?

(b) You then return all the fish to the lake (alive), and start fishing again. You now catch m fish. What is the probability that exactly k bream are caught twice?

Solution (a) The question clearly intends us to assume that all possible selections of n fish are equally likely to occur. Then the number of ways of selecting n fish without repetition is $\binom{b+c}{n}$. The number of ways of catching x bream (and hence also $n-x$ chub) is $\binom{b}{x}\binom{c}{n-x}$, so the required probability is

(1) $$p_x = \frac{\binom{b}{x}\binom{c}{n-x}}{\binom{b+c}{n}}, \qquad \max\{0, n-c\} \leqslant x \leqslant \min\{b, n\}.$$

(b) We assume that fish do not learn from experience, so that all $\binom{b+c}{m}$ selections of m fish are still equally likely. If x bream were in the first catch, where $k \leqslant x \leqslant b$, then the number of ways of selecting m fish, of which k are bream being caught for the second time, is $\binom{x}{k}\binom{b+c-x}{m-k}$. Therefore the required conditional probability of catching k bream twice, given a first catch of x bream, is

$$p_{k|x} = \binom{x}{k}\binom{b+c-x}{m-k}\bigg/\binom{b+c}{m}$$

by the same argument as in (a). Hence by Theorem 2.1.3 the required unconditional probability is

$$p_k = \sum_x p_{k|x}p_x = \sum_{x=k}^{b} \frac{\binom{b}{x}\binom{c}{n-x}}{\binom{b+c}{n}} \frac{\binom{x}{k}\binom{b+c-x}{m-k}}{\binom{b+c}{m}}$$

(2) Exercise What is the probability of catching x bream if you catch n fish and
(a) you throw bream back but not chub?
(b) you throw both species back?
Part (a) considers sampling with partial replacement, and part (b) considers sampling with replacement. Discuss the difference in your answers.

(3) Exercise Show that $\sum_{k=0}^{\min\{b,n\}}\binom{b}{k}\binom{c}{n-k} = \binom{b+c}{n}$. [You may wish to recall that $(1+x)^b(1+x)^c = (1+x)^{b+c}$.]

(4) Exercise Suppose that as b and c approach ∞, $b/(b+c) \to p$ and $c/(b+c) \to 1-p$. Find the limit of the probabilities in (1) and (2) as b and $c \to \infty$. Discuss.

3.16 Example: Colouring

Let $K(b, c)$ be the number of different ways in which b indistinguishable balls may be coloured with c different colours. Show that $K(b, c) = K(b-1, c) +$

$K(b, c-1)$ and deduce that

$$\sum_{b=0}^{\infty} x^b K(b, c) = (1-x)^{-c}.$$

Use this to show $K(b, c) = \binom{b+c-1}{c-1}$.

Solution Pick any colour and call it grurple. The number of colourings is the number of ways of colouring the balls which do not colour any grurple, plus the number of ways of colouring which do use grurple. Hence

(1) $$K(b, c) = K(b-1, c) + K(b, c-1).$$

Also $K(1, c) = c$, and $K(0, c) = 1$, because there are c colours for one ball, and only one way of colouring no balls. Now let

$$g_c(x) = \sum_{b=0}^{\infty} x^b K(b, c).$$

Multiply (1) by x^b and sum from $b = 0$, to get

(2) $$g_c(x) = x g_c(x) + g_{c-1}(x).$$

Now using Theorem 3.5.1 we solve (2) to find $g_c(x) = (1-x)^{-c}$. Furthermore, we may write

$$(1-x)^{-c} = (1 + x + x^2 + \ldots)(1 + x + x^2 + \ldots) \ldots (1 + x + x^2 + \ldots)$$

where the right side is the product of c brackets. We get $K(b, c)$ by picking a term from each bracket, and we can say that picking x^k from the ith bracket is like picking k objects of type i. The coefficient $K(b, c)$ of x^b is thus obtained by choosing b objects from c different types of objects with repetition. By Theorem 3.3.4 we have

(3) $$K(b, c) = \binom{c+b-1}{b}.$$

(4) **Exercise** Let $C(n, k)$ be the number of ways of choosing a set of k objects from n distinct objects. Show that $\sum_{k=0}^{n} C(n, k) x^k = (1+x)^n$.

(5) **Exercise** How many non-negative integer valued solutions for x_1, x_2 and x_3 does $x_1 + x_2 + x_3 = 20$ have? [For example, $x_1 = 0, x_2 = 4, x_3 = 16$.]

(6) **Exercise** How many positive integer valued solutions does $x_1 + x_2 + x_3 = 20$ have for x_1, x_2 and x_3? [For example, $x_1 = 5, x_2 = 6, x_3 = 9$.]

(7) **Exercise** Show that $K(b, c) = \binom{b+c-1}{c-1}$ by a method different from that in the above solution.

3.17 Example: Matching (Rencontres)

Suppose n different letters are typed with their corresponding envelopes. If the letters are placed at random in the envelopes, show that the probability that exactly r letters match their envelopes is

(1) $$p(n, r) = \frac{1}{r!} \sum_{k=0}^{n-r} \frac{(-1)^k}{k!}.$$

(This problem first surfaced in France during the eighteenth century as a question about coincidences when turning over cards from packs (a kind of French snap).)

Solution We can suppose that the order of the envelopes is fixed. Let the number of permutations of the letters in which r out of the n letters match their envelopes be $a(n, r)$. Then

$$p(n, r) = \frac{a(n, r)}{n!}.$$

Suppose we have another letter sealed in its correct envelope. Consider the number A of arrangements that there are of this letter and n letters of which r match their envelopes. We can get this number A in two ways. Either:

(i) we place the sealed letter in any one of $n+1$ positions among the n letters to get $(n+1)a(n, r)$ arrangements;

or:

(ii) we permute $n+1$ unsealed letters of which $r+1$ match and then choose one of the $r+1$ matching letters to seal, giving $(r+1)a(n+1, r+1)$ arrangements.

The two numbers must both be equal to A, so $(r+1)a(n+1, r+1) = (n+1)a(n, r)$ and hence, dividing by $(n+1)!$ we obtain

(2)
$$(r+1)p(n+1, r+1) = p(n, r)$$

with $p(n, n) = 1/n!$.

This is a rather interesting recurrence relation, which is solved by standard methods. First iterating (2) we have

(3)
$$p(n, r) = \frac{1}{r!}p(n-r, 0).$$

Now define the probability generating function

$$g_n(x) = \sum_{r=0}^{n} x^r p(n, r); \quad n \geq 1.$$

Multiplying (2) by x^r and summing over r gives

$$g_n(x) = \frac{x^n}{n!} + \sum_{r=1}^{n} p(r, 0)\frac{x^{n-r}}{(n-r)!}.$$

The sum on the right is a convolution as in Theorem 3.6.6, so multiplying by y^n and summing over n, by Theorem 3.6.6

$$\sum_{1}^{\infty} y^n g_n(x) = e^{xy} - 1 + e^{xy}\sum_{1}^{\infty} y^n p(n, 0).$$

Setting $x = 1$ and using Theorem 3.6.10 gives

$$\frac{y}{1-y} = e^y - 1 + e^y \sum_{1}^{\infty} y^n p(n, 0)$$

so that

(4) $$\sum_{1}^{\infty} y^n p(n, 0) = \frac{e^{-y}}{1 - y} - 1 = \sum_{1}^{\infty} y^n \sum_{k=0}^{n} \frac{(-1)^k}{k!}.$$

Hence by (4) and (3) we get (1).

(5) Exercise Find the probability that exactly $r + s$ matches occur given that at least r matches occur. Show that for large n, it is approximately

$$\frac{1}{(r+s)!} \Bigg/ \left(\sum_{r}^{\infty} \frac{1}{k!} \right).$$

(6) Exercise Show that the probability that the first letter matches its envelope, given that there are exactly r such matches, is $\frac{r}{n}$.

(7) Exercise If a cheque is written for each addressee and these are also placed at random in the envelopes, find:
(a) the probability that exactly r envelopes contain the correct letter and cheque.
(b) the probability that no envelope contains the correct letter and cheque.
(c) the probability that every letter contains the wrong letter and the wrong cheque.

(8) Exercise Find the limit of each probability in 7(a), 7(b) and 7(c) as $n \to \infty$.

PROBLEMS

1 You have two pairs of red socks, three pairs of mauve socks, and four pairs with a rather attractive rainbow motif. If you pick two socks at random, what is the probability that they match?

2 A keen student has a algebra books, b books on boundary layers, and c calculus books. If he places them on one shelf at random, what is the probability that:
(a) books on the same subject are not separated?
(b) books on the same subject are in the usual alphabetical order, but not necessarily adjacent?
(c) books on the same subject are adjacent and in alphabetical order?

3 A pack of cards is well-shuffled and one hand of thirteen cards is dealt to each of four players. Find the probability that:
(a) each player has an ace;
(b) at least one player has a complete suit;
(c) my hand is void in at least one suit;
(d) some player has all the aces.
What is the most likely distribution among suits in the dealer's hand?

4 **Poker** You are dealt five cards in your hand at poker. What is the probability that you hold:
(a) one pair?
(b) two pairs?
(c) a straight?
(d) a flush?
(e) a full house?

5 **Birthdays** Assume people are independently equally likely to be born on any day of the year. Given a randomly selected group of r people, of whom it is known that none were born on February 29th, show that the probability that at least two of them have their birthdays

either on consecutive days or on the same day is p_r where

$$p_r = 1 - \frac{(365 - r - 1)!}{(365 - 2r)!} 365^{-r+1}, \quad (2r < 365).$$

Deduce that if $r = 13$ then the probability of at least two such contiguous birthdays is approximately $\frac{1}{2}$, while if $r = 23$ then the probability of at least two such contiguous birthdays is approximately $\frac{9}{10}$.

6 You pick an integer at random between zero and 10^5 inclusive. What is the probability that its digits are all different?

7 One hundred light bulbs are numbered consecutively from 1 to 100, and are off. They are wired to 100 switches in such a way that the nth switch changes the state (off to on, or on to off) of all the bulbs numbered kn; $k \geqslant 1$. If the switches are all thrown successively, how many light bulbs are on? What is the answer if you start with M light bulbs and M switches?

8 (a) Show that the product of any r consecutive integers is divisible by $r!$.
(b) Show that $(k!)!$ is divisible by $(k!)^{(k-1)!}$.

9 **Poker dice** Each die bears the symbols A, K, Q, J, 10, 9. If you roll five such dice, what is the probability that your set of five symbols includes:

(a) four aces? (b) four of a kind? (c) A, K, Q?

10 Eight rooks are placed randomly on a chess board (with at most one on each square). What is the probability that:
(a) they are all in a straight line?
(b) no two are in the same row or column?

11 An urn contains $4n$ balls, n of which are coloured black, n pink, n blue and n brown. Now r balls are drawn from the urn without replacement, $r \geqslant 4$. What is the probability that:
(a) at least one of the balls is black?
(b) exactly two balls are black?
(c) there is at least one ball of each colour?

12 Find the number of distinguishable ways of colouring the faces of a solid regular tetrahedron with:
(a) at most three colours (red, blue and green);
(b) exactly four colours (red, blue, green and yellow);
(c) at most four colours (red, blue, green and yellow).

13 An orienteer runs on the rectangular grid through the grid points (m, n), $m, n = 0, 1, 2, \ldots$ of a Cartesian plane. On reaching (m, n) the orienteer must next proceed either to $(m + 1, n)$ or $(m, n + 1)$.
(a) Show the number of different paths from $(0, 0)$ to (n, n) equals the number from $(1, 0)$ to $(n + 1, n)$ and that this equals $\binom{2n}{n}$, where $\binom{k}{r} = \frac{k!}{r!(k-r)!}$.
(b) Show that the number of different paths from $(1, 0)$ to $(n + 1, n)$ passing through at least one of the grid points (r, r) with $1 \leqslant r \leqslant n$ is equal to the total number of different paths from $(0, 1)$ to $(n + 1, n)$ and that this equals $\binom{2n}{n-1}$.
(c) Suppose that at each grid point the orienteer is equally likely to choose to go to either of the two possible next grid points. Let A_k be the event that the first of the grid points (r, r), $r \geqslant 1$, to be visited is (k, k). Show that

$$\mathbf{P}(A_k) = \frac{4^{-k}}{2k - 1}\binom{2k - 1}{k}.$$

14 A bag contains b black balls and w white balls. If balls are drawn from the bag without replacement, what is the probability P_k that exactly k black balls are drawn before the first white ball?

By considering $\sum_{k=0}^{b} P_k$, or otherwise, prove the identity

$$\sum_{k=0}^{b} \binom{b}{k} \bigg/ \binom{b+w-1}{k} = \frac{b+w}{w}$$

for positive integers b, w.

15 (a) Show that N'£' symbols and m'.' symbols may be set out in a line with a'.' at the right-hand end in $\binom{N+m-1}{m-1}$ ways, provided $m \geqslant 1$.

(b) A rich man decides to divide his fortune, which consists of N one-pound coins, among his m friends. Happily $N > m \geqslant 1$.

(i) In how many ways can the coins be so divided?

(ii) In how many ways can the coins be so divided if every friend must receive at least one?

(c) Deduce, or prove otherwise, that whenever $N > m \geqslant 1$,

$$\sum_{k=1}^{m} \binom{m}{k} \binom{N-1}{k-1} = \binom{N+m-1}{m-1}.$$

16 Let N balls be placed independently at random in n boxes, where $n \geqslant N > 1$, each ball having an equal chance $1/n$ of going into each box. Obtain an expression for the probability P that no box will contain more than one ball. Prove that $N(N-1) < Kn$ where $K = -2\log P$, and hence that $N < \frac{1}{2} + \sqrt{(Kn + \frac{1}{4})}$.

Now suppose that $P \geqslant e^{-1}$. Show that $N - 1 < 4n/5$ and hence that $Kn < N(N+1)$.

Prove finally that N is the integer nearest to $\sqrt{(Kn + \frac{1}{4})}$ when $P \geqslant e^{-1}$.

[You may assume that $\log(1-x) < -x$ for $0 < x < 1$, that $\log(1-x) > -x - \frac{3}{2}x^2$ for $0 < x < \frac{4}{5}$, and that $\sum_{r=1}^{N-1} r^2 = N(N-1)(2N-1)/6$.]

17 Consider sequences of n integers $a_1, a_2, \ldots, a_n$ such that $0 \leqslant a_i < k$ for each i, where k is a positive integer.

(a) How many such sequences are there?

(b) How many sequences have all a_i distinct?

(c) How many sequences have the property that $a_1 \leqslant a_2 \leqslant \ldots \leqslant a_n$?

18 Let a_n ($n = 2, 3, \ldots$) denote the number of distinct ways the expression $x_1 x_2 \ldots x_n$ can be bracketed so that only two quantities are multiplied together at any one time. (For example when $n = 2$ there is only one way, $(x_1 x_2)$, and when $n = 3$ there are two ways, $(x_1(x_2 x_3))$ and $((x_1 x_2)x_3)$.)

Prove that

$$a_{n+1} = a_n + a_2 a_{n-1} + a_3 a_{n-2} + \ldots + a_{n-2} a_3 + a_{n-1} a_2 + a_n.$$

Defining $A(x) = x + a_2 x^2 + a_3 x^3 + \ldots$ prove that $(A(x))^2 = A(x) - x$.

Deduce that $A(x) = \frac{1}{2}(1 - (1-4x)^{\frac{1}{2}})$, and show that

$$a_n = \frac{1.3 \ldots (2n-3)}{n!} 2^{n-1}.$$

19 **Coupons** Each packet of some harmful and offensive product contains one of a series of r different types of object. Every packet is equally likely to contain one of the r types. If you buy $n \geqslant r$ packets, show that the probability that you are then the owner of a set of all r types is

$$\sum_{k=0}^{r} (-)^k \binom{r}{k} \left(1 - \frac{k}{r}\right)^n.$$

20 **Tennis** Suppose that $2n$ players enter for two consecutive tennis tournaments. If the draws for each tournament are random, what is the probability that no two players meet in the first round of both tournaments? If n is large, show that this probability is about $e^{-\frac{1}{2}}$.

21 **Lotteries again** Suppose that n balls numbered from 1 to n are drawn randomly from an urn. Show that the probability that no two consecutive numbers are actually carried by consecutive balls drawn is

$$1-\frac{1}{1!}\left(1-\frac{1}{n}\right)+\frac{1}{2!}\left(1-\frac{2}{n}\right)-\frac{1}{3!}\left(1-\frac{3}{n}\right)+\ldots+\frac{(-)^{n-1}}{n!}.$$

[**Hint:** show that the number of arrangements of 1, 2, . . . , n such that at least j pairs of consecutive integers occur is $(n-j)!$.]

22 **Runs** A fair coin is tossed n times yielding a heads and $n-a$ tails. Show that the probability that there are k head runs and k tail runs (see Example 3.14 for definitions) is $2\binom{a-1}{k-1}\binom{n-a-1}{k-1}\frac{a!(n-a)!}{n!}$. Deduce that

$$\sum_{k=1}^{a\wedge(n-a)}\binom{a-1}{k-1}\binom{n-a-1}{k-1}\bigg/\binom{n-2}{a-1}=1$$

(where $x\wedge y$ denotes the smaller of x and y).

23 **Camelot** For obvious reasons Arthur would rather not sit next to Mordred or Lancelot at the round table. (There are n seats, and n knights including these three.)

(a) If the n knights sit at random, what is the probability that Arthur sits next to neither? Does it make any difference whether Arthur sits at random or not?

(b) If the n knights sit at random on two occasions, what is the probability that no one has the same left-hand neighbour on the two occasions?

24 By considering $(x+x^2+\ldots+x^n)^r$ show that n indistinguishable objects may be divided into r distinct groups with at least one object in each group in $\binom{n-1}{r-1}$ ways.

25 There are $2n$ balls in an urn; the balls are numbered 1, 2, . . . , $2n$. They are withdrawn at random without replacement. What is the probability that

(a) for no integer j, the $2j$th ball drawn bears the number $2j$?

(b) for no integer j, the ball bearing the number $j+1$ is removed next after the ball bearing the number j?

Find the limit as $n\to\infty$ of the probabilities in (a) and (b).

26 A chandelier has seven light bulbs arranged around the circumference of a circle. By the end of a given year each will have burnt out with probability $\frac{1}{2}$. Assuming that they do so independently, what is the probability that four or more bulbs will have burnt out?

If three bulbs burn out, what is the probability that no two are adjacent?

I decide that I will replace all the dead bulbs at the end of the year only if at least two are adjacent. Find the probability that this will happen. If it does, what is the probability that I will need more than 2 bulbs?

27 A biased coin is tossed $2n$ times. Show that the probability that the number of heads is the same as the number of tails is $\binom{2n}{n}(pq)^n$. Find the limit of this as $n\to\infty$.

28 Show that:

(a) $\sum_0^n(-)^k\binom{n}{k}=0$;

(b) $\sum_0^n\binom{n}{k}=2^n$;

(c) $\sum_{k=0}^{n/2}\binom{n}{2k}=2^{n-1}$ if n is even;

(d) $\sum_{k=0}^{n/2}\binom{n}{k}=2^{n-1}$ if n is even.

29 Observe that

(i) $(1+x)^m(1+x)^n=(1+x)^{m+n}$

(ii) $(1-x)^m(1-x)^{-n-2}=(1-x)^{m-n-2}$.

Now show that

(a) $\sum_{j=0}^{k}\binom{m}{j}\binom{n}{k-j}=\binom{m+n}{k}$ and (b) $\sum_{k=1}^{m}(-)^{m-k}\binom{m}{k}\binom{n+k}{n+1}=\binom{n}{m-1}$.

30 Show that for $j \leqslant n/2$, $n^{-n}\sum_{k=0}^{j}\binom{n}{k} \leqslant j^{-j}(n-j)^{-(n-j)}$.

31 Show that for fixed n, $\binom{n}{k}$ is largest when k is the integer nearest to $\frac{n}{2}$.

32 Show that

$$\binom{n}{r}\binom{n+k}{r+2k}\binom{n+2k}{r+k}=\binom{n}{r+k}\binom{n+k}{r}\binom{n+2k}{r+2k}$$

and interpret this in Pascal's triangle.

33 Show that $\sum_{k=0}^{n}\binom{a-k}{b}=\binom{a+1}{b+1}-\binom{a-n}{b+1}$, and deduce that $\sum_{k=0}^{n}\binom{k+a-1}{a-1}=\binom{n+a}{a}$.

34 An urn contains b blue balls and a aquamarine balls. The balls are removed successively at random from the urn without replacement. If $b > a$, show that the probability that at all stages until the urn is empty there are more blue than aquamarine balls in the urn is $(b-a)/(a+b)$.

Why is this result called the ballot theorem?

[**Hint:** use conditional probability and induction.]

35 The points $A_0, A_1, \ldots, A_n$ lie, in that order, on a circle. Let $a_1 = 1$, $a_2 = 1$ and for $n > 2$, let a_n denote the number of dissections of the polygon $A_0A_1 \ldots A_n$ into triangles by a set of non-crossing diagonals, A_iA_j.

(a) Check that $a_3 = 2$ and $a_4 = 5$.

(b) Show that in each dissection there is a unique i $(1 \leqslant i \leqslant n-1)$ such that cuts are made along both A_0A_i and A_nA_i.

(c) Show that

$$a_n = a_1a_{n-1} + a_2a_{n-2} + \ldots + a_{n-2}a_2 + a_{n-1}a_1.$$

(d) If $f(x) = \sum_1^\infty a_nx^n$, show (by considering the coefficient of each power of x) that $(f(x))^2 - f(x) + x = 0$, and show that $f(x) = \frac{1}{2} - \frac{1}{2}\surd(1-4x)$.

36 Let A, B, C, D be the vertices of a tetrahedron. A beetle is initially at A; it chooses any of the edges leaving A and walks along it to the next vertex. It continues in this way; at any vertex it is equally likely to choose to go to any other vertex next. What is the probability that it is at A when it has traversed n edges?

37 Suppose that n sets of triplets form a line at random. What is the probability that no three triplets from one set are adjacent?

38 Suppose a group of N objects may each have up to r distinct properties $b_1, \ldots, b_r$. With the notation of (3.4.2) show that the number possessing exactly m of these properties is

$$M_m = \sum_{k=0}^{r-m}(-)^k\binom{m+k}{k}N(b_1, \ldots, b_{m+k}).$$

39 **The ménages problem revisited** Use the result of Problem 38 to show that the probability that exactly m couples are seated in adjacent seats is

$$p_m = \frac{2}{m!}\sum_{k=0}^{n-m}(-)^k\frac{k(n-m-k)!(2n-m-k-1)!}{k!n!(2n-2m-2k)!}.$$

40 Suppose that N objects are placed in a row. The operation S_k is defined as follows: "Pick one of the first k objects at random and swap it with the object in the kth place". Now perform $S_N, S_{N-1}, \ldots, S_1$. Show that the final arrangement is equally likely to be any one of the $N!$ permutations of the objects.

41 Suppose that n contestants are to be placed in order of merit, and ties are possible. Let $r(n)$ be the number of possible distinct such orderings of the n contestants. (Thus $r(0) = 0$, $r(1) = 1$, $r(2) = 3$, $r(3) = 13$, and so on.) Show that $r(n)$ has exponential generating function

$$E_r(x) = \sum_{n=0}^{\infty} \frac{x^n}{n!} r(n) = \frac{1}{2 - e^x}.$$

[**Hint:** remember the multinomial theorem, and consider the coefficient of x^n in $(e^x - 1)^k$.]

4
Random variables: distribution and expectation

4.1 Random variables

In many experiments outcomes are defined in terms of numbers (for example, the number of heads in n tosses of a coin), or they may be associated with numbers, if we so choose. In either case we want to assign probabilities directly to these numbers, as well as to the underlying events. This requires the introduction of some new functions.

(1) Definition Given a sample space Ω, (with $\mathcal{F}$ and $\mathbf{P}(.)$), a *discrete random variable* X is a function such that for each outcome ω in Ω, $X(\omega)$ is one of a countable set D of real numbers. Formally $X(.)$ is a function with domain Ω and range D, and so for each $\omega \in \Omega$, $X(\omega) = x \in D$, where D is a countable (denumerable) subset of the real numbers. ▲

(2) Example: Pairs in poker How many distinct pairs are there in your poker hand of five cards? Your hand is one outcome ω in the sample space Ω of all possible hands; if you are playing with a full deck then $|\Omega| = \binom{52}{5}$. The number of pairs X depends on the outcome ω, and obviously $X(\omega) \in \{0, 1, 2\}$, as you can have no more than two pairs. Notice that this holds regardless of how the hand is selected, and whether or not the pack is shuffled. However, this information will be required later to assign probabilities. ●

We shall always use upper case letters (such as X, Y, T, R, and so on), to denote random variables, and lower case letters (x, y, z, etc.), to denote their possible numerical values. You should do the same. Since the possible values of X are countable, we can denote them by $\{x_i; i \in I\}$, where the index set I is a subset of the integers. Very commonly all the possible values of X are integers, in which case we may denote them simply by x,· r, k, j, or any other conventional symbol for integers.

(3) Definition (a) If X takes only the values 0 or 1 it is called an *indicator*, or sometimes a *Bernoulli trial*.
(b) If X takes one of only a finite number of values then it is called *simple*. ▲

(4) Example Suppose n coins are tossed. Let X_j be the number of heads shown by the jth coin. Then X_j is obviously zero or one; we may write $X_j(H) = 1$, and $X_j(T) = 0$.

Let Y be the total number of heads shown by the n coins. Clearly, for each outcome $\omega \in \Omega$, $Y(\omega) \in \{0, 1, 2, \ldots, n\}$. Thus X_j is an indicator and Y is simple. ●

It is intuitively clear also that

$$Y = \sum_1^n X_j.$$

We discuss the meaning of this and its implications in Chapter 5.

Finally, note that the sample space Ω need not be countable, even though $X(\omega)$ takes one of a countable set of values.

(5) Example: Darts You throw one dart at a conventional dartboard. A natural sample space Ω is the set of all possible points of impact. This is of course uncountable, since it includes every point of the dartboard, much of the wall, and even parts of the floor or ceiling if you happen not to be especially adroit.

However your score $X(\omega)$ is one of a finite set of integers lying between 0 and 60, inclusive. ●

4.2 Distributions

Next we need a function, defined on the possible values x of X, to tell us how likely they are. For each such x, there is an event $A_x \subseteq \Omega$, such that

(1) $$\omega \in A_x \Leftrightarrow X(\omega) = x.$$

Hence, just as the probability that any event A in Ω occurs is given by the probability function $\mathbf{P}(A) \in [0, 1]$, the probability that $X(\omega)$ takes any value x is given by a function $\mathbf{P}(A_x) \in [0, 1]$. (We assume that A_x is in $\mathcal{F}$.)

For example, if a coin is tossed, and X is the number of heads shown, then $X \in \{0, 1\}$ and $A_1 = H$; $A_0 = T$. Hence $\mathbf{P}(X = 1) = \mathbf{P}(A_1) = \mathbf{P}(H) = \frac{1}{2}$, if the coin is fair.

This function has its own special name and notation. Given Ω, $\mathcal{F}$, and $\mathbf{P}(.)$:

(2) Definition A discrete random variable X has a *probability mass function* $f_X(x)$ given by

$$f_X(x) = \mathbf{P}(A_x).$$

This is also denoted by $\mathbf{P}(X = x)$, which can be thought of as an obvious shorthand for $\mathbf{P}(\{\omega: X(\omega) = x\})$. ▲

For example, let X be the number of pairs in a poker hand, as discussed in Example 4.1.2. If the hand is randomly selected then

$$f_X(2) = \mathbf{P}(\{\omega: X(\omega) = 2\}) = |\{\omega: X(\omega) = 2\}| \bigg/ \binom{52}{5} = \binom{13}{2}\binom{4}{2}\binom{4}{2}\binom{44}{1} \bigg/ \binom{52}{5}$$
$$\simeq 0.048.$$

Likewise

$$f_X(1) = \binom{13}{1}\binom{4}{2}\binom{12}{3}\binom{4}{1}^3 \Big/ \binom{52}{5} \simeq 0.42,$$

and hence

$$f_X(0) = 1 - f_X(1) - f_X(2) \simeq 0.53.$$

Returning to Example 4.1.4 gives an example of great theoretical and historical importance.

(3) **Example 4.1.4 revisited: Binomial distribution** The random variable Y takes the value r, if exactly r heads appear in the n tosses. The probability of this event is $\binom{n}{r}p^r q^{n-r}$, where $p = \mathbf{P}(H) = 1 - q$. Hence Y has probability mass function

$$f_Y(r) = \binom{n}{r} p^r q^{n-r}, \quad 0 \leqslant r \leqslant n.$$ ●

The suffix in $f_X(x)$ or $f_Y(y)$ is included to stress the role of X or Y. Where this is unnecessary, or no confusion can arise, we omit it. In the interests of brevity, $f(x)$ is often called simply the *mass function* of X, or even more briefly the p.m.f.

The p.m.f., $f(x) = f_X(x)$, has the following properties: firstly

(4)
$$f(x) \geqslant 0 \quad \text{for } x \in \{x_i : i \in \mathbb{Z}\}$$
$$f(x) = 0 \quad \text{elsewhere.}$$

That is to say it is positive for a countable number of values of x, and is zero elsewhere.

Secondly, if $X(\omega)$ is finite with probability one, then it is called a *proper* random variable and we have

(5)
$$\sum_i f(x_i) = \sum_i \mathbf{P}(A_{x_i}) = \mathbf{P}(\Omega) = 1.$$

If

(6)
$$\sum_i f(x_i) < 1,$$

then X is said to be defective or improper. It is occasionally useful to allow X to take values in the extended real line, so that $f_X(\infty)$ has a meaning. In general it does not.

We remark that any function satisfying (4) and (5) can be regarded as a mass function, in that given such an $f(.)$ it is quite simple to construct a sample space, probability function and random variable X, such that $f(x) = \mathbf{P}(X = x)$.

Here are two famous mass functions.

(7) **Example: Poisson distribution** Let X be a random variable with mass function

$$f_X(x) = \frac{\lambda^x e^{-\lambda}}{x!}, \quad x \in \{0, 1, 2, \ldots\}, \lambda > 0.$$

Then

$$\sum_{x=0}^{\infty} f(x) = e^{-\lambda} \sum_{x=0}^{\infty} \frac{\lambda^x}{x!} = 1 \quad \text{by Theorem 3.6.9.}$$

Hence X is proper. This mass function is called the *Poisson distribution*, and X is said to be Poisson (or a Poisson random variable), with parameter λ. ●

(8) **Example: Negative binomial distribution** By the negative binomial theorem, for any number q such that $0 < q < 1$, we have

$$(1-q)^{-n} = \sum_{r=0}^{\infty} \binom{n+r-1}{r} q^r.$$

Hence the function $f(r)$ defined by

$$f(r) = \binom{n+r-1}{r} q^r (1-q)^n, \quad r \geq 0,$$

is a probability mass function. Commonly we let $1 - q = p$. ●

The following function is also very useful.

Definition A discrete random variable X has a *cumulative distribution function* $F_X(x)$, where

(9) $$F_X(x) = \sum_{i: x_i \leq x} f(x_i).$$ ▲

This is also denoted by

$$\mathbf{P}(X \leq x) = \mathbf{P}(\{\omega: X(\omega) \leq x\});$$

it may be referred to simply as the distribution function (or rarely as the c.d.f.), and the suffix X may be omitted. The following properties of $F(x)$ are trivial consequences of the definition (9):

(10) $$F(x) \leq F(y) \quad \text{for } x \leq y.$$

(11) $$1 - F(x) = \mathbf{P}(X > x).$$

(12) $$\mathbf{P}(a < X \leq b) = F(b) - F(a) \quad \text{for } a < b.$$

Some further useful properties are not quite so trivial, in that they depend on Theorem 1.5.2. Thus, if we define the event $B_n = \{X \leq x - 1/n\}$, we find that

$$\mathbf{P}(X < x) = \mathbf{P}\left(\bigcup_{n=1}^{\infty} B_n\right) = \mathbf{P}(\lim_{n\to\infty} B_n) = \lim_{n\to\infty} \mathbf{P}(B_n) \quad \text{by Theorem 1.5.2,}$$

$$= \lim_{n\to\infty} F\left(x - \frac{1}{n}\right) = \lim_{y \uparrow x} F(y).$$

If the random variable X is not defective then, again from (9) (and Theorem 1.5.2), $\lim_{x\to\infty} F(x) = 1$, and $\lim_{x\to-\infty} F(x) = 0$.

The c.d.f. is obtained from the p.m.f. by (9). Conversely the p.m.f. is obtained

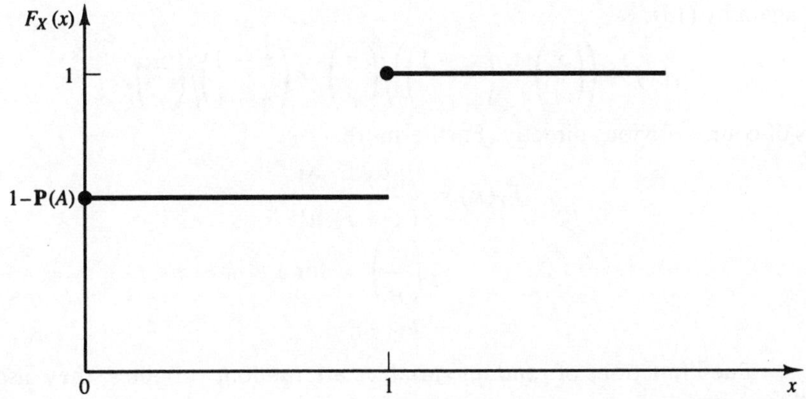

Figure 4.1 The distribution function $F_X(x)$ of the random variable X which is the indicator of the event A. Thus the jump at zero is $\mathbf{P}(X=0)=\mathbf{P}(A^c)=1-\mathbf{P}(A)$ and the jump at $x=1$ is $\mathbf{P}(X=1)=\mathbf{P}(A)$.

from the c.d.f. by

$$f(x) = F(x) - \lim_{y \uparrow x} F(y) \quad \text{where } y < x.$$

When X takes only integer values this relationship has the following simpler more attractive form: for integer x

(13) $$f(x) = F(x) - F(x-1).$$

(14) Example: Lottery An urn contains n tickets bearing numbers from 1 to n inclusive. Of these r are withdrawn at random. Let X be the largest number removed if the tickets are replaced in the urn after each drawing, and let Y be the largest number removed if the drawn tickets are not replaced. Find $f_X(x)$, $F_X(x)$, $f_Y(x)$ and $F_Y(x)$. Show that $F_Y(k) < F_X(k)$, for $0 < k < n$.

Solution The number of ways of choosing r numbers less than or equal to x, with repetition allowed, is x^r. Since there are n^r outcomes,

$$F_X(x) = \left(\frac{x}{n}\right)^r, \quad \text{for } 1 \leqslant x \leqslant n,$$

when x is an integer. For any real x, $F_X(x) = ([x]/n)^r$, (where $[x]$ denotes the largest integer which is not greater than x). Hence for integer x, and $1 \leqslant x \leqslant n$, by (13),

$$f_X(x) = \left(\frac{x}{n}\right)^r - \left(\frac{x-1}{n}\right)^r;$$

and elsewhere $f_X(x)$ is zero.

Without replacement, the number of ways of choosing r different numbers less than or equal to x is $\binom{x}{r}$. Hence for integer x, and $1 \leqslant x \leqslant n$,

$$F_Y(x) = \binom{x}{r} \bigg/ \binom{n}{r}.$$

Hence, again by (13),

$$f_Y(x) = \left(\binom{x}{r} - \binom{x-1}{r}\right)\Big/\binom{n}{r} = \binom{x-1}{r-1}\Big/\binom{n}{r},$$

which is of course obvious directly. Furthermore

$$\begin{aligned} F_Y(k) &= \frac{k!(n-r)!}{(k-r)!n!} \\ &< \left(\frac{k}{n}\right)^r \quad \text{for } 1 < k < n \\ &= F_X(k). \end{aligned}$$ ●

Since real valued functions of random variables are random variables, they also have probability mass functions.

(15) Theorem If X and Y are random variables such that $Y = g(X)$, then Y has p.m.f. given by

$$\sum_{x:g(x)=y} f_X(x).$$

Proof

$$f_Y(y) = \mathbf{P}(g(X) = y) = \sum_{x:g(x)=y} \mathbf{P}(X = x) = \sum_{x:g(x)=y} f_X(x).$$ ■

(16) Example Let X have mass function $f(x)$. Find the mass functions of the following functions of X.

(a) $-X$
(b) $X^+ = \max\{0, X\}$
(c) $X^- = \max\{0, -X\}$
(d) $|X| = X^+ + X^-$
(e) $\operatorname{sgn} X = \begin{cases} \dfrac{X}{|X|}, & X \neq 0, \\ 0, & X = 0. \end{cases}$

Solution Using Theorem 15 repeatedly we have:

(a) $f_{-X}(x) = f_X(-x)$.

(b) $f_{X^+}(x) = \begin{cases} f_X(x) & ; x > 0. \\ \sum_{x \leqslant 0} f_X(x) & ; x = 0. \end{cases}$

(c) $f_{X^-}(x) = \begin{cases} f_X(-x) & ; x > 0 \\ \sum_{x \geqslant 0} f_X(x) & ; x = 0. \end{cases}$

(d) $f_{|X|}(x) = \begin{cases} f_X(x) + f_X(-x) & ; x \neq 0 \\ f_X(0) & ; x = 0. \end{cases}$

(e) $f_{\operatorname{sgn} X}(x) = \begin{cases} \sum_{x>0} f_X(x) & ; x = 1 \\ f_X(0) & ; x = 0 \\ \sum_{x<0} f_X(x) & ; x = -1. \end{cases}$ ●

Finally we note that any number m such that $\lim_{x \uparrow m} F(x) \leq \frac{1}{2} \leq F(m)$ is called a *median* of F (or a median of X, if X has distribution F)

4.3 Expectation

(1) **Definition** Let X be a random variable with probability mass function $f(x)$ such that

$$\sum_x |x| f(x) < \infty.$$

The *expected value* of X is then denoted by $\mathbf{E}(X)$ and defined by

$$\mathbf{E}(X) = \sum_x x f(x).$$

This is also known as the *expectation*, or *mean*, or *average* or *first moment* of X. ▲

The importance of $\mathbf{E}(X)$ will become progressively more apparent.

(2) **Example** Suppose that X is an indicator random variable, so that $X(\omega) \in \{0, 1\}$. Define the event $A = \{\omega\colon X(\omega) = 1\}$. Then X is the indicator of the event A; we have $f_X(1) = \mathbf{P}(A)$, and $\mathbf{E}(X) = 0.f_X(0) + 1.f_X(1) = \mathbf{P}(A)$. ●

(3) **Example** Let X have mass function

$$f_X(x) = \frac{4}{x(x+1)(x+2)}, \quad x = 1, 2, \ldots$$

and let Y have mass function

$$f_Y(x) = \frac{1}{x(x+1)}, \quad x = 1, 2, \ldots$$

Show that X does have an expected value, and that Y does not have an expected value.

Solution For any $m < \infty$

$$\sum_{x=1}^{m} |x| f_X(x) = \sum_{x=1}^{m} \frac{4}{(x+1)(x+2)} = 4\sum_{x=1}^{m}\left(\frac{1}{x+1} - \frac{1}{x+2}\right) = 2 - 4(m+2)^{-1},$$

by successive cancellations in the sum. Hence the sum converges as $m \to \infty$, and so because $X > 0$, $\mathbf{E}(X) = 2$. However, for the random variable Y,

$$\sum_x |x| f_Y(x) = \sum_{x=1}^{\infty} \frac{1}{x+1}$$

which is not finite. ●

Notice that the condition $\sum|x|f(x) < \infty$ amounts to $\mathbf{E}(X^+) + \mathbf{E}(X^-) < \infty$ (use Example 4.2.16 to see this). A little extension of Definition 1 is sometimes useful. Thus if $\mathbf{E}(X^-) < \infty$, but $\mathbf{E}(X^+)$ diverges, then we may define $\mathbf{E}(X) = +\infty$. With this extension in Example 3, $\mathbf{E}(Y) = \infty$. Likewise if $\mathbf{E}(X^+) < \infty$, but $\mathbf{E}(X^-)$ diverges, then $\mathbf{E}(X) = -\infty$. If both $\mathbf{E}(X^+)$ and $\mathbf{E}(X^-)$ diverge, then $\mathbf{E}(X)$ is undefined.

In general, real valued functions of random variables are random variables having a mass function given by Theorem 4.2.15. They may therefore have an expected value. In accordance with Example 3, if $Y = g(X)$, then by definition

$$\mathbf{E}(g(X)) = \sum_i y_i f_Y(y_i).$$

We used this with Example 4.2.16(b) in observing that

$$\mathbf{E}(X^+) = \sum_{x>0} x f_X(x).$$

This was easy, because it was easy to find the mass function of X^+ in terms of that of X. It is not such an immediately attractive prospect to calculate (for example) $\mathbf{E}(\cos(\theta X))$ by first finding its mass function. The following theorem is therefore extremely useful.

(4) **Theorem** Let X be a random variable with mass function $f(x)$, and let $g(.)$ be a real valued function defined on $\mathbb{R}$. Then

$$\mathbf{E}(g(X)) = \sum_x g(x) f(x)$$

whenever $\sum_x |g(x)| f(x) < \infty$.

Proof Let (g_j) denote the possible values of $g(X)$, and for each j define the set $A_j = \{x\colon g(x) = g_j\}$. Then $\mathbf{P}(g(X) = g_j) = \mathbf{P}(X \in A_j)$, and therefore, provided all the following summations converge absolutely, we have

$$\mathbf{E}(g(X)) = \sum_j g_j \mathbf{P}(g(X) = g_j) = \sum_j g_j \sum_{x \in A_j} f(x)$$

$$= \sum_j \sum_{x \in A_j} g(x) f(x), \quad \text{since } g(x) = g_j \text{ for } x \in A_j,$$

$$= \sum_x g(x) f(x), \quad \text{since } A_j \cap A_k = \phi \text{ for } j \neq k. \qquad ■$$

(5) **Example** Let X be Poisson with parameter λ. Find $\mathbf{E}(\cos(\theta X))$.

Solution First recall de Moivre's Theorem that $e^{i\theta} = \cos\theta + i\sin\theta$, where i is an imaginary square root of -1. Now, by Theorem 4

$$\begin{aligned}\mathbf{E}(\cos(\theta X)) &= \sum_{k=0}^{\infty} \cos(k\theta)e^{-\lambda}\lambda^k/k! \\ &= \text{Re}\left(\sum_{k=0}^{\infty} e^{ik\theta}e^{-\lambda}\lambda^k/k!\right), \quad \text{where Re}(z) \text{ is the real part of } z, \\ &= \text{Re}(\exp(\lambda e^{i\theta} - \lambda)) \\ &= e^{\lambda(\cos\theta - 1)}\cos(\lambda\sin\theta), \quad \text{using de Moivre's Theorem again.}\end{aligned}$$ ●

Now we can use Theorem 4 to establish some important properties of $\mathbf{E}(.)$.

(6) **Theorem** Let X be a random variable with finite mean $\mathbf{E}(X)$, and let a and b be constants. Then:

(i) $\mathbf{E}(aX + b) = a\mathbf{E}(X) + b$;
(ii) if $\mathbf{P}(X = b) = 1$, then $\mathbf{E}(X) = b$;
(iii) if $\mathbf{P}(a < X \leqslant b) = 1$, then $a < \mathbf{E}(X) \leqslant b$;
(iv) if $g(X)$ and $h(X)$ have finite mean, then

$$\mathbf{E}(g(X) + h(X)) = \mathbf{E}(g(X)) + \mathbf{E}(h(X)).$$

Proof (i) First we establish the necessary absolute convergence:

$$\sum_x |ax + b|f(x) \leqslant \sum_x (|a|\,|x| + |b|)f(x) = |a|\sum_x |x|f(x) + |b| < \infty,$$

as required. Hence, by Theorem 4,

$$\mathbf{E}(aX + b) = \sum_x (ax + b)f(x) = a\sum_x xf(x) + b = a\mathbf{E}(X) + b.$$

(ii) Here X has mass function $f(b) = 1$, so by definition $\mathbf{E}(X) = bf(b) = b$.
(iii) In this case $f(x) = 0$ for $x \notin (a, b]$, so

$$\mathbf{E}(X) = \sum_x xf(x) \begin{cases} \leqslant \sum_x bf(x) = b; \\ > \sum_x af(x) = a. \end{cases}$$

(iv) Since $|g(x) + h(x)| \leqslant |g(x)| + |h(x)|$, absolute convergence is quickly established. Hence by Theorem 4,

$$\begin{aligned}\mathbf{E}(g(X) + h(X)) &= \sum_x (g(x) + h(x))f(x) = \sum_x g(x)f(x) + \sum_x h(x)f(x) \\ &= \mathbf{E}(g(X)) + \mathbf{E}(h(X)).\end{aligned}$$ ■

The following simple corollary is of some importance.

(7) **Theorem** If $\mathbf{E}(X)$ exists then

$$(\mathbf{E}(X))^2 \leq (\mathbf{E}(|X|))^2 \leq \mathbf{E}(X^2).$$

Proof First note that $(|X| - \mathbf{E}(|X|))^2 \geq 0$. Hence by Theorem 6(iii),

$$\begin{aligned} 0 \leq \mathbf{E}((|X| - \mathbf{E}(|X|))^2 &= \mathbf{E}(|X|^2) - (\mathbf{E}(|X|))^2, \quad \text{by theorem 6(iv) and 6(ii)} \\ &= \mathbf{E}(X^2) - (\mathbf{E}(|X|))^2, \end{aligned}$$

which proves the second inequality. Also $|X| - X \geq 0$, so by 6(iv) and 6(iii)

$$\mathbf{E}(X) \leq \mathbf{E}(|X|),$$

which proves the first inequality. ■

(8) **Example: Uniform distribution** Recall that an urn contains n tickets numbered from 1 to n. You take one ticket at random; it bears the number X. Find $\mathbf{E}(X)$ and $\mathbf{E}(X^2)$, and verify that Theorem 7 holds explicitly.

Solution The mass function of X is $\mathbf{P}(X = k) = 1/n$. (Because it distributes probability evenly over the values of X, it is called the *uniform distribution*.) Hence

$$\begin{aligned} \mathbf{E}(X) &= \sum_{x=1}^{n} \frac{x}{n} = \frac{1}{n} \sum_{x=1}^{n} \frac{1}{2}(x(x+1) - x(x-1)) \\ &= \frac{1}{2}(n+1) \quad \text{by successive cancellation.} \end{aligned}$$

Likewise, using Theorems 4 and 6(iv),

$$\begin{aligned} \mathbf{E}(X^2) + \mathbf{E}(X) &= \sum_{x=1}^{n} \frac{x^2 + x}{n} = \frac{1}{n} \sum_{x=1}^{n} \frac{1}{3}(x(x+1)(x+2) - (x-1)x(x+1)) \\ &= \frac{1}{3}(n+1)(n+2) \quad \text{by successive cancellation.} \end{aligned}$$

Hence

$$\mathbf{E}(X^2) = \frac{1}{6}(n+1)(2n+1) \geq \frac{1}{4}(n+1)^2 = (\mathbf{E}(X))^2. \quad ●$$

In practice we shall often be interested in the expectations of two particularly important collections of functions of X; namely $(X^k; k \geq 1)$ and $([X - \mathbf{E}(X)]^k; k \geq 1)$.

(9) **Definition** Let X have mass function $f(x)$ such that $\sum_x |x|^k f(x) < \infty$. Then

(a) the kth *moment* of X is $\mu_k = \mathbf{E}(X^k)$
(b) the kth *central moment* of X is $\sigma_k = \mathbf{E}((X - \mathbf{E}(X))^k)$
(c) the kth *factorial moment* of X is $\mu^{(k)} = \mathbf{E}(X(X-1)\dots(X-k+1))$.

In particular, σ_2 is called the *variance* of X and is denoted by σ^2, σ_X^2 or var(X). Thus

$$\operatorname{var}(X) = \mathbf{E}((X - \mathbf{E}(X))^2). \quad ▲$$

Example: Indicators Let X be the indicator of the event A (recall Example 4.3.2). Since $X^k(\omega) = X(\omega)$ for all k and ω, we have

$$\mu_k = \mathbf{E}(X^k) = \mathbf{P}(A).$$

Also $\operatorname{var}(X) = \mathbf{P}(A)\mathbf{P}(A^c)$, and

$$\mu^{(k)} = \begin{cases} \mathbf{P}(A) & ; k = 1 \\ 0 & ; k > 1. \end{cases}$$

●

(10) Example Show that if $\mathbf{E}(X^2) < \infty$, and a and b are constants then

$$\operatorname{var}(aX + b) = a^2 \operatorname{var}(X).$$

Solution Using Theorem 6(i), and the definition of variance,

$$\operatorname{var}(aX + b) = \mathbf{E}((a(X - \mathbf{E}(X)) + b - b)^2) = \mathbf{E}(a^2(X - \mathbf{E}(X))^2) = a^2 \operatorname{var}(X).$$

●

Sometimes the tail of a distribution, $\mathbf{P}(X > x)$, has a simpler form than the mass function $f(x)$. In these and other circumstances the following theorems are useful.

(11) Theorem If $X \geqslant 0$, and X takes integer values, then $\mathbf{E}(X) = \sum_{x=1}^{\infty} \mathbf{P}(X \geqslant x)$.

Proof By definition

$$\mathbf{E}(X) = \sum_{x=1}^{\infty} x f(x) = \sum_{x=1}^{\infty} f(x) \sum_{r=1}^{x} 1.$$

Since all terms are non-negative we may interchange the order of summation to obtain

$$\sum_{x=1}^{\infty} \sum_{r=x}^{\infty} f(r) = \sum_{x=1}^{\infty} \mathbf{P}(X \geqslant x).$$

■

This theorem has various generalizations; we state one.

(12) Theorem If $X \geqslant 0$ and $k \geqslant 2$, then

$$\mu^{(k)} = \mathbf{E}(X(X-1)\dots(X-k+1)) = k \sum_{x=k}^{\infty} (x-1)\dots(x-k+1)\mathbf{P}(X \geqslant x).$$

Proof This is proved in the same way as Theorem 11, by changing the order of summation on the right hand side. ■

(13) Example: Waiting–the geometric distribution A biased coin shows a head with probability p, or a tail with probability $q = 1 - p$. How many times do you expect to toss the coin until it first shows a head? Find the various second moments of this waiting time.

Solution Let the required number of tosses until the first head be T. Then because they are independent, $\mathbf{P}(T = x) = q^{x-1}p;\ x \geq 1$. ($T$ is said to have the *geometric* distribution.)
Hence

$$\mathbf{E}(T) = \sum_{x=1}^{\infty} xq^{x-1}p = \frac{p}{(1-q)^2}, \quad \text{by (3.6.12) with } n = 2,$$
$$= \frac{1}{p}.$$

Alternatively we can use Theorem 11 as follows. Using the independence of tosses again gives $\mathbf{P}(T > x) = q^x$, so

$$\mathbf{E}(T) = \sum_{x=0}^{\infty} \mathbf{P}(T > x) = \sum_{x=0}^{\infty} q^x = \frac{1}{p}.$$

For the second factorial moment, by Theorem 12

$$\mu^{(2)} = 2\sum_{x=2}^{\infty}(x-1)q^{x-1} = \frac{2q}{p^2}, \quad \text{by (3.6.12) again.}$$

Hence the second moment is

$$\mathbf{E}(T^2) = \mathbf{E}(T(T-1)) + \mathbf{E}(T) = \frac{2q}{p^2} + \frac{1}{p} = \frac{1+q}{p^2}.$$

And, finally, the second central moment is

$$\sigma_2 = \mathbf{E}((T - \mathbf{E}(T))^2) = \mathbf{E}(T^2) - (\mathbf{E}(T))^2 = \frac{1+q}{p^2} - \frac{1}{p^2} = \frac{q}{p^2}.$$ ●

(14) Example Let X have mass function

$$f_X(x) = \frac{a}{x^2}; \quad x = 1, 2, 3, \ldots$$

and Y have mass function

$$f_Y(y) = \frac{b}{y^2}; \quad y = \pm 1, \pm 2, \ldots$$

(a) Find a and b.
(b) What can you say about $\mathbf{E}(X)$ and $\mathbf{E}(Y)$?

Solution (a) Since $f_X(x)$ is a mass function

$$1 = \sum_x f_X(x) = a\sum_{x=1}^{\infty}\frac{1}{x^2} = a\frac{\pi^2}{6}.$$

Hence $a = 6\pi^{-2}$. Likewise $b = 3\pi^{-2}$.
(b) We have

$$\mathbf{E}(X) = a\sum_{x=1}^{\infty} xf_X(x) = a\sum_{x=1}^{\infty}\frac{1}{x} = \infty.$$

Since $\mathbf{E}(Y^+)$ and $\mathbf{E}(Y^-)$ both diverge, $\mathbf{E}(Y)$ does not exist. ●

(15) Example: Coupons Each packet of an injurious product is equally likely to contain any one of n different types of coupon, independently of every other packet. What is the expected number of packets you must buy to obtain at least one of each type of coupon?

Solution Let A_n^r be the event that the first r coupons you obtain do not include a full set of n coupons. Let C_k^r be the event that you have not obtained one of the kth coupon in the first r. Then

$$A_n^r = \bigcup_{k=1}^{n} C_k^r.$$

We may calculate:

$$\mathbf{P}(C_1^r) = \left(\frac{n-1}{n}\right)^r,$$

$$\mathbf{P}(C_i^r \cap C_j^r) = \left(\frac{n-2}{n}\right)^r; \quad i \neq j,$$

and in general, for any set S_j of j distinct coupons

$$\mathbf{P}\left(\bigcap_{i \in S_j} C_i^r\right) = \left(\frac{n-j}{n}\right)^r.$$

Hence, by (1.4.8)

$$\mathbf{P}(A_n^r) = \mathbf{P}\left(\bigcup_{k=1}^{n} C_k^r\right) = \sum_{j=1}^{n} (-1)^{j+1} \binom{n}{j} \left(1 - \frac{j}{n}\right)^r,$$

since for each j there are $\binom{n}{j}$ sets S_j.

Now let R be the number of packets required to complete a set of n distinct coupons. Since A_n^r occurs if and only if $R > r$, we have $\mathbf{P}(R > r) = \mathbf{P}(A_n^r)$. Hence by Theorem 11,

$$\mathbf{E}(R) = \sum_{r=0}^{\infty} \mathbf{P}(R > r) = \sum_{r=0}^{\infty} \sum_{j=1}^{n} (-1)^{j+1} \binom{n}{j} \left(1 - \frac{j}{n}\right)^r$$

$$= \sum_{j=1}^{n} (-1)^{j+1} \binom{n}{j} \frac{n}{j} = nu_n, \quad \text{say.}$$

Now

(16)
$$u_{n+1} - u_n = \sum_{j=1}^{n+1} (-1)^{j+1} \binom{n+1}{j} \frac{1}{j} - \sum_{j=1}^{n} (-1)^{j+1} \binom{n}{j} \frac{1}{j}$$

$$= \frac{(-1)^{n+2}}{n+1} + \sum_{j=1}^{n} \frac{(-1)^{j+1}}{j} \left(\binom{n+1}{j} - \binom{n}{j} \right)$$

$$= \frac{(-1)^{n+2}}{n+1} + \sum_{j=1}^{n} \frac{(-1)^{j+1}}{n+1} \binom{n+1}{j}$$

$$= \frac{1}{n+1},$$

because

$$\sum_{j=0}^{n+1}(-1)^{j+1}\binom{n+1}{j} = (1-1)^{n+1} = 0.$$

Hence, iterating (16)

$$u_n = \sum_{j=1}^{n}\frac{1}{n-j+1},$$

so that

$$\mathbf{E}(R) = \sum_{j=1}^{n}\frac{n}{n-j+1}.$$

In Chapter 5 we discover a much easier method of obtaining this result. ●

4.4 Conditional distributions

Let Ω be some sample space, X some random variable defined on Ω, and $\mathbf{P}(.)$ a probability function defined on Ω. Now suppose that we are given that some event $B \subseteq \Omega$ occurs, with $\mathbf{P}(B) > 0$. Just as we argued in Chapter 2 that this gives rise to a conditional probability function, so we now conclude that this gives rise to a conditional distribution of X given B.

In fact, using (4.2.1) we write

(1) $$\mathbf{P}(X(\omega) = x|B) = \mathbf{P}(A_x|B) = \mathbf{P}(A_x \cap B)/\mathbf{P}(B) \geqslant 0,$$

where as usual $A_x = \{\omega: X(\omega) = x\}$. Furthermore, since $\bigcup_x A_x = \Omega$, and $A_x \cap A_y = \phi$, whenever $x \neq y$, we have

(2) $$\sum_x \mathbf{P}(A_x|B) = \sum_x \mathbf{P}(A_x \cap B)/\mathbf{P}(B) = \mathbf{P}(\Omega \cap B)/\mathbf{P}(B) = 1.$$

Hence the function $f(x|B)$ defined by

(3) $$f(x|B) = \mathbf{P}(A_x|B) = \mathbf{P}(X = x|B)$$

is a probability mass function, in that $f(x|B) \geqslant 0$, and $\sum_x f(x|B) = 1$. It is the conditional mass function of X given B.

(4) **Example** Let X be uniformly distributed on $\{1, 2, \ldots, n\}$, and let B be the event that $a \leqslant X \leqslant b$, where $1 \leqslant a < b \leqslant n$. Find the mass function of X given B.

Solution Obviously

$$\mathbf{P}(B) = \sum_{i=a}^{b}\frac{1}{n} = (b-a+1)/n$$

and

$$\mathbf{P}(\{X = k\} \cap B) = \begin{cases} \dfrac{1}{n}; & a \leqslant k \leqslant b \\ 0; & \text{otherwise.} \end{cases}$$

Hence

$$f(x|B) = \begin{cases} \dfrac{1}{b-a+1}; & a \leqslant x \leqslant b \\ 0 \quad ; & \text{otherwise} \end{cases}$$

Thus, given that X lies in B, it is uniformly distributed over B. ●

Since $f(x|B)$ is a probability mass function it may have an expectation. In line with Definition 4.3.1 we require that $\sum_x |x| f(x|B) < \infty$. If this condition is satisfied, then the conditional expectation of X given B is denoted by $\mathbf{E}(X|B)$, and defined by

(5) $$\mathbf{E}(X|B) = \sum_x x f(x|B).$$

Expectation and conditional expectation are related by the following exceptionally important result.

(6) Theorem Let X be a random variable with mean $\mathbf{E}(X)$, and let B be an event such that $\mathbf{P}(B)\mathbf{P}(B^c) > 0$. Then

$$\mathbf{E}(X) = \mathbf{E}(X|B)\mathbf{P}(B) + \mathbf{E}(X|B^c)\mathbf{P}(B^c).$$

Proof By conditional probability

$$\begin{aligned} f(x) = \mathbf{P}(X = x) &= \mathbf{P}(\{X = x\} \cap B) + \mathbf{P}(\{X = x\} \cap B^c) \\ &= f(x|B)\mathbf{P}(B) + f(x|B^c)\mathbf{P}(B^c). \end{aligned}$$

Hence

$$\mathbf{E}(X) = \sum_x x f(x) = \mathbf{P}(B)\sum_x x f(x|B) + \mathbf{P}(B^c)\sum_x x f(x|B^c)$$

as required. ■

More generally, it is shown in exactly the same way that if $(B_i; i \geqslant 1)$ is a collection of events such that

(i) $\bigcup_i B_i = \Omega$,

(ii) $B_i \cap B_j = \phi; \quad i \neq j$, and

(iii) $\mathbf{P}(B_i) > 0$,

then

(7) $$\mathbf{E}(X) = \sum_i \mathbf{E}(X|B_i)\mathbf{P}(B_i)$$

whenever the summation is absolutely convergent.

Finally we make the small but useful observation that if $A \subseteq B$, then

(8) $$\mathbf{E}(X|A \cap B) = \mathbf{E}(X|A).$$

(9) **Example** A coin is tossed repeatedly. As usual, for each toss $\mathbf{P}(H) = p = 1 - q = 1 - \mathbf{P}(H^c)$. The outcome is a sequence of runs of heads alternating with runs of tails; the first run can be of either heads or tails. Let the length of the nth run be R_n. For all k and j, show that $\mathbf{E}(R_{2k+1}) \geqslant \mathbf{E}(R_{2j})$ and $\text{var}\,(R_{2k+1}) \geqslant \text{var}\,(R_{2j})$, with equality in each case if and only if $p = q = \frac{1}{2}$.

Solution Let X be the number of heads shown before the first appearance of a tail. We know that

(10) $$\mathbf{P}(X = k) = p^k q, \quad k \geqslant 0.$$

Let us consider the mass function of X conditional of the first toss. Given that the first toss is H, let X' be the further number of tosses before the first tail. By independence

$$\mathbf{P}(X' = k) = p^k q = \mathbf{P}(X = k).$$

Hence, conditional on H, we have $X = 1 + X'$, and conditional on H^c, we have $X = 0$. Therefore, by Theorem 6

(11) $$\mathbf{E}(X) = p\mathbf{E}(X|H) + q\mathbf{E}(X|H^c) = p(1 + \mathbf{E}(X')) + 0 = p + p\mathbf{E}(X)$$

since $\mathbf{E}(X) = \mathbf{E}(X')$.

Thus $\mathbf{E}(X) = p/q$, which of course we could have obtained directly from (10); we chose to do it this way to display the new technique. Likewise, if Y is the number of tails before the first head $\mathbf{E}(Y) = q/p$.

Now R_{2k+1} is a run of heads if and only if the first toss is a head. Hence, again using independence,

$$\mathbf{E}(R_{2k+1}) = \mathbf{E}(R_{2k+1}|H)p + \mathbf{E}(R_{2k+1}|H^c)q = \mathbf{E}(1 + X')p + \mathbf{E}(1 + Y')q$$
$$= \left(1 + \frac{p}{q}\right)p + \left(1 + \frac{q}{p}\right)q = \frac{p}{q} + \frac{q}{p}.$$

Likewise R_{2k} is a run of heads if and only if the first toss yields a tail, so

$$\mathbf{E}(R_{2k}) = \left(1 + \frac{p}{q}\right)q + \left(1 + \frac{q}{p}\right)p = 2 \leqslant \frac{p}{q} + \frac{q}{p},$$

with equality if and only if $p = \frac{1}{2} = q$. (Because $(p - q)^2 > 0$, for $p \neq q$.) Now $\text{var}\,(R_{2k}) = \mathbf{E}(R_{2k}^2) - 4$. Arguing as above, and using conditional probability again, yields

$$\mathbf{E}(R_{2k}^2) = q\mathbf{E}((1 + X')^2) + p\mathbf{E}((1 + Y')^2) = q\left(\frac{1 + p}{q^2}\right) + p\left(\frac{1 + q}{p^2}\right)$$

and so

(12) $$\text{var}\,(R_{2k}) = \frac{1 + (p - q)^2 - 2pq}{pq}.$$

Likewise

$$\mathbf{E}(R_{2k+1}^2) = q\left(\frac{1 + q}{p^2}\right) + p\left(\frac{1 + p}{q^2}\right)$$

and so

(13) $$\operatorname{var}(R_{2k+1}) = \frac{q}{p^2} + \frac{p}{q^2} - 2.$$

Now

$$\operatorname{var}(R_{2k+1}) - \operatorname{var}(R_{2j}) = \frac{p^4 + q^4 + 2p^2q^2 - pq}{p^2q^2} = \frac{(p^3 - q^3)(p - q)}{p^2q^2} \geqslant 0$$

with equality if and only if $p = q = \frac{1}{2}$. ●

4.5 Sequences of distributions

If an experiment is repeated indefinitely it may give rise to a sequence $(F_n(x); n \geqslant 1)$ of distributions.

(1) Definition Let $f(x)$ be a probability mass function which is non-zero for $x \in D$, and zero for $x \in \mathbb{R}\backslash D = C$. Let $F(x)$ be the corresponding distribution function

$$F(x) = \sum_{x_i \leqslant x} f(x_i).$$

A sequence of distribution functions $F_n(x)$ is said to converge to $F(x)$ if, as $n \to \infty$,

$$F_n(x) \to F(x) \quad \text{for } x \in C.$$ ▲

One special case is important to us; if D is included in the integers then $F_n(x)$ converges to $F(x)$ if, for all x, $f_n(x) \to f(x)$ as $n \to \infty$.

(2) Example: Matching revisited In Example 3.17 we showed that the probability of exactly r matches in n random assignments of letters is

$$p(n, r) = \frac{1}{r!} \sum_{k=0}^{n-r} \frac{(-)^k}{k!} \to \frac{e^{-1}}{r!}$$

as $n \to \infty$. This shows that as $n \to \infty$ the number of matches has a Poisson distribution (with parameter 1) in the limit. ●

(3) Example: Ménages revisited In Problem 3.38 we found the probability that exactly m couples were adjacent when seated randomly at a circular table (alternating the sexes) is

$$p_m = \frac{2}{m!} \sum_{k=0}^{n-m} \frac{(-)^k (n - m - k)!(2n - m - k - 1)!}{k!(2n - 2m - 2k)!n!} \to \frac{2^m}{m!} \sum_{k=0}^{\infty} (-)^k \frac{2^k}{k!} = \frac{2^m e^{-2}}{m!}$$

as $n \to \infty$. Thus the number of adjacent couples is Poisson with parameter 2 in the limit as $n \to \infty$. ●

Finally we note that the appearance of the Poisson distribution in Examples 2 and 3 is significant. This distribution commonly arises in limits of this type, and that is one of the reasons for its major importance.

4.6 Inequalities

Calculating the exact probability that X lies in some set of interest is not always easy. However, simple bounds on these probabilities will often be sufficient for the task in hand. We start with a basic inequality.

(1) **Theorem: Basic inequality** If $h(x)$ is a non-negative function then, for $a > 0$,

$$\mathbf{P}(h(X) \geqslant a) \leqslant \mathbf{E}(h(X))/a).$$

Proof Define the following function of X:

$$I(h \geqslant a) = \begin{cases} 1 & \text{whenever } h(X) \geqslant a \\ 0 & \text{otherwise} \end{cases}$$

Observe that I is an indicator, and so by Example 4.3.2 $\mathbf{E}(I) = \mathbf{P}(h(X) \geqslant a)$. Now, by its construction I satisfies $h(X) - aI \geqslant 0$, and so by Theorem 4.3.6 (parts (iii) and (iv)),

$$\mathbf{E}(h(X)) \geqslant a\mathbf{E}(I) = a\mathbf{P}(h(X) \geqslant a). \qquad \blacksquare$$

The following useful inequalities can all be proved using Theorem 1, or by essentially the same method. You should do some as exercises. For any $a > 0$ we have:

Markov's inequality

(2)
$$\mathbf{P}(|X| \geqslant a) \leqslant \mathbf{E}(|X|)/a.$$

† Chebyshov's inequality

(3)
$$\mathbf{P}(|X| \geqslant a) \leqslant \mathbf{E}(X^2)/a^2.$$

One-sided Chebyshov's inequality

$$\mathbf{P}(X - \mathbf{E}(X) \geqslant a) \leqslant \frac{\operatorname{var}(X)}{a^2 + \operatorname{var}(X)}.$$

Generalized Markov inequality

If $h(x)$ is increasing for $x > 0$, even, and non-negative then

(5)
$$\mathbf{P}(|X| \geqslant a) \leqslant \mathbf{E}(h(X))/h(a).$$

If X is non-negative then

(6)
$$\mathbf{P}(X > a) \leqslant \frac{\mathbf{E}(X)}{a}.$$

If $c > 0$, then

(7)
$$\mathbf{P}(X > a) \leqslant \frac{\mathbf{E}((X + c)^2)}{(a + c)^2},$$

† Some writers use the transliteration 'Chebyshev'. They then have to remember that the second 'e' is pronounced as 'o'.

and

(8) $$\mathbf{P}(X > a) \leqslant \mathbf{E}(\exp(c(X - a))).$$

Here is one important application.

(9) Example Let X be a random variable such that $\text{var}(X) = 0$. Show that X is constant with probability one.

Solution By (3), for any integer $n \geqslant 1$,

(10) $$\mathbf{P}\left(|X - \mathbf{E}(X)| > \frac{1}{n}\right) \leqslant n^2 \text{var}(X) = 0.$$

Hence, defining the events $C_n = \{|X - \mathbf{E}(X)| > 1/n\}$, we have

$$\mathbf{P}(X \neq \mathbf{E}(X)) = \mathbf{P}\left(\bigcup_{n=1}^{\infty} C_n\right) = \mathbf{P}(\lim_{n\to\infty} C_n) = \lim_{n\to\infty} \mathbf{P}(C_n) \quad \text{by Theorem 1.5.2}$$
$$= 0.$$
●

An important concept that crops up in many areas of pure and applied mathematics is that of *convexity*. We shall be interested in the following manifestation of this.

(11) Definition A function $g(x)$ (from $\mathbb{R}$ to $\mathbb{R}$), is called *convex* if, for all a, there exists $\lambda(a)$ such that

(12) $$g(x) \geqslant g(a) + \lambda(a)(x - a), \quad \text{for all } x.$$
▲

If $g(x)$ is differentiable, then a suitable λ is given by $\lambda(a) = g'(a)$ and (12) takes the form

(13) $$g(x) \geqslant g(a) + g'(a)(x - a).$$

This says that a convex function lies above all its tangents. If g is not differentiable then there may be many choices for λ; draw a picture of $g(x) = |x|$ at $x = 0$ to see this. (There are several other definitions of a convex function, all equivalent to Definition 11.) We are interested in the following property of convex functions.

(14) Theorem: Jensen's inequality Let X be a random variable with finite mean, and $g(x)$ a convex function. Then

(15) $$\mathbf{E}(g(X)) \geqslant g(\mathbf{E}(X)).$$

Proof Choosing $a = \mathbf{E}(X)$ in (12), we have

$$g(X) \geqslant g(\mathbf{E}(X)) + \lambda(X - \mathbf{E}(X)).$$

Taking the expected value of each side gives (15). ■

For example, $g(x) = |x|$ and $g(x) = x^2$ are both convex, so $\mathbf{E}(|X|) \geqslant |\mathbf{E}(X)|$ and $\mathbf{E}(X^2) \geqslant (\mathbf{E}(X))^2$. This is Theorem 4.3.7. Here is a less trivial example.

(16) Example Let X be a positive random variable. Show that $\mathbf{E}(\log X) \leqslant \log \mathbf{E}(X)$.

Solution This follows immediately from Jensen's inequality if we can show that $-\log x$ is convex. Fortunately this is easy, as follows. By definition, for $x > 0$,

$$-\log x = \int_x^1 y^{-1}\,dy = \int_a^1 y^{-1}\,dy + \int_x^a y^{-1}\,dy, \quad \text{for } a > 0,$$

$$= -\log a + \int_x^a y^{-1}\,dy \geqslant -\log a + \int_x^a a^{-1}\,dy = -\log a - \frac{1}{a}(x-a),$$

and this is (12) with $\lambda(a) = -a^{-1}$. ●

Example 16 has many important applications, of which we see more later. Here is one to begin with.

(17) Example: Arithmetic–geometric means inequality Let $(x_i; 1 \leqslant i \leqslant n)$ be any collection of positive numbers, and $(p_i; 1 \leqslant i \leqslant n)$ any collection of positive numbers such that $\sum_i p_i = 1$. Show that

(18) $$p_1x_1 + p_2x_2 + \ldots + p_nx_n \geqslant {x_1}^{p_1}{x_2}^{p_2} \ldots {x_n}^{p_n}.$$

Solution Let X be the random variable with probability mass function $\mathbf{P}(X = x_i) = p_i; 1 \leqslant i \leqslant n$. Then from (16)

$$\log \mathbf{E}(X) = \log(p_1x_1 + \ldots + p_nx_n) \geqslant \mathbf{E}(\log X)$$
$$= p_1 \log x_1 + \ldots + p_n \log x_n = \log({x_1}^{p_1}{x_2}^{p_2} \ldots {x_n}^{p_n}).$$

The result (18) follows since $\log x$ is an increasing function. ●

(19) Example: AM/GM inequality In the special case when $p_i = 1/n$, $1 \leqslant i \leqslant n$, then (18) takes the form

$$\frac{1}{n}\sum_1^n x_i \geqslant \left(\prod_{i=1}^n x_i\right)^{1/n}.$$ ●

WORKED EXAMPLES AND EXERCISES

4.7 Example: Royal Oak Lottery

This eighteenth century lottery paid winners 28 to 1; the chance of winning at any given bet was 2^{-5}, independently of other bets. Gamesters (as usual) complained that the odds were unfair. It is reported by de Moivre (in *Doctrine of Chances*, 1756) that the Master of the Ball maintained that any particular point of the Ball should come up once in 22 throws; he offered to bet on this (at evens) at any time, and did so when required. The seeming contradiction between the 2^{-5} chance at any bet, with 22 throws for any chance to come up, so perplexed the gamesters that they began to think they had the advantage; so they played on and continued to lose.

Explain why there is no contradiction.

Solution Let P be a point of the Ball. Let T be the number of trials required to yield P for the first time. At each trial P fails to appear with probability 31/32, and

$T > k$ if and only if the first k trials do not yield P. Hence by independence

$$\mathbf{P}(T > k) = \left(\frac{31}{32}\right)^k.$$

Now

$$\left(\frac{31}{32}\right)^{22} \simeq 0.49 < 0.5.$$

Hence

$$\mathbf{P}(T \leqslant 22) > 0.5.$$

(However note that $\mathbf{E}(T) = \sum_0^\infty \mathbf{P}(T > k) = (1 - \frac{31}{32})^{-1} = 32$.)

Thus, by betting on the event $T \leqslant 22$, the Master of the Ball was giving himself a better than evens chance of winning. However, if we let W be the profit to the gambler of a \$1 stake wagered on P turning up, we have $\mathbf{P}(W = 28) = \frac{1}{32}$ and $\mathbf{P}(W = -1) = \frac{31}{32}$. Hence $\mathbf{E}(W) = \frac{28}{32} - \frac{31}{32} = -\frac{3}{32}$. A *loss*.

Thus in the long run the gambler will surely lose at a rate of nearly 10% of his stake each play. (See Example 4.17 for a proof of this.)

Remark The Master of the Ball was exploiting the fact that the median of the distribution of T is less than its mean.

Note that T has a *geometric* distribution.

(1) Exercise Give an example of a distribution for which the median is larger than the mean.

(2) Exercise Find: (a) var (T) and (b) $\mu_T^{(k)}$.

(3) Exercise Which of the following strategies gives the gambler a better chance of winning if she takes up the offer of a bet on P not occurring in 22 trials:

(a) making such a bet immediately? or

(b) waiting for a run of 22 trials during which P has not appeared? or

(c) waiting until P has appeared in consecutive trials and then betting on its non-appearance in the following 22?

(4) Exercise Calculate $\mathbf{P}(T > j + k | T > j)$. Explain the significance of your answer.

4.8 Example: Misprints

Each printed character in a book is misprinted independently with probability p, or is correct with probability $1 - p$. Let n be the number of characters in the book, and let X be the number of misprinted characters.

(a) Find $\mathbf{P}(X = r)$.

(b) Show that $\mathbf{E}(X) = np$.

(c) Suppose that $\mathbf{E}(X)$ is fixed, and let A be the event that $X \neq 0$. Find $\mathbf{E}(X|A)$, and show that as $n \to \infty$, $\mathbf{E}(X|A) \to \mathbf{E}(X)/(1 - \exp[-\mathbf{E}(X)])$.

Solution (a) We give two methods of solution.

Method I Because characters are misprinted independently, the probability that r given characters are misprinted and the remaining $n - r$ are correct is

$p^r(1-p)^{n-r}$. Since there are $\binom{n}{r}$ distinct ways of fixing the positions of the r misprints, it follows that

(1) $$\mathbf{P}(X=r)=\binom{n}{r}p^r(1-p)^{n-r}.$$

Remark This is the *binomial* distribution, which we met in Example 4.2.3. We shall sometimes denote it by $\mathrm{B}(n, p)$.

Method II Consider the first character and let M be the event that it is misprinted. Then

$$\mathbf{P}(X=r)=\mathbf{P}(X=r|M)\mathbf{P}(M)+\mathbf{P}(X=r|M^c)\mathbf{P}(M^c).$$

We write $\mathbf{P}(X=r)=p(n,r)$ and observe that if M occurs, then $X=r$ if and only if there are $r-1$ misprints in the remaining $n-1$ characters. Hence

(2) $$p(n,r)=p(n-1,r-1)p+p(n-1,r)(1-p)$$

where $p(n,0)=(1-p)^n$ and $p(n,n)=p^n$, $n\geqslant 0$. Now the substitution $p(n,r)=p^r(1-p)^{n-r}c(n,r)$ gives $c(n,r)=c(n-1,r-1)+c(n-1,r)$ where $c(n,0)=c(n,n)=1$, $n\geqslant 0$.

We already know that this difference equation has the solution $c(n,r)=\binom{n}{r}$ as required; (recall Pascal's triangle).

(b) We consider two methods here also.

Method I Let $m(n)$ be the expected number of misprints in n characters. Then by Theorem 4.4.6

$$m(n)=\mathbf{E}(X|M)\mathbf{P}(M)+\mathbf{E}(X|M^c)\mathbf{P}(M^c)=(1+m(n-1))p+m(n-1)(1-p),$$

where we have used

$$\mathbf{E}(X|M)=\sum_r r\mathbf{P}(X=r|M)=\sum_r rp(n-1,r-1),$$

since misprints are independent,

$$=\sum_r p(n-1,r-1)+\sum_r(r-1)p(n-1,r-1)=1+m(n-1).$$

Hence

$$m(n)=m(n-1)+p.$$

Obviously $m(0)=0$, so this difference equation has solution

(3) $$m(n)=np.$$

Method II Using (a)

$$m(n)=\sum_{r=0}^{n}r\binom{n}{r}p^r(1-p)^{n-r}=\sum_{r=1}^{n}np\frac{(n-1)!}{(r-1)!(n-r)!}p^{r-1}(1-p)^{n-r}$$

$$=np\sum_{r=1}^{n}\binom{n-1}{r-1}p^{r-1}(1-p)^{n-r}=np \qquad \text{by the binomial theorem.}$$

(c) By definition

$$\mathbf{E}(X|A) = \sum_{r=1}^{n} \frac{\mathbf{P}(X=r)}{\mathbf{P}(X>0)} = np/(1-(1-p)^n)$$

$$= np\Big/\left(1-\left(1-\frac{np}{n}\right)^n\right) = \mathbf{E}(X)\Big/\left(1-\left(1-\frac{\mathbf{E}(X)}{n}\right)^n\right)$$

$$\to \mathbf{E}(X)/(1-\exp(-\mathbf{E}(X))$$

as $n \to \infty$.

(4) Exercise Show that $\operatorname{var}(X) = np(1-p)$ by two different methods.

(5) Exercise Show that as $n \to \infty$, if $\mathbf{E}(X)$ is fixed $\mathbf{P}(X=0) \to \exp(-\mathbf{E}(X))$.

(6) Exercise Let X have the binomial distribution $\mathbf{P}(X=k) = \binom{n}{k} p^k (1-p)^{n-k}$.

(a) For fixed n and p, for what value of k is $\mathbf{P}(X=k)$ greatest?

(b) For fixed k and p, for what value of n is $\mathbf{P}(X=k)$ greatest?

(7) Exercise If X has a binomial distribution with parameters n and p find:

(a) the probability that X is even,

(b) $\mathbf{E}(\sin^2(\frac{1}{2}\pi X))$

(c) $\mu_X^{(k)}$.

4.9 Example: Dog bites; Poisson distribution

(a) Let X be a binomial random variable with parameters n and p, such that $np = \lambda$. Show that for fixed k, as $n \to \infty$, with λ fixed,

$$\mathbf{P}(X=k) \to \frac{1}{k!}\lambda^k e^{-\lambda}.$$

(b) During 1979–1981, in Bristol, 1103 postmen sustained 215 dog bites. A total of 191 postmen were bitten, of whom 145 were bitten just once. Which should be the postman's motto: 'Once bitten, twice shy' or 'Once bitten, twice bitten'?

Solution (a) Since X is binomial

$$\mathbf{P}(X=k) = \binom{n}{k} p^k (1-p)^{n-k} = \frac{n(n-1)\ldots(n-k+1)}{n^k}.\frac{\lambda^k}{k!}.\left(1-\frac{\lambda}{n}\right)^{n-k}.$$

Now for fixed k, as $n \to \infty$ with λ fixed

$$\frac{n-k+j}{n} \to 1; \quad 1 \le j \le k,$$

$$\left(1-\frac{\lambda}{n}\right)^k \to 1,$$

and

$$\left(1-\frac{\lambda}{n}\right)^n \to e^{-\lambda}.$$

Hence as $n \to \infty$

$$\mathbf{P}(X=k) \to \frac{\lambda^k}{k!} e^{-\lambda}.$$

Remark This is the *Poisson* distribution, which we met in Example 4.2.7.

(b) Suppose you were a postman, and let X be the number of your bites. If dogs bite any postman at random, then X is a binomial random variable with parameters 215 and $(1103)^{-1}$, since it may be thought of as a series of 215 trials in which a 'success' is being bitten with probability $(1103)^{-1}$ at each trial, independently of the rest.

Hence

$$\mathbf{P}(X=0)=\left(1-\frac{1}{1103}\right)^{215}$$

and

$$\mathbf{P}(X=1)=\frac{215}{1103}\left(1-\frac{1}{1103}\right)^{214}.$$

You may either compute these directly, or recognise from (a) that the number of bites you get is approximately Poisson, with parameter $\lambda=\frac{215}{1103}\simeq 0.195$. So

$$\mathbf{P}(X=0)\simeq e^{-\lambda}\simeq 0.82$$
$$\mathbf{P}(X=1)\simeq \lambda e^{-\lambda}\simeq 0.16$$
$$\mathbf{P}(X>1)\simeq 1-e^{-\lambda}-\lambda e^{-\lambda}\simeq 0.02.$$

However, if we pick a postman at random and let X' be the number of bites he sustained we find that in fact

$$\mathbf{P}(X'=0)=\frac{912}{1103}\simeq 0.83$$
$$\mathbf{P}(X'=1)=\frac{145}{1103}\simeq 0.13$$
$$\mathbf{P}(X'>1)=\frac{46}{1103}\simeq 0.04.$$

It seems that 'once bitten, twice bitten' should be the postman's motto.

Remark Our conclusion may be given more substance by investigating the extent to which the observed distribution differs from the expected Poisson distribution. Such techniques are known to statisticians as 'goodness-of-fit tests', and an appropriate procedure here would use the χ^2 test. This may be found in textbooks of elementary statistics; the motto is the same.

(1) **Exercise** If bites are Poisson with parameter λ, what is the probability that you get more than one bite, given that you get at least one?

(2) **Exercise** If bites are Poisson, what is your expected total number of bites given that you get at least one?

(3) **Exercise** Let X_n have a binomial distribution with parameters n and p, such that $np=\lambda$, and let A_n be the event that $X_n\geqslant 1$. If Y is a Poisson random variable with parameter λ, show that as $n\to\infty$, $\mathbf{P}(X_n=k|A_n)\to\mathbf{P}(Y=k|Y\geqslant 1)$.

(4) **Exercise** Let X have a Poisson distribution with parameter λ. Show that $\mathbf{E}(X)=\lambda$.

(a) For fixed λ, what value of k maximizes $\mathbf{P}(X=k)$?

(b) For fixed k, what value of λ maximizes $\mathbf{P}(X=k)$?

(5) Exercise If X has a Poisson distribution with parameter λ, find:
(a) $\mathbf{E}(e^X)$ (b) $\mathbf{E}(\cos(\pi X))$ (c) $\operatorname{var}(X)$ (d) $\mu_X^{(k)}$.

(6) Exercise If X has a Poisson distribution with parameter λ, show that:
(a) the probability that X is even is $e^{-1}\cosh\lambda$.
(b) $\mathbf{E}(|X-\lambda|) = 2\lambda^\lambda e^{-\lambda}/(\lambda-1)!$, when λ is an integer greater than zero.

4.10 Example: Guesswork

You are trying to guess the value of a proper integer valued random variable X, with probability mass function $f(x)$ (which you know). If you underestimate by y, it will cost you $\$by$; if you overestimate by y it will cost you $\$ay$. Your guess is an integer; what guess minimizes your expected loss?

Solution If you guess t, then your expected loss is

(1) $$L(t) = a\sum_{x\le t}(t-x)f(x) + b\sum_{x>t}(x-t)f(x).$$

Substituting $t+1$ for t in (1) gives an expression for $L(t+1)$, and subtracting this from (1) gives

(2) $$L(t) - L(t+1) = a\sum_{x\le t}f(x) + b\sum_{x>t}f(x) = -aF(t) + b(1-F(t)). = D(t) \quad \text{(say)}.$$

Now $\lim_{x\to-\infty} D(x) = b$, and $\lim_{x\to\infty} D(t) = -a$, and both $-F(t)$ and $1-F(t)$ are non-increasing. Therefore there is a smallest t such that

$$D(t) = L(t) - L(t+1) \le 0,$$

and this is the guess which minimizes your expected loss. Hence, denoting this guess by $\hat{t}$,

$$\hat{t} = \min\left\{t \colon F(t) \ge \frac{b}{a+b}\right\}, \quad \text{by (2)}.$$

(3) Exercise Suppose that if you underestimate X you incur a fixed loss £b, whereas if you overestimate X by y it will cost you £ay. Find an expression that determines the guess which minimizes your expected loss.
Find this best guess when

(a) $\mathbf{P}(X = x) = pq^{x-1}; \quad x \ge 1, p = 1-q > 0.$
(b) $\mathbf{P}(X = x) = 1/(x(x+1)); \quad x \ge 1$
(c) $\mathbf{P}(X = x) = 1/(2n+1); \quad -n \le x \le n.$

(4) Exercise What is your best guess if

(a) $L(t) = \mathbf{E}(|X-t|)$?
(b) $L(t) = \mathbf{E}((X-t)^2)$?

(5) Exercise Icarus Airways sells $m+n$ tickets for its n-seat aeroplane. Passengers fail to show up with probability p independently. Empty seats cost $\$c$, and a passenger with a ticket who cannot fly is paid $\$b$ for being bumped. What choice of m minimizes the airline's expected losses on booking errors?

What level of compensation b would be sufficient to ensure that it was not worthwhile for the airline to overbook at all (for fixed p)? For fixed b, what value of p would entail no overbooking by the airline?

4.11 Example: Gamblers ruined again

Alberich and Brunnhilde have a and b gold coins respectively. They play a series of independent games in which the loser gives a gold piece to the winner; they stop when one of them has no coins remaining. If Alberich wins each game with probability p, find the expected number of games played before they stop. (Assume $p \neq q = 1 - p$.)

Solution Let X_k be the number of games they will play when Alberich's fortune is k, and let $m_k = \mathbf{E}(X_k)$. Clearly $m_0 = m_{a+b} = 0$, because in each case one player has no coins. If A is the event that Alberich wins the first game, then for $0 < k < a + b$,

$$\mathbf{E}(X_k|A) = 1 + \mathbf{E}(X_{k+1}) = 1 + m_{k+1}$$

because his fortune is then $k + 1$, and succeeding games are independent of A. Likewise it follows that

$$\mathbf{E}(X_k|A^c) = 1 + \mathbf{E}(X_{k-1}).$$

Hence, by Theorem 4.4.6,

$$m_k = \mathbf{E}(X_k|A)\mathbf{P}(A) + \mathbf{E}(X_k|A^c)\mathbf{P}(A^c) = 1 + pm_{k+1} + qm_{k-1}.$$

Setting

$$m_k = \frac{k}{q - p} + u_k,$$

gives

$$u_k = pu_{k+1} + qu_{k-1}, \quad \text{for } 0 < k < a + b.$$

In particular $u_0 = 0$, and $u_{a+b} = -(a + b)(q - p)^{-1}$. Proceeding as in Example 2.10, using Theorem 2.3.1, shows that

(1) $$u_k = + \frac{(a + b)}{p - q} \frac{1 - \left(\dfrac{q}{p}\right)^k}{1 - \left(\dfrac{q}{p}\right)^{a+b}}.$$

(2) **Exercise** What is the expected number of games played when $p = \frac{1}{2}$?

(3) **Exercise** Let B be the event that Brunnhilde wins the entire contest. Find a difference equation satisfied by $\mathbf{E}(X_k|B)$. Solve this in the case when $p = \frac{1}{2}$.

(4) **Exercise** When the first game is over they re-divide the $a + b$ coins as follows. All the coins are tossed, one player gets those showing a head, the other gets all those showing a tail. Now they play a series of games as before. What is the expected number to be played until one or other player again has all the coins? What if $p = \frac{1}{2}$?

(5) Exercise Alberich is blackmailing Fafner, so each time he loses his last gold coin, he immediately demands (and gets) one replacement coin, with which to continue gambling. What now is the expected number of games played? What if $p = \frac{1}{2}$?

4.12 Example: Postmen

A and B are postmen. They start work on day 1. The probability that A sustains a dog bite on day n, given that he has not been bitten on any of the preceding days is $p_A(n)$. The corresponding probability for B is $p_B(n)$. Let X_A and X_B respectively be the number of days until each sustains his first bite.

(a) Find $\mathbf{P}(X_A = n)$ and $\mathbf{P}(X_B = n)$.
(b) A is wary, so $p_A(n)$ decreases as n increases. If

$$p_A(n) = \frac{1}{n+1}; \quad n \geqslant 1,$$

find $\mathbf{P}(X_A = n)$, and show that $\mathbf{E}(X_A) = \infty$, while $\mathbf{P}(X_A < \infty) = 1$.
(c) B is complacent, so $p_B(n)$ increases as n increases. If, for some $\lambda < 0$,

$$p_B(n) = 1 - e^{-\lambda n}$$

find $\mathbf{P}(X_B = n)$ and $\mathbf{E}(X_B)$.

Solution (a) Let H_k be the event that A is bitten on the kth day. Then the event that he is bitten for the first time on the nth day is $\bigcap_1^{n-1} H_k^c \cap H_n$. Hence

$$\begin{aligned}\mathbf{P}(X_A = n) &= \mathbf{P}\left(\bigcap_1^{n-1} H_k^c \cap H_n\right) = \mathbf{P}\left(H_n \Big| \bigcap_1^{n-1} H_k^c\right)\mathbf{P}\left(\bigcap_1^{n-1} H_k^c\right)\\ &= p_A(n)\mathbf{P}\left(H_{n-1}^c \Big| \bigcap_1^{n-2} H_k^c\right)\mathbf{P}\left(\bigcap_1^{n-2} H_k^c\right) = p_A(n)\prod_1^{n-1}(1 - p_A(k)),\end{aligned}$$

on iterating.

Likewise

$$\mathbf{P}(X_B = n) = p_B(n)\prod_1^{n-1}(1 - p_B(k)).$$

(b) Employing a similar argument

$$\begin{aligned}\mathbf{P}(X_A > n) &= \mathbf{P}\left(\bigcap_{k=1}^{n} H_k^c\right) = \prod_{k=1}^{n}(1 - p_A(k))\\ &= \prod_{k=1}^{n}\left(1 - \frac{1}{k+1}\right) = \frac{1}{n+1} \to 0 \quad \text{as } n \to \infty.\end{aligned}$$

Hence $\mathbf{P}(X_A < \infty) = 1$. Also

$$\mathbf{P}(X_A = n) = \mathbf{P}(X_A > n - 1) - \mathbf{P}(X_A > n) = \frac{1}{n(n+1)},$$

and finally $\mathbf{E}(X_A) = \sum_1^\infty 1/(n+1)$, which diverges. The expected time until A is first bitten is infinite.

(c) By the same argument,

$$\mathbf{P}(X_B \geqslant n) = \prod_{j=1}^{n}(1 - p_B(j)) = \prod_{j=1}^{n} e^{-\lambda j} = e^{-(\lambda/2)n(n+1)}.$$

Hence $\mathbf{P}(X_B = n) = (1 - e^{-\lambda n})e^{-(\lambda/2)n(n-1)}$, and

$$\mathbf{E}(X_B) = \sum_{1}^{\infty} e^{-(\lambda/2)n(n+1)} < \infty.$$

B expects to be bitten in a finite time.

(1) **Exercise** In both cases (b) and (c) find the probability that the postman is first bitten on the jth day, given that he is bitten on or before day M.

(2) **Exercise** If A is less wary, so that $p_A(n) = 2/(n+2)$, show that $\mathbf{E}(X_A)$ is now finite, but $\text{var}\,(X_A)$ diverges.

(3) **Exercise** In each case (b) and (c), given that the postman has not been bitten during the first m days, find the expected further time until he is bitten.

(4) **Exercise** If A is extremely wary, and $p_A(n) = 1/(n+1)^2$ show that with probability $\frac{1}{2}$, he is never bitten. What is the median of the distribution of X_A in this case? Find the expectation of X_A, given that X_A is finite.

4.13 Example: Acme Gadgets

This company has developed a new product. The demand for it is unknown, but it is assumed to be a random variable X, which is distributed uniformly on $\{0, 1, \ldots, N\}$. The gadgets have to be made in advance; each one sold makes a profit of \$$b$, and each one made and left unsold represents a net loss of \$$c$.

How many should be made, to maximize the expected profit?

Solution Suppose that m items are made. Then the total profit (negative profits are interpreted as losses) is

$$Y_m = \begin{cases} bm; & X \geqslant m \\ bX - c(m - X); & m > X. \end{cases}$$

The expected profit is

$$\mathbf{E}(Y_m) = bm\mathbf{P}(X \geqslant m) + b\sum_{x=1}^{m-1}\frac{x}{N+1} - c\sum_{x=0}^{m-1}\frac{m-x}{N+1},$$

$$= \frac{m}{N+1}\left(b\left(N + \frac{1}{2}\right) - \frac{1}{2}c - \frac{1}{2}m(b+c)\right).$$

Now

$$2(N+1)(\mathbf{E}(Y_{m+1}) - \mathbf{E}(Y_m)) = (2N+1)b - c - (2m+1)(b+c),$$

so that the expected profit is largest when $m = \hat{m}$, where

$$\hat{m} = \max\left\{0, \left[\frac{Nb - c}{b + c}\right]\right\}.$$

(1) Exercise Suppose that an unsatisfied customer represents a loss of $\$d$. What now is the choice of m which maximizes expected profit?

(2) Exercise Suppose that the unknown demand X is assumed to have a geometric distribution with parameter p. Find the choice of m which maximizes the expected profit.

(3) Exercise Suppose the unknown demand X is a Poisson random variable with parameter λ. Show that the expected profit if they make m items is

$$\lambda(b+c) - \frac{(b+c)\lambda^{m+1}}{m!} \Bigg/ \sum_0^m \frac{\lambda^k}{k!} - mc,$$

and that this is maximized by the value of m which minimizes

$$\frac{(b+c)\lambda^{m+1}}{m!} \Bigg/ \sum_0^m \frac{\lambda^k}{k!} + mc.$$

4.14 Example: Roulette and the martingale

Suppose you are playing roulette; the wheel has a zero. The chance of winning on red is $p < \frac{1}{2}$, and you bet at evens: if you win you gain an amount equal to your stake.

Your first bet is \$1 on red, if it wins you quit. If it loses, your second bet is \$2 on red, if it wins you quit. And so on; your nth bet is $\$2^{n-1}$ so long as you lose; as soon as you win you quit.

(a) Show that you are certain to win \$1 every time you use this system.
(b) Find the expected size of your winning bet.

Now suppose the house limit is $\$2^L$, so this must be your last bet if you have not already won.

(c) What is your expected gain when you stop?
(d) Would you prefer large or small house limits?

Remark This gambling system is the martingale. Avoid it unless you seek ruin!

Solution Let T be the number of spins of the wheel until the outcome is first red.

(a) Your bet on that spin is $\$2^{T-1}$, and since you win, you gain $\$2^{T-1}$. However your losses on the previous $T-1$ spins are $\$\sum_1^{T-1} 2^{k-1} = \$2^{T-1} - 1$. Since

$$\sum_{k=1}^{\infty} \mathbf{P}(T = k) = \sum_{k=1}^{\infty} p(1-p)^{k-1} = 1$$

this means you are certain to win \$1.

(b) Since your winning bet is $\$2^{T-1}$, it has expected value

$$\mathbf{E}(2^{T-1}) = \sum_{k=1}^{\infty} 2^{k-1} p(1-p)^{k-1} = \infty, \quad \text{since } 2(1-p) > 1.$$

(c) You win \$1 if $1 \leqslant T \leqslant L+1$, otherwise you lose $\$\sum_{k=0}^{L} 2^k$. Hence your expected gains are $\$\gamma$ where

$$\gamma = \mathbf{P}(T \leqslant L+1) - (2^{L+1} - 1)\mathbf{P}(T > L+1) = 1 - (2(1-p))^{L+1}.$$

(d) Since your expected losses increase exponentially fast with L, you must hope the casino is sufficiently generous to have low limits.

(1) **Exercise** What difference does it make to these results if the wheel is fair? (That is, $p = \frac{1}{2}$.)

(2) **Exercise** With house limit $\$2^L$, what is the expected size of your winning bet, given that you do indeed win? What happens as $L \to \infty$? [Remember to consider all three cases, $p > \frac{1}{2}$, $p = \frac{1}{2}$, $p < \frac{1}{2}$.]

4.15 Example: Searching

(a) Let X be a positive integer valued random variable such that $f(n) = \mathbf{P}(X = n)$ is non-increasing as n increases. Suppose that $(g(x); x = 1, 2, \ldots)$ is a function, taking positive integer values, such that for any k, $g(x) = k$ for at most one positive integer $x = r_k$.

Show that $\mathbf{E}(g(X)) \geq \mathbf{E}(X)$.

(b) You have lost a key. There are n places in which you might have mislaid it with respective probabilities $(p_k;\ 1 \leq k \leq n)$. If you search the kth place once, you find the key with probability d_k, if it is indeed there. (You can search any place any number of times.) How do you arrange your searching in order to minimize the expected time until you find the key? [Searches are successful independently of each other.]

Solution (a) Consider the distribution of $g(X)$. Since $g(X) = k$ for at most one value r_k of X,

$$\mathbf{P}(g(X) \leq n) = \sum_k \mathbf{P}(g(X) = k) = \sum_k \mathbf{P}(X = r_k) = \sum_k f(r_k)$$

where the final sum contains $m \leq n$ non-zero terms. If these are arranged in decreasing order as $f(r_{k_1}) \geq f(r_{k_2}) \geq \ldots \geq f(r_{k_m})$, then

$$f(r_{k_1}) \leq f(1)$$
$$f(r_{k_2}) \leq f(2)$$

and so on. Hence, summing these inequalities yields

$$\mathbf{P}(g(X) \leq n) \leq \mathbf{P}(X \leq n),$$

and so

$$\mathbf{E}(g(X)) = \sum_0^\infty \mathbf{P}(g(X) > n) = \sum_0^\infty (1 - \mathbf{P}(g(X) \leq n))$$

$$\geq \sum_0^\infty (1 - \mathbf{P}(X \leq n)) = \mathbf{E}(X).$$

(b) The probability that you find the key on the sth search of the rth room is

$$m_{rs} = (1 - d_r)^{s-1} d_r p_r.$$

To see this, note that the key has to be there (with probability p_r) and you have to fail to find it $s - 1$ times before you succeed. Let p_k be the kth largest of the numbers $(m_{rs};\ r \geqslant 1,\ s \geqslant 1)$. Then p_k is a probability mass function and $(p_k;\ k \geqslant 1)$ is non-increasing. Take this ordering as an order of search; that is if $m_{rs} \geqslant m_{uv}$ then the sth search of the rth place precedes the vth search of the uth place. This search is consistent (the mth search of a given place precedes the $(m+1)$st for every m), and $\sum kp_k$ is the expected number of searches required to find the key.

By part (a) any other order yields greater expected duration of the search time, since the function $g(x)$ is a permutation, and thus one–one.

(1) Exercise Show that you can arrange your searches so that the expected time to find the key is finite.

(2) Exercise The key is upstairs with probability $\frac{2}{3}$, or downstairs with probability $\frac{1}{3}$. Any search upstairs is successful with probability $\frac{1}{4}$ if the key is there, any search downstairs is successful with probability $\frac{3}{4}$ if the key is there. How do you arrange your searches to minimize the expected number of searches?

(3) Exercise Suppose the sth search of the rth room (conditional on $s-1$ previous unsuccessful searches of this room) discovers the key with probability d_{rs}. How do you order your searches to minimize the expected number of searches?

4.16 Example: Duelling

Pascal and Brianchon fight a series of independent bouts. At each bout either Pascal is awarded a hit with probability p, or Brianchon is awarded a hit with probability $q = 1 - p$. The first to be awarded two consecutive hits is declared the winner and the duel stops. Let X be the number of bouts fought. Find the distribution and expected value of X. For what value of p is $\mathbf{E}(X)$ greatest?

Solution Let B be the event that Brianchon wins. Then

$$f_X(n) = \mathbf{P}(\{X = n\} \cap B) + \mathbf{P}(\{X = n\} \cap B^c).$$

For B to occur at the nth bout, he must win the nth and $(n-1)$th bouts (with probability q^2), and the preceding $n-2$ bouts must be awarded alternately to each contestant. The probability of this is $p^{(n/2)-1}q^{(n/2)-1}$ if n is even, or $p^{(n/2)-(1/2)}q^{(n/2)-(3/2)}$ if n is odd, since bouts are independent.

A similar argument applies if B^c occurs, yielding

$$f_X(n) = \begin{cases} p^{(n/2)-1}q^{(n/2)-1}(q^2 + p^2) & \text{if } n \text{ is even} \\ p^{(n/2)-(1/2)}q^{(n/2)-(1/2)}(q + p) & \text{if } n \text{ is odd.} \end{cases}$$

The expected value of X is then, by definition,

$$\mathbf{E}(X) = \sum_{j=1}^{\infty} p^{j-1}q^{j-1}(q^2 + p^2)2j + \sum_{j=1}^{\infty} p^j q^j (q + p)(2j + 1).$$

Summing this series is elementary and boring. To get a solution in closed form it is more fun to argue as follows. Let A_k be the event that Pascal is awarded the kth bout. Then

(1) $$\mathbf{E}(X) = \mathbf{E}(X|A_1)p + \mathbf{E}(X|A_1^c)q,$$

by conditioning on the outcome of the first bout. Now if Pascal is awarded the first bout but not the second, the state of the duel in respect of the final outcome is exactly the same as if he had lost the first bout, except of course that one bout extra has been fought. Formally this says

$$\mathbf{E}(X|A_1 \cap A_2^c) = 1 + \mathbf{E}(X|A_1^c).$$

Hence

(2) $$\mathbf{E}(X|A_1) = \mathbf{E}(X|A_1 \cap A_2)p + \mathbf{E}(X|A_1 \cap A_2^c)q = 2p + q(1 + \mathbf{E}(X|A_1^c)).$$

Likewise

(3) $$\mathbf{E}(X|A_1^c) = 2q + p(1 + \mathbf{E}(X|A_1)).$$

Solving (2) and (3) and substituting into (1) yields

$$\mathbf{E}(X) = \frac{2 + qp}{1 - qp}.$$

Since $qp = \frac{1}{4} - (p - \frac{1}{2})^2$ this is greatest when $p = \frac{1}{2}$, and then $\mathbf{E}(X) = 3$.

(4) Exercise What is $\mathbf{P}(B)$?

(5) Exercise What is $\mathbf{P}(B^c)$?

(6) Exercise Given that Pascal wins, find the distribution and expected value of the number of bouts.

(7) Exercise Find $\mathbf{P}(A_1|B)$ and $\mathbf{P}(A_2|B)$.

(8) Exercise Find the median number of bouts when $p = \frac{1}{2}$.

(9) Exercise Find $\mathbf{P}(B)$, and the expectation of the number of bouts fought, if the winner is required to win *three* consecutive bouts.

(10) Exercise Brianchon suggests that they adopt a different rule for deciding the winner, *viz*: when first a player has been awarded a total number of bouts two greater than the number of bouts awarded to his opponent, then the match stops and the leading player wins. If $p > q$, do you think Brianchon was wise to suggest this? (Assume he wants to win.) What is the expected duration of this game when $p = q$?

4.17 Binomial distribution; the long run

Let X have a binomial distribution with parameters n and p, where $p = 1 - q$. Show that for $\lambda > 0$ and $\epsilon > 0$,

$$\mathbf{P}(X - np > n\epsilon) \leq \mathbf{E}(\exp[\lambda(X - np - n\epsilon)]).$$

Deduce that as $n \to \infty$,

$$\mathbf{P}(|X - np| \leq n\epsilon) \to 1.$$

[You may assume without proof that for any x, $0 < e^x \leq x + e^{x^2}$.]

Solution For $k > np + n\epsilon$ we have, when $\lambda > 0$, $\exp(\lambda(k - np - n\epsilon)) > 1$. Hence

$$\sum_{k>n(p+\epsilon)} \mathbf{P}(X = k) < \sum_{k>n(p+\epsilon)} \exp(\lambda(k - np - n\epsilon))\mathbf{P}(X = k)$$

$$< \sum_k \exp(\lambda(k - np - n\epsilon))\mathbf{P}(X = k), \quad \text{since } e^x > 0,$$

$$= \mathbf{E}(\exp(\lambda(X - np - n\epsilon))).$$

Now the left side is just $\mathbf{P}(X > np + n\epsilon)$, and

$$\mathbf{E}(e^{\lambda X}) = \sum_0^n \binom{n}{k}(pe^{\lambda})^k q^{n-k} = (q + pe^{\lambda})^n,$$

so the right side is

$$(pe^{\lambda q} + qe^{-\lambda p})^n e^{-\lambda n\epsilon} \leq (pe^{\lambda^2 q^2} + qe^{\lambda^2 p^2})^n e^{-\lambda n\epsilon}, \quad \text{since } e^x \leq x + e^{x^2}$$
$$\leq \exp(n\lambda^2 - \lambda n\epsilon).$$

Now choosing $\lambda = \epsilon/2$ gives

$$\mathbf{P}(X - np > n\epsilon) \leq \exp(-n\epsilon^2/4).$$

Likewise

$$\mathbf{P}(X - np < -n\epsilon) \leq \exp(-n\epsilon^2/4),$$

so

(1) $$\mathbf{P}(|X - np| > n\epsilon) \leq 2\exp(-n\epsilon^2/4) \to 0 \quad \text{as } n \to \infty,$$

as required.

(2) Exercise You wish to ask each of a large number n of people a question to which the answer 'yes' is so embarrassing that many individuals would falsely answer 'no'. The answer 'no' is not embarrassing. The following procedure is proposed to determine the embarrassed fraction of the population. As the question is asked, a coin is tossed out of sight of the questioner. If the true answer would have been 'no' and the coin shows heads, then the answer 'yes' is given. Otherwise people should respond truthfully. If the number responding 'yes' is now Y_n and 'yes' is the true answer for a proportion p of the whole population, show that for $\epsilon > 0$

$$\mathbf{P}\left(\left|\frac{Y_n}{n} - \frac{1}{2}(1 + p)\right| > \epsilon\right) \leq 2\exp(-n\epsilon^2/4).$$

Explain the advantages of this procedure.

(3) Exercise Suppose a coin shows a head with probability p, and let X_n be the number of heads in n tosses, and $A_n(\epsilon)$ the event that $|X_n/n - p| > \epsilon$, where $2 > \epsilon > 0$. Show that as $n \to \infty$,

$$\mathbf{P}\left(\bigcup_n^\infty A_k(\epsilon)\right) \to 0.$$

(4) Exercise Suppose a gambler wins \$28 with probability $\frac{1}{32}$, or loses his stake of \$1 with probability $\frac{31}{32}$ at each trial. Let W_n be his fortune after n such independent trials, and $A_n(\epsilon)$ the event that $|W_n/n + 3/32| > \epsilon$. Show that as $n \to \infty$, $\mathbf{P}(\bigcup_{n=m}^\infty A_n(\epsilon)) \to 0$. Deduce that his fortune is equal to its initial value on only finitely many occasions, with probability one. [*Hint*: recall Problem 1.24.]

Note: in the following exercises X is a binomially distributed random variable with parameters n and p.

(5) Exercise Show that for any fixed finite a and b, as $n \to \infty$, $\mathbf{P}(a < X \leq b) \to 0$.

(6) Exercise Show that for $a > 0$,

$$\mathbf{P}\left(\left|\frac{X}{n} - p\right| > a\right) \leq \frac{(p(1-p))^{1/2}}{a^2 n} \min\{(p(1-p))^{1/2}, an^{1/2}\}.$$

(7) Exercise (a) Show that if $p = (m-1)/n$ where m is an integer, then

$$\mathbf{E}\left(\left|\frac{X}{n}-p\right|\right)=2\binom{n-1}{m-1}p^m(1-p)^{n-m+1}.$$

(b) Find var $(|X/n-p|)$.

(8) **Exercise** If $n=2m$ and $p=\frac{1}{2}$, show that

$$\mathbf{P}(X-m=k)=\binom{2m}{m}\frac{1}{4^m}a(m,k)$$

where, as $m\to\infty$, $(a(m,k))^m\to e^{-k^2}$.
Also, show that

$$\frac{1}{2m^{\frac{1}{2}}}<\binom{2m}{m}\frac{1}{4^m}<\frac{1}{(2m+1)^{\frac{1}{2}}}.$$

[You may assume that $|\log(1+x)-x|<x^2$ for small enough x.]

4.18 Example: Uncertainty and entropy

Let X and Y be simple random variables taking values in the same set $\{x_1,\ldots,x_n\}$ with respective probability mass functions $f_X(.)$ and $f_Y(.)$. Show that

(1) $$-\mathbf{E}(\log f_Y(X))\geqslant-\mathbf{E}(\log f_X(X)),$$

and

(2) $$-\mathbf{E}(\log f_X(X))\leqslant\log n,$$

with equality in (1) if and only if $f_Y(.)\equiv f_X(.)$, and equality in (2) if and only if $f_X(x_i)=n^{-1}$ for all x_i. [*Hint*: show first that $\log x\leqslant x-1$.]

Solution By definition, for $x>0$,

(3) $$-\log x=\int_x^1 y^{-1}\,dy\geqslant\int_x^1 dy=1-x,$$

with equality if and only if $x=1$. Hence

$$\begin{aligned}\mathbf{E}(\log f_X(X))-\mathbf{E}(\log f_Y(X))&=\sum_i f_X(x_i)\log f_X(x_i)-\sum_i f_X(x_i)\log f_Y(x_i)\\&=-\sum_i f_X(x_i)\log[f_Y(x_i)/f_X(x_i)]\\&\geqslant-\sum_i f_X(x_i)[1-f_Y(x_i)/f_X(x_i)]\quad\text{by (3)}\\&=0,\end{aligned}$$

with equality iff $f_X(x_i)=f_Y(x_i)$ for all x_i, which proves (1). In particular, setting $f_Y(x_i)=n^{-1}$ yields (2).

Remark It is conventional to denote $-\mathbf{E}(\log f_X(X))$ by $H(X)$ (or alternatively $h(X)$) and the logarithms are taken to base 2. The number $H(X)$ is known as the *uncertainty* or *entropy* of X, and is an essential tool in information theory and communication theory. The result (1) is sometimes called the Gibbs inequality.

(4) Exercise Let $f_X(x) = \binom{n}{x} p^x (1-p)^{n-x}$; $0 \le x \le n$. Show that

$$H(X) \le -n(p \log p + (1-p)\log(1-p)),$$

with equality if $n = 1$.

(5) Exercise Let $f_X(x) = pq^{x-1}/(1-q^M)$, for $1 \le x \le M$, where $p = 1 - q$. Show that

$$\lim_{M\to\infty} H(X) = -p^{-1}[p \log p + (1-p)\log(1-p)].$$

(6) Exercise Let $Y = g(X)$ be a function of the random variable X. Show that for any $c > 0$

$$H(Y) \le H(X) \le c\mathbf{E}(Y) + \log\left(\sum_i e^{-cg(x_i)}\right).$$

When does equality hold?

PROBLEMS

1 A box contains 12 sound grapefruit and 4 that are rotten. You pick three at random.
(a) Describe the sample space.
(b) Let X be the number of sound grapefruit you pick. Find $f_X(x)$ and $\mathbf{E}(X)$.

2 Show that the expected number of pairs in your poker hand is about 0.516.

3 You roll a die once. What is the variance of your score?

4 What is the variance of a uniform random variable?

5 For each of the following functions $f(x)$ (defined on the positive integers $x = 1, 2, \ldots$), find:
(a) the value of c for which $f(x)$ is a probability mass function, and
(b) the expectation
(i) $f(x) = c.2^x/x!$
(ii) $f(x) = cp^x$; $0 \le p \le 1$
(iii) $f(x) = cp^x x^{-1}$; $0 \le p \le 1$
(iv) $f(x) = cx^{-2}$
(v) $f(x) = c[x(x+1)]^{-1}$.

6 If X is a random variable, explain whether or not it is true that $X + X = 2X$, and $X - X = 0$.
Are 0 and $2X$ random variables?

7 For what value of c is $f(x) = c(x(x+1)(x+2))^{-1}$; $1 \le x \le M$, a probability mass function? Find its expectation $\mathbf{E}(X)$. Find the limit of c and $\mathbf{E}(X)$ as $M \to \infty$.

8 A fair coin is tossed repeatedly. Let A_n be the event that three heads have appeared in consecutive tosses for the first time on the nth toss. Let T be the number of tosses required until three consecutive heads appear for the first time. Find $\mathbf{P}(A_n)$ and $\mathbf{E}(T)$.

Let U be the number of tosses required until the sequence HTH appears for the first time. Can you find $\mathbf{E}(U)$?

9 You choose a random number X as follows. Toss a coin repeatedly and count the number of tosses until it shows a head, N say. Then pick an integer at random in $1, 2, \ldots, 10^N$. Show that

$$\mathbf{P}(X = k) = \frac{1}{19} \cdot \frac{1}{20^{d-1}}$$

where d is the number of digits in the decimal expansion of k. What is $\mathbf{E}(X)$?

10 Let X have a Poisson distribution $f(k)$, with parameter λ. Show that the largest term in this distribution is $f([\lambda])$.

11 Show that if $\mathbf{E}(X^2) < \infty$, $\min_a \mathbf{E}((X-a)^2) = \operatorname{var}(X)$.

12 Let $f_1(x)$ and $f_2(x)$ be probability mass functions. Show that if $0 \leqslant p \leqslant 1$, then $f_3(x) = pf_1(x) + (1-p)f_2(x)$ is a probability mass function. Interpret this result.

13 Let X be a geometric random variable. Show that, for $n > 0$ and $k > 0$, $\mathbf{P}(X > n+k | X > n) = \mathbf{P}(X > k)$.

14 Let X be a random variable uniform on $1 \leqslant x \leqslant m$. What is $\mathbf{P}(X = k | a \leqslant X \leqslant b)$? In particular find $\mathbf{P}(X > n + k | X > n)$.

15 A random variable is symmetric if for some a and all k, $f(a-k) = f(a+k)$. Show that the mean and a median are equal for symmetric random variables. Find a non-symmetric random variable for which the mean and median are equal.

16 If X is symmetric about zero, and takes integer values, find $\mathbf{E}(\cos(\pi X))$ and $\mathbf{E}(\sin(\pi X))$.

17 Let X have distribution function F. Find the distribution of $Y = aX + b$ and of $Z = |X|$.

18 Let X have a geometric distribution such that $\mathbf{P}(X = k) = q^{k-1}p;\ k \geqslant 1$. Show that $\mathbf{E}(X^{-1}) = \log(p^{(1/p)-1})$.

19 (a) Let X have a Poisson distribution with parameter λ. Show that $\mathbf{E}(1/(X+1)) = \lambda^{-1}(1 - e^{-\lambda})$, and deduce that for all λ, $\mathbf{E}(1/(X+1)) \geqslant (\mathbf{E}(X+1))^{-1}$. When does equality hold?
(b) Find $\mathbf{E}(1/(X+1))$ when $\mathbf{P}(X = k) = (-k^{-1}p^k)/\log(1-p);\ k \geqslant 1$.

20 **Fingerprints** It is assumed that the number X of individuals in a population, whose fingerprints are of a given type, has a Poisson distribution with some parameter λ.
(a) Explain when and why this is a plausible assumption.
(b) Show that $\mathbf{P}(X = 1 | X \geqslant 1) = \lambda(e^{\lambda} - 1)^{-1}$.
(c) A careless miscreant leaves a clear fingerprint of type t. It is known that the probability that any randomly selected person has this type of fingerprint is 10^{-6}. The city has 10^7 inhabitants and a citizen is produced who has fingerprints of type t. Do you believe him to be the miscreant on this evidence alone? In what size of city would you be convinced?

21 Initially urn I contains n red balls and urn II contains n blue balls. A ball selected randomly from urn I is placed in urn II, and a ball selected randomly from urn II is placed in urn I. This whole operation is repeated indefinitely. Given that r of the n balls in urn I are red, find the mass function of R, the number of red balls in urn I after the next repetition.

Show that the mean of this is $r + 1 - 2r/n$, and hence find the expected number of red balls in urn I in the long run.

22 A monkey has a bag with four apples, three bananas and two pears. He eats fruit at random until he takes a fruit of a kind he has eaten already. He throws that away and the bag with the rest. What is the mass function of the number of fruit eaten, and what is its expectation?

23 **Matching** Consider the matching problem of Example 3.17. Let $\mu^{(k)}$ be the kth factorial moment of the number X of matching letters, $\mu^{(k)} = \mathbf{E}(X(X-1)\ldots(X-k+1))$. Show that

$$\mu^{(k)} = \begin{cases} 1; & k \leqslant n \\ 0; & k > n. \end{cases}$$

24 Suppose an urn contains m balls which bear the numbers from 1 to m inclusive. Two balls are removed with replacement. Let X be the difference between the two numbers they bear.
(a) Find $\mathbf{P}(X \leqslant n)$.
(b) Show that if $n/m = x$ is fixed as $m \to \infty$ then $\mathbf{P}(|X| \leqslant n) \to 1 - (1-x)^2;\ 0 \leqslant x \leqslant 1$.
(c) Show that $\mathbf{E}|X|/m \to \frac{1}{3}$.

25 Suppose the probability of an insect laying n eggs is given by the Poisson distribution with mean $\mu > 0$, that is by the probability distribution over all the non-negative integers defined by

$p_n = e^{-\mu}\mu^n/n!$ $(n = 0, 1, 2, \ldots)$ and suppose further that the probability of an egg developing is p. Assuming mutual independence of the eggs, show that the probability distribution q_m for the probability that there are m survivors is of the Poisson type and find the mean.

26 Preparatory to a camping trip, you can buy 6 cans of food, all of the same size, 2 each of meat, vegetables and fruit. Assuming that cans with the same contents have indistinguishable labels, in how many distinguishable ways can the cans be arranged in a row?

On the trip there is heavy rain and all the labels are washed off. Show that if you open 3 of the cans at random the chance that you will open one of each type is $\frac{2}{5}$. If you do not succeed, you continue opening cans until you have one of each type; what is the expected number of open cans?

27 A belt conveys tomatoes to be packed. Each tomato is defective with probability p, independently of all the others. Each is inspected with probability r; inspections are also mutually independent. If a tomato is defective and inspected, it is rejected.

(a) Find the probability that the nth tomato is the kth defective tomato.
(b) Find the probability that the nth tomato is the kth rejected tomato.
(c) Given that the $(n+1)$th tomato is the first to be rejected, let X be the number of its predecessors which were defective. Find $\mathbf{P}(X = k)$, the probability that X takes the value k, and $\mathbf{E}(X)$.

28 Mr Smith must site his widget warehouse in either Acester or Beeley. Initially he assesses the probability as p that the demand for widgets is greater in Acester, and as $1 - p$ that it is greater in Beeley. The ideal decision is to site the warehouse in the town with the larger demand. The cost of the wrong decision, because of increased transport costs, may be assumed to be £1000 if Acester is the correct choice and £2000 if Beeley is the correct choice. Find the expectations of these costs for each of the two possible decisions, and the values of p for which Acester should be chosen on the basis of minimum expected cost.

Mr Smith could commission a market survey to assess the demand. If Acester has the higher demand the survey will indicate this with probability $\frac{3}{4}$ and will indicate Beeley with probability $\frac{1}{4}$. If Beeley has the higher demand the survey will indicate this with probability $\frac{2}{3}$, and will indicate Acester with probability $\frac{1}{3}$. Show that the probability that the demand is higher in Acester is $9p/(4 + 5p)$ if the survey indicates Acester. Find also the expected cost for each of the two possible decisions if the survey indicates Acester.

If the survey indicates Acester and $p < 8/17$, where should Mr Smith site the warehouse?

29 A coin is tossed repeatedly and on each occasion the probability of obtaining a head is p and the probability of obtaining a tail is $1 - p$ $(0 < p < 1)$.

(a) What is the probability of not obtaining a tail in the first n tosses?
(b) What is the probability p_n of obtaining the first tail at the nth toss?
(c) What is the expected number of tosses required to obtain the first tail?

30 The probability of a day being fine is p if the previous day was fine, and is p' if the previous day was wet. Show that, in a consecutive sequence of days, the probability u_n that the nth is fine satisfies $u_n = (p - p')u_{n-1} + p'$, $n \geqslant 2$.
Show that as $n \to \infty$, $u_n \to p'(1 - p + p')^{-1}$.

By considering the alternative possibilities for tomorrow's weather, or otherwise, show that if today is fine the expected number of future days up to and including the next wet day is $1/(1 - p)$.

Show that (today being fine) the expected number of future days up to and including the next two consecutive wet days is $(2 - p)/((1 - p)(1 - p'))$.

31 Cars are parked in a line in a parking lot in order of arrival and left there. There are two types of cars, small ones requiring only one unit of parking length (say 15 ft) and large ones requiring two units of parking length (say 30 ft). The probability that a large car turns up to park is p and the probability that a small car turns up is $q = 1 - p$. It is required to find the expected maximum number of cars that can park in a parking length of n units, where n is an integer. Denoting this number by $M(n)$ show that:

(a) $M(0) = 0$
(b) $M(1) = 1 - p$
(c) $M(n) - qM(n-1) - pM(n-2) = 1, \quad (n \geq 2)$.

Show that the equations are satisfied by a solution of the form $M(n) = A\alpha^n + B\beta^n + Cn$, where α, β are the roots of the equation $x^2 - qx - p = 0$, and A, B, C are constants to be found. What happens to $M(n)$ as $n \to \infty$?

32 The probability that the postman delivers at least one letter to my house on any day (including Sundays) is p. Today is Sunday, the postman has passed my house and no letter has been delivered.

(a) What is the probability that at least one letter will be delivered during the next week (including next Sunday)?
(b) Given that at least one letter is delivered during the next week, let X be the number of days until the first is delivered. What is $f_X(x)$?
(c) What is the expected value of X?
(d) Suppose that all the conditions in the first paragraph hold, except that it is known that a letter will arrive on Thursday. What is the expected number of days until a letter arrives?

33 A gambler plays two games, in each of which the probability of her winning is 0.4. If she loses a game she loses her stake, but if she wins she gets double her stake. Suppose that she stakes a in the first game and b in the second, with $a + b = 1$. Show that her expected loss after both games is 0.2.

Suppose she plays again, but now the stake in the first game buys knowledge of the second, so that the chance of winning in the second is ap (≤ 1). Show that the value of a which gives the greatest expected gain is $0.5 + 0.2/p$.

34 Let $f_1(X)$ and $f_2(X)$ be functions of the random variable X. Show that (when both sides exist) $[\mathbf{E}(f_1 f_2)]^2 \leq \mathbf{E}(f_1^2)\mathbf{E}(f_2^2)$. Deduce that $\mathbf{P}(X = 0) \leq 1 - [\mathbf{E}(X)]^2/\mathbf{E}(X^2)$.
[Recall that $at^2 + 2bt + c$ has distinct real roots if and only if $b^2 > ac$.]

35 Any oyster contains a pearl with probability p independently of its fellows. You have a tiara which requires k pearls, and are opening a sequence of oysters until you find exactly k pearls. Let X be the number of oysters you have opened which contain no pearl.

(a) Find $\mathbf{P}(X = r)$ and show that $\sum_r \mathbf{P}(X = r) = 1$.
(b) Find the mean and variance of X.
(c) If $p = 1 - \lambda/k$, find the limit of the distribution of X as $k \to \infty$.

36 A factory produces 100 zoggles a day. Each is defective independently with probability p. If a defective zoggle is sold it costs the factory £100 in fines and replacement charges. Therefore each day 10 are selected at random and tested. If they all pass, all 100 zoggles are sold. If more than one is defective then all 100 zoggles are scrapped. If one is defective it is scrapped and a further sample of size 10 is taken. If any are defective, the day's output is scrapped, otherwise 99 zoggles are sold.

(a) Show that the probability r of not scrapping the day's output is $(1-p)^{10}(1 + 10p(1-p)^9)$.
(b) If testing one zoggle costs £10, find the expected cost of a day's testing.
(c) Find the expected returns on a day's output in terms of the profit b of a sold zoggle and cost c of a scrapped zoggle.

37 An urn contains 2 blue balls and $n - 2$ red balls; they are removed without replacement.

(a) Show that the probability of removing exactly one blue ball in $r-1$ removals is
$$\frac{2(r-1)(n-r+1)}{n(n-1)}.$$
(b) Show that the probability that the urn first contains no blue balls after the rth removal is
$$\frac{2(r-1)}{n(n-1)}.$$
(c) Find the expected number of removals required to remove both blue balls.

38 Suppose that n dice are rolled once; let X be the number of sixes shown. These X dice are rolled again, let Y be the number of sixes shown after this second set of rolls.
(a) Find the distribution and mean of Y.
(b) Given that the second set of rolls yielded r sixes, find the distribution and mean of X.

39 Pascal and Brianchon now play a series of games which may be drawn (that is, tied) with probability r. Otherwise Pascal wins with probability p or Brianchon wins with probability q, where $p+q+r=1$.
(a) Find the expected duration of the match if they stop when one or other wins two consecutive games. Also find the probability that Pascal wins.
(b) Find the expected duration of the match if they stop when one or other wins two successive games of the games that are won. (That is, draws are counted but ignored.) Find the probability that Pascal wins.
If you were Brianchon and $p>q$, which rules would you rather play by?

40 Let the random variable X have a geometric distribution, $\mathbf{P}(X=k)=q^{k-1}p;\ k\geqslant 1$. Show that for $t>0$, $\mathbf{P}(X\geqslant a)\leqslant pe^{-ta}(1-qe^{t})^{-1}$. Deduce that $\mathbf{P}(X\geqslant a)\leqslant (a+1)p[q(a+1)a^{-1}]^{a}$ and compare this with the exact value of $\mathbf{P}(X\geqslant a)$.

41 An archer shoots arrows at a circular target of radius 1 where the central portion of the target inside radius $\frac{1}{4}$ is called the bull. The archer is as likely to miss the target as she is to hit it, and when the archer does hit the target she is as likely to hit any one point on the target as any other. What is the probability that the archer will hit the bull? What is the probability that the archer will hit k bulls in n attempts? Prove that the mean number of bulls that the archer hits in n attempts is $n/32$.

Show that if the archer shoots 96 arrows in a day, the probability of her hitting no more than one bull is approximately $4e^{-3}$. Show that the average number of bulls the archer hits in a day is 3, and that the variance is approximately $(63\sqrt{3}/64)^2$.

42 Prove Chebyshov's inequality that, for a random variable X with mean μ and variance σ^2,
$$\mathbf{P}(|X-\mu|\leqslant h\sigma)\geqslant 1-\frac{1}{h^2}$$
for any $h>0$.

When an unbiased coin is tossed n times, let the number of tails obtained be m. Show that
$$\mathbf{P}\left(0.4\leqslant\frac{m}{n}\leqslant 0.6\right)\geqslant 0.75$$
when $n\geqslant 100$. Given that $n=100$, show that
$$\mathbf{P}\left(0.49\leqslant\frac{m}{n}\leqslant 0.51\right)\simeq 3(5\sqrt{(2\pi)})^{-1}.$$
[You may assume Stirling's formula that $n!\simeq\sqrt{(2\pi)}n^{n+1/2}e^{-n}$ when n is large.]

43 An ambidextrous student has a left and a right pocket, each initially containing n humbugs. Each time she feels hungry she puts a hand into one of her pockets and if it is not empty, takes a humbug from it and eats it. On each occasion, she is equally likely to choose either the left or the right pocket. When she first puts her hand into an empty pocket the other pocket contains H humbugs.

Show that if p_h is the probability that $H = h$, then

$$p_h = \binom{2n-h}{n} \frac{1}{2^{2n-h}},$$

and find the expected value of H, by considering $\sum_{h=0}^{n} (n-h) p_h$, or otherwise.

44 You insure your car. You make a claim in any year with probability q independently of events in other years. The premium in year j is a_j (where $a_j < a_k$ for $k < j$), so long as no claim is made. If you make a claim in year k, then the premium in year $k + j$ is a_j, so long as no further claim is made; and so on. Find the expected total payment of premiums until the first claim.

45 A Scotch die has faces bearing tartan patterns: three are McDiarmid, two are Meldrum and one is Murray. Show that the expected number of times you must roll the die before all three patterns have appeared is 7.3.

5

Random vectors; independence and dependence

5.1 Joint distributions

Commonly each outcome of an experiment generates two (or more) real numbers of interest. We can treat these as individual random variables $(X_i; 1 \leq i \leq n)$, say, but it is often important to consider their joint behaviour. For example, if the experiment is your visit to your doctor, you may find out your height H and weight W. These are separate random variables, but are often informative when considered jointly. Thus the outcome $H = 150$ cm and $W = 150$ kg might disturb your physician, whereas the outcome $H = 190$ cm and $W = 80$ kg probably would not. Likewise the random vector comprising height, weight, age, sex, blood pressure and heart rate is of more use considered jointly than separately.

As another example, complicated systems (such as space shuttles) have several on-board computers which work together to run the system. If one fails, or makes an error, the others can over-ride it; thus the system fails only when a majority of the computers fail. If X_i is the time until the ith processor fails, then the time until the system fails depends jointly on the collection of random variables, $X_1, \ldots, X_n$. It is natural to refer to such a collection as a random vector, and write $\mathbf{X} = (X_1, X_2, \ldots, X_n)$. Formally, as before, we have $\mathbf{X} = \mathbf{X}(\omega)$; $\omega \in \Omega$, and $A_{\mathbf{x}} = \{\omega: \mathbf{X}(\omega) = \mathbf{x}\} \in \mathcal{F}$, but we shall not often refer to the underlying sample space Ω. Since $\mathbf{X}$ maps Ω into a countable subset S of $\mathbb{R}^n$, we think of S as the sample space. (You may well have already been doing this instinctively in Chapter 4.)

For simplicity we summarize the properties of random vectors in two dimensions; the appropriate generalizations in more dimensions are straightforward.

Definition Let X and Y be two discrete random variables taking values $(x_i; i \geq 1)$ and $(y_j; i \geq 1)$ respectively. Their *joint probability mass function* $f(x, y)$ is defined by

$$f(x, y) = \mathbf{P}(X = x, Y = y)$$

as x and y range over all possible values x_i and y_j, of X and Y. ▲

The mass function $f(x, y)$ is zero except at a countable set of points in $\mathbb{R}^2$. In fact

(1) $$f(x, y) \geq 0 \quad \text{for all } x \text{ and } y,$$

and further if

(2) $$\sum_{i,j} f(x_i, y_j) = 1;$$

then the joint distribution $f(x, y)$ is not defective.

(3) Example Suppose that a coin is tossed twice; let X be the total number of heads shown, and Y the total number of tails. Then (X, Y) takes values in

$$S = \{0, 1, 2\} \times \{0, 1, 2\} = \{(i, j): i \in \{0, 1, 2\}, j \in \{0, 1, 2\}\}.$$

Clearly $f(x, y)$ is zero except at the points $(0, 2)$, $(1, 1)$ and $(2, 0)$. Furthermore

$$f(0, 2) + f(1, 1) + f(2, 0) = (1 - p)^2 + 2p(1 - p) + p^2 = 1,$$

where we have denoted the probability of a head by p, as usual. ●

Any real function $g(X, Y)$ of two such jointly distributed random variables is itself a random variable. If we set $Z = g(X, Y)$ then Z has a probability mass function given by

(4) $$f_Z(z) = \mathbf{P}(g(X, Y) = z) = \sum f(x, y)$$

where the summation is over all x and y such that $g(x, y) = z$.
In particular, if $g(x, y) = x$, we have

(5) $$f_X(x) = \sum_y f(x, y),$$

and if $g(x, y) = y$, we have

(6) $$f_Y(y) = \sum_x f(x, y).$$

Thus we have shown the important result that, if we know the joint mass function of several random variables, we can find all their separate mass functions. When obtained in this way, $f_X(x)$ and $f_Y(y)$ are sometimes called *marginal* mass functions.
Here are some examples illustrating joint mass functions.

(7) Example A row of n numbered machines are producing components which are identical except for the serial number. On any day the kth component produced by the jth machine bears the serial number (j, k). On the day in question the rth machine produces c_r $(1 \leq r \leq n)$ components, and at the end of the day one component C is picked at random from all those produced. Let its serial number be (X, Y). Find $f(x, y)$, $f_X(x)$ and $f_Y(y)$.

Solution Since C is picked at random from all $\sum_{r=1}^{n} c_r$ components, we have

$$f(x, y) = \left(\sum_1^n c_r\right)^{-1}, \quad 1 \leq x \leq n; 1 \leq y \leq c_x,$$
$$= a \quad \text{(say)}.$$

Then by (5)

$$f_X(x) = \sum_y f(x, y) = ac_x.$$

Now define the function $H(i, j) = \begin{cases} 1 & \text{if } c_i \geq c_j, \\ 0 & \text{otherwise.} \end{cases}$

Then by (6)

$$f_Y(y) = \sum_x f(x, y) = a \sum_{x=1}^{n} H(x, y).$$ ●

(8) Example: Cutting for the deal It is customary, before engaging in a card game, to cut for the deal; each player removes a portion of the deck in turn, and then each reveals the bottom card of his segment. The highest card wins. For these to be random variables we need to assign numerical values to the court cards, so we set J = 11, Q = 12, K = 13, A = 14, when aces are high.

(a) Art and Bart cut for deal, aces high. Let X be Art's card, and Y be Bart's card. Find the joint mass function of X and Y. Does it make any difference how many cards Art removes from the deck?
(b) Let V be the loser's card, and W the dealer's (winning) card. Find the joint mass function of V and W, and the separate mass functions of V and W.
(c) Find the mass function of the dealer's winning margin (namely, $W - V$).
(d) What is the mass function of the dealer's card when three players cut for deal?

Note that in the event of a tie, the deck is shuffled and the players cut again to choose the dealer.

Solution (a) Each random variable takes values in $\{2, 3 \ldots, 14\}$. Cutting the deck twice amounts to selecting two cards at random, and since ties are not allowed, $X \neq Y$. By symmetry any two unequal values are equally likely to occur, so

$$f_{X,Y}(x, y) = \mathbf{P}(X = x, Y = y) = \left\{\begin{matrix} \frac{1}{12} \cdot \frac{1}{13} & x \neq y \\ \\ 0 & x = y \end{matrix}\right\} \quad 2 \leq x, y \leq 14.$$

It makes no difference how many cards Art removes with his cut.

(b) Of course $W > V$, so

$$f_{V,W}(v, w) = \mathbf{P}(V = v, W = w) = \mathbf{P}(X = v, Y = w) + \mathbf{P}(X = w, Y = v)$$

$$= \begin{cases} \frac{2}{12} \cdot \frac{1}{13} = \frac{1}{78}; & 2 \leq v < w \leq 14 \\ 0\,. \quad ; & \text{otherwise.} \end{cases}$$

This is otherwise obvious, because the experiment amounts to choosing an unordered pair of unequal cards at random, with equal probability of choosing any pair. Hence, for $v < w$, $f(v, w) = \binom{13}{2}^{-1}$ as above. Now by (5)

$$f_V(v) = \sum_{w=v+1}^{14} \frac{1}{78} = \frac{14-v}{78}; \quad 2 \leqslant v \leqslant 13.$$

And by (6)

$$f_W(w) = \sum_{2}^{w-1} \frac{1}{78} = \frac{w-2}{78}; \quad 3 \leqslant w \leqslant 14.$$

(c) By (4)

$$f_Z(z) = \mathbf{P}(W - V = z) = \sum \frac{1}{78}$$

where the summation is over all v and w such that $w - v = z$. Since $z \leqslant v < w \leqslant 14$, there are exactly $13 - z$ terms in this sum, so

$$\mathbf{P}(W - V = z) = \frac{13-z}{78}; \quad 1 \leqslant z \leqslant 12.$$

(d) Arguing as we did for (b), where now $u < v < w$, we have

$$\mathbf{P}(U = u, V = v, W = w) = \binom{13}{3}^{-1}.$$

Hence

$$f_W(w) = \sum_{2 \leqslant u < v < w} \binom{13}{3}^{-1}; \quad 4 \leqslant w \leqslant 14,$$

$$= \sum_{3 \leqslant v < w} (v-2)\binom{13}{3}^{-1} = \frac{1}{2}(w-3)(w-2)\binom{13}{3}^{-1}. \quad \bullet$$

(9) **Example** For what value of c is the function

$$f(x, y) = c\binom{x+y-1}{x}\lambda^x \mu^y, \quad x \geqslant 0;\ y \geqslant 1;\ 1 > 1 - \lambda > \mu > 0$$

a joint mass function? For this value of c, find the mass functions of X and Y.

Solution By (2)

$$c^{-1} = \sum_{y=1}^{\infty}\sum_{x=0}^{\infty}\binom{x+y-1}{x}\lambda^x\mu^y = \sum_{y=1}^{\infty}\frac{\mu^y}{(1-\lambda)^y} \quad \text{by (3.6.12)},$$

$$= \frac{\mu}{1-\lambda-\mu}.$$

Then by (5)

$$f_X(x) = c\sum_{y=1}^{\infty}\mu\lambda^x\binom{x+1+y-2}{y-1}\mu^{y-1} = \frac{(1-\lambda-\mu)\lambda^x}{(1-\mu)^{x+1}}, \quad x \geqslant 0.$$

Likewise

$$f_Y(y) = c\sum_{x=0}^{\infty}\binom{x+y-1}{x}\lambda^x\mu^y = \frac{(1-\lambda-\mu)\mu^{y-1}}{(1-\lambda)^y}, \quad y \geqslant 1.$$

Thus $X+1$ and Y are both geometric, with parameters $\frac{\lambda}{1-\mu}$ and $\frac{\mu}{1-\lambda}$ respectively, X taking values in the non-negative integers and Y in the positive integers. ●

(10) Example Leif and Rolf are bored with fair games. They want to play a game in which the probability of winning (for Leif) is λ, where λ is an arbitrary number in $[0, 1]$. Also, they want the game to be of finite duration with probability 1. Unfortunately the only gaming aid they have is a fair coin. Can you supply them with a game?

Solution Let λ have binary expansion

$$\lambda = 0.b_1b_2b_3\ldots = \sum_{n=1}^{\infty} b_n 2^{-n}.$$

Now toss the coin repeatedly and let I_n be the indicator of the event that the nth toss is a head. Let T be the first toss such that $I_n \neq b_n$, $T = \min\{n: I_n \neq b_n\}$. If $I_T < b_T$, then Leif wins, otherwise Rolf wins. Now $\mathbf{P}(T = n) = (\frac{1}{2})^n$ so that

$$\mathbf{P}(T < \infty) = \sum_{1}^{\infty}\left(\frac{1}{2}\right)^n = 1$$

and indeed $\mathbf{E}(T) = 2$. Also Leif can only win at the nth toss if $b_n = 1$ so

$$\mathbf{P}(\text{Leif wins}) = \sum_n b_n \mathbf{P}(T = n) = \sum_n b_n 2^{-n} = \lambda,$$

as required. ●

5.2 Independence

Given the joint mass function of X and Y, equations (5.1.5) and (5.1.6) yield the marginal mass functions of X and Y. However, to be given the marginal distributions does not in general uniquely determine a joint distribution.

(1) Example Let X and Y have joint mass function given by

$$f(0,0) = \frac{1}{6}, \quad f(0,1) = \frac{1}{3}, \quad f(1,0) = \frac{1}{12}, \quad f(1,1) = \frac{5}{12},$$

and let U and V have joint mass function given by

$$f(0,0) = \frac{1}{4}, \quad f(0,1) = \frac{1}{4}, \quad f(1,0) = 0, \quad f(1,1) = \frac{1}{2}.$$

Then summing to get the marginal mass functions shows that:

$$f_X(0) = f_U(0) = \frac{1}{2}, \quad f_X(1) = f_U(1) = \frac{1}{2};$$

and

$$f_Y(0) = f_V(0) = \frac{1}{4}, \quad f_Y(1) = f_V(1) = \frac{3}{4}.$$

These marginal mass functions are the same, but the joint mass functions are different. ●

There is one exceptionally important special case when marginal mass functions do determine the joint mass function uniquely.

(2) **Definition** Random variables X and Y are independent if, for all x and y,

$$f(x, y) = f_X(x)f_Y(y).$$

This is equivalent to $\mathbf{P}(A \cap B) = \mathbf{P}(A)\mathbf{P}(B)$ where $A = \{\omega: X(\omega) = x\}$ and $B = \{\omega: X(\omega) = y\}$, which is the definition of independence for the events A and B. More generally, a collection $(X_i; 1 \leq i \leq n)$ with mass function f is independent if for all $\mathbf{x} = (x_1, \ldots, x_n)$

(3) $$f(\mathbf{x}) = \prod_{i=1}^{n} f_{X_i}(x_i). \qquad \blacktriangle$$

Note that if X or Y (or both) are improper random variables (so that $f(x, y)$ is defective), then to say they are independent is interpreted as meaning

(4) $$\mathbf{P}(X = x, Y = y) = \mathbf{P}(X = x)\mathbf{P}(Y = y)$$

for all finite x and y. This may seem odd, but such random variables occur quite naturally in simple random walks and other topics.

Example 5.1.7 revisited Recall that n machines produce components. Suppose that all the machines produce the same number c of components, and as before we pick one at random and let its serial number be (X, Y), where X is the machine number and Y is the component index. Then

$$f(x, y) = (nc)^{-1}; \quad 1 \leq x \leq n; 1 \leq y \leq c$$

and

$$f_X(x) = \frac{1}{n}; \qquad 1 \leq x \leq n$$

and

$$f_Y(y) = \frac{1}{c}; \qquad 1 \leq y \leq c.$$

Obviously $f(x, y) = f_X(x)f_Y(y)$ and so X and Y are independent in this case. ●

(5) **Example 1 revisited** Observe that the mass functions of X and Y, and of U and V, do not satisfy Definition 2. Let W and Z be independent random variables such that $f_W(0) = \frac{1}{2}$, $f_W(1) = \frac{1}{2}$, and $f_Z(0) = \frac{1}{4}$, $f_Z(1) = \frac{3}{4}$.

Then by Definition 2, their joint mass function is

$$f(0, 0) = \frac{1}{8}, \quad f(0, 1) = \frac{3}{8}, \quad f(1, 0) = \frac{1}{8}, \quad f(1, 1) = \frac{3}{8}. \qquad ●$$

(6) **Example 5.1.9 revisited** Observe that X and Y are not independent because $\binom{x+y-1}{x}\lambda^x\mu^y$ cannot be expressed in the form $f_X(x)f_Y(y)$. If $X+1$ and Y were independent geometric random variables with parameters $\frac{\lambda}{1-\mu}$ and $\frac{\mu}{1-\lambda}$, then the joint mass function would be

$$f(x, y) = \left(1 - \frac{\lambda}{1-\mu}\right)\left(1 - \frac{\mu}{1-\lambda}\right)\left(\frac{\lambda}{1-\mu}\right)^x\left(\frac{\mu}{1-\lambda}\right)^{y-1}, \quad x \geqslant 0;\ y \geqslant 1. \qquad \bullet$$

The apparently simple Definition 2 implies a great deal more about independent random variables, as the following result shows.

(7) **Theorem** Let X and Y be independent random variables. Then:
(a) for arbitrary countable sets A and B

(8) $$\mathbf{P}(X \in A, Y \in B) = \mathbf{P}(X \in A)\mathbf{P}(Y \in B),$$

and
(b) for any real functions $g(.)$ and $h(.)$, $g(X)$ and $h(Y)$ are independent.

Proof (a) The left hand side of (8) is

$$\sum_{x\in A}\sum_{y\in B}\mathbf{P}(X = x, Y = y) = \sum_{x\in A}\sum_{y\in B}\mathbf{P}(X = x)\mathbf{P}(Y = y) \quad \text{by independence}$$

$$= \sum_{x\in A}\mathbf{P}(X = x)\sum_{y\in B}\mathbf{P}(Y = y) = \mathbf{P}(X \in A)\mathbf{P}(Y \in B)$$

as required. For (b) let $A = \{x\colon g(X) = \xi\}$ and $B = \{y\colon h(Y) = \eta\}$. Then by part (a), for any ξ and η,

$$\mathbf{P}(g(X) = \xi, h(Y) = \eta) = \mathbf{P}(X \in A, Y \in B) = \mathbf{P}(X \in A)\mathbf{P}(Y \in B)$$
$$= \mathbf{P}(g(X) = \xi)\mathbf{P}(h(Y) = \eta),$$

as required. ■

(9) **Example** Independent random variables X and Y take the values -1 or $+1$ only, and $\mathbf{P}(X = 1) = a$, $\mathbf{P}(Y = 1) = \alpha$. A third random variable Z is defined by $Z = \cos\left((X + Y)\frac{\pi}{2}\right)$. If $0 < a, \alpha < 1$, show that there are unique values of a and α such that X and Z are independent, and Y and Z are independent. In this case, are X, Y and Z independent?

Solution First, for Z,

$$\mathbf{P}(Z = 1) = \mathbf{P}(X + Y = 0) = a(1 - \alpha) + \alpha(1 - a)$$

and likewise $\mathbf{P}(Z = -1) = a\alpha + (1 - a)(1 - \alpha)$. Now

$$\mathbf{P}(Z = 1, X = 1) = \mathbf{P}(X = 1, Y = -1) = a(1 - \alpha).$$

Hence if

(10) $$a(1 - \alpha) = a(a(1 - \alpha) + \alpha(1 - a))$$

we have $\mathbf{P}(Z = 1, X = 1) = \mathbf{P}(Z = 1)\mathbf{P}(X = 1)$. Simplifying (10) yields $\alpha = \frac{1}{2}$. Now plodding through three similar constraints shows that X and Z are independent iff $\alpha = a = \frac{1}{2}$. By symmetry the same condition holds iff Y and Z are independent.

However, X, Y and Z are not independent because

$$\mathbf{P}(X = 1, Y = 1, Z = -1) = 0 \neq \mathbf{P}(X = 1)\mathbf{P}(Y = 1)\mathbf{P}(Z = -1).$$ ●

Independent random variables often have interesting and useful properties.

(11) **Example** Let X and Y be independent geometric random variables having respective mass functions $f_X(x) = (1 - \lambda)\lambda^x$ and $f_Y(y) = (1 - \mu)\mu^y$ for $x \geqslant 0$ and $y \geqslant 0$. What is the mass function of $Z = \min\{X, Y\}$?

Solution By independence

$$\mathbf{P}(Z > n) = \mathbf{P}(X > n \cap Y > n) = \mathbf{P}(X > n)\mathbf{P}(Y > n) = \lambda^{n+1}\mu^{n+1} = (\lambda\mu)^{n+1}.$$

Hence

$$\mathbf{P}(Z = n) = \mathbf{P}(Z > n - 1) - \mathbf{P}(Z > n) = (1 - \lambda\mu)(\lambda\mu)^n$$

and Z is also geometric with parameter $\lambda\mu$. ●

5.3 Expectation

Let the random variable $Z = g(X, Y)$ be a function of X and Y. Using (5.1.4) and the definition of expectation (4.3.1), we have

$$\mathbf{E}(Z) = \sum_z z f_Z(z) = \sum_z z\mathbf{P}(g(X, Y) = z).$$

This expression for $\mathbf{E}(Z)$ is not always simple or convenient for use in calculation. The following generalization of Theorem 4.3.4 is therefore very useful.

(1) **Theorem** Let X and Y have joint mass function $f(x, y)$. Whenever the sum on the right hand side is absolutely convergent, we have

$$\mathbf{E}(g(X, Y)) = \sum_{x,y} g(x, y)f(x, y).$$

Proof The proof is essentially the same as that of Theorem 4.3.4. ■

(2) **Corollary** For any real numbers a and b,

$$\mathbf{E}(aX + bY) = a\mathbf{E}(X) + b\mathbf{E}(Y)$$

when both sides exist and are finite.

Proof Since the sum is absolutely convergent, by (1)

$$\mathbf{E}(aX+bY)=\sum_{x,y}(ax+by)f(x,y)=\sum_{x,y}axf(x,y)+\sum_{x,y}byf(x,y)$$

$$=\sum_x axf_X(x)+\sum_y byf_Y(y) \qquad \text{by (5.1.5) and (5.1.6)}$$

$$=a\mathbf{E}(X)+b\mathbf{E}(Y).$$ ■

(3) **Example: Coupons** Recall Example 4.3.15 in which you were collecting coupons; we can now find $\mathbf{E}(R)$ more quickly. Let T_1 be the number of packets required to obtain the first coupon, T_2 the *further* number of packets required to obtain a second type of coupon, T_3 the *further* number required for the third type and so on. Then

$$R=\sum_{k=1}^{n}T_k.$$

Obviously, $T_1=1$. Also

$$\mathbf{P}(T_2=r)=\left(\frac{1}{n}\right)^{r-1}\left(1-\frac{1}{n}\right)$$

so that T_2 is geometric with mean $\frac{n}{n-1}$. Likewise T_k is geometric with mean

$$\mathbf{E}(T_k)=\frac{n}{n-k+1};\quad 1\leqslant k\leqslant n.$$

Hence, by (2)

$$\mathbf{E}(R)=\sum_{k=1}^{n}\mathbf{E}(T_k)=\sum_{k=1}^{n}\frac{n}{n-k+1},$$

the same as the answer obtained with somewhat more effort in Example 4.3.15. ●

Corollary (2) is often useful when considering sums of indicators. For example, let $\{A_1, A_2, \ldots, A_n\}$ be any collection of events, and let

$$I_i=\begin{cases}1 & \text{if } A_i \text{ occurs}\\ 0 & \text{if } A_i \text{ does not occur}\end{cases}$$

be the indicator of A_i. Now let X be the number of the A_i which occur. Then by construction $X=\sum_{i=1}^{n}I_i$, and by (2)

$$\mathbf{E}(X)=\sum_{i=1}^{n}\mathbf{E}(I_i)=\sum_{i=1}^{n}\mathbf{P}(A_i).$$

We use this result in the following example.

Example: Binomial distribution Let X be binomial with parameters n and p. Show that the factorial moments of X are given by

$$\mu^{(k)}=p^k n(n-1)\ldots(n-k+1).$$

Solution Suppose a coin which shows heads with probability p is tossed n times. Then the number of heads has the mass function of X. Let Y be the number of distinct sets of k such that all k tosses show heads. Then $Y = \binom{X}{k}$. However each of the $\binom{n}{k}$ distinct sets of k tosses shows k heads with probability p^k. Hence $\mathbf{E}(Y) = \binom{n}{k}p^k$. Therefore we have

$$\mathbf{E}\left(\binom{X}{k}\right) = \binom{n}{k}p^k$$

which is the desired result. ●

We single certain expectations out for special notice. Just as random variables have moments, jointly distributed random variables have joint moments.

(4) **Definition** The joint moments of X and Y are

$$\mu_{ij} = \mathbf{E}(X^i Y^j) \;;\quad i, j \geqslant 1.$$ ▲

(5) **Definition** The covariance of X and Y is

$$\operatorname{cov}(X, Y) = \mathbf{E}[(X - \mathbf{E}(X))(Y - \mathbf{E}(Y))] = \mathbf{E}(XY) - \mathbf{E}(X)\mathbf{E}(Y).$$ ▲

This is the most important of the *central joint moments* which are

$$\sigma_{ij} = \mathbf{E}[(X - \mathbf{E}(X))^i (Y - \mathbf{E}(Y))^j]; \quad i, j \geqslant 1.$$

Here are two interesting properties of $\operatorname{cov}(X, Y)$.

(6) **Theorem** For jointly distributed random variables X and Y, and constants a, b, c, d, we have:

(i) $\operatorname{cov}(aX + b, cY + d) = ac \operatorname{cov}(X, Y)$,
(ii) $\operatorname{var}(X + Y) = \operatorname{var}(X) + \operatorname{var}(Y) + 2\operatorname{cov}(X, Y)$.

Proof
(i)
$$\begin{aligned}\operatorname{cov}(aX + b, cY + d) &= \mathbf{E}[(aX + b - a\mathbf{E}(X) - b)(cY + d - c\mathbf{E}(Y) - d)] \\ &= \mathbf{E}[ac(X - \mathbf{E}(X))(Y - \mathbf{E}(Y))] = ac\operatorname{cov}(X, Y).\end{aligned}$$

(ii)
$$\begin{aligned}\operatorname{var}(X + Y) &= \mathbf{E}(X + Y - \mathbf{E}(X) - \mathbf{E}(Y))^2 \\ &= \mathbf{E}[(X - \mathbf{E}(X))^2 + (Y - \mathbf{E}(Y))^2 + 2(X - \mathbf{E}(X))(Y - \mathbf{E}(Y))]\end{aligned}$$

as required. ■

Let us find $\operatorname{cov}(X, Y)$ for the simple examples we have met above.

Examples 5.2.1 and 5.2.5 revisited Find the covariance for each of the three joint mass functions given in these two examples.

Solution In every case

$$\mathbf{E}(XY) = 1^2 f(1, 1)$$

and

$$\mathbf{E}(X) = f(1,0) + f(1,1)$$

and

$$\mathbf{E}(Y) = f(0,1) + f(1,1).$$

Hence

$$\operatorname{cov}(X,Y) = f(1,1) - (f(1,0) + f(1,1))(f(0,1) + f(1,1)).$$

Evaluating this in the three given instances shows that:

(i) $$\operatorname{cov}(X,Y) = \frac{5}{12} - \left(\frac{1}{3} + \frac{5}{12}\right)\left(\frac{1}{12} + \frac{5}{12}\right) = \frac{1}{24}$$

(ii) $$\operatorname{cov}(U,V) = \frac{1}{2} - \left(\frac{1}{4} + \frac{1}{2}\right).\frac{1}{2} = \frac{1}{8}$$

(iii) $$\operatorname{cov}(W,Z) = \frac{3}{8} - \left(\frac{3}{8} + \frac{3}{8}\right)\left(\frac{1}{8} + \frac{3}{8}\right) = 0.$$ ●

(7) Example 5.1.8 revisited Recall that Art and Bart are cutting for the deal. Find $\operatorname{cov}(X,Y)$ and $\operatorname{cov}(V,W)$.

Solution

$$\mathbf{E}(X) = \mathbf{E}(Y) = \sum_{2}^{14} \frac{x}{13} = 8.$$

Also, using (1),

$$\mathbf{E}(XY) = \sum_{2\le x\ne y\le 14} \frac{x}{12}.\frac{y}{13} = \frac{1}{12}.\frac{1}{13}\sum_{y}(105 - y - 1)y = \frac{1}{12}\cdot\frac{1}{13}(105 \times 104 - 1118)$$

$$= 64 - \frac{7}{6}.$$

Hence $\operatorname{cov}(X,Y) = -\frac{7}{6}$.

Likewise, using the expressions in Example 5.1.8 for the marginal mass functions of V and W, we find

$$\mathbf{E}(V) = \sum_{v=2}^{13} \frac{v(14-v)}{78} = \frac{1}{78}\sum_{v=2}^{13}(13v - v(v-1))$$

$$= +\frac{1}{78}\sum_{v=2}^{13}\left(\frac{13}{2}[v(v+1) - v(v-1)] - \frac{1}{3}[(v+1)v(v-1) - v(v-1)(v-2)]\right)$$

$$= \frac{17}{3}$$

after successive cancellation of the terms in the sum.

Similarly, we find $\mathbf{E}(W) = \frac{31}{3}$.

Now for all ω, $X(\omega)Y(\omega) = V(\omega)W(\omega)$, so $\mathbf{E}(VW) = \mathbf{E}(XY)$, and finally

$$\operatorname{cov}(V,W) = 62 + \frac{5}{6} - \frac{17}{3}\cdot\frac{31}{3} = +\frac{77}{18}.$$ ●

Just as joint mass functions have a simple form when random variables are independent, so too do joint moments simplify.

(8) Theorem If X and Y are independent random variables with finite expectations, then $\mathbf{E}(XY)$ exists, and $\mathbf{E}(XY) = \mathbf{E}(X)\mathbf{E}(Y)$. It follows that $\text{cov}(X, Y) = 0$ in this case.

Proof By independence and Theorem 5.3.1,

$$\mathbf{E}(|XY|) = \sum_{x,y} |xy| f_X(x) f_Y(y) = \sum_x |x| f_X(x) \sum_y |y| f_Y(y) = \mathbf{E}(|X|)\mathbf{E}(|Y|)$$

so $\mathbf{E}(|XY|) < \infty$. Thus $\mathbf{E}(XY)$ exists, and the same argument shows that $\mathbf{E}(XY) = \mathbf{E}(X)\mathbf{E}(Y)$. ■

(9) Definition If $\text{cov}(X, Y) = 0$ then X and Y are said to be *uncorrelated*. If $\mathbf{E}(XY) = 0$ then X and Y are said to be *orthogonal*. ▲

It follows that independent random variables are uncorrelated, but the converse is not true, as the following example shows.

(10) Example A random variable X is said to be symmetric if $\mathbf{P}(X = -x) = \mathbf{P}(X = x)$ for all x. Let X be symmetric with $\mathbf{E}(X^3) < \infty$, and let $Y = X^2$. Then since X has an expectation it is zero, by symmetry, and

$$\mathbf{E}(XY) = \mathbf{E}(Y^3) = \sum_{x>0} x^3(f(x) - f(-x)) = 0 = \mathbf{E}(X)\mathbf{E}(Y).$$

Thus $\text{cov}(X, Y) = 0$ even though X and Y are not independent. In this case X and Y are uncorrelated and orthogonal, but dependent. ●

Thus up to a point, and in a way which we carefully leave unspecified, $\text{cov}(X, Y)$ can be an indication of the dependence of X and Y. It has the drawback that it depends on the scale of X and Y. Thus, if a is a constant, aX and Y have the same 'dependence' as X and Y (whatever we mean by that), but $\text{cov}(aX, y) = a\,\text{cov}(X, Y)$. For this reason statisticians more commonly use the following.

(11) Definition The *correlation coefficient* of random variables X and Y is

$$\rho(X, Y) = \frac{\text{cov}(X, Y)}{(\text{var}(X)\,\text{var}(Y))^{\frac{1}{2}}},$$

whenever the right hand side exists. ▲

Example 5.1.3 revisited Here X and Y are the number of heads and tails respectively when a coin is tossed twice. What are $\text{cov}(X, Y)$ and $\rho(X, Y)$?

Solution Trivially

$$\mathbf{E}(XY) = 1^2\mathbf{P}(X = 1, \mathbf{Y} = 1) = 2p(1 - p).$$

Likewise $\mathbf{E}(X) = 2p$, $\mathbf{E}(Y) = 2(1-p)$, $\text{var}(X) = 2p(1-p)$ and $\text{var}(Y) = 2p(1-p)$. Hence

$$\text{cov}(X, Y) = 2p(1-p) - 4p(1-p) = -2p(1-p)$$

and

$$\rho(X, Y) = \frac{-2p(1-p)}{(4p^2(1-p)^2)^{\frac{1}{2}}} = -1. \qquad \bullet$$

The correlation coefficient ρ has the following interesting properties; we assume that X and Y are not constant, and have finite variance.

(12) Theorem If X and Y have correlation $\rho(X, Y)$, then:

(i) $-1 \leq \rho(X, Y) \leq 1$;
(ii) $|\rho| = 1$ if and only if $\mathbf{P}(X = aY) = 1$ for some constant a;
(iii) $\rho(aX + b, cY + d) = \text{sgn}(ac)\rho(X, Y)$, where $\text{sgn}(x)$ denotes the sign of x;
(iv) $\rho = 0$ if X and Y are independent.

The proof of this theorem relies on the following important and useful result.

(13) Lemma: Cauchy–Schwarz inequality If $\mathbf{E}(X^2)\mathbf{E}(Y^2) < \infty$ then

(14) $$(\mathbf{E}(XY))^2 \leq \mathbf{E}(X^2)\mathbf{E}(Y^2).$$

Proof Suppose $0 < \mathbf{E}(X^2)\mathbf{E}(Y^2)$. By Theorem 4.3.6 (iii)

(15) $$\begin{aligned} 0 &\leq \mathbf{E}[(X\mathbf{E}(Y^2) - Y\mathbf{E}(XY))^2] \\ &= \mathbf{E}(X^2)(\mathbf{E}(Y^2))^2 - 2\mathbf{E}(XY)^2\mathbf{E}(Y^2) + \mathbf{E}(Y^2)[\mathbf{E}(XY)]^2 \\ &= \mathbf{E}(Y^2)[\mathbf{E}(X^2)\mathbf{E}(Y^2) - (\mathbf{E}(XY))^2]. \end{aligned}$$

Since $\mathbf{E}(Y^2) > 0$, (14) follows. Of course, (14) is trivially true if $\mathbf{E}(Y^2) = 0$, for then $Y = XY = 0$ with probability one. ■

Proof of (12) (i) Applying Lemma 13 to the random variables $X - \mathbf{E}(X)$ and $Y - \mathbf{E}(Y)$ shows that $(\rho(X, Y))^2 \leq 1$, and so $-1 \leq \rho \leq 1$, as required.

(ii) If $|\rho| = 1$ then from (15)

$$\mathbf{E}[(X\mathbf{E}(Y^2) - Y\mathbf{E}(XY))^2] = 0,$$

and so from Example 4.6.10, with probability one $X = (\mathbf{E}(XY)/\mathbf{E}(Y^2))Y$.

(iii) Expanding, and using Theorem 6(i)

$$\rho(aX + b, cY + d) = \frac{ac\,\text{cov}(X, Y)}{(a^2\,\text{var}(X)c^2\,\text{var}(Y))^{\frac{1}{2}}} = \frac{ac}{\sqrt{(ac)^2}}\rho(X, Y)$$

as required

(iv) This follows immediately from Theorem 8. ■

(16) Example (5.1.9) revisited Recall that X and Y have joint mass function

$$f(x, y) = \frac{1-\lambda-\mu}{\mu}\binom{x+y-1}{x}\lambda^x\mu^y.$$

Show that $\rho(X, Y) = \left(\dfrac{\lambda\mu}{(1-\lambda)(1-\mu)}\right)^{\frac{1}{2}}$.

Solution First we calculate $\mathbf{E}(XY)$ as

$$\frac{1-\lambda-\mu}{\mu}\sum_{x,y} xy\binom{x+y-1}{x}\lambda^x\mu^y = \frac{1-\lambda-\mu}{\mu}\sum_{y=1}^{\infty} y^2\mu^y\lambda\sum_{x=1}^{\infty}\binom{x+y-1}{x-1}\lambda^{x-1}$$

$$= \frac{(1-\lambda-\mu)\lambda}{(1-\lambda)\mu}\sum_{y=1}^{\infty} y^2\left(\frac{\mu}{1-\lambda}\right)^y = \frac{\lambda(1-\lambda+\mu)}{(1-\lambda-\mu)^2}.$$

Now we have already discovered in Example 5.1.9 that X and Y have geometric mass functions, so by Example 4.3.13

$$\mathbf{E}(X) = \frac{1-\mu}{1-\lambda-\mu} - 1, \quad \mathbf{E}(Y) = \frac{1-\lambda}{1-\lambda-\mu}$$

$$\operatorname{var}(X) = \frac{\mu(1-\lambda)}{(1-\lambda-\mu)^2}, \quad \operatorname{var}(Y) = \frac{\lambda(1-\mu)}{(1-\lambda-\mu)^2}$$

and plugging all this into (11) yields

$$\rho = \frac{\lambda(1-\lambda+\mu) - \lambda(1-\lambda)}{(\mu(1-\lambda)\lambda(1-\mu))^{\frac{1}{2}}}$$

as required. ●

Finally we remark that $\operatorname{cov}(X, Y)$ and $\rho(X, Y)$ are not the only functions used to measure dependence between X and Y. Another such function is

(17) $$I(X, Y) = \sum_x\sum_y f(x, y)\log\left(\frac{f(x, y)}{f_X(x)f_Y(y)}\right) = \mathbf{E}\left(\log\left(\frac{f(X, Y)}{f_X(X)f_Y(Y)}\right)\right).$$

See Example 5.15 for more on this.

5.4 Sums and products of random variables

These arise in many ways. For example, it is often useful to write a random variable as a sum of simpler random variables.

(1) Example: Binomial random variable The random variable X with mass function

$$f_X(k) = \binom{n}{k}p^k(1-p)^{n-k}$$

has arisen in many ways, classically it is the number of heads in n tosses of a biased coin. We now see that we can think about X in a different way. Let I_k be the indicator of the event that the kth toss of the coin shows a head. Then

$$X = I_1 + I_2 + \ldots + I_n = \sum_{k=1}^{n} I_k.$$

We have written X as a sum of Bernoulli trials, or indicators. Hence

$$\mathbf{E}(X) = \mathbf{E}\left(\sum_{k=1}^{n} I_k\right) = \sum_{k=1}^{n} \mathbf{E}(I_k) = np.$$

Likewise

$$\mathbf{E}(X^2) = \mathbf{E}\left(\sum_{k=1}^{n} I_k\right)^2 = \sum_{k=1}^{n} \mathbf{E}(I_k^2) + \sum_{j \neq k} \mathbf{E}(I_j I_k) = np + n(n-1)p^2.$$

Hence $\operatorname{var}(X) = np(1-p)$. You should compare this with your earlier methods using

$$\mathbf{E}(X^2) = \sum_{k=1}^{n} k^2 \binom{n}{k} p^k (1-p)^{n-k} = \sum_{k=1}^{n} (k(k-1)+k) \frac{n!}{k!(n-k)!} p^k (1-p)^{n-k}$$

and so on. ●

(2) Theorem Any discrete random variable X can be written as a linear combination of indicator random variables, thus

$$X = \sum_i a_i I(A_i)$$

for some collection of events $(A_i; i \geq 1)$ and real numbers $(a_i; i \geq 1)$.

Proof Just let $(a_i; i \geq 1)$ include the set of possible values of X, and set $A_i = \{\omega: X(\omega) = a_i\}$. ■

(3) Example: Matching Suppose that n distinct numbered keys ordinarily hang on n hooks bearing corresponding distinct numbers. On one occasion an inebriated turnkey hangs the keys at random on the hooks (one to each hook). Let X be the number of keys which are then on the correct hooks. Find $\mathbf{E}(X)$ and $\operatorname{var}(X)$.

Solution Let I_j be the indicator of the event that the jth key does hang on the jth hook. Then

$$X = \sum_{j=1}^{n} I_j.$$

Now by symmetry $\mathbf{P}(I_j = 1) = 1/n$, and for $j \neq k$

(4) $$\mathbf{P}(I_j I_k = 1) = \frac{1}{n(n-1)}.$$

Hence

$$\mathbf{E}(X) = \mathbf{E}\left(\sum_{j=1}^{n} I_j\right) = \sum_{j=1}^{n} \mathbf{E}(I_j) = \sum_{j=1}^{n} \mathbf{P}(I_j = 1) = 1.$$

Also

$$\mathbf{E}(X^2) = \mathbf{E}\left(\sum_{j=1}^{n} I_j^2 + \sum_{j\neq k} I_j I_k\right) = 1 + 1,$$

using (4), and the fact that $I_j^2 = I_j$. Hence $\operatorname{var}(X) = 1$. ●

Indicators can also be useful when multiplied together; here is an illustration.

(5) Example Let us prove (1.4.8). Recall that we have events $A_1, \ldots, A_n$, and we seek the probability that at least one of them occurs, namely

$$\mathbf{P}\left(\bigcup_{j=1}^{n} A_j\right) = t_n \quad \text{(say)}.$$

For economy of notation we set

(6) $$s_r = \sum_{i_1<i_2<\ldots<i_r} \mathbf{P}(A_{i_1} \cap \ldots \cap A_{i_r}); \; i \leqslant r \leqslant n.$$

We let I_j be the indicator of the event that A_j occurs, and set

$$S_r = \sum_{i_1<i_2<\ldots<i_r} I_{i_1} I_{i_2} \ldots I_{i_r}.$$

Next observe that

(7) $$1 - \prod_{j=1}^{n}(1 - I_j) = \begin{cases} 1 & \text{if at least one } A_j \text{ occurs} \\ 0 & \text{otherwise.} \end{cases}$$

Hence this is the indicator of the event whose probability we seek, and

$$\begin{aligned} \mathbf{P}\left(\sum_{j=1}^{n} A_j\right) &= \mathbf{E}\left(1 - \prod_{j=1}^{n}(1 - I_j)\right) = \mathbf{E}(S_1 - S_2 + \ldots + (-)^{n+1} S_n) \quad \text{on multiplying out,} \\ &= s_1 - s_2 + \ldots + (-)^{n+1} s_n \quad \text{by (6)} \end{aligned}$$

as required. ●

The same expansion (7) can be used to prove the following interesting inequalities.

Theorem: Inclusion–exclusion inequalities With the notation of Example 5, for $1 \leqslant r \leqslant n$,

$$(-)^r\left(\mathbf{P}\left(\bigcup_{j=1}^{n} A_j\right) - s_1 + s_2 - \ldots (-)^r s_r\right) \geqslant 0.$$

Proof First we prove a simple identity. Obviously $(1 + x)^k(1 + \frac{1}{x})^{-1} = x(1 + x)^{k-1}$. Hence equating the coefficient of x^{r+1} on each side gives

(8) $$\binom{k}{r+1} - \binom{k}{r+2} + \ldots + (-)^{k-r+1}\binom{k}{k} = \binom{k-1}{r} \geqslant 0.$$

Furthermore we have

$$1 - \prod_{1}^{n}(1 - I_j) - S_1 + S_2 - \ldots + (-)^r S_r = (-)^r (S_{r+1} - S_{r+2} + \ldots + (-)^{n-r+1} S_n). \tag{9}$$

Now suppose exactly k of $A_1, \ldots, A_n$ occur. If $k \leqslant r$, the right hand side of (9) is zero. If $k > r$, the contribution in the bracket on the right hand side is

$$\binom{k}{r+1} - \binom{k}{r+2} + \ldots + (-)^{k-r+1}\binom{k}{k} = \binom{k-1}{r} \quad \text{by (8)},$$
$$> 0.$$

Hence no matter how many A_j's occur

$$(-)^r \mathbf{E}\left(1 - \prod_{1}^{n}(1 - I_j) - S_1 + S_2 + \ldots + (-)^r S_r\right)$$
$$= (-)^{2r}\mathbf{E}(S_{r+1} - \ldots + (-)^{t-r+1} S_n) \geqslant 0$$

as required. ■

It can similarly be shown (or deduced from the above) that if

$$t_r = \sum_{i_1 < i_2 < \ldots < i_r} \mathbf{P}(A_{i_1} \cup \ldots \cup A_{i_r})$$

then $t_n = \sum_{r=1}^{n}(-)^{r-1} s_r$, and

$$(-)^r \left(\mathbf{P}\left(\bigcap_{1}^{n} A_i\right) - t_1 + t_2 - \ldots + (-)^r t_r\right) \geqslant 0.$$

Very often it is natural and important to consider the sum Z of two variables X and Y. This is itself a random variable, and so we may require the distribution of $Z = X + Y$. This is given by Example 5.1.3, so we have proved that

$$f_Z(z) = \sum_x f_{X,Y}(x, z - x). \tag{10}$$

One special case of this result must be singled out.

(11) Theorem If X and Y are independent discrete random variables, then $Z = X + Y$ has probability mass function

$$f_Z(z) = \sum_x f_X(x) f_Y(z - x). \tag{12}$$

Proof Substitute Definition 5.2.2 into (10). A summation of this form is called a *convolution*. ■

(13) Example: Sum of geometric random variables (a) Let X_1 and X_2 be independent random variables, each having a geometric distribution

$$\mathbf{P}(X_i = k) = (1 - p)^k p; \quad k \geqslant 0. \tag{14}$$

Show that the mass function of $X_1 + X_2$ is $f(z) = (z+1)(1-p)^z p^2;\ z \geq 0$.

(b) If $(X_i; i > 1)$ are independent random variables each having the geometric distribution (14), find the mass function of $Z = \sum_1^n X_i$.

Solution (a) Using (12)

$$f(z) = \sum_{k=0}^{z} (1-p)^k p(1-p)^{z-k} p = \sum_{k=0}^{z} p^2(1-p)^z = (z+1)p^2(1-p)^z.$$

Alternatively, suppose we have a coin which, when tossed, shows a head with probability p. Toss this coin repeatedly until a head first appears, and let T_1 be the number of tails shown up to that point. Continue tossing the coin until the second head appears and let T_2 be the further number of tails shown up to that point. Then T_1 and T_2 are independent and have the geometric distribution (14), and $T_1 + T_2$ is the number of tails shown before the second head appears. But we know from Example 4.2.8 that

$$\mathbf{P}(T_1 + T_2 = z) = \binom{z+1}{1} p^2(1-p)^z.$$

Thus $X_1 + X_2$ has this mass function also.

(b) Extending the second argument above, we see that $\sum_1^n X_i$ has the same distribution as the number of tails shown up to the point where n heads have first appeared in successive tosses of our biased coin. But the probability that z tails have appeared is just

$$\binom{z+n-1}{n-1} p^n(1-p)^z \tag{15}$$

and so this is the mass function of Z. It is negative binomial.

Now we are in possession of the answer, we can use (12) to verify it by induction. Assume that $\sum_1^n X_i$ has mass function (15). Then by (12) the mass function of $\sum_1^{n+1} X_i$ is

$$\mathbf{P}\left(\sum_{i=1}^{n+1} X_i = z\right) = \sum_{k=0}^{z} p(1-p)^k \binom{z+n-k-1}{n-1} (1-p)^{z-k} p^n$$

$$= \sum_{k=0}^{z} p^{n+1}(1-p)^z \binom{z+n-k-1}{n-1} = p^{n+1}(1-p)^z \binom{z+n}{k}$$

as required, where we have used the identity

$$\sum_{r=0}^{z} \binom{n-1+r}{n-1} = \binom{n+z}{n}. \tag{16}$$

Since (15) holds for $n = 1$, the result follows by induction. Note that (16) can be derived immediately by equating the coefficient of x^z on both sides of the trivial identity

$$\frac{1}{1+x} \cdot \frac{1}{(1+x)^n} = \frac{1}{(1+x)^{n+1}}.$$

●

Example: Sum of binomial random variables Let X and Y be independent random variables with respective mass functions

$$f_X(x) = \binom{m}{x} p^x (1-p)^{m-x},$$

and

$$f_Y(y) = \binom{n}{y} p^y (1-p)^{n-y}.$$

Show that $X + Y$ has a binomial distribution.

Solution The expeditious method of doing this is to use Example 5.4.1 to write $X = \sum_{i=1}^{m} I_i$ and $Y = \sum_{i=m+1}^{m+n} I_i$. Hence

$$X + Y = \sum_{i=1}^{m+n} I_i;$$

this has the $\mathrm{B}(m+n, p)$ mass function by Example 5.4.1.

Alternatively this may be shown by using (5.4.12); this is an exercise for you. ●

Turning to expected values, we recall that from Definition 5.2.2, for any random variables $(X_i; i \geqslant 1)$ with finite expectation, *whether or not they are independent*,

(17) $$\mathbf{E}\left(\sum_{i=1}^{n} X_i\right) = \sum_{i=1}^{n} \mathbf{E}(X_i).$$

If in addition the X_i are independent, then

(18) $$\mathbf{E}\left(\prod_{i=1}^{n} X_i\right) = \prod_{i=1}^{n} \mathbf{E}(X_i).$$

5.5 Dependence; conditional expectation

Let X and Y be jointly distributed random variables. We may be given the value of Y, either in fact, or as a supposition. What is the effect on the distribution of X?

(1) Definition If X and Y have joint probability mass function $f(x, y)$, then given $Y = y$, the random variable X has a *conditional probability mass function* given by

$$f_{X|Y}(x|y) = \frac{f(x, y)}{f_Y(y)}$$

for all y such that $f_Y(y) > 0$. ▲

Example Let X and Y be independent geometric random variables each having mass function $f(x) = (1-\lambda)\lambda^x$; $x \geqslant 0$, $0 < \lambda < 1$. Let $Z = X + Y$. Show that for $0 \leqslant x \leqslant z$, $f_{X|Z}(x|z) = 1/(z+1)$.

Solution From Example 5.4.13 we know that Z has mass function $f_Z(z) = (z+1)(1-\lambda)^2\lambda^z$ and so

$$f_{X|Z}(x|z) = \frac{\mathbf{P}(X = x,\ Z = z)}{(z+1)(1-\lambda)^2\lambda^z} = \frac{(1-\lambda)^2\lambda^x\lambda^{z-x}}{(1-\lambda)^2\lambda^z(z+1)} = (z+1)^{-1}.$$ ●

(2) **Example 5.1.8 revisited: Cutting for the deal** Find the conditional mass function of the loser's card conditional on $W = w$; find also $f_{W|V}(w|v)$.

Solution According to Example 5.1.8, $f(v, w) = \frac{1}{78}$; $2 \leqslant v < w \leqslant 4$, and

$$f_W(w) = \frac{w-2}{78}; \qquad 3 \leqslant w \leqslant 14.$$

Hence, using Definition 1,

$$f_{V|W}(v|w) = \frac{1}{w-2}; \quad 2 \leqslant v < w.$$

The loser's score is uniformly distributed given W. Likewise

$$f_{W|V}(w|v) = \frac{1}{78}\bigg/\frac{14-v}{78} = \frac{1}{14-v}; \quad v < w \leqslant 14,$$

also a uniform distribution. ●

(3) **Example** Let X and Y be independent. Show that the conditional mass function of X given Y is $f_X(x)$, the marginal mass function.

Solution Since X and Y are independent, $f(x, y) = f_X(x)f_Y(y)$. Hence, applying Definition 1,

$$f_{X|Y}(x|y) = f(x, y)/f_Y(y) = f_X(x).$$ ●

(4) **Theorem** $f_{X|Y}(x|y)$ is a probability mass function, which is to say that

(i) $f_{X|Y}(x|y) \geqslant 0$

and

(ii) $\sum_x f_{X|Y}(x|y) = 1$.

Proof Part (i) is trivial. Part (ii) follows immediately from (5.1.5). ■

Recall that two events A and B are said to be conditionally independent given C, if $\mathbf{P}(A \cap B|C) = \mathbf{P}(A|C)\mathbf{P}(B|C)$. Likewise it is possible for two random variables X and Y to be conditionally independent given Z, if

$$f_{X,Y|Z} = f_{X|Z}f_{Y|Z}.$$

Example: Cutting for the deal (Example 5.1.8 revisited) Suppose three players cut for the deal (with ties not allowed, as usual). Let X be the lowest card, Y the highest card and Z the intermediate card. Clearly X and Y are dependent.

However, conditional on $Z = z$, X and Y are independent. The mass function $f_{X|Z}$ is uniform on $\{2, \ldots, z-1\}$ and $f_{Y|Z}$ is uniform on $\{z+1, \ldots, 14\}$. ●

Being a mass function, $f_{X|Y}$ may have an expectation; it has a special name and importance.

(5) Definition The *conditional expectation* of X, given that $Y = y$ where $f_Y(y) > 0$, is

$$\mathbf{E}(X|\{Y = y\}) = \sum_x x f(x, y)/f_Y(y),$$

when the sum is absolutely convergent. ▲

As y varies over the possible values of Y, this defines a function of Y, denoted by $\mathbf{E}(X|Y)$. Since it is a function of Y, it is itself a random variable, which may have an expectation.

(6) Theorem If both sides exist

(7) $$\mathbf{E}(\mathbf{E}(X|Y)) = \mathbf{E}(X).$$

Proof Assuming the sums are absolutely convergent we have, by Theorem 4.3.4,

$$\begin{aligned}\mathbf{E}(\mathbf{E}(X|Y)) &= \sum_y \mathbf{E}(X|\{Y = y\}) f_Y(y) = \sum_y \sum_x \frac{x f(x, y)}{f_Y(y)} f_Y(y) && \text{by Example 2}\\ &= \sum_x x f_X(x) && \text{by (5.1.4)}\\ &= \mathbf{E}(X).\end{aligned}$$

■

This is an exceptionally important and useful result. Judicious use of Theorem 6 can greatly simplify many calculations; we give some examples.

(8) Example: Eggs A hen lays X eggs where X is Poisson with parameter λ. Each hatches with probability p, independently of the others, yielding Y chicks. Show that $\rho(X, Y) = \sqrt{p}$.

Solution Conditional on $X = k$, the number of chicks is binomial $\mathrm{B}(k, p)$, with mean kp. Hence

$$\mathbf{E}(XY) = \mathbf{E}(\mathbf{E}(XY|X)) = \mathbf{E}(X^2 p) = (\lambda^2 + \lambda)p.$$

Likewise

$$\begin{aligned}\mathbf{E}(Y^2) &= \mathbf{E}(\mathbf{E}(Y^2|X)) = \mathbf{E}(Xp(1-p) + X^2p^2)\\ &= \lambda p^2(1-p) + (\lambda + \lambda^2)p^2 = \lambda p + \lambda^2 p^2.\end{aligned}$$

Hence

$$\rho(X, Y) = \frac{\mathbf{E}(XY) - \mathbf{E}(X)\mathbf{E}(Y)}{(\operatorname{var}(X)\operatorname{var}(Y))^{\frac{1}{2}}} = \frac{\lambda^2 p + \lambda p - \lambda.\lambda p}{(\lambda(\lambda p + \lambda^2 p^2 - \lambda^2 p^2))^{\frac{1}{2}}} = \sqrt{p}.$$ ●

(9) **Example: Variance of a random sum** Let $X_1, X_2, \ldots$ be a collection of independent identically distributed random variables, and let Y be an integer valued random variable independent of all the X_i. Let $S_Y = \sum_{i=1}^{Y} X_i$. Show that

$$\text{var}(S_Y) = [\mathbf{E}(X_1)]^2 \text{var}(Y) + \mathbf{E}(Y) \text{var}(X_1).$$

Solution By (7),

$$\mathbf{E}(S_Y) = \mathbf{E}(\mathbf{E}(S_Y|Y)) = \mathbf{E}\left(\mathbf{E}\left(\sum_1^Y X_i|Y\right)\right) = \mathbf{E}(Y\mathbf{E}(X_1)) = \mathbf{E}(Y)\mathbf{E}(X_1).$$

Likewise

$$\begin{aligned}\mathbf{E}(S_Y^2) &= \mathbf{E}(\mathbf{E}(S_Y^2|Y)) = \mathbf{E}(Y\mathbf{E}(X_1^2) + Y(Y-1)[\mathbf{E}(X_1)]^2) \\ &= \mathbf{E}(Y)(\mathbf{E}(X_1^2) - [\mathbf{E}(X_1)]^2) + \mathbf{E}(Y^2)[\mathbf{E}(X_1)]^2,\end{aligned}$$

and so substituting into $\text{var}(S_Y) = \mathbf{E}(S_Y^2) - (\mathbf{E}(S_Y))^2$, gives the result. ●

(10) **Example 4.11 revisited: Gamblers' ruin** Two gamblers, A and B, have n coins. They divide this hoard by tossing each coin; A gets those which show heads, X say, B gets the rest, totalling $n - X$.

They then play a series of independent fair games; each time A wins he gets a coin from B, each time he loses he gives a coin to B. They stop when one or other has all the coins.

Let D_X be the number of games played. Find $\mathbf{E}(D_X)$, and show that $\rho(X, D_X) = 0$.

Solution Conditional on $X = k$, as in Example 4.11,

$$D_k = \frac{1}{2}D_{k+1} + \frac{1}{2}D_{k-1} + 1$$

with solution $D_k = k(n-k)$. Hence, observing that X is $\text{B}(n, p)$ (where p is the chance of a head), we have

$$\mathbf{E}(D_X) = \mathbf{E}(\mathbf{E}(D_X|X)) = \mathbf{E}(X(n-X)) = n(n-1)p(1-p).$$

Finally,

$$\text{cov}(X, D_X) = \mathbf{E}(X^2(n-X)) - \frac{n}{2}.\frac{n(n-1)}{4} = 0,$$

whence $\rho = 0$. ●

(11) **Example** Show that if X and Y are jointly distributed, then

$$f_X(x) = \sum_y f_Y(y) f_{X|Y}(x|y).$$

Solution This is just Theorem 6 in the special case when we take X to be I_x, the indicator of the event $\{X = x\}$. Then $\mathbf{E}(I_x) = f_X(x)$, and $\mathbf{E}(I_x|Y = y) = f_{X|Y}(x|y)$. The result follows from (7). Alternatively you can substitute from Definition 1. ●

Recall that we have already defined $\mathbf{E}(X|B)$ for any event B, in Chapter 4. It is convenient occasionally to consider quantities such as $\mathbf{E}(X|Y;B)$ This is defined to be the expected value of the conditional distribution

(12) $$\mathbf{P}(X = x|\{Y = y\} \cap B) = \frac{\mathbf{P}(\{X = x\} \cap \{Y = y\} \cap B)}{\mathbf{P}(\{Y = y\} \cap B)}$$

for any value y of Y such that $\mathbf{P}(\{Y = y\} \cap B) > 0$.

We give some of the more important properties of conditional expectation.

(13) Theorem Let a and b be constants, $g(.)$ an arbitrary function, and suppose that X, Y and Z are jointly distributed. Then (assuming all the expectations exist):

(i) $\mathbf{E}(a|Y) = a$
(ii) $\mathbf{E}(aX + bZ|Y) = a\mathbf{E}(X|Y) + b\mathbf{E}(Z|Y)$
(iii) $\mathbf{E}(X|Y) \geqslant 0$ if $X \geqslant 0$
(iv) $\mathbf{E}(X|Y) = \mathbf{E}(X)$ if X and Y are independent
(v) $\mathbf{E}(Xg(Y)|Y) = g(Y)\mathbf{E}(X|Y)$
(vi) $\mathbf{E}(X|Y; g(Y)) = \mathbf{E}(X|Y)$
(vii) $\mathbf{E}(\mathbf{E}(X|Y;Z)|Y) = \mathbf{E}(X|Y)$.

Property (vii) is called the tower property. It enables us to consider multiple conditioning by taking the random variables in any convenient order.

Proof We shall prove the odd parts of Theorem 13, the even parts are left as exercises for you.

(i) $f(a, y) = f_Y(y)$, so

$$\mathbf{E}(a|Y) = af_Y(y)/f_Y(y) = a.$$

(iii) If $X \geqslant 0$, then every term in the sum in Theorem 6 is non-negative. The result follows.

(v) $$\mathbf{E}(Xg(Y)|Y = y) = \sum_x xg(y)f(x, y)/f_Y(y) = g(y)\sum_x f(x, y)/f_Y(y)$$
$$= g(y)\mathbf{E}(X|Y = y).$$

(vii) For arbitrary values $Y = y$ and $Z = z$ of Y and Z, we have $\mathbf{E}(X|Y;Z) = \sum_x xf(x, y, z)/f_{Y,Z}(y, z)$. Hence by definition

(14) $$\mathbf{E}(\mathbf{E}(X|Y;Z)|Y) = \sum_z \mathbf{E}(X|Y;Z)f_{Y,Z}(y, z)/f_Y(y) = \sum_z\sum_x xf(x, y, z)/f_Y(y)$$
$$= \sum_x xf(x, y)/f_Y(y) = \mathbf{E}(X|Y).$$ ■

(15) Example Three children (Aelhyde, Beowulf and Canute) roll a die in the order A, B, C, A, ..., etc. until one of them rolls a six and wins. Find the expected number of rolls, given that Canute wins.

Solution We use a form of the tower property, (14). Let C be the event that Canute wins, let X be the duration of the game, and let Y denote the first roll to show a six in the first three rolls, with $Y = 0$ if there is no six. Then

(16) $$\mathbf{E}(X|C) = \mathbf{E}(\mathbf{E}(X|Y; C)|C).$$

Now if $Y = 0$, then with the fourth roll the game stands just as it did initially except that three rolls have been made. So

$$\mathbf{E}(X|Y = 0; C) = 3 + \mathbf{E}(X|C).$$

Obviously Y is otherwise 3, and

$$\mathbf{E}(X|Y = 3; C) = 3.$$

Therefore, substituting in (16) we have

$$\mathbf{E}(X|C) = 3 + \mathbf{E}(X|C)\mathbf{P}(Y = 0|C)$$

and so

$$\mathbf{E}(X|C) = \frac{3}{1 - \left(\frac{5}{6}\right)^3} = \frac{648}{91}.$$

Of course there are other ways of doing this. ●

Finally we remark that conditional expectation arises in another way.

(17) Theorem Let $h(Y)$ be any function of Y such that $\mathbf{E}(h(Y)^2) < \infty$. Then

(18) $$\mathbf{E}((X - h(Y))^2) \geq \mathbf{E}((X - \mathbf{E}(X|Y))^2).$$

Further, if $h(Y)$ is any function of Y such that

(19) $$\mathbf{E}((X - h(Y))^2) = \mathbf{E}((X - \mathbf{E}(X|Y))^2)$$

then

$$\mathbf{E}((h(Y) - \mathbf{E}(X|Y))^2) = 0.$$

Proof

(20) $$\begin{aligned}\mathbf{E}((X - h(Y))^2) &= \mathbf{E}((X - \mathbf{E}(X|Y) + \mathbf{E}(X|Y) - h(Y))^2)\\ &= \mathbf{E}((X - \mathbf{E}(X|Y))^2) + \mathbf{E}((\mathbf{E}(X|Y) - h(Y))^2)\\ &\quad + 2\mathbf{E}((X - \mathbf{E}(X|Y))(\mathbf{E}(X|Y) - h(Y))).\end{aligned}$$

However, by (7) we can write

$$\begin{aligned}\mathbf{E}((X - \mathbf{E}(X|Y))(\mathbf{E}(X|Y) - h(Y))) &= \mathbf{E}(\mathbf{E}((X - \mathbf{E}(X|Y))(\mathbf{E}(X|Y) - h(Y))|Y))\\ &= \mathbf{E}(\mathbf{E}(X|Y) - h(Y))\mathbf{E}((X - \mathbf{E}(X|Y))|Y)\\ &\qquad \text{by Theorem 13(v)}\\ &= 0,\end{aligned}$$

because $\mathbf{E}((X - \mathbf{E}(X|Y))|Y) = 0$. The result (18) follows because $\mathbf{E}((\mathbf{E}(X|Y) -$

$h(Y))^2) \geq 0$. Finally if (19) holds then from (20) we have

$$\mathbf{E}((\mathbf{E}(X|Y) - h(Y))^2) = 0,$$

as required to complete the proof of Theorem 17. ∎

Recall that if $\mathbf{E}(X^2) = 0$, then $\mathbf{P}(X = 0) = 1$; and hence if $\mathbf{E}((X - Y)^2) = 0$, then $X = Y$ with probability one. This suggests that the smaller $\mathbf{E}((X - Y)^2)$ is, then the 'closer' X is to Y, in some sense. The point of the theorem is then that among all functions of Y, $\mathbf{E}(X|Y)$ is the one which is 'closest' to X.

It is thus possible, and in later work desirable, to define $\mathbf{E}(X|Y)$ by this property. However, to explore all these ideas at this moment would take us too far afield.

5.6 Simple random walk

The ideas above now enable us to consider an exceptionally famous and entertaining collection of random variables.

(1) **Definition** Let $(X_i; i \geq 1)$ be a collection of independent identically distributed random variables with mass function

$$\mathbf{P}(X_1 = 1) = p; \mathbf{P}(X_1 = -1) = q = 1 - p.$$

Then the collection $(S_n; n \geq 0)$ where

$$S_n = S_0 + \sum_{i=1}^{n} X_i \tag{2}$$

is called a simple random walk. If $p = q$ it is called a symmetric simple random walk. ▲

The nomenclature follows from the visualization of S_n as representing the position of a particle which is initially at S_0, and then takes a series of independent unit steps; each step being positive with probability p, or negative with probability q.

It is conventional to display the walk in Cartesian coordinates as the sequence of points (n, S_n) for $n \geq 0$. Any particular such sequence is called a path of the random walk.

Much of the effort of the first probabilists (Fermat, Pascal, Bernoulli, de Moivre, Laplace) was devoted to discovering the properties of the simple random walk, and more general random walks are still being investigated by modern probabilists.

It is easy to see that the celebrated gambler's ruin problem of Example 2.10 is just a simple random walk in which S_0 is interpreted as the initial capital of the gambler, and the walk stops on the first occasion D when either $S_n = 0$ (the gambler is ruined) or $S_n = K$ (his opponent is ruined). That is the random variable

$$D = \min\{n: \{S_n = 0\} \cup \{S_n = K\}\} \tag{3}$$

is the duration of the game. In the context of random walks, D is called the first passage time of the walk to $\{0, K\}$.

The first thing to find is the mass function of S_n (when it is not stopped).

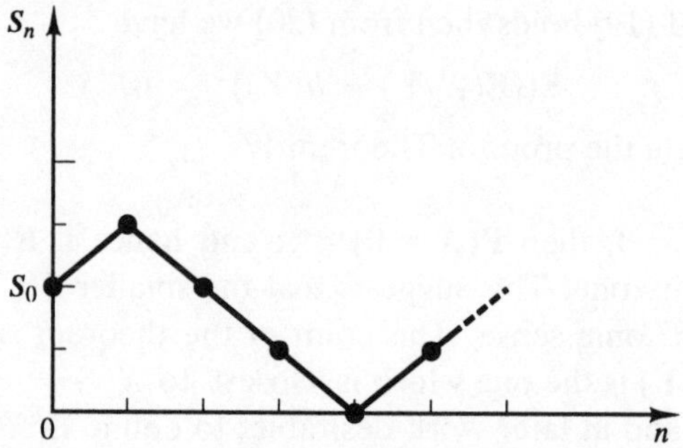

Figure 5.1 A path (or realization) of a simple random walk. The points represent successive positions (or values) of the walk; these are joined by steps of the walk. In this case $S_0 = 2$, $S_1 = 3$, $S_2 = 2$, and so on. The walk visits the origin at the fourth step.

(4) Theorem
$$\mathbf{P}(S_n - S_0 = k) = \binom{n}{\frac{1}{2}(n+k)} p^{\frac{1}{2}(n+k)} q^{\frac{1}{2}(n-k)}.$$

Proof Consider a path of the walk from $(0, S_0)$ to (n, S_n) with r positive steps and s negative steps. If $S_n - S_0 = k$, then $r - s = k$, and $r + s = n$. Hence $r = \frac{1}{2}(n+k)$ and $s = \frac{1}{2}(n-k)$. There are $\binom{n}{r}$ such paths, and each has the same probability, namely $p^r q^s$. Hence
$$\mathbf{P}(S_n - S_0 = k) = \binom{n}{r} p^r q^s$$
which is (4).

Alternatively, we can simply observe that $\frac{1}{2}(S_n - S_0 + n)$ is a binomial random variable with parameters n and p, and (4) follows. ■

As suggested by the gambler's ruin problem, we are interested in the first passage times of random walks.

(5) Definition Let $(S_n; n \geqslant 0)$ be a simple random walk with $S_0 = i$. Then the *first passage time* from i to k is the random variable
$$T_{ik} = \min\{n > 0; S_n = k\}.$$ ▲

When $i = k$, the random variable T_{kk} is called the *recurrence time* of k. We often denote T_{kk} by T_k. One obvious but important property of a first passage time T_{ik} is that steps after the first passage to k are independent of those before. It follows that we can write, for example,

(6)
$$T_{02} = T_{01} + T_{12}$$

where T_{01} and T_{12} are independent. Furthermore T_{12} and T_{01} have the same distribution because the X_i are identically distributed. These simple remarks are of great importance in examining the properties of the walk.

Our first result is very striking.

(7) **Theorem** If $p \geqslant q$ then T_{01} is certainly finite. If $p < q$ then T_{01} is finite with probability p/q.

Proof Let $\mathbf{P}(T_{jk} < \infty) = r_{jk}$. Now let us condition on the first step of the walk from $S_0 = 0$, giving

(8) $$r_{01} = \mathbf{P}(T_{01} < \infty) = p\mathbf{P}(T_{01} < \infty | X_1 = 1) + q\mathbf{P}(T_{01} < \infty | X_1 = -1).$$

On the one hand if $X_1 = 1$ then $T_{01} = 1$. On the other hand, if $X_1 = -1$, then the walk has to go from -1 to 0 and then from 0 to $+1$ in a finite number of steps for $T_{01} < \infty$. Hence, as $r_{01} = r_{-10}$,

(9) $$r_{01} = p + qr_{01}^2.$$

This has two roots $r_{01} = 1$ and $r_{01} = p/q$. If follows that if $p \geqslant q$, then the only root which is a probability is $r_{01} = 1$, as required.

If $p < q$ then it certainly seems plausible that $r_{01} = p/q$, but a little work is needed to prove it. This can be done in several ways. We choose to use the following interesting fact, that is for all i

(10) $$\mathbf{E}\left(\left(\frac{q}{p}\right)^{X_i}\right) = p\left(\frac{q}{p}\right)^{+1} + q\left(\frac{q}{p}\right)^{-1} = 1.$$

Hence, if $S_0 = 0$,

(11) $$\mathbf{E}\left(\left(\frac{q}{p}\right)^{S_n}\right) = \prod_{i=1}^{n}\mathbf{E}\left(\left(\frac{q}{p}\right)^{X_i}\right) = 1.$$

Now denote T_{01} by T, and suppose that

(12) $$p < q \quad \text{and} \quad r_{01} = 1.$$

Then conditioning on whether $T \leqslant n$ or $T > n$ gives

(13) $$1 = \mathbf{E}\left(\left(\frac{q}{p}\right)^{S_n}\right) = \mathbf{E}\left(\left(\frac{q}{p}\right)^{S_n}\middle| T \leqslant n\right)\mathbf{P}(T \leqslant n) + \mathbf{E}\left(\left(\frac{q}{p}\right)^{S_n}\middle| T > n\right)\mathbf{P}(T > n).$$

Now if $T \leqslant n$, then $S_T = 1$ and $S_n = 1 + X_{T+1} + \ldots + X_n$. Hence

$$\mathbf{E}\left(\left(\frac{q}{p}\right)^{S_n}\middle| T \leqslant n\right) = \frac{q}{p}\mathbf{E}\left(\left(\frac{q}{p}\right)^{X_{T+1}+\ldots+X_n}\right) = \frac{q}{p} \quad \text{by (10).}$$

Furthermore, if $T > n$, then $S_n \leqslant 0$, and so

$$\mathbf{E}\left(\left(\frac{q}{p}\right)^{S_n}\middle| T > n\right) \leqslant 1.$$

Hence, allowing $n \to \infty$ in (13) gives $1 = q/p + 0$. But this is impossible when $p < q$, so (12) must be impossible. Hence when $p < q$ we must have $r_{01} = p/q$. ■

In the case when $p > q$, so that T_{01} is finite, it is natural to ask what is $\mathbf{E}(T_{01})$? If we knew that $\mathbf{E}(T_{01}) < \infty$, then we could write

$$\begin{aligned}\mathbf{E}(T_{01}) &= \mathbf{E}(\mathbf{E}(T_{01}|X_1)) = p\mathbf{E}(T_{01}|X_1 = 1) + q\mathbf{E}(T_{01}|X_1 = -1)\\ &= p + q(1 + \mathbf{E}(T_{-1,1})) = 1 + 2q\mathbf{E}(T_{01}) \quad \text{by (6).}\end{aligned}$$

Hence

(14) $$\mathbf{E}(T_{01}) = \frac{1}{p-q}; \quad p > q,$$

as required.

It is not too difficult to show that $\mathbf{E}(T_{01}) < \infty$, as we now demonstrate.

(15) Theorem $\mathbf{E}(T_{01}) < \infty$ when $p > q$.

Proof

$$\begin{aligned}\mathbf{P}(T_{01} > n) &= \mathbf{P}(S_i \leqslant 0 \text{ for } 0 \leqslant i \leqslant n)\\ &\leqslant \mathbf{P}(S_n \leqslant 0) = \mathbf{P}\left(\tfrac{1}{2}(S_n + n) \leqslant \frac{n}{2}\right)\\ &= \mathbf{P}\left(\left(\frac{q}{p}\right)^{\frac{1}{2}(S_n+n)} \geqslant \left(\frac{q}{p}\right)^{n/2}\right) \quad \text{since } p > q,\\ &\leqslant \mathbf{E}\left(\left(\frac{q}{p}\right)^{\frac{1}{2}(S_n+n)}\right)\left(\frac{q}{p}\right)^{-n/2}\end{aligned}$$

by the basic inequality, Theorem 4.6.1.
Now we recall the observation in Theorem 4 that $\frac{1}{2}(S_n + n)$ has the $\mathrm{B}(n, p)$ mass function. Hence

(16) $$\mathbf{E}\left(\left(\frac{q}{p}\right)^{\frac{1}{2}(S_n+n)}\right) = \left(q + p.\frac{q}{p}\right)^n = (2q)^n.$$

Therefore, finally,

$$\mathbf{E}(T_{01}) = \sum_{n=0}^{\infty}\mathbf{P}(T_{01} > n) \leqslant \sum_{n=0}(2q)^n\left(\frac{p}{q}\right)^{n/2} = \frac{1}{1-2(pq)^{\frac{1}{2}}} \quad \text{since } pq < \frac{1}{4}$$
$$< \infty.$$

This establishes (14), as required. ■

There are one or two gaps in the above; for example we do not yet know either the mass function of T_{01}, or $\mathbf{E}(T_{01})$ in the case $p = q = \frac{1}{2}$. Both of these can be filled by the following beautiful theorem.

(17) Hitting time theorem Let $(S_n; n \geqslant 0)$ be a simple random walk with $S_0 = 0$. Let T_{0b} be the first passage time from 0 to $b > 0$, with mass function $f_{0b}(n)$. Then

$$f_{0b}(n) = \mathbf{P}(T_{0b} = n) = \frac{b}{n}\mathbf{P}(S_n = b) = \frac{b}{n}\binom{n}{\frac{1}{2}(n+b)}p^{\frac{1}{2}(n+b)}q^{\frac{1}{2}(n-b)}.$$

The proof of (17) relies on the following lemma, which is of considerable interest in its own right. First we observe that the number of paths of the walk from $(0, 0)$ to $(n-1, b+1)$ is denoted by $N_{n-1}(0, b+1)$ and we have

(18) $$N_{n-1}(0, b+1) = \binom{n-1}{\frac{1}{2}(n-b)-1}.$$

(19) **Lemma: The reflection principle** Let $N^b_{n-1}(0, b-1)$ be the number of paths from $(0,0)$ to $(n-1, b-1)$ which pass through b at least once. Then

$$N^b_{n-1}(0, b-1) = N_{n-1}(0, b+1). \tag{20}$$

Proof Let π be a path which visits b on its journey from $(0,0)$ to $(n-1, b-1)$. Let L be the occasion of its last visit. Now reflect that part of the walk after L in the line $y = b$. This yields a path π' from $(0,0)$ to $(n-1, b+1)$. Conversely, for any path from $(0,0)$ to $(n-1, b+1)$, we may reflect the segment in $y = b$ after its last visit to b, to give a path π from $(0,0)$ to $(n-1, b-1)$. These two sets are thus in one–one correspondence, and (20) follows. ■

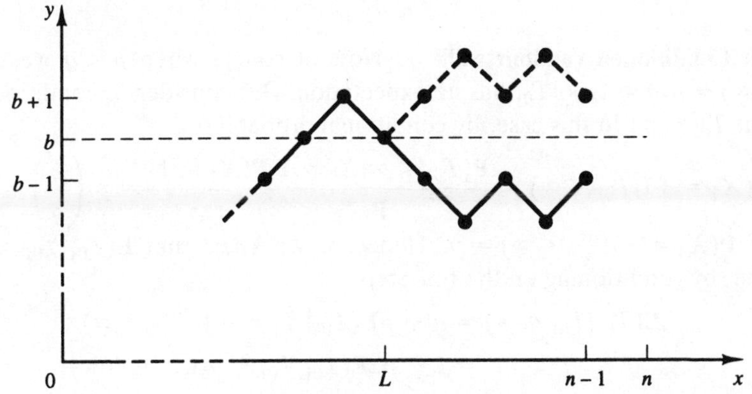

Figure 5.2 The solid line is the path of the walk; the dashed line is the reflection in $y = b$ of that part of the walk after its last visit to b before n, at time L.

Proof of (17): Hitting time theorem If $T_{0b} = n$, then we must have $X_n = +1$ and $S_{n-1} = b - 1$. Now there are $N_{n-1}(0, b-1)$ paths from $(0,0)$ to $(n-1, b-1)$ of which $N^b_{n-1}(0, b-1)$ visit b on route. Each such path has probability $p^{\frac{1}{2}(n+b)-1}q^{\frac{1}{2}(n-b)}$. Hence, using the reflection principle,

$$\begin{aligned}
\mathbf{P}(T_{0b} = n) &= p(N_{n-1}(0, b-1) - N_{n-1}(0, b+1))p^{\frac{1}{2}(n+b)-1}q^{\frac{1}{2}(n-b)} \\
&= \left(\binom{n-1}{\frac{1}{2}(n+b)-1} - \binom{n-1}{\frac{1}{2}(n+b)}\right)p^{\frac{1}{2}(n+b)}q^{\frac{1}{2}(n-b)} \\
&= \frac{b}{n}\binom{n}{\frac{1}{2}(n+b)}p^{\frac{1}{2}(n+b)}q^{\frac{1}{2}(n-b)} \\
&= \frac{b}{n}\mathbf{P}(S_n = b), \qquad \text{by (4).}
\end{aligned}$$

■

Since a similar argument works for negative values of b, we have

$$\mathbf{P}(T_{0b} = n) = \frac{|b|}{n}\mathbf{P}(S_n = b) \tag{21}$$

and $\mathbf{E}(T_{0b}) = \sum_{n=1}^{\infty} |b|\mathbf{P}(S_n = b)$.

(22) Example: Symmetric random walk When $p = q$, the simple random walk is said to be symmetric. In this case

$$\mathbf{E}(T_{01}) = \sum_{m=0}^{\infty} \mathbf{P}(S_{2m+1} = 1) = \sum_{m=0}^{\infty} \binom{2m+1}{m+1} 2^{-(2m+1)}$$

$$\geqslant \sum_{m=0}^{\infty} \frac{2m(2m-2)\ldots 2.1}{(m+1)(m)\ldots 1} 2^{-m} = \sum_{m=0}^{\infty} \frac{1}{m+1} = \infty.$$

Hence the symmetric random walk has the interesting property that $\mathbf{P}(T_{01} < \infty) = 1$, but

(23) $$\mathbf{E}(T_{01}) = \infty.$$ ●

(24) Example: Conditioned random walk Now of course when $p < q$ we know that $\mathbf{P}(T_{01} < \infty) = p/q < 1$, so T_{01} has no expectation. But consider S_n conditional on the event that $T_{01} < \infty$. In this case, by conditional probability

$$\mathbf{P}(X_1 = +1 | T_{01} < \infty) = \frac{\mathbf{P}(T_{01} < \infty | X_1 = 1)\mathbf{P}(X_1 = 1)}{\mathbf{P}(T_{01} < \infty)} = p \Big/ \left(\frac{p}{q}\right) = q.$$

Likewise $\mathbf{P}(X_1 = -1 | T_{01} < \infty) = p$. Hence, if we knew that $\mathbf{E}(T_{01} | T_{01} < \infty)$ were finite then, by conditioning on the first step

$$\mathbf{E}(T_{01} | T_{01} < \infty) = q + p\mathbf{E}(T_{01} | X_1 = -1;\, T_{01} < \infty)$$
$$= 1 + p2\mathbf{E}(T_{01} | T_{01} < \infty), \quad \text{by (6)}.$$

Hence, when $q > p$,

(25) $$\mathbf{E}(T_{01} | T_{01} < \infty) = \frac{1}{q - p}.$$

It is straightforward to use Theorem 17 to establish that $\mathbf{E}(T_{01} | T_{01} < \infty) < \infty$ (an exercise for you) and (25) is proved. Together with (14) and Example 22, this shows that for any value of p, and $b > 0$,

(26) $$\mathbf{E}(T_{0b} | T_{0b} < \infty) = \frac{b}{|p - q|}.$$ ●

We may also consider recurrence times.

(27) Example Let T_0 be the recurrence time of 0. Show that $\mathbf{P}(T_0 < \infty) = 1 - |p - q|$ and

$$\mathbf{E}(T_0 | T_0 < \infty) = 1 + \frac{1}{|p - q|}.$$

Solution Just consider T_0 conditional on the outcome of the first step, and then use what we know about T_{01} and T_{10}. You fill in the details. ●

Finally we prove a famous result, the so-called

(28) Ballot theorem Let $S_n = \sum_1^n X_i$ be a simple random walk with $S_0 = 0$. Then

$$\mathbf{P}\left(\prod_1^{2n-1} S_i \neq 0 | S_{2n} = 2r\right) = \frac{r}{n}.$$

Proof We count paths as we did in the hitting time theorem. What is the number $N_{2n-1}^0(1, 2r)$ of paths from $(1, 1)$ to $(2n, 2r)$ which visit the origin? We can reflect the walk before its first zero in the x-axis, and this shows that $N_{2n-1}^0(1, 2r) = N_{2n-1}(-1, 2r)$. Since all $N_{2n}(0, 2r)$ paths from $(0, 0)$ to $(2n, 2r)$ are equally likely it follows that the required probability is

$$\frac{N_{2n-1}(1, 2r) - N_{2n-1}^0(1, 2r)}{N_{2n}(0, 2r)} = \frac{N_{2n-1}(1, 2r) - N_{2n-1}(-1, 2r)}{N_{2n}(0, 2r)}$$
$$= \frac{\binom{2n-1}{n+r-1} - \binom{2n-1}{n+r}}{\binom{2n}{n+r}} = \frac{2r}{2n}. \qquad \blacksquare$$

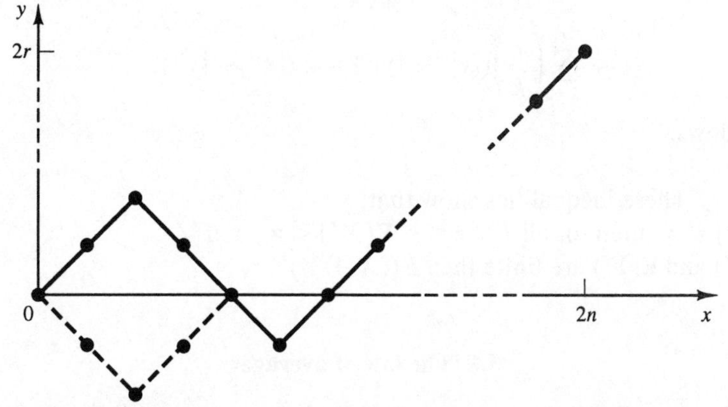

Figure 5.3 The ballot theorem The solid line is a path of the walk; the dashed line is the reflection in the x-axis of that part of the walk before its first visit to zero.

The following application explains the name of the theorem.

Example: Ballot In an election, candidate A secures a votes and candidate B secures b votes. What is the probability that A is ahead throughout the count? By just the above argument this probability is $(a - b)/(a + b)$ when $a > b$. ●

5.7 Inequalities

Very many inequalities for probabilities and moments are useful when dealing with random vectors. Most of them are beyond our scope at this stage, but we give a few simple examples with some applications.

(1) **Basic inequality** If $X \leqslant Y$ with probability one, then $\mathbf{E}(X) \leqslant \mathbf{E}(Y)$.

Proof This follows immediately from Theorem 4.3.6. ■

Corollary (a) For $1 < r < s$,

(2) $$\mathbf{E}(|X|^r) \leqslant (\mathbf{E}(|X|^s)) + 1.$$

(b) For $r \geqslant 1$,

(3) $$\mathbf{E}(|X+Y|^r) \leqslant 2^r(\mathbf{E}(|X|^r) + \mathbf{E}(|Y|^r)).$$

Proof (a) If $|x| \leqslant 1$ then $|x|^r \leqslant 1$, and if $|x| > 1$, and $r < s$, then $|x|^r \leqslant |x|^s$. Hence in any case, when $r \leqslant s$, $|x|^r \leqslant |x|^s + 1$. Thus

$$\mathbf{E}(|X|^r) = \sum_x |x|^r f(x) \leqslant \sum_x |x|^s f(x) + 1 = \mathbf{E}(|X|^s) + 1.$$

(b) For any real numbers x and y, if $k \leqslant r$, $|x|^k|y|^{r-k} \leqslant |x^r + |y|^r$, because either $(|x|^k/|y|^k) \leqslant 1$ or $(|y|^{r-k}/|x|^{r-k}) < 1$. Hence

$$|x+y|^r \leqslant (|x| + |y|)^r \leqslant \sum_{k=0}^{r} \binom{r}{k} |x|^k|y|^{r-k}$$

$$\leqslant \sum_{k=0}^{r} \binom{r}{k} (|x|^r + |y|^r) = 2^r(|x|^r + |y|^r),$$

and (3) follows. ■

(4) **Corollary** These inequalities show that:
(a) if $\mathbf{E}(X^s) < \infty$, then for all $1 \leqslant r \leqslant s$, $\mathbf{E}(X^r) < \infty$, and
(b) if $\mathbf{E}(X^r)$ and $\mathbf{E}(Y^r)$ are finite then $\mathbf{E}((X+Y)^r) < \infty$. ■

5.8 The law of averages

Suppose that, as a result of some experiment, the event A occurs with probability p (or the event A^c occurs with probability $1-p$). Typically, A might be an event such as: 'the patient was cured' or 'the dart hit the target' or 'the molecule split'. Let N_n be the number of times A occurs in n independent repetitions of this experiment. Now we have shown that N_n is a binomial random variable, and in Example 4.17 we proved that for $\epsilon > 0$

(1) $$\mathbf{P}\left(\left|\frac{1}{n}N_n - p\right| > \epsilon\right) \leqslant 2\exp(-n\epsilon^2/4)$$
$$\to 0 \quad \text{as } n \to \infty.$$

Roughly speaking this says that the proportion of experiments in which A occurs approaches the probability of A as n increases. (It is pleasing that this agrees with our intuitive notions about events and their probabilities.)

We now develop this simple idea a little. Statements like (1) are common in probability, so we make a formal definition.

(2) **Definition** Let $(X_n; n \geqslant 1)$ be a sequence of random variables. We say the sequence X_n converges in probability to X if, for any $\epsilon > 0$, as $n \to \infty$

(3)
$$\mathbf{P}(|X_n - X| > \epsilon) \to 0.$$

For brevity this is often written as $X_n \xrightarrow{\text{P}} X$. ▲

In this notation (1) becomes

(4)
$$\frac{1}{n} N_n \xrightarrow{\text{P}} p.$$

Here the limit p is a constant random variable of course.

Now in Section 5.5 above we observed that we could write

$$N_n = S_n = \sum_{k=1}^{n} I_k,$$

where I_k is the indicator of the event that A occurs in the kth experiment, and $\mathbf{E}(I_k) = p$. In the notation of Definition 2, we can thus write (4) as

(5)
$$\frac{1}{n} S_n \xrightarrow{\text{P}} \mathbf{E}(I_1).$$

It is natural to wonder whether this result may also hold for sequences other than indicators. The following celebrated result shows that in many cases it does.

(6) **Theorem: Weak law of large numbers** Let $(X_n; n \geqslant 1)$ be a sequence of independent random variables having the same finite mean and variance, $\mu = \mathbf{E}(X_1)$ and $\sigma^2 = \operatorname{var}(X_1)$. Then as $n \to \infty$

(7)
$$\frac{1}{n}(X_1 + \ldots + X_n) \xrightarrow{\text{P}} \mu.$$

It is customary to write $S_n = \sum_{i=1}^{n} X_i$ for the partial sums of the X_i.

Proof Recall Chebyshov's inequality: for any random variable Y, and $\epsilon > 0$, $\mathbf{P}(|Y| > \epsilon) \leqslant \mathbf{E}(Y^2)\epsilon^{-2}$. Hence letting $Y = n^{-1}(S_n - n\mu)$, we have

$$\begin{aligned}\mathbf{P}\left(\left|\frac{1}{n}S_n - \mu\right| > \epsilon\right) &\leqslant \frac{1}{n^2\epsilon^2}\mathbf{E}\left(\left[\sum_{i=1}^{n}(X_i - \mu)\right]^2\right) \\ &= n^{-2}\epsilon^{-2}\sum_{i=1}^{n}\operatorname{var}(X_i) = \sigma^2/(n\epsilon^2) \\ &\to 0 \quad \text{as } n \to \infty,\end{aligned}$$

and (6) is proved. ■

Actually, when S_n is a binomial random variable we have already shown in Exercise 4.17.3 that more can be said.

(8) Theorem If $(X_i; i \geq 1)$ are independent indicator random variables with $\mathbf{E}(X_i) = p$ then as $n \to \infty$, for $0 < \epsilon < 1$,

$$\mathbf{P}\left(\left|\frac{1}{m}S_m - p\right| > \epsilon \quad \text{for any } m \geq n\right) \to 0.$$

Proof Remember that for any events $(A_i; i \geq 1)$ we have

$$\mathbf{P}\left(\bigcup_i A_i\right) \leq \sum_i \mathbf{P}(A_i).$$

It follows (using (1)) that

$$\mathbf{P}\left(\left|\frac{1}{m}S_m - p\right| > \epsilon \quad \text{for any } m \geq n\right) \leq \sum_{m=n}^{\infty} \mathbf{P}\left(\left|\frac{1}{m}S_m - p\right| > \epsilon\right) \leq \sum_{m=n}^{\infty} 2e^{-m\epsilon^2/4}$$

$$= 2e^{-n\epsilon^2/4}(1 - e^{-\epsilon^2/4})^{-1} \to 0 \quad \text{as } n \to \infty. \quad \blacksquare$$

Roughly speaking, this says that not only does the chance of finding $n^{-1}S_n$ far from p vanish, but also the chance that any of $(m^{-1}S_m; m \geq n)$ are far from p vanishes as $n \to \infty$. Results of this type are called strong laws of large numbers.

Results like those above go some way towards justifying our intuitive feelings about averages in the long run. Such laws of large numbers have other applications also; the following two examples are typical.

(9) Example: Monte Carlo integration Suppose $f(x)$ is a non-negative function, and we require $I = \int_a^b f(x)\,dx$. If f is sufficiently nasty to defy basic methods, a surprisingly effective method of finding I is as follows. Let R be the rectangle $\{x, y: a \leq x \leq b;\ c \leq y \leq d\}$ where $0 \leq c < f(x) < d$ for $a \leq x < b$. The curve $y = f(x)$ divides R into two disjoint regions, A lying above f and B lying below f.

Now pick a point P_1 at random in R, by which we mean uniformly in R. Then

(10) $$\mathbf{P}(P_1 \in B) = \frac{1}{|R|}\int_a^b f(x)\,dx = p \quad \text{say.}$$

Now we pick a series $(P_j; j \geq 1)$ of such points independently in R, and let I_j be the indicator of the event that P_i lies in B. Then by the weak law above

$$\frac{1}{n}\sum_{j=1}^{n} I_j \xrightarrow{\text{P}} p,$$

as $n \to \infty$. Hence for large n we may expect that $(|R|/n)\sum_{j=1}^{n} I_j$ is a reasonable approximation to I, and that it improves as n increases.

(We have glossed over one or two details in this simple account; more discussion is provided in Chapter 7.) ●

Here is a similar example.

(11) Example: Estimation of mass functions Let $(X_i; i \geq 1)$ be independent and identically distributed with an unknown mass function $f(x)$, and suppose we wish to know $f(x)$ for some given x. Let I_k be the indicator of the event that $X_k = x$. Obviously

$\mathbf{E}(I_k) = f(x)$, so by the weak law

$$\frac{1}{n}S_n = \frac{1}{n}\sum_{k=1}^{n} I_k \xrightarrow{\mathrm{P}} f(x).$$

Thus $n^{-1}S_n$ should be a good guess at $f(x)$ for large n. ●

Notice that in both these cases we applied the weak law of large numbers (WLLN) to a binomial random variable arising as a sum of indicators. Despite its simplicity this is a very important special case, so we return to Theorem 6 and note that the proof shows something a little stronger than the WLLN. In fact in using Chebyshov's inequality we showed that as $n \to \infty$

(12)
$$\mathbf{E}\left(\left|\frac{S_n}{n} - \mu\right|^2\right) \to 0.$$

This type of statement is also widespread in probability and warrants a formal emphasis.

(13) Definition Let $X_1, X_2, \ldots$ be a sequence of random variables. If there is a random variable X such that

$$\lim_{n\to\infty} \mathbf{E}(X_n - X)^2 = 0$$

then X_n is said to converge in mean square to X. We sometimes write this as $X_n \xrightarrow{\text{m.s.}} X$. ▲

5.9 Convergence

Though important, this section may be omitted at a first reading.

In the preceding section, and at various earlier times, we have introduced a number of ideas about the long run behaviour of random variables and their distributions. This seems an appropriate moment to point out the connections between these concepts.

First we recall our earlier definitions. Here $(X_n;\, n \geqslant 1)$ is a sequence of random variables with corresponding distributions $(F_n(x);\; n \geqslant 1)$. Also, $F(x)$ is the distribution of a random variable X. Then as $n \to \infty$ we say:

(1) X_n converges in distribution if $F_n(x) \to F(x)$ whenever $F(x)$ is continuous;
(2) X_n converges in probability if $\mathbf{P}(|X_n - X| > \epsilon) \to 0$, for any $\epsilon > 0$;
(3) X_n converges in mean square if $\mathbf{E}(|X_n - X|^2) \to 0$, where $\mathbf{E}(X_n^2)\mathbf{E}(X^2) < \infty$.

These are clearly not equivalent statements. For example, let X be integer valued and symmetrically distributed about zero, then let $Y = -X$. By symmetry X and Y have the same distribution, so $|F_Y(x) - F_X(x)| = 0$. However $\mathbf{P}(|X - Y| > \epsilon) = \mathbf{P}(2|X| > \epsilon) = 1 - \mathbf{P}(X = 0)$. Hence convergence in distribution does not necessarily imply convergence in probability. This in turn does not necessarily imply convergence in mean square, because (2) may hold for random variables without a variance.

What we can say is the following.

Theorem Let $(X_n; n \geq 1)$ be a sequence of random variables having corresponding distributions $(F_n(x); n \geq 1)$. We have the following two results:

(i) if $X_n \xrightarrow{\text{m.s.}} X$ then $X_n \xrightarrow{\text{P}} X$;

(ii) if $X_n \xrightarrow{\text{P}} X$ then $F_n(x) \to F(x)$ at all points x where $F(x)$ is continuous.

Proof (i) By Chebyshov's inequality, for $\epsilon > 0$

$$\mathbf{P}(|X_n - X| > \epsilon) \leq \epsilon^{-2}\mathbf{E}(X_n - X)^2 \to 0 \quad \text{by hypothesis.}$$

(ii) For any $\epsilon > 0$

$$\begin{aligned} F_n(x) &= \mathbf{P}(X_n \leq x) \\ &= \mathbf{P}(X_n \leq x, X \leq x + \epsilon) + \mathbf{P}(X_n \leq x, X > x + \epsilon) \\ &\leq \mathbf{F}(x + \epsilon) + \mathbf{P}(|X_n - X| > \epsilon). \end{aligned}$$

Likewise

$$\begin{aligned} F(x - \epsilon) &= \mathbf{P}(X \leq x - \epsilon, X_n \leq x) + \mathbf{P}(X \leq x - \epsilon, X_n > x) \\ &\leq F_n(x) + \mathbf{P}(|X_n - X| > \epsilon). \end{aligned}$$

Hence

$$F(x - \epsilon) - \mathbf{P}(|X_n - X| > \epsilon) \leq F_n(x) \leq F(x + \epsilon) + \mathbf{P}(|X_n - X| > \epsilon).$$

Now allowing $n \to \infty$ and $\epsilon \to 0$ yields the result. ■

Much more can be said about these matters, but we refrain from doing so.

WORKED EXAMPLES AND EXERCISES

5.10 Example: Golf

Arnold and Bobby play a complete round of 18 holes at golf. Holes are independent, and any hole is won by Arnold with probability p, or won by Bobby with probability q, or it is halved with probability r. Of the 18 Arnold wins X, Bobby wins Y, and Z are halved.

(a) Find the joint mass function of X, Y and Z.
(b) What is the marginal distribution of X?
(c) Show that the correlation coefficient of X and Y is

$$\rho(X, Y) = -\left(\frac{pq}{(1 - p)(1 - q)}\right)^{\frac{1}{2}}.$$

Solution Let A_k be the event that Arnold wins the kth hole, B_k the event that he loses the kth hole, and H_k the event that the kth hole is halved. We illustrate the possibilities by giving more than one method of solution.

(a)
Method I A typical outcome is a string of xA's, yB's and zH's. Such a sequence has probability $p^x q^y r^z$ of occurring, by independence, and by Theorem 3.3.2 there are $18!/(x!y!z!)$ such sequences. Hence

(1) $$\mathbf{P}(X = x, Y = y, Z = z) \frac{18!p^x q^y r^z}{x!y!z!}; \quad x + y + z = 18.$$

Method II By Definition 5.5.1

$$\mathbf{P}(X = x, Y = y, Z = z) = \mathbf{P}(X = x, Y = y|Z = z)\mathbf{P}(Z = z).$$

Now the number of holes halved is just the number of successes in 18 Bernoulli trials with $\mathbf{P}(\text{success}) = r$. Hence by Example 5.4.1 (or Example 4.2.3),

$$\mathbf{P}(Z = z) = r^z(1 - r)^{18-z}\binom{18}{z}.$$

Now for any given hole, $\mathbf{P}(A|H^c) = p/(p + q)$. Hence, given $Z = z$, the number of holes won by Arnold is just the number of successes in $18 - z$ Bernoulli trials with $\mathbf{P}(\text{success}) = p/(p + q)$. Therefore

$$\mathbf{P}(X = x, Y = y|Z = z) = \binom{18 - z}{x}\left(\frac{p}{p + q}\right)^x\left(\frac{q}{p + q}\right)^y$$

where $x + y = 18 - z$. Thus

$$\mathbf{P}(X = x, Y = y, Z = z) = \frac{(x + y)!}{x!y!}\frac{p^x q^y}{(1 - r)^{x+y}}r^z(1 - r)^{18-z}\frac{18!}{(x + y)!z!}$$
$$= \frac{18!p^x q^y r^z}{x!y!z!}.$$

(b)
Method I As in (5.1.5) we have

(2) $$\mathbf{P}(X = x) = \sum_{y,z}\mathbf{P}(X = x, Y = y, Z = z)$$
$$= \sum_{y=0}^{18-x}\frac{18!p^x}{x!(18 - x)!}\frac{(18 - x)!}{y!(18 - x - y)!}q^y r^{18-x-y}$$
$$= \binom{18}{x}p^x(q + r)^{18-x}$$

which is binomial with parameters 18 and p.

Method II Either Arnold succeeds with probability p in winning each hole, or he fails with probability $1 - p = q + r$. Hence by Example by 5.4.1 the mass function of X is binomial as in (2).

(c) Let I_k be the indicator of the event A_k that Arnold wins the kth hole, and J_k the indicator of the event B_k that Bobby wins it. Then $I_k J_k = 0$, and I_j is independent of J_k for $j \neq k$. Hence $\mathbf{E}(I_j I_k) = pq$ for $j \neq k$, and therefore

$$\mathbf{E}(XY) = \mathbf{E}\left(\sum_{k=1}^{18} I_k \sum_{j=1}^{18} J_j\right) = 18 \times 17pq.$$

Thus

$$\operatorname{cov}(X, Y) = 18 \times 17pq - 18p \times 18q = -18pq.$$

Finally we note that since X and Y are binomial, we have $\operatorname{var}(X) = 18p(1-p)$ and $\operatorname{var}(Y) = 18q(1-q)$. Therefore

$$\rho(X, Y) = \frac{\operatorname{cov}(X, Y)}{(\operatorname{var}(X)\operatorname{var}(Y))^{\frac{1}{2}}} = \frac{-18pq}{(18p(1-p).18q(1-q))^{\frac{1}{2}}}$$

as required.

(3) Exercise What is $\rho(Y, Z)$?
(4) Exercise What is the conditional mass function of X, given $X + Y = m$?
(5) Exercise What is the probability that the match is halved?
(6) Exercise What is $\mathbf{E}(X|Y)$?
(7) Exercise What is $\mathbf{E}(X|Y, Z)$?

5.11 Example: Joint lives

Suppose that $2m$ individuals constitute m married couples at some given initial date. We wish to consider the survivors at some given later date. Suppose that each individual is alive at the later date with probability p independently of the others. Let A be the number of individuals then alive, and let S be the number of surviving couples in which both the partners are alive. Show that

$$\mathbf{E}(S|A) = \frac{A(A-1)}{2(2m-1)}.$$

Remark This problem was discussed by Daniel Bernoulli in 1768.

Solution Let S_a be the number of surviving couples given that $A = a$. We shall give several methods of solution; you can choose your favourite, or of course find a better one.

Method I Let I_j be the indicator of the event that the jth couple survives. Then

$$\mathbf{E}(S_a) = \mathbf{E}\left(\sum_1^m I_j\right) = m\mathbf{E}(I_1) = m\mathbf{P}(I_1 = 1),$$

because the chance of survival is the same for every couple. Now we can choose the a survivors in $\binom{2m}{a}$ ways, and the number of these in which the first couple remain alive is $\binom{2m-2}{a-2}$. (This is the number ways of choosing $a-2$ other survivors from the other $m-1$ couples.)

Since these are equally likely outcomes

$$\mathbf{P}(I_1 = 1) = \binom{2m-2}{a-2}\Big/\binom{2m}{a} = \frac{a(a-1)}{2m(2m-1)}.$$

Hence $\mathbf{E}(S_a) = a(a-1)/(2(2m-1))$.

Method II Suppose that the a individuals remaining alive include x couples. If one more individual were to die, then the expected number of couples remaining would be $\mathbf{E}(S_{a-1}|S_a = x)$. To evaluate this, observe that if a widow/er dies then there are still x couples; however, if a survivor's spouse dies, there are now $x - 1$ couples. The probability of a widow/er's death is $(a - 2x)/a$; the probability of the death of one individual of the x couples is $(2x/a)$. Hence

$$\mathbf{E}(S_{a-1}|S_a = x) = \frac{x(a-2x)}{a} + \frac{(x-1)2x}{a} = \frac{(a-2)}{a}x.$$

Hence by Theorem 5.5.6

$$\mathbf{E}(S_{a-1}) = \mathbf{E}(\mathbf{E}(S_{a-1}|S_a)) = \frac{a-2}{a}\mathbf{E}(S_a).$$

This relation may be iterated on the left or the right, to give either

$$\mathbf{E}(S_a) = \frac{a(a-1)}{2}\mathbf{E}(S_2) = \frac{a(a-1)}{2(2m-1)}$$

or

$$\mathbf{E}(S_a) = \frac{a(a-1)}{2m(2m-1)}\mathbf{E}(S_{2m}) = \frac{a(a-1)}{2(2m-1)}.$$

Method III Number the couples $1, \ldots, m$. Let Y_i be the indicator of the event that the male of the ith couple survives and X_i the indicator of the event that the female of the ith couple survives. Then $S_a = \sum_1^m X_i Y_i$.

Now $\mathbf{P}(Y_1 = 1|X_1 = 1) = (a-1)/(2m-1)$ and $\mathbf{P}(X_1 = 1) = a/(2m)$. Hence

$$\mathbf{E}(S_a) = \sum_1^m \mathbf{E}(Y_i X_i) = m\mathbf{E}(Y_1|X_1 = 1)\mathbf{P}(X_1 = 1) = m\frac{a}{2m}\frac{a-1}{2m-1}$$

as required.

(1) **Exercise** Find the mass function of S, and write down its mean and variance.

(2) **Exercise** Show that $\mathbf{E}(AS) = 2m((m-1)p^3 + p^2)$.

(3) **Exercise** Show that $\mathbf{E}(A|S) = 2mp + 2(1-p)S$.

(4) **Exercise** Show that the correlation $\rho(A, S)$ is given by $\rho(A, S) = (2p/(1+p))^{\frac{1}{2}}$.

(5) **Exercise** Suppose that males and females have different death rates, so the probability of a male surviving is μ and the probability of a female surviving is ϕ. Show that S has a $\mathrm{B}(m, \mu\phi)$ mass function. What is the mass function of A? What is $\mathbf{E}(A)$?

(6) **Exercise** When males and females have different survival rates μ and ϕ, find $\mathbf{E}(A|S)$ and hence show that in this case

$$\rho(A, S) = \frac{(2-\phi-\mu)(\phi\mu)^{\frac{1}{2}}}{((1-\phi\mu)(\phi(1-\phi)+\mu(1-\mu)))^{\frac{1}{2}}}.$$

5.12 Example: Tournament

Suppose that 2^n tennis players enter a knock-out singles tournament, and the players are completely ranked (with no ties). The draw for the tournament is at random, and we suppose that in any match the higher ranked player always wins. Let R_n be the rank of the losing finalist; find $\mathbf{E}(R_n)$, and show that as $n \to \infty$

$$\mathbf{E}(R_n) \to 3.$$

Solution The losing finalist comes from the half of the draw not containing the top-ranked player. These 2^{n-1} players have ranks $N_1 < N_2 < \ldots < N_{2^{n-1}}$, which are drawn at random from the $2^n - 1$ integers $\{2, 3, 4, \ldots, 2^n\}$, and $R_n = N_1$.

Let $X_1, X_2, \ldots$ be the numbers of players drawn with the top-ranked player, between successive players drawn for the other half. That is to say

$$X_1 = N_1 - 2$$

$$X_k = N_k - N_{k-1} - 1, \quad 2 \leq k \leq 2^{n-1},$$

$$X_{2^{n-1}+1} = 2^n - N_{2^{n-1}}.$$

By symmetry, for all j and k, $\mathbf{E}(X_k) = \mathbf{E}(X_j)$. Hence

(1) $$(2^{n-1} + 1)\mathbf{E}(X_1) = \mathbf{E}\left(\sum_1^{2^{n-1}+1} X_k\right) = 2^n - 1 - 2^{n-1}.$$

Thus

$$\mathbf{E}(R_n) = \mathbf{E}(N_1) = \mathbf{E}(X_1) + 2$$

$$= \frac{2^{n+1} - 2^{n-1} + 1}{2^{n-1} + 1} \to 3, \quad \text{as } n \to \infty.$$

(2) Exercise Suppose that the ranking allows ties (so that, for example, a possible ranking is 1, 1, 3, 3, 3, 6, ...). Show that as $n \to \infty$, $\lim \mathbf{E}(R_n) \leq 3$.

(3) Exercise Suppose that there are $3 \times 2^{n-1}$ entrants, and these are divided at random into a group of size 2^{n-1} and a group of size 2^n who then knock each other out in the usual way to provide two finalists. Find $\mathbf{E}(R_n)$ and show that $\mathbf{E}(R_n) \to \frac{7}{2}$.

(4) Exercise An urn contains b blue and r red balls. Balls are removed at random until the first blue ball is drawn. Show that the expected number drawn is $(b + r + 1)/(b + 1)$.

The balls are replaced, and then removed at random until all the balls remaining are of the same colour. Show that the expected number remaining is $r/(b + 1) + b/(r + 1)$. What is the probability p_r that they are all red?

(5) Exercise Let $X_1, X_2, \ldots$ be independent and identically distributed. What is

$$\mathbf{E}\left(\frac{\sum_1^m X_i}{\sum_1^n X_i}\right) \quad \text{when } m \leq n?$$

5.13 Example: Congregations

Suppose that n initially separate congregations of people are then united, and one person is picked at random from the united group. Let the size of the congregation of which she was originally a member be Y. If the respective sizes of the original congregations are the random variables $(X_i;\ 1 \leqslant i \leqslant n)$, show that

(1) $$\mathbf{E}(Y) = \mathbf{E}\left(\frac{\sum_1^n X_i^2}{\sum_1^n X_i}\right)$$

(2) $$\geqslant \frac{1}{n}\sum_1^n \mathbf{E}(X_i).$$

Solution Let us use conditional expectation. Given that $X_i = x_i$ for $1 \leqslant i \leqslant n$, the probability that the selected individual was in the rth congregation initially is $x_r/(\sum_1^n x_i)$. Hence

$$\mathbf{E}(Y|X_1 = x_1, \ldots, X_n = x_n) = \sum_1^n x_r \mathbf{P}(Y = x_r) = \sum_{r=1}^n x_r^2 \bigg/ \left(\sum_{i=1}^n x_i\right).$$

Therefore

$$\mathbf{E}(Y) = \mathbf{E}(\mathbf{E}(Y|X_1, \ldots, X_n)) = \mathbf{E}\left(\frac{\sum_1^n X_i^2}{\sum_1^n X_i}\right).$$

Now recall Cauchy's inequality for real numbers $(x_i;\ 1 \leqslant i \leqslant n)$ and $(y_i;\ 1 \leqslant i \leqslant n)$, namely

(3) $$\left(\sum_1^n x_i y_i\right)^2 \leqslant \left(\sum_1^n x_i^2\right)\left(\sum_1^n y_i^2\right).$$

Setting $y_i = 1$, for all i, yields $(\sum_1^n x_i)^2/(\sum_1^n x_i) \geqslant \sum_1^n x_i/n$, and the required inequality (2) follows.

Remark Observe that if a congregation is picked at random by choosing a number in $\{1, 2, \ldots, n\}$ at random, then the expected size of the chosen congregation is $\frac{1}{n}\sum_1^n \mathbf{E}(X_i)$. The fact that a member picked at random was in a larger expected congregation is a form of sampling 'paradox'.

(4) Exercise For what distributions of X_i, if any, does the expected size of a randomly selected individual's group actually equal the mean size of groups?

(5) Exercise Family sizes are independent and identically distributed with mean μ. If you pick an individual at random find the probability that she is the kth born of her family, and the expectation of her order of birth in her family. Compare this with μ.

(6) Exercise Use the Cauchy–Schwarz inequality Lemma 5.3.13 to prove Cauchy's inequality (3).

5.14 Example: Propagation

A plant sheds N seeds, where N is a binomial random variable with parameters n and p. Each seed germinates with probability γ independently of all the others. Let S denote the number of resulting seedlings. Find:

(a) the conditional mass function of S given N;
(b) the joint mass function of S and N;
(c) the probability mass function of S;
(d) the conditional mass function of N given S.

Solution (a) Given that there are i seeds, that is $N = i$, the germination of any one can be regarded as a Bernoulli trial with $\mathbf{P}(\text{success}) = \gamma$. Then by Example 5.4.1 (also discussed in Example 4.2.3), the total number of successes is a binomial random variable with parameters i and γ. So

$$\mathbf{P}(S = j|N = i) = \binom{i}{j}\gamma^j(1-\gamma)^{i-j}; \quad 0 \leq j \leq i.$$

(b) Now for the joint mass function

$$\mathbf{P}(N = i \cap S = j) = \mathbf{P}(S = j|N = i)\mathbf{P}(N = i) = \binom{i}{j}\gamma^j(1-\gamma)^{i-j}\binom{n}{i}p^i(1-p)^{n-i}.$$

(c) Now we require the marginal mass function of S, which is given by

$$\begin{aligned}
\mathbf{P}(S = j) &= \sum_{i=j}^{n}\mathbf{P}(N = i, S = j) \quad \text{using (5.1.6)}\\
&= \sum_{i=j}^{n}\frac{\gamma^j(1-\gamma)^{i-j}p^i(1-p)^{n-i}}{(n-i)!j!(i-j)!}\\
&= (1-p)^n\left(\frac{\gamma}{1-\gamma}\right)^j\binom{n}{j}\sum_{i=j}^{n}\left(\frac{(1-\gamma)p}{1-p}\right)^i\frac{(n-j)!}{(n-i)!(i-j)!}\\
&= \binom{n}{j}(\gamma p)^j(1-\gamma p)^{n-j}.
\end{aligned}$$

Thus S is binomial with parameters n and γp.

(d) Finally

$$\mathbf{P}(N = i|S = j) = \mathbf{P}\frac{(N = i \cap S = j)}{\mathbf{P}(S = j)} = \binom{n-j}{n-i}\left(\frac{1-p}{1-\gamma p}\right)^{n-i}\left(1 - \frac{1-p}{1-\gamma p}\right)^{i-j}; \quad j \leq i \leq n.$$

Thus the variable $N - S$ given that $S = j$ germinate is binomial with parameters $n - j$ and $(p - p\gamma)/(1 - \gamma p)$.

(1) Exercise Find $\mathbf{E}(N|S)$.
(2) Exercise Find $\mathbf{E}(S|N)$.
(3) Exercise Find $\operatorname{cov}(N, S)$ and $\rho(N, S)$.
(4) Exercise Each seedling independently succeeds in growing into a tree with probability τ, or succumbs to wilt with probability $1 - \tau$. Let T be the number of resulting trees.

Find the joint probability mass function of N, S and T, and also the conditional mass function of N given T.

(5) **Exercise** Find the joint mass function of N and T given that $S = s$.

(6) **Exercise** Find the conditional covariance of N and T given that $S = s$.

5.15 Example: Information and entropy

(a) Let the random variable X take a finite number of values $(x_i;\ 1 \leqslant i \leqslant n)$, with mass function $\mathbf{P}(X = x) = f(x)$. Suppose that $(a_i; 1 \leqslant i \leqslant n)$ are such that $a_i > 0$ for $1 \leqslant i \leqslant n$ and $\sum_{i=1}^n a_i = 1$. Show that

$$-\sum_{i=1}^{n} f(i)\log a_i \geqslant -\sum_{i=1}^{n} f(i)\log f(i)$$

with equality if and only if $a_i = f(i)$ for all i.

(b) The random variables X and Y take a finite number of values and have joint mass function $f(x, y)$. Define

$$I(X, Y) = \sum_x \sum_y f(x, y)\log\left(\frac{f(x, y)}{f_X(x) f_Y(y)}\right).$$

Show that $I \geqslant 0$, with equality if and only if X and Y are independent.

Solution (a) By definition

$$\begin{aligned}\log y &= \int_1^y x^{-1}\,dx \\ &\leqslant \int_1^y dx \quad \text{with equality if } y = 1, \\ &= y - 1.\end{aligned}$$

Hence

(1) $$\log y \leqslant y - 1$$

with equality if and only if $y = 1$. Therefore

$$\begin{aligned}-\sum_i f(i)\log f(i) + \sum_i f(i)\log a_i &= \sum_i f(i)\log\left(\frac{a_i}{f(i)}\right) \\ &\leqslant \sum_i f(i)\left(\frac{a_i}{f(i)} - 1\right) \quad \text{by (1)} \\ &= 0,\end{aligned}$$

with equality if and only if $f(i) = a_i$ for all i.

(b) The positive numbers $f_X(x) f_Y(y)$ satisfy $\sum f_X(x) f_Y(y) = 1$. Therefore by part (a)

$$\sum_{x,y} f(x, y)\log f(x, y) \geqslant \sum_{x,y} f(x, y)\log\left(f_X(x) f_Y(y)\right)$$

with equality if and only if for all x and y $f(x, y) = f_X(x) f_Y(y)$. But this is a necessary and sufficient condition for the independence of X and Y.

(2) Exercise Show that $I = \mathbf{E}(\log f(X, Y)) - \mathbf{E}(\log f_X(X)) - \mathbf{E}(\log f_Y(Y))$.

(3) Exercise Show that if the conditional mass function of X given that $Y = y$ is $f(x|y)$, we have

$$I = \sum_{x,y} f_Y(y) f(x|y) \log f(x|y) - \mathbf{E}(\log f_X(X))$$
$$= \sum_{x,y} f_X(x) f(y|x) \log f(y|x) - \mathbf{E}(\log f_Y(Y)).$$

(4) Exercise A die is rolled twice, yielding the respective scores X and Y. Let $Z = \max\{X, Y\}$. Find $I(X, Z)$ and $I(Z, X)$.

Remark The quantity I is sometimes said to be the information about X conveyed by Y. It is interesting that this is equal to the information about Y conveyed by X.

The quantity $H(X) = \mathbf{E}(-\log f_X(X))$ is known as the entropy or uncertainty of X, and $H(X|Y) = H(X) - I(X, Y)$ is known as the conditional entropy (or uncertainty) of X given Y. It is interpreted as the uncertainty of X, reduced by the information conveyed about X by Y.

(5) Exercise Show that $H(X|X) = 0$.

(6) Exercise Show that if $H(X|Y) = 0 = H(Y|X)$ and $H(Y|Z) = H(Z|Y) = 0$, then $H(X|Z) = H(Z|X) = 0$.

5.16 Example: Cooperation

Achilles and his two friends, Briseis and Chryseis, play a cooperative game. They possess a die with n faces, and each of them rolls it once. Then $I(AC)$ is the indicator of the event that Achilles and Chryseis each turn up the same face of the die. $I(AB)$ and $I(BC)$ are defined similarly. Show that $I(AB)$, $I(AC)$ and $I(BC)$ are pairwise independent if and only if the die is unbiased.

Solution Let the die, when rolled, show its kth face with probability $f(k)$. Then

$$\mathbf{P}(I(AB) = 1, I(BC) = 1) = \mathbf{P}(\text{all three rolls show the same face}) = \sum_{k=1}^{n} (f(k))^3$$

and

$$\mathbf{P}(I(AB) = 1) = \mathbf{P}(\text{two rolls show the same face}) = \sum_{k=1}^{n} (f(k))^2.$$

Pairwise independence then requires that

(1) $$\left(\sum_k (f(k))^2\right)^2 = \sum_k (f(k))^3.$$

Now let X be a random variable which takes the value $f(k)$ with probability $f(k)$. Then (1) states that

$$0 = \mathbf{E}(X^2) - (\mathbf{E}(X))^2 = \mathbf{E}(X - \mathbf{E}(X))^2 = \operatorname{var}(X).$$

Hence X must be constant by Example 4.6.10, and so the die is unbiased because $f(k) = \frac{1}{n}$; $1 \leqslant k \leqslant n$. In this case it is easy to check that

$$\mathbf{P}(I(AB) = 1, I(BC) = 0) = \sum_{k=1}^{n} \frac{1}{n} . \frac{n-1}{n} . \frac{1}{n} = \frac{1}{n} . \frac{n-1}{n}$$
$$= \mathbf{P}(I(AB) = 1)\mathbf{P}(I(BC) = 0)$$

and two other conditions are satisfied. The indicators are pairwise independent only in this case.

(2) **Exercise** Are the indicators independent?

(3) **Exercise** Find the mean and variance of $Z = I(AB) + I(BC) + I(AC)$.

(4) **Exercise** If n women play a similar game, and $I(A_i, A_j)$ is the indicator of the event that the ith and jth women turn up the same face of the die, find the mean and variance of $\sum_{i \neq j} I(A_i A_j)$.

5.17 Example: Strange but true

Let $(S_n; n \geqslant 0)$ be a simple symmetric random walk with $S_0 = 0$. Let $f_0(n) = \mathbf{P}(T_0 = n)$ be the probability that the walk first returns to zero at the nth step, and let $u(n) = \mathbf{P}(S_n = 0)$. Let V_n be the number of values which a walk of n steps has visited exactly once.

(a) Show that

(1) $$f_0(2k) = u(2k-2) - u(2k).$$

(b) Deduce that

(2) $$\mathbf{P}\left(\prod_{k=1}^{2n} S_k \neq 0\right) = \mathbf{P}(T_0 > 2n) = u(2n).$$

(c) Hence show that for all $n \geqslant 1$

(3) $$\mathbf{E}(V_n) = 2.$$

Solution (a) By symmetry and the reflection principle, using Theorem 5.6.17

$$f_0(2k) = \frac{1}{2k-1}\mathbf{P}(S_{2k-1} = 1) = \frac{2^{-(2k-1)}}{2k-1}\binom{2k-1}{k}$$

$$= \frac{2^{-2k}}{2k-1}\binom{2k}{k} = 2^{-2k+2}\binom{2k-2}{k-1} - 2^{-2k}\binom{2k}{k} = u(2k-2) - u(2k).$$

(b)

$$\mathbf{P}(T_0 > 2n) = 1 - \sum_{k=1}^{n} f_0(2k) = 1 - \sum_{k=1}^{n}(u(2k-2) - u(2k)) \quad \text{by (1)},$$

$$= u(2n).$$

(c) Clearly $V_1 = 2$, so it suffices to show that $\mathbf{E}(V_n) = \mathbf{E}(V_{n-1})$ for $n \geqslant 2$. Let V'_n be the number of points visited just once by $S_1, S_2, \ldots, S_n$. This has the same distribution as V_{n-1}, and so also the same expectation. Let T_0 be the time of first return to the origin.

Now

(4)
$$V_n = \begin{cases} V'_n + 1 & \text{if } T_0 > n \\ V'_n - 1 & \text{if } S_1, \ldots, S_n \text{ revisits zero exactly once} \\ V'_n & \text{otherwise.} \end{cases}$$

Hence

(5)
$$\begin{aligned} \mathbf{E}(V_n) - \mathbf{E}(V_{n-1}) &= \mathbf{E}(V_n) - \mathbf{E}(V'_n) \\ &= \mathbf{P}(T_0 > n) - \mathbf{P}(S_1, \ldots, S_n \text{ revisits 0 exactly once}) \quad \text{by (4)}, \\ &= \mathbf{P}(T_0 > n) - \sum_{k=1}^{[\frac{n}{2}]} \mathbf{P}(T_0 = 2k)\mathbf{P}\left(\prod_{i=2k+1}^{n} S_i \neq 0\right) \\ &= \mathbf{P}(T_0 > n) - \sum_{k=1}^{[\frac{n}{2}]} \mathbf{P}(T_0 = 2k)\mathbf{P}(S_{n-2k} = 0) \quad \text{by (2)} \\ &= \mathbf{P}(T_0 > n) - \mathbf{P}(S_n = 0) \\ &= 0 \quad \text{by (2)}. \end{aligned}$$

(6) Exercise Show that $2kf_0(2k) = u(2k-2)$.

(7) Exercise Show that $\mathbf{P}(S_{2n} = 0) = (1/2n)\mathbf{E}(|S_{2n}|)$.

(8) Exercise Let L_{2n} be the time of the last visit to 0 up to time $2n$. Show that $\mathbf{P}(L_{2n} = 2k) = u(2k)u(2n-2k)$.

Show that if k and n increase in such a way that $k/n = x$ then

$$\mathbf{P}\left(\frac{L_{2n}}{2n} \leqslant x\right) \to \frac{2}{\pi}\sin^{-1}x; \quad 0 \leqslant x \leqslant 1$$
$$= \frac{2}{\pi}\arcsin x.$$

[Stirling's formula says that $n! \simeq e^{-n}n^{n+\frac{1}{2}}(2\pi)^{\frac{1}{2}}$ for large n.] This is an *arc-sine law*.

5.18 Example: Capture–recapture

A population of b animals has had a number a of its members captured, marked and released.

(a) Let Y_m be the number of animals that it is necessary to capture (without re-release) in order to obtain m which have been marked. Find $\mathbf{P}(Y_m = n)$ and $\mathbf{E}(Y_m)$.

(b) If, instead, it had been decided just to capture $[\mathbf{E}(Y_m)]$ animals, what would have been the expected number of marked animals among them? Compare this with m.

Solution (a) *Method* I For the event $\{Y_m = n\}$ to occur it is necessary that:

(i) the nth animal is marked, which can occur in a ways, and

(ii) the preceding $n-1$ animals include exactly $m-1$ marked and $n-m$ unmarked animals, which may occur in $\binom{a-1}{m-1}\binom{b-a}{n-m}$ ways.

The total number of ways of first selecting a distinct animal to fill the nth place, and then choosing $n-1$ animals to fill the remaining $n-1$ places is $b.\binom{b-1}{n-1}$. Since these are assumed to be equally likely, the required probability is

(1) $$\mathbf{P}(Y_m = n) = \frac{a}{b}\binom{a-1}{m-1}\binom{b-a}{n-m}\bigg/\binom{b-1}{n-1}; \quad m \leqslant n \leqslant b-a+m.$$

To calculate $\mathbf{E}(Y_m)$ you may write

$$\mathbf{E}(Y_m) = \sum_{n=m}^{b-a+m} n\mathbf{P}(Y_m = n) = \sum_{n=m}^{b-a+m} m\binom{a}{m}\binom{b-a}{n-m}\bigg/\binom{b}{n}$$

$$= m\left(\frac{b+1}{a+1}\right)\sum_{n+1=m+1}^{b+1-(a+1)+m} \frac{a+1}{b+1}\binom{a}{m}\binom{b+1-(a+1)}{n+1-(m+1)}\bigg/\binom{b}{n}$$

$$= m\left(\frac{b+1}{a+1}\right)\sum_{n'=m'}^{b'-a'+m'} \frac{a'}{b'}\binom{a'-1}{m'-1}\binom{b'-a'}{n'-m'}\bigg/\binom{b'-1}{n'-1}$$

where $a' = a+1$ and so on,

$$= m\left(\frac{b+1}{a+1}\right),$$

because (1) is a probability distribution with sum equal to unity.

Method II Alternatively, suppose that you were to capture them all, and let X_0 be the number of unmarked animals captured before the first marked animal, X_r the number of unmarked animals captured between the rth and the $(r+1)$st marked animals and X_a the number captured after the last marked animal. Then

$$\sum_0^a X_i = b-a,$$

and by symmetry, for all i and j, $\mathbf{E}(X_i) = \mathbf{E}(X_j)$. Hence

$$\mathbf{E}(X_r) = \frac{b-a}{a+1}, \quad \text{and} \quad \mathbf{E}(Y_m) = \sum_0^{m-1} \mathbf{E}(X_r) + m = m\frac{b+1}{a+1}.$$

(b) It is possible to write down the distribution of the number of marked animals captured, and then evaluate the mean by a method similar to the first method of (a). It is easier to let I_j be the indicator of the event that the jth captured animal is marked. Then the required expectation is

(2) $$\mathbf{E}\left(\sum_1^{[\mathbf{E}(Y_m)]} I_j\right) = [\mathbf{E}(Y_m)]\mathbf{E}(I_j) = [\mathbf{E}(Y_m)]\frac{a}{b} = \left[\frac{b+1}{a+1}.m\right]\frac{a}{b} < m.$$

Remark The distribution of Y_m is called the negative hypergeometric distribution, by analogy with the relation between the negative binomial distribution and the binomial distribution. The hypergeometric p.m.f. is (3.15.1).

(3) Exercise If you capture and keep a fixed number n of animals, find the variance of the number which are marked.

(4) Exercise Your pen will only hold m animals, so you return the unmarked ones. Now if Z_m is the number of captures required to secure m marked animals, find $\mathbf{E}(Z_m)$.

(5) Exercise Let X and Y be independent binomial random variables with the same parameters n and p. Find $\mathbf{P}(X = k | X + Y = j)$ and explain why the answer takes the form you find.

5.19 Example: Visits of a random walk

Let $(S_n; n \geqslant 0)$ be a simple symmetric random walk with $S_0 = 0$.

(a) Let V_r be the number of visits to r before the walk revisits the origin. Show that $\mathbf{E}(V_r) = 1$.

(b) Show that the expected number of visits to the origin is infinite.

Solution Let I_n be the indicator of a visit to the point r at the nth step before any return to 0. Then

(1) $$\mathbf{E}(V_r) = \mathbf{E}\left(\sum_{n=1}^{\infty} I_n\right) = \sum_{n=1}^{\infty} \mathbf{E}(I_n)$$

$$= \sum_{n=1}^{\infty} \mathbf{P}(S_n = r, S_1 \neq 0, \ldots, S_{n-1} \neq 0).$$

Now we make two important observations:

(2) if $S_n = r$, then $X_1 + \ldots + X_k \neq 0$ if and only if $X_{k+1} + \ldots + X_n \neq r$; and

(3) since the X_i are independent and identically distributed, $X_{k+1} + \ldots + X_n$ has the same distribution as $X_1 + \ldots + X_{n-k}$, and this remains true when $S_n = r$.

Hence we can write (1) as

(4) $$\mathbf{E}(V_r) = \sum_{n=1}^{\infty} \mathbf{P}\left(\sum_{i=1}^{n} X_i = r, X_1 \neq 0, \ldots, X_1 + \ldots + X_{n-1} \neq 0\right)$$

$$= \sum_{n=1}^{\infty} \mathbf{P}\left(\sum_{i=1}^{n} X_i = r, X_2 + \ldots + X_n \neq r, \ldots, X_n \neq r\right) \quad \text{by (2)},$$

$$= \sum_{n=1}^{\infty} \mathbf{P}\left(\sum_{i=1}^{n} X_i = r, X_1 + \ldots + X_{n-1} \neq r, \ldots, X_1 \neq r\right) \quad \text{by (3)},$$

$$= \sum_{n=1}^{\infty} f_r(n) = 1 \quad \text{, using Theorem 5.6.7.}$$

(b) Let J_n be the indicator of a visit to the origin at the nth step, and let R be the total number of returns to the origin. Then

$$\mathbf{E}(R) = \mathbf{E}\left(\sum_{n=1}^{\infty} J_n\right) = \sum_{n=1}^{\infty} \mathbf{E}(J_n) = \sum_{n=1}^{\infty} \mathbf{P}(S_n = 0)$$

$$= \sum_{k=1}^{\infty} \frac{1}{2^{2k}} \binom{2k}{k} \quad \text{where } n = 2k,$$

$$= \sum_{k=1}^{\infty} \frac{(2k-1)(2k-3)\dots 3.1}{2^k k(k-1)\dots 2.1}$$

$$\geqslant \sum_{k=1}^{\infty} (2k-1).\frac{(2k-2)}{(2k-1)}.(2k-3).\frac{(2k-4)}{(2k-3)}\dots 3.\frac{2}{3}.\frac{1}{2^k k!}$$

$$= \sum_{k=1}^{\infty} \frac{1}{2k} = \infty.$$

Remark Result (a) is indeed remarkable. Interpreted as a game, it says that if a coin is tossed repeatedly and you get \$1 every time the total number of heads is r more than the total number of tails, until heads and tails are equal, then your expected gain is \$1, independently of r.

(5) Exercise For a symmetric simple random walk with $S_0 = 0$, let R_r be the total number of returns to r. What is $\mathbf{E}(R_r)$?

(6) Exercise For a symmetric simple random walk with $S_0 = 0$, show that the probability that the first visit to S_{2n} takes place at time $2k$ is $\mathbf{P}(S_{2k} = 0)\mathbf{P}(S_{2n-2k} = 0)$; $0 \leqslant k \leqslant n$.

(7) Exercise What is $\mathbf{E}(V)$, the expected number of visits of an asymmetric simple random walk to r?

(8) Exercise Consider a two-dimensional symmetric random walk (S_X, S_Y) on the points (i, j) where i and j are integers. From (i, j) the walk steps to any one of $(i \pm 1, j)$ or $(i, j \pm 1)$ with equal probability $\frac{1}{4}$. Show that the expected number $\mathbf{E}(V)$ of visits to the origin is infinite.

5.20 Example: Ordering

Let X and Y be random variables such that for all x

(1) $$F_X(x) \leqslant F_Y(x).$$

Show that $\mathbf{E}(X) \geqslant \mathbf{E}(Y)$, and deduce that $F_X(x) \leqslant F_Y(x)$ if and only if, for all increasing functions $h(.)$,

(2) $$\mathbf{E}(h(X)) \geqslant \mathbf{E}(h(Y)).$$

Solution From Example 4.3.3 we have

(3) $$\mathbf{E}(X) = \sum_0^{\infty} \mathbf{P}(X > k) - \sum_{k=0}^{-\infty} \mathbf{P}(X < k) = \sum_0^{\infty}(1 - F_X(k)) - \sum_0^{-\infty} F_X(k)$$

$$\geqslant \sum_0^{\infty}(1 - F_Y(k)) - \sum_0^{-\infty} F_Y(k) \qquad \text{by (1)}$$

$$= \mathbf{E}(Y).$$

Now if $h(.)$ is an increasing function

$$\mathbf{P}(h(X) > z) = \mathbf{P}(X > \inf\{t: h(t) > z\}) \geqslant \mathbf{P}(Y > \inf\{t: h(t) > z\}) \quad \text{by (1)}$$

$$= \mathbf{P}(h(Y) > z).$$

Hence $\mathbf{P}(h(X) \leqslant z) \leqslant \mathbf{P}(h(Y) \leqslant z)$ and (2) follows on using (3).

Conversely, if we choose $h(Z)$ to be the indicator of the event that $Z \leq x$, then

(4) $$\mathbf{E}(h(X)) = \mathbf{P}(X \leq x) \leq \mathbf{P}(Y \leq x) = \mathbf{E}(h(Y)).$$

(5) Exercise If X and Y are independent and for all x $F_X(x) \leq F_Y(x)$, show that $\mathbf{P}(X \geq Y) \geq \frac{1}{2}$.

(6) Exercise If X, Y and Z are independent show that X, Y and Z can be distributed in such a way that $\mathbf{P}(X > Y) > \frac{1}{2}$; $\mathbf{P}(Y > Z) > \frac{1}{2}$; $\mathbf{P}(Z > X) > \frac{1}{2}$.

(7) Exercise Let $X(n, p)$ have binomial distribution with parameters n and p. Show that

$$\mathbf{P}(X(m, p) \leq x) \geq \mathbf{P}(X(n, p) \leq x) \quad \text{for } m \leq n$$

and

$$\mathbf{P}(X(n, p_1) \leq x) \geq \mathbf{P}(X(n, p_2) \leq x) \quad \text{for } p_1 \leq p_2.$$

PROBLEMS

1 You roll two fair dice. Let X be the number of 2's shown, and Y the number of 4's. Write down the joint probability mass function of X and Y, and find $\operatorname{cov}(X, Y)$ and $\rho(X, Y)$.

2 Let the random variables X and Y have joint probability mass function $f(x, y)$ such that:

$$f(1, 2) = \frac{1}{8}, \quad f(1, 3) = \frac{1}{16}, \quad f(1, 4) = \frac{1}{4},$$
$$f(2, 2) = \frac{1}{16}, \quad f(2, 3) = \frac{1}{8}, \quad f(2, 4) = \frac{3}{8}.$$

Find the probability of the following:

(a) $X > Y$ (c) $X + Y$ is odd
(b) $X \geq Y$ (d) $X - Y \leq 1$.

3 Find two random variables X and Y that are uncorrelated, but not independent.

4 Show that if $\mathbf{E}((X - Y)^2) = 0$, then $X = Y$ with probability one.

5 Show that if $\mathbf{E}((X - Y)^2) = \mathbf{E}(X^2) + \mathbf{E}(Y^2)$ then X and Y are orthogonal.

6 Let X be uniformly distributed on $\{0, 1, 2, \ldots, n\}$. Let $Y = \sin(\frac{1}{2}\pi X)$ and $Z = \cos(\frac{1}{2}\pi X)$.
(a) What is the joint probability mass function of Y and Z?
(b) What is the distribution of $Y + Z$?
Show that Y and Z are orthogonal.

7 Let X and Y be jointly distributed with finite second moments and unit variance. Show that for some non-zero constants a, b, c, d, the random variables U and V are uncorrelated where $U = aX + bY$, $V = cX + dY$. Are a, b, c and d unique?

8 A source produces a message forming a sequence of zeros and ones. In being transmitted it passes through two independent channels, each of which transmits the wrong symbol with probability $1 - p$, or the correct symbol with probability p. Show that a symbol is least likely to be transmitted correctly when $p = \frac{1}{2}$.

Find the probability of correct transmission of a symbol when the message passes through three similar independent channels.

9 Let $(X_n; n \geq 1)$ be a sequence of independent random variables such that $\mathbf{P}(X_n = 1) = p = 1 - q = 1 - \mathbf{P}(X_n = -1)$. Let U be the number of terms in the sequence before the first change of sign, and V the further number of terms before the second change of sign. [In other words, $X_1, X_2, \ldots$ is made up of runs of $+1$'s and runs of -1's; U is the length of the first run and V the length of the second.]
(a) Show that $\mathbf{E}(U) = pq^{-1} + qp^{-1}$ and $\mathbf{E}(V) = 2$.
(b) Write down the joint distribution of U and V, and find $\operatorname{cov}(U, V)$ and $\rho(U, V)$.

10 An urn contains n balls numbered individually with the integers from 1 to n. Two balls are drawn at random without replacement, and the numbers they bear are denoted by X and Y. Find $\operatorname{cov}(X, Y)$, $\rho(X, Y)$, and the limit of $\rho(X, Y)$ as $n \to \infty$.

11 Let X and Y have joint distribution defined by $f(0,0) = 1 - 3a$; and $f(0,1) = f(1,0) = f(1,1) = a$; $a \leqslant \frac{1}{3}$. Find:
(a) the p.m.f.s of X and Y
(b) $\operatorname{cov}(X, Y)$
(c) $\mathbf{E}(X|Y)$ and $\mathbf{E}(Y|X)$
(d) whether X and Y can be independent, and if so, when.

12 You roll two fair dice and they show X and Y respectively. Let $U = \min\{X, Y\}$, $V = \max\{X, Y\}$. Write down the joint distributions of:
(a) $\{U, X\}$
(b) $\{U, V\}$
(c) $\{X, Y, V\}$.
Find $\operatorname{cov}(U, V)$ and $\mathbf{E}(XYV)$.

13 (a) If X and Y are independent with finite expectation, show that $\mathbf{E}(XY)$ exists.
(b) Find a sufficient condition on the moments of X and Y, for $\mathbf{E}(XY)$ to exist in general.

14 Which of the following functions $f(i, j)$ can be a joint probability mass function of two random variables X and Y?
(a) $\theta^{|i|+|j|}$; $\quad |i| + |j| \neq 0$
(b) θ^{i+j}; $\quad 0 \leqslant i \leqslant j < \infty$
(c) θ^{i+j+2}; $\quad 0 \leqslant i < j < \infty$
(d) θ^{i+j+1}; $\quad 0 \leqslant i, j < \infty$
(e) $(i^j - (i-1)^j)\alpha\left(\dfrac{\beta}{c}\right)^j$; $\quad 1 \leqslant i \leqslant c,\ j \geqslant 1,\ $ c an integer
(f) $\alpha(i^n - (i-1)^n)j^{-n-2}$; $\quad 1 \leqslant i \leqslant j < \infty$.
Are X and Y independent in any case?

15 For each function in problem 11 that is a joint probability mass function of X and Y, find the marginal mass functions of X and Y.

16 Suppose that random variables X and Y are such that $\mathbf{P}(|X - Y| \leqslant M) = 1$, where M is finite. Show that if $\mathbf{E}(X) < \infty$, then $\mathbf{E}(Y) < \infty$ and $|\mathbf{E}(X) - \mathbf{E}(Y)| \leqslant M$.

17 Show that the following are joint p.m.f.s and find the marginal distributions.

(a)
$$f(x_1, \ldots, x_k) = \frac{x!}{x_1! \ldots x_k!} p_1^{x_1} \ldots p_k^{x_k};$$

where $\sum_1^k p_j = 1$ and $\sum_1^k x_j = x$.

(b)
$$f(x, x_1, \ldots, x_k) = \frac{(x + r - 1)!}{(r-1)!} p_0^r \frac{p_1^{x_1}}{x_1!} \ldots \frac{p_k^{x_k}}{x_k!},$$

where $\sum_0^k p_j = 1$ and $\sum_1^k x_j = x \geqslant 0$.

(c)
$$f(x_1, \ldots, x_k) = \binom{a_1}{x_1} \ldots \binom{a_k}{x_k} \Bigg/ \binom{\sum_1^k a_i}{\sum_1^k x_i}.$$

18 Let X and Y be independent geometric random variables with parameters p_1 and p_2 respectively.
(a) If c is an integer and $Z = \min\{c, X\}$ find $\mathbf{E}(Z)$.
(b) Find the distribution and expectation of $\min\{X, Y\}$.

19 Let X and Y have joint p.m.f.

$$f(x, y) = \frac{C}{(x+y-1)(x+y)(x+y+1)}; \quad m \geqslant 1, n \geqslant 1.$$

Find the p.m.f. of X, the p.m.f. of Y, and the value of C.

20 Let X and Y be independent random variables each with a geometric distribution, so

$$f_X(x) = \alpha\beta^{x-1}; \quad x \geqslant 1, \alpha + \beta = 1,$$

and

$$f_Y(y) = pq^{y-1}; \quad y \geqslant 1, p + q = 1.$$

Let $R = X/Y$.

(a) Find $\mathbf{P}(R > 1)$.

(b) If $r = m/n$ where m and n are integers with no common factor except unity, find $\mathbf{P}(R = r)$, and show that when $\alpha = p = \frac{1}{2}$, $\mathbf{P}(R = r) = 1/(2^{m+n} - 1)$.

21 **Triangular distribution** Let X and Y be independent random variables each uniformly distributed on $\{0, 1, \ldots, n\}$. Find the p.m.f. of

(a) $X + Y$

(b) $X - Y$.

22 Let X have binomial distribution with parameters n and p, and Y a binomial distribution with parameters m and p. Show that if X and Y are independent then $X + Y$ has a binomial distribution with parameters $m + n$ and p. Deduce that

$$\sum_{k=0}^{r} \binom{m}{r-k}\binom{n}{k} = \binom{m+n}{r}.$$

Find the conditional distribution of X given that $X + Y = k$.

23 If X has a binomial distribution with parameters n and p, show that

$$\mathbf{E}\left(\frac{1}{1+X}\right) = \frac{1-(1-p)^{n+1}}{(n+1)p}.$$

24 Let X and Y be independent Poisson random variables. Show that $Z = X + Y$ has a Poisson distribution.

Show also that for some p, $\mathbf{P}(X = k|z = n) = \binom{n}{k}p^k(1-p)^{n-k}$, which is to say that the conditional distribution of X given Z is binomial.

25 Let X and Y be independent geometric random variables such that for $m \geqslant 0$

$$\mathbf{P}(X = m) = (1-\lambda)\lambda^m \quad \text{and} \quad \mathbf{P}(Y = m) = (1-\mu)\mu^m.$$

(a) Show that

$$\mathbf{P}(X + Y = n) = \frac{(1-\lambda)(1-\mu)}{\lambda - \mu}(\lambda^{n+1} - \mu^{n+1}), \quad \lambda \neq \mu.$$

Find $\mathbf{P}(X = k|X + Y = n)$.

(b) Find the distribution of $Z = X + Y$ when $\lambda = \mu$, and show that in this case $\mathbf{P}(X = k|Z = n) = 1/(n+1)$.

26 **Bell's inequality** Let X, Y and Z be jointly distributed random variables such that each can take either of the values ± 1. Show that $\mathbf{E}(XY) \leqslant 1 - |\mathbf{E}((X - Y)Z)|$.

[This inequality is interesting because it has been claimed that there are experiments in quantum mechanics for which it does not hold true.]

27 Show that for any c such that $|c| \leqslant 4$ the function

$$f(i, j) = \frac{1}{(2m+1)(2n+1)} + \frac{c(i-m)(j-n)}{((2n+1)(2m+1))^2}; \quad \begin{array}{l} 0 \leqslant i \leqslant 2m, \\ 0 \leqslant j \leqslant 2n, \end{array}$$

is a joint probability mass function with marginal distributions which do not depend on c.

Show that the covariance of this distribution is

$$\frac{cmn(m+1)(n+1)}{9(2n+1)(2m+1)}.$$

28 Construct two identically distributed random variables X and Y such that

$$\mathbf{P}(X < Y) \neq \mathbf{P}(Y < X).$$

29 **Bernoulli's urn** Initially an urn contains U umber balls and a vase contains V viridian balls. From each container a ball is removed at random and placed in the other container. Let U_r be the number of umber balls in the urn after r repetitions of this operation.

(a) Find $\mathbf{E}(U_r)$ and show that $\lim_{r\to\infty} \mathbf{E}(U_r) = U^2/(U+V)$.

(b) Just before each time balls are exchanged, a coin is tossed (which shows a head with probability p); find the expected number of umber balls in the urn when the coin first shows a head. Show that if $U = V$, and $Up = 1$, this expectation is about $\frac{2}{3}U$ for large U.

30 Let $(X_i; i \geq 1)$ be a random walk with $S_0 = 0$ and $S_r = \sum_1^r X_i$. Define the maximum $M_n = \max_{1\leq k\leq n}\{S_k\}$. Show that $\mathbf{P}(M_n \geq x) \leq (\frac{p}{q})^x$, $x \geq 0$, and deduce that

$$\lim_{n\to\infty} \mathbf{E}(M_n) \leq q/(q-p), \text{ for } p < q.$$

31 An urn contains n balls such that each of the n consecutive integers $1, 2, \ldots, n$ is carried by one ball. If k balls are removed at random find the mean and variance of the total of their number in the two cases:

(a) they are not replaced,

(b) they are replaced.

What is the distribution of the largest number removed in each case?

32 Let the random variables X and Y have joint distribution

$$\mathbf{P}(X = a, Y = 0) = \mathbf{P}(X = 0, Y = a) = \mathbf{P}(X = -a, Y = 0) = \mathbf{P}(X = 0, Y = -a) = \tfrac{1}{4}.$$

Show that $X - Y$ and $X + Y$ are independent.

33 The random variables U and V each take the values ± 1. Their joint distribution is given by

$$\mathbf{P}(U = +1) = \mathbf{P}(U = -1) = \tfrac{1}{2},$$
$$\mathbf{P}(V = +1 | U = 1) = \tfrac{1}{3} = \mathbf{P}(V = -1 | U = -1),$$
$$\mathbf{P}(V = -1 | U = 1) = \tfrac{2}{3} = \mathbf{P}(V = +1 | U = -1).$$

(a) Find the probability that $x^2 + Ux + V = 0$ has at least one real root.

(b) Find the expected value of the larger root given that there is at least one real root.

(c) Find the probability that $x^2 + (U+V)x + U + V = 0$ has at least one real root.

34 Let $S_n = \sum_1^n X_i$ be a random walk with $S_0 = a > 0$, such that $\mathbf{P}(X_i = -1) = q$, $\mathbf{P}(X_i = +2) = p$, $p + q = 1$. Let T_{a0} be the time at which the walk first visits zero. Show that if $p \leq \frac{1}{3}$ then $\mathbf{P}(T_{a0} < \infty) = 1$, but if $p > \frac{1}{3}$ then $\mathbf{P}(T_{a0} < \infty) = r^a < 1$. What is r?

35 Casualties arriving at a certain hospital require surgery, independently of one another, with probability $\frac{1}{4}$. What is the probability that, on a day when n casualties arrive, exactly r require surgery?

The number X of casualties arriving on weekdays follows a Poisson distribution with mean 8, that is, for each day, $\mathbf{P}\{X = n\} = e^{-8}8^n/n!$ $n = 0, 1, 2, \ldots$

Show that the number requiring surgery each day also follows a Poisson distribution and find its mean.

Suppose that the situation is identical on Saturdays and Sundays except that there are on average only 4 casualties arriving per day. Find the mean and variance of the number of patients requiring surgery each week. [Assume that each day's arrivals are independent, and recall Problem 24.]

36 An urn contains m white balls and $M - m$ black balls. Balls are chosen at random without replacement. Show that the probability p_k of choosing exactly k white balls in n choices $(0 \leq k \leq m)$ is given by

$$p_k = \binom{M}{n}^{-1}\binom{m}{k}\binom{M-m}{n-k}.$$

Define a random variable $X = X_1 + X_2 + \ldots + X_n$, where $X_i = 0$ or 1 according as the ith ball is black or white. Show that

$$\mathbf{P}(X = k) = p_k,$$
$$\mathbf{P}(X_i = 1) = m/M,$$
$$\mathbf{P}(X_i = 1, X_j = 1) = \frac{m(m-1)}{M(M-1)}, \quad i \neq j.$$

By considering $\mathbf{E}(X)$, $\mathbf{E}(X^2)$, or otherwise, find the mean and variance of the distribution given by p_k.

37 **Conditional gambler's ruin** An optimistic gambler seeks to know the expected duration of the game assuming that he wins. As usual he plays a sequence of fair wagers losing or gaining \$1 each time. The game stops as soon as he has \$0 or \$$K$. Initially his fortune is \$$k$ (< \$$K$), the event that he stops with \$$K$ is V_k, and D_k is the duration of the game. Let $\delta_k = \mathbf{E}(D_k|V_k)$. Show that for $1 < k < K$,

$$(k+1)\delta_{k+1} - 2k\delta_k + (k-1)\delta_{k-1} + 2k = 0.$$

Write down two boundary conditions at $k = 1$ and $k = K$, and deduce that

$$\mathbf{E}(D_k|V_k) = \tfrac{1}{3}(K^2 - k^2), \quad 1 \leq k \leq K.$$

38 Let $(S_n; n \geq 1)$ be a simple random walk, and let M be its maximum, $M = \max_{n\geq 1}\{S_n\}$.

(a) If $S_0 = 0$, and $p < q$, show that m has a geometric distribution and find its mean.

(b) If S_0 is a random variable with distribution $\mathbf{P}(S_0 = -k) = \alpha\beta^k$; $k = 0, 1, 2, \ldots$ find the distribution of M.

In this case, what is the conditional distribution of S_0 given M?

39 Let X_1, X_2 and X_3 be independent geometric random variables with parameters $1 - p_1$, $1 - p_2$ and $1 - p_3$, respectively.

(a) Show that

$$\mathbf{P}(X_1 < X_2 < X_3) = \frac{(1-p_1)(1-p_2)p_2p_3^2}{(1-p_2p_3)(1-p_1p_2p_3)}.$$

(b) Find $\mathbf{P}(X_1 \leq X_2 \leq X_3)$.

(c) Three players, A, B and C, roll a fair die in turn, that is, in the order $ABCABCA\ldots$ Show that the probability that A throws the first six, B the second six, and C the third six, is $\frac{216}{1001}$.

40 **Matching** Once again n letters with n matching envelopes are inadvertently placed at random in the envelopes. Let X be the number of letters that are in their matching envelope. Find $\mathbf{E}(X)$ and $\operatorname{var}(X)$, and show that

$$\mathbf{E}(X(X-1)\ldots(X-k+1)) = \begin{cases} 1 & k \leq n \\ 0 & k > n. \end{cases}$$

41 Let n be a prime number greater than two, and let X and Y be independently and uniformly distributed on $\{0, 1, \ldots, n-1\}$. For all r such that $0 \leq r \leq n-1$, define $Z_r = X + rY$. Show that the random variables $(Z_r; 0 \leq r \leq n-1)$ are pairwise independent.

Is this true if n is not prime?

6

Generating functions and their applications

This chapter deals with a special subject, and may be omitted on a first reading. Its contents are important and useful, but are not a prerequisite for most of the following chapters.

6.1 Introduction

In Chapter 3 we found that generating functions can provide very elegant and concise methods for handling collections of real numbers. The mass function of an integer valued random variable is such a collection, and so we may anticipate (correctly as it turns out) that the following generating function will be very useful.

(1) **Definition** The probability generating function $G(s)$ of the integer valued random variable X is defined by

$$G(s) = \sum_k \mathbf{P}(X = k)s^k. \qquad \blacktriangle$$

Since all random variables in this chapter will be integer valued, this will not again be mentioned explicitly.

(2) **Example** Let X be uniformly distributed in $\{-a, -a+1, \ldots, b-1, b\}$, where $a, b > 0$. Then provided $s \neq 1$,

$$G(s) = \sum_{k=-a}^{b} \frac{1}{a+b+1} s^k = \frac{s^{-a} - s^{b+1}}{(a+b+1)(1-s)}. \qquad \bullet$$

Notice that by Theorem 4.3.4 we have from Definition 1 of $G(s)$ that

(3) $$G(s) = \mathbf{E}(s^X);$$

this is a particularly useful representation of $G(s)$, and we shall use it a great deal in what follows.

For example, suppose we seek the probability generating function of $Y = X + a$, where a is constant. Using (3) we can write

$$G_Y(s) = \mathbf{E}(s^Y) = \mathbf{E}(s^{X+a}) = s^a G_X(s).$$

We shall see many other applications of (3) later. When X is defective (that is when

$\mathbf{P}(|X| < \infty) < 1$), the representation (3) can still be used, provided that we remember that the expectation is taken over the finite part of the distribution of X.

We write $G_X(s)$ when we wish to stress the role of X; and for brevity $G_X(s)$ is sometimes known as the p.g.f. of X.

Obviously, if $\mathbf{P}(|X| < \infty) = 1$, then

$$G_X(1) = \sum_k \mathbf{P}(X = k) = 1.$$

To sum up, if X is finite with probability 1, then $G_X(s)$ is a power series in s with non-negative coefficients such that $G_X(1) = 1$.

Conversely, if $G(s)$ is a power series with non-negative coefficients such that $G(1) = 1$, then G is the p.g.f. of some integer valued random variable X, which is finite with probability 1.

(4) **Example** Let $G(s) = (a + bs)/(1 - cs)$. When is G the p.g.f. of a finite integer valued random variable X?

Solution First we note that if X is finite then $G(1) = 1$, and so $a + b + c = 1$. Now we need to consider various cases.

(i) If $0 \leqslant c < 1$, then we can write, for any n

$$G(s) = (a + bs)(1 + cs + \ldots + (cs)^n) + \frac{a + bs}{1 - cs}(cs)^{n+1}$$
$$= a + (b + ac)s + (b + ac)cs^2 + \ldots + (b + ac)c^{n-1}s^n + bc^n s^{n+1}$$
$$+ \frac{a + bs}{1 - cs}(cs)^{n+1}.$$

For $|s| < c^{-1}$ we can let $n \to \infty$ to obtain a series expansion of $G(s)$. This has the required properties of a p.g.f. if $1 \geqslant a \geqslant 0$, and $1 \geqslant b + ac \geqslant 0$. In this case X is a non-negative random variable.

(ii) If $c = 1$, then $a = -b = 1$. In this case X is zero with probability 1.

(iii) If $c > 1$, then we can use a method similar to that of (i) to obtain a different series expansion of $G(s)$, that is:

$$G = -\frac{b}{c} - \frac{ac + b}{c^2 s} - \frac{ac + b}{c^3 s^2} - \ldots$$

This is a p.g.f. if $-c \leqslant b \leqslant 0$, and $-c^2 \leqslant b + ac \leqslant 0$. In this case X is non-positive.

(iv) If $c < 0$, then $a = 1$ and $b = c$. In this case X is zero with probability 1.

See Example 14 for more insight into the nature of this probability generating function. ●

Another useful theorem is 3.6.7 which we re-state here.

(5) Theorem Let X be a random variable with mass function $f(k)$, and suppose that $a \leqslant X \leqslant b$. Let

$$t_n = \mathbf{P}(X > n) = \sum_{k=n+1}^{b} f(k).$$

Define the tail generating function

$T(s) = \sum_{a}^{b-1} s^n t_n$. Then whenever both sides exist

$$(1 - s)T(s) = s^a - G(s). \tag{6}$$

In particular, if $X \geqslant 0$, then

$$(1 - s)T(s) = 1 - G(s). \tag{7}$$

Proof The left hand side of (7) may be written as

$$
\begin{aligned}
(1 - s)\sum_{n=a}^{b-1} \mathbf{P}(X > n)s^n
&= \sum_{n=a}^{b-1} s^n \mathbf{P}(X > n) - \sum_{n=a}^{b-1} s^{n+1}\mathbf{P}(X > n) \\
&= \sum_{n=a+1}^{b-1} s^n(\mathbf{P}(X > n) - \mathbf{P}(X > n - 1)) + s^a\mathbf{P}(X > a) - s^b\mathbf{P}(X > b - 1) \\
&= s^a - \sum_{a}^{b} \mathbf{P}(X = n)s^n = s^a - G(s),
\end{aligned}
$$

as required. ■

(8) Example 2 revisited Here X is uniform on $\{-a, \ldots, b\}$, and so

$$
\begin{aligned}
T_X(s) &= \frac{s^{-a}}{1 - s} - \frac{(s^{-a} - s^{b+1})}{(a + b + 1)(1 - s)^2} \\
&= \frac{(1 - s)(a + b)s^{-a} + s^{b+1} - s^{-a+1}}{(a + b + 1)(1 - s)^2}.
\end{aligned}
$$

●

More generally, we can show that the identity (7) holds for unbounded non-negative random variables. One way of doing this is to observe that the coefficients of s^n on each side are equal for all n, and then use a standard theorem about power series.

(9) Example Let X be geometric with mass function $f(k) = (1 - q)q^{k-1}$; $k \geqslant 1$, $0 < q < 1$. Then

$$G(s) = \sum_{k=1}^{\infty} (1 - q)q^{k-1}s^k = \frac{(1 - q)s}{1 - qs},$$

and

$$(1-s)T(s) = 1 - \frac{(1-q)s}{1-qs} = \frac{1-s}{1-qs}.$$

Thus

$$T(s) = \frac{1}{1-qs}. \qquad \bullet$$

For future reference we record the following trivial corollary of (7), that is, if $\mathbf{P}(0 \leq X < \infty) = 1$, then

(10)
$$\sum_{j=0}^{\infty} s^j \mathbf{P}(X \leq j) = \frac{G_X(s)}{1-s}.$$

It is very useful to bear in mind that conditional probability mass functions also have generating functions. Thus if A is some event, we can write $\mathbf{P}(X = k|A) = f(k|A)$, and define the generating function

$$G_{X|A}(s) = \sum_k f(k|A)s^k = \mathbf{E}(s^X|A),$$

in the usual notation.

If $(A_i;\ i \geq 1)$ is a collection of disjoint events with $\bigcup_i A_i = \Omega$, then it is easy to show that

(11)
$$\mathbf{E}(s^X) = \sum_i \mathbf{E}(s^X|A_i)\mathbf{P}(A_i).$$

This result is often useful in finding $\mathbf{E}(s^X)$.

If the random variables X and Y are jointly distributed, then in like manner we have

(12)
$$\mathbf{E}(s^X|Y = y) = \sum_k s^k \mathbf{P}(X = k|Y = y) = \sum_k s^k \frac{f(k, y)}{f_Y(y)}; \quad f_Y(y) > 0.$$

As y runs over all the possible values of Y, this yields the conditional p.g.f. of X given Y

(13)
$$G_{X|Y}(s) = \mathbf{E}(s^X|Y).$$

We therefore have the useful result:

$$G_X(s) = \mathbf{E}(G_{X|Y}(s)) = \mathbf{E}(\mathbf{E}(s^X|Y)).$$

(14) Example 4 revisited Suppose we have two biased coins; the first shows a head with probability a, and the second shows a head with probability $1-c$. The first coin is tossed and, if it shows a tail then the second coin is tossed repeatedly until a head is shown. Let X be the number of times the second coin is tossed. What is $G_X(s)$?

Solution Let H be the event that the first coin shows a head. If H occurs then $X = 0$, so $\mathbf{E}(s^X|H) = 1$. If H^c occurs then X is geometric with $f_X(k) = (1-c)c^{k-1}$;

$k \geqslant 1$. Hence by Example 9

$$\mathbf{E}(s^X | H^c) = \frac{(1-c)s}{1-cs}.$$

Therefore by (11)

$$\mathbf{E}(s^X) = a + \frac{(1-a)(1-c)s}{1-cs} = \frac{a+(1-a-c)s}{1-cs}.$$

Looking back, we see that this is the generating function considered in Example 4, case (i). It follows that we can think of $(a + bs)/(1 - cs)$ as being the p.g.f. of a random variable X which is either zero with probability a, or with probability $1 - a$ is a geometric random variable with parameter c. Such random variables arise quite naturally in applications. ●

(15) Example A biased coin is tossed repeatedly until the first occasion when r consecutive heads have resulted. Let X be the number of tosses required. Find $\mathbf{E}(s^X)$.

Solution We suppose that the chance of a head is p, and note that if the first i tosses are $i - 1$ heads followed by a tail, then the further number of tosses required has the same mass function as X. Hence, with an obvious notation:

$$\mathbf{E}(s^X | H^{i-1}T) = s^i \mathbf{E}(s^X); \quad 1 \leqslant i \leqslant r.$$

Also

$$\mathbf{E}(s^X | H^r) = s^r.$$

It follows that

$$\mathbf{E}(s^X) = \sum_{i=1}^{r} qp^{i-1} s^i \mathbf{E}(s^X) + p^r s^r$$

and so

$$\mathbf{E}(s^X)\Big(1 - qs \sum_{i=0}^{r-1} (ps)^i\Big) = p^r s^r.$$

Hence

$$\mathbf{E}(s^X) = \frac{p^r s^r (1-ps)}{1 - s + qp^r s^{r+1}}.$$

We shall discover different methods for proving this later on. ●

6.2 Moments and the probability generating function

For the remainder of this chapter, random variables will be assumed to be non-negative unless stated otherwise. In this case, whenever $|s| \leqslant 1$,

$$|G_X(s)| = \left| \sum_0^\infty f(k) s^k \right| \leqslant \sum_0^\infty f(k) |s^k| \leqslant \sum_0^\infty f(k) = 1.$$

This simple property has enormous consequences for the p.g.f. $G(s)$. These are fully explored in textbooks on calculus and analysis, so we merely state the most important relevant results here. Firstly we state without proof:

(1) Theorem The function $G(s)$ is differentiable for $|s| < 1$ and its derivative is

$$G'(s) = \sum_{n=1}^{\infty} nf(n)s^{n-1} < \infty.$$

At $s = 1$

(2) $$G'(1) = \lim_{s\uparrow 1} \sum_{n=1}^{\infty} nf(n)s^{n-1}$$

whether or not the limit is finite. More generally, it follows that for $k \geqslant 1$,

(3) $$G^{(k)}(s) = \sum_{n=k}^{\infty} \frac{n!}{(n-k)!} f(n)s^{n-k}, \quad |s| < 1$$

and

(4) $$G^{(k)}(1) = \lim_{s\uparrow 1} \sum_{n=k}^{\infty} \frac{n!}{(n-k)!} f(n)s^{n-k}.$$

Secondly it follows that $G(s)$ determines the collection $(f(k);\ k \geqslant 0)$.

(5) Theorem (Uniqueness) Let X and Y have generating functions $G_X(s)$ and $G_Y(s)$. If for some $G(s)$ we have

$$G_X(s) = G_Y(s) = G(s) \text{ for } |s| < 1,$$

then X and Y have the same mass function.

Proof This follows from (3), because both $f_X(k)$ and $f_Y(k)$ are given by

$$f_X(k) = \frac{G^{(k)}(0)}{k!} = f_Y(k)$$

for all k. ■

Thirdly it follows that we can obtain all the moments of X from $G(s)$.

(6) Theorem If X has p.g.f. $G(s)$, then

(7) $$\mathbf{E}(X) = G'(1);$$

more generally, the kth factorial moment is

(8) $$\mu^{(k)} = \mathbf{E}(X(X-1)\dots(X-k+1)) = G^{(k)}(1);$$

and in particular

(9) $$\operatorname{var}(X) = G''(1) + G'(1) - (G'(1))^2.$$

Proof Equation (7) is a trivial consequence of (2), and (8) follows from (4). To see (9), write

(10)
$$\begin{aligned}\operatorname{var}(X) &= \mathbf{E}(X - \mathbf{E}(X))^2 = \mathbf{E}(X^2) - (G'(1))^2 \\ &= \mathbf{E}(X(X-1) + X) - (G'(1))^2\end{aligned}$$

as required. ■

Just as (10) gives the second moment and second central moment in terms of the first two factorial moments, likewise σ_k and μ_k may be obtained in principle in terms of $(\mu^{(k)}; k \geqslant 1)$.

(11) Example: Binomial p.g.f. Let X have a binomial distribution with parameters n and p. Then, with $q = 1 - p$ as usual,

$$G(s) = \sum_0^n \binom{n}{k} q^{n-k} p^k s^k = (q + ps)^n.$$

Now using (8) we have

$$\mu^{(k)} = \begin{cases} \dfrac{n!}{(n-k)!} p^k & 1 \leqslant k \leqslant n, \\ 0 & k > n. \end{cases}$$

Hence, by (9), $\operatorname{var}(X) = n(n-1)p^2 + np - (np)^2) = npq$. ●

(12) Example: Poisson p.g.f. Let X have a Poisson distribution with parameter λ. Then

$$G(s) = \sum_0^\infty e^{-\lambda} \frac{\lambda^k}{k!} s^k = e^{+\lambda(s-1)}.$$

Hence we find that $\mu^{(k)} = \lambda^k$, for $k \geqslant 1$. ●

Moments can also be obtained from the tail generating function $T(s)$ defined in Theorem 6.1.5.

(13) Theorem Let X be a random variable with

$$T(s) = \sum_0^\infty s^k \mathbf{P}(X > k).$$

Then

(14)
$$\mathbf{E}(X) = T(1)$$

and, if $\mathbf{E}(X) < \infty$,

(15)
$$\operatorname{var}(X) = 2T'(1) + T(1) - T(1)^2.$$

Proof By L'Hôpital's rule

$$T(1) = \lim_{s \uparrow 1} \frac{1 - G(s)}{1 - s} = G'(1) = \mathbf{E}(X), \quad \text{by (7).}$$

Likewise, differentiating (6.1.7) yields

$$T'(1) = \lim_{s\uparrow 1}\left(\frac{1-G(s)}{(1-s)^2} - \frac{G'(s)}{1-s}\right) = \frac{G''(1)}{2}, \quad \text{by L'Hôpital's rule,}$$

and the result follows using Theorem 6. ■

More generally, a straightforward extension of this theorem shows that

(16) $$\mu^{(k)} = kT^{(k-1)}(1).$$

(17) Example: Geometric p.g.f. Let X have a geometric distribution with mass function $f(k) = (1-q)q^{k-1}$, $k \geqslant 1$; $0 < q < 1$. Then by Example 6.1.9

$$T(s) = \frac{1}{1-qs}.$$

Hence by Theorem 13

$$\mathbf{E}(X) = T(1) = \frac{1}{1-q},$$

and likewise

$$\operatorname{var}(X) = 2T'(1) + T(1) - T(1)^2 = \frac{q}{(1-q)^2}.$$

From (16)

$$\mu^{(k)} = \frac{kq^{k-1}}{(1-q)^k}.$$ ●

We conclude this section with a note about defective probability mass functions. If X is a non-negative random variable such that $\sum_{k=0}^{\infty} f(k) < 1$, then it still makes sense to define the generating function $G(s) = \sum_{k=0}^{\infty} s^k f(k)$. Furthermore, if $\sum_k kf(k) < \infty$, then $G'(1) = \sum_k kf(k)$. However this is not now the expectation $\mathbf{E}(X)$, but rather the 'defective' expectation

$$\mathbf{E}(XI\{X<\infty\}) = \mathbf{E}(X;\, X<\infty) = \mathbf{E}(X|X<\infty)\mathbf{P}(X<\infty)$$

where $I\{X<\infty\}$ is the indicator of the event that X is finite.

In the general case we have likewise

$$G'(1) = \mathbf{E}(X\,|\,|X|<\infty)\mathbf{P}(|X|<\infty)$$

when the expectation exists. In such cases it can be of interest to calculate

$$\mathbf{E}(X|X<\infty) = \frac{G'(1)}{G(1)}.$$

6.3 Sums of independent random variables

If X and Y are independent then the mass function of their sum $Z = X + Y$ is

(1) $$f_Z(k) = \sum_j f_X(j) f_Y(k-j).$$

Practical folk (such as statisticians and the like) are frequently interested in the sum of n independent random variables:

(2)
$$S = \sum_1^n X_i.$$

The prospect of performing the summation in (1) on $n-1$ occasions to find $f_S(k)$ is not an attractive one. The next theorem renders it unnecessary in many important cases.

(3) **Theorem** (a) Let X_1 and X_2 be independent with generating functions $G_1(s)$ and $G_2(s)$ respectively. Then the sum $Z = X_1 + X_2$ has generating function

(4)
$$G(s) = G_1(s)G_2(s).$$

(b) More generally, if $(X_i;\ 1 \leqslant i \leqslant n)$ are independent with generating functions $(G_i(s);\ 1 \leqslant i \leqslant n)$, then the sum $Z = \sum_1^n X_i$ has generating function

$$G_Z(s) = \prod_{i=1}^n G_i(s).$$

Proof (a) Since X_1 and X_2 are independent, s^{X_1} and s^{X_2} are also independent. Hence

$$\begin{aligned} G_Z(s) &= \mathbf{E}(s^{X_1+X_2}) = \mathbf{E}(s^{X_1})\mathbf{E}(s^{X_2}) \quad \text{by Theorem 5.3.8} \\ &= G_1(s)G_2(s). \end{aligned}$$

Part (b) is proved similarly. ■

Example: Binomial sum Let X and Y be independent and binomially distributed with parameters (m, p) and (n, p) respectively. Then recalling Example 6.2.11 we have

$$\begin{aligned} G_{X+Y}(s) &= \mathbf{E}(s^{X+Y}) = \mathbf{E}(s^X)\mathbf{E}(s^Y) \quad \text{by independence} \\ &= (1 - p + ps)^{m+n}. \end{aligned}$$

Hence $X + Y$ is binomially distributed with parameters $m + n$ and p, using Theorem 6.2.5, the uniqueness theorem. ●

(5) **Example** Let $(X_i;\ i \geqslant 1)$ be independent Poisson random variables having respective parameters $(\lambda_i;\ i \geqslant 1)$. Find the mass function of $Z = \sum_{i=1}^n X_i$.

Solution Reproducing the argument of the above theorem we have:

$$\begin{aligned} G_Z(s) &= \mathbf{E}(s^Z) = \prod_{i=1}^n \mathbf{E}(s^{X_i}) \quad \text{by independence} \\ &= \exp\left(\sum_{i=1}^n \lambda_i(s-1)\right) \quad \text{by Example 6.2.12.} \end{aligned}$$

Thus Z is Poisson with parameter $\sum_{i=1}^n \lambda_i$, by the uniqueness theorem. ●

(6) **Example** Let $(X_i;\ i \geqslant 1)$ be independently and uniformly distributed on $\{1, 2, \ldots, n\}$. Let $S_k = \sum_{i=1}^{k} X_i$, and define

$$T_n = \min\{k\colon S_k > n\}.$$

(Thus T_n is the smallest number of the X_i required to achieve a sum exceeding n.) Find the mass function and p.g.f. of T_n, and hence calculate $\mathbf{E}(T_n)$ and $\operatorname{var}(T_n)$.

Solution First we observe that $T_n \geqslant j+1$ if and only if $S_j \leqslant n$. Therefore

(7) $$\mathbf{P}(T_n \geqslant j+1) = \mathbf{P}(S_j \leqslant n).$$

Now, by independence

$$\mathbf{E}(z^{S_j}) = (\mathbf{E}(z^{X_1}))^j = \frac{1}{n^j}\left(\frac{z - z^{n+1}}{1-z}\right)^j \quad \text{by Example 6.1.2.}$$

Hence, by Example 6.1.9

(8) $$\sum_{k=j}^{\infty} z^k \mathbf{P}(S_j \leqslant k) = \left(\frac{z}{n}\right)^j \frac{(1-z^n)^j}{(1-z)^{j+1}}.$$

Equating coefficients of z^n on each side of (8) gives

(9) $$\mathbf{P}(S_j \leqslant n) = \frac{1}{n^j}\binom{n}{j} = \mathbf{P}(T_n \geqslant j+1) \quad \text{by (7).}$$

Hence

$$\mathbf{P}(T_n = j) = \frac{1}{n^{j-1}}\binom{n}{j-1} - \frac{1}{n^j}\binom{n}{j}.$$

From (9), T_n has tail generating function

$$\sum_{j=0}^{n} z^j \mathbf{P}(T_n > j) = \sum_{j=0}^{n}\left(\frac{z}{n}\right)^j\binom{n}{j} = \left(1 + \frac{z}{n}\right)^n.$$

Hence from Theorem 6.2.13

(10) $$\mathbf{E}(T_n) = \left(1 + \frac{1}{n}\right)^n$$

and

(11) $$\operatorname{var}(T_n) = 2\left(1 + \frac{1}{n}\right)^{n-1} + \left(1 + \frac{1}{n}\right)^n - \left(1 + \frac{1}{n}\right)^{2n}.$$

Finally T_n has p.g.f.

(12) $$G(z) = 1 + (z-1)\left(1 + \frac{z}{n}\right)^n. \qquad \bullet$$

Generating functions become even more useful when you are required to consider the sum of a random number of random variables.

(13) **Theorem** Let N and $(X_i;\ i \geqslant 1)$ be independent random variables, and suppose that N is nonnegative and that for all i

(14) $$\mathbf{E}(s^{X_i}) = G(s).$$

Then the sum $Z = \sum_{i=1}^{N} X_i$ has generating function

$$G_Z(s) = G_N(G(s)).$$

Proof By conditional expectation

$$\begin{aligned}\mathbf{E}(s^Z) &= \mathbf{E}(\mathbf{E}(s^Z|N)) = \mathbf{E}(\mathbf{E}(s^{X_1}) \ldots \mathbf{E}(s^{X_N})) \quad \text{by independence}\\ &= \mathbf{E}(G(s)^N) \quad \text{by (14)}\\ &= G_N(G(s))\end{aligned}$$

by Definition 6.1.3, and the result follows. ■

(15) **Example** You toss a fair coin repeatedly. Each time it shows a tail you roll a fair die, when the coin first shows a head you stop. What is the p.g.f. of the total sum of the scores shown by the rolls of the die?

Solution As you know by now, the number N of tails shown has mass function $f_N(k) = (\frac{1}{2})^{k+1}$; $k \geqslant 0$, with generating function

$$G_N(s) = \frac{1}{2-s}.$$

The score shown by each die has p.g.f.

$$G_X(s) = \frac{1}{6}\frac{s(1-s^6)}{1-s},$$

and so the p.g.f. of the total is given by Theorem 13 as

$$G(s) = \left(2 - \frac{1}{6}\frac{s(1-s^6)}{1-s}\right)^{-1}. \quad \bullet$$

(16) **Example** Let $Z = \sum_{i=1}^{N} X_i$ where

$$f_X(k) = \frac{k^{-1}p^k}{\log(1-p)}; \quad k \geqslant 1, 0 < p < 1$$

and $f_N(k) = \lambda^k e^{-\lambda}/k!$; $k \geqslant 1$, $0 < \lambda$. Show that Z has a negative binomial mass function.

Solution It is easy to show that

$$G_X(s) = \frac{\log(1-sp)}{\log(1-p)}; \quad G_N(s) = e^{\lambda(s-1)}.$$

Hence

$$G_Z(s) = e^{-\lambda}\exp(\lambda G_X(s)) = \left(\frac{1-p}{1-ps}\right)^{-\lambda(\log(1-p))^{-1}},$$

which is the p.g.f. of a negative binomial mass function. ●

(17) **Example: Branching** A collection of particles behaves in the following way. At time $n = 0$ there is one particle. At time $n = 1$ it is replaced by a random number X

of particles, where X has mass function $f(k)$, $k \geq 0$. At every subsequent time $n = 2, 3, \ldots$ each particle in existence at that time is replaced by a random number of new particles, called its *family*. All family sizes are independent, and they all have the same mass function as the first family X.

Let the number of particles in existence at time n be Z_n. Find $\mathbf{E}(s^{Z_n})$ and $\lim_{n\to\infty} \mathbf{P}(Z_n = 0)$.

Solution Let $G(s) = \mathbf{E}(s^X)$ and $G_n(s) = \mathbf{E}(s^{Z_n})$. Let the family sizes of the particles existing at time n be $(X_j;\ 0 \leq j \leq Z_n)$. Then we obtain the attractive and useful representation

$$Z_{n+1} = \sum_{j=0}^{Z_n} X_j$$

and by Theorem 13

$$G_{n+1}(s) = G_n(G(s)).$$

(This basic argument is used repeatedly in the theory of branching processes.) Hence $G_{n+1}(s)$ is the $(n+1)$th iterate of $G(.)$, that is to say:

(18) $$G_{n+1}(s) = G(G(\ldots G(s)\ldots)), \quad n \geq 0.$$

Now let $\mathbf{P}(Z_n = 0) = \eta_n$, and define η to be the smallest non-negative root of the equation

(19) $$G(s) = s$$

We shall now show that

(20) $$\lim_{n\to\infty} \eta_n = \eta.$$

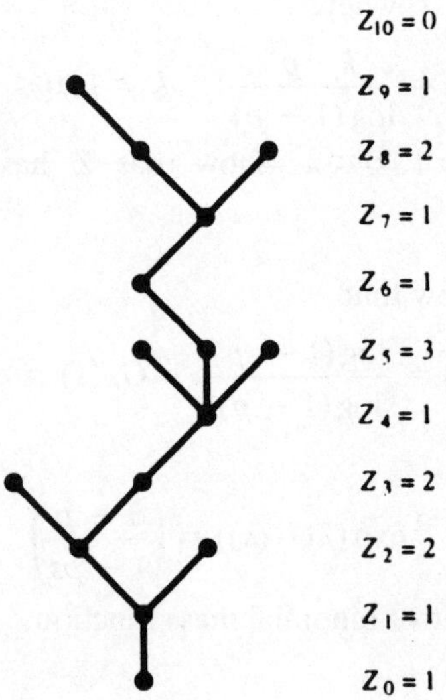

Figure 6.1 A realization of a branching process. The orientation of the diagram explains the name.

First we consider three trivial cases:

(i) if $f(0) = 0$ then $\eta_n = G_n(0) = 0 = \eta$.
(ii) if $f(0) = 1$ then $\eta_n = G_n(0) = 1 = \eta$.
(iii) if $f(0) + f(1) = 1$, with $f(0)f(1) \neq 0$, then

$$\eta_n = G_n(0) = 1 - (f(1))^n \to 1 = \eta.$$

Thus (20) is true in each case. In what follows we exclude these cases by requiring that $0 < f(0) < f(0) + f(1) < 1$.

Now note that $\{Z_n = 0\} \subseteq \{Z_{n+1} = 0\}$, and so by Example 1.4.11

$$\eta_n \leqslant \eta_{n+1} \leqslant 1.$$

Hence $\lim_{n\to\infty} \eta_n$ exists; let us denote it by λ. By (18) $G_{n+1}(0) = G(G_n(0))$; now letting $n \to \infty$ and using the continuity of $G(s)$, we find that λ is a root of (19):

$$\lambda = G(\lambda).$$

However, if for some n, $\eta_n < \eta$ then, because $G(s)$ is increasing,

$$\eta_{n+1} = G(\eta_n) \leqslant G(\eta) = \eta.$$

But $\eta_0 = G(0) \leqslant G(\eta) = \eta$, and so $\eta_n < \eta$ for all n. Hence $\lambda \leqslant \eta$ and so $\lambda = \eta$. ●

Once again we conclude with a note about defective random variables. If X and Y are defective then they are independent if $\mathbf{P}(X = i, Y = j) = \mathbf{P}(X = i)\mathbf{P}(Y = j)$ for all finite X and Y. Hence we can still write in this case

$$G_{X+Y}(s) = G_X(s)G_Y(s),$$

and we can denote this by

$$\mathbf{E}(s^{X+Y}) = \mathbf{E}(s^X)\mathbf{E}(s^Y),$$

if we remember that the expectation is taken only over the finite part of the distribution.

6.4 Moment generating functions

The moments $(\mu_k;\ k \geqslant 1)$ of a random variable X also form a collection of real numbers, so we may expect their generating functions to be useful also. In this case it is convenient to use the exponential generating function of the collection $(\mu_k;\ k \geqslant 1)$.

(1) **Definition** Let the random variable X have finite moments $\mu_k = \mathbf{E}(X^k)$ for all $k \geqslant 1$. Then the moment generating function (or m.g.f.) of X is the function $M_X(t)$ given by

(2)
$$M_X(t) = \sum_{k=0}^{\infty} \frac{\mu_k t^k}{k!}. \qquad ▲$$

If X takes only a finite number of values then we easily obtain the very convenient representation

(3) $$M_X(t) = \sum_{k=0}^{\infty} \mathbf{E}\frac{(X^k)t^k}{k!} = \mathbf{E}\left(\sum_{k=0}^{\infty} \frac{(Xt)^k}{k!}\right) = \mathbf{E}(e^{Xt}).$$

More generally, (3) holds provided the moments μ_k do not get too large as k increases. For example, if $\sum_{k=0}^{\infty}|\mu_k|/k! < \infty$, then $M_X(t)$ exists for $|t| < 1$, and we can use the equivalent of Theorem 6.2.1. This yields

(4) $$\mu_k = M_X^{(k)}(0).$$

From (3) we find that

(5) $$M_X(t) = G_X(e^t)$$

where $G_X(s)$ is the p.g.f. of X.

(6) Example Let X have a negative binomial distribution with mass function $f(k) = \binom{n+k-1}{k}q^k p^n$, $k \geqslant 0$. By the negative binomial expansion

$$G(s) = \sum_{k=0}^{\infty} p^n \binom{n+k-1}{k} q^k s^k = \left(\frac{p}{1-qs}\right)^n, \quad |s| < q^{-1}.$$

Then X has moment generating function

$$M(t) = \left(\frac{p}{1-qe^t}\right)^n, \quad t < -\log q.$$ ●

Let us consider an example in which X may take negative integer values.

(7) Example Let X have mass function

$$f(k) = \frac{1}{2}q^{|k|-1}(1-q); \quad k = \pm 1, \pm 2, \ldots,$$

where $0 < q < 1$. Then X has p.g.f.

$$G(s) = \sum_{k \neq 0} s^k \frac{1}{2} q^{|k|-1}(1-q)$$

$$= \frac{1}{2}(1-q)\left(\frac{s}{1-qs} + \frac{1}{s-q}\right), \quad q < |s| < q^{-1}.$$

Hence X has m.g.f.

$$M(t) = \frac{1}{2}(1-q)\left(\frac{e^t}{1-qe^t} + \frac{1}{e^t - q}\right), \quad \log q < t < -\log q.$$

In this case $M(t)$ exists in an interval including the origin, and (4) holds. ●

The factorial moments $(\mu^{(k)}; k \geqslant 1)$ also have a generating function, which is related to the probability generating function as follows:

(8) $$\sum_k \frac{\mu^{(k)} t^k}{k!} = \sum_k \mathbf{E}\frac{(X(X-1)\ldots(X-k+1))t^k}{k!}$$

$$= \sum_k \sum_{n=k}^{\infty} \frac{n!}{(n-k)!} \frac{\mathbf{P}(X=n)}{k!} t^k$$

$$= \sum_{n=0}^{\infty} \mathbf{P}(X=n) \sum_{k=0}^{n} \binom{n}{k} t^k$$

$$= \sum_n \mathbf{P}(X=n)(1+t)^n = G_X(1+t).$$

The change in the order of summation is justified since the terms in the sum are all non-negative.

Now let us make the important observation that both of these moment generating functions are useful for dealing with sums of independent random variables, for essentially the same reasons that made the p.g.f. so useful. To see this, let X and Y be independent, and set $Z = X + Y$. Then

(9) $$M_Z(t) = \mathbf{E}(e^{t(X+Y)}) = M_X(t)M_Y(t) \quad \text{by independence.}$$

Likewise

(10) $$G_Z(1+t) = \mathbf{E}((1+t)^Z) = G_X(1+t)G_Y(1+t).$$

Finally we record the existence of yet another function that generates the moments of X, albeit indirectly.

(11) Definition If the function

$$\kappa(t) = \log(\mathbf{E}(e^{Xt})) = \log(M_X(t))$$

can be expanded in powers of t, in the form

(12) $$\kappa(t) = \sum_{r=1}^{\infty} \kappa_r t^r / r!$$

then it is called the *generating function of the cumulants* (κ_r; $r \geqslant 1$). ▲

(13) Example If X is Poisson with parameter λ then

$$\log(M_X(t)) = \log(\exp[\lambda(e^t - 1)]) = \lambda(e^t - 1) = \sum_{r=1}^{\infty} \frac{\lambda}{r!} t^r.$$

Hence for all r, $\kappa_r = \lambda$. ●

6.5 Joint generating functions

Generating functions can be equally useful when we wish to consider the joint behaviour of a number of random variables. Not surprisingly, we need a joint generating function.

(1) Definition A random vector $(X_1, \ldots, X_n)$, with joint mass function $f(x_1, \ldots, x_n)$, has a *joint probability generating function*

$$G_{\mathbf{X}}(\mathbf{s}) = G_{X_1,\ldots,X_n}(s_1, \ldots, s_n) = \sum_{x_1, x_2, \ldots, x_n} s^{x_1} s^{x_2} \ldots s^{x_n} f(x_1, \ldots, x_n). \quad ▲$$

By Theorem 5.3.1 we obtain the following useful representation of G,

$$G_X(s) = \mathbf{E}(s_1^{X_1} s_2^{X_2} \ldots s_n^{X_n}) = \mathbf{E}\left(\prod_{i=1}^{n} s_i^{X_i}\right).$$

(2) Example A coin shows heads with probability p or tails with probability $q = 1 - p$. If it is tossed n times, then the joint p.g.f. of the number X of heads and the number of tails is

$$G(s, t) = \mathbf{E}(s^X t^Y) = t^n \mathbf{E}\left(\left(\frac{s}{t}\right)^X\right) = t^n\left(q + p\frac{s}{t}\right)^n \quad \text{because } X \text{ is binomial,}$$

$$= (qt + ps)^n. \quad \bullet$$

The fact that G is the nth power of $(qt + ps)$ suggests that independence could have been used to get this result. We use this idea in the next example.

(3) Example Each of a sequence of n independent trials results in a win, a loss or a draw, with probabilities α, β and γ respectively. Find the joint p.g.f. of the wins, losses and draws.

Solution Let W_i, L_i and D_i be the respective indicators on the ith trial of a win, a loss or a draw. Then

$$\mathbf{E}(x^{W_i} y^{L_i} z^{D_i}) = \alpha x + \beta y + \gamma z.$$

But the required joint p.g.f. is

$$G(x, y, z) = \mathbf{E}(x^{\sum_1^n W_i} y^{\sum_1^n L_i} z^{\sum_1^n D_i}) = [\mathbf{E}(x^{W_i} y^{L_i} z^{D_i})]^n \quad \text{by independence,}$$

$$= (\alpha x + \beta y + \gamma z)^n. \quad \bullet$$

Knowledge of the joint p.g.f. entails knowledge of all the separate p.g.f.s because, for example, if X and Y have joint p.g.f. $G(s, t)$, then

(4) $$G_X(s) = \mathbf{E}(s^X) = \mathbf{E}(s^X 1^Y) = G(s, 1).$$

Likewise

(5) $$G_Y(t) = G(1, t)$$

Indeed we can quickly obtain the p.g.f. of any linear combination of X and Y; for example let $Z = aX + bY$, then

(6) $$\mathbf{E}(s^Z) = \mathbf{E}(s^{aX+bY}) = \mathbf{E}(s^{aX} s^{bY}) = G(s^a, s^b).$$

Further, the joint generating function also provides us with the joint moments when they exist, in the same way as $G_X(s)$ provides the moments of X.

(7) Example Let X and Y have joint p.g.f. $G(s, t)$ and suppose that X and Y have finite variance. Then $\mathbf{E}(XY)$ exists (by the Cauchy–Schwarz inequality) and

$$\frac{\partial^2 G}{\partial s \partial t} = \frac{\partial^2}{\partial s \partial t}\mathbf{E}(s^X s^Y) = \mathbf{E}(XY s^{X-1} t^{Y-1}).$$

Hence

(8) $$\mathbf{E}(XY) = \left.\frac{\partial^2 G}{\partial s \partial t}\right|_{s=t=1}.$$

Likewise

$$\mathbf{E}(X) = \left.\frac{\partial G}{\partial s}\right|_{s=t=1}, \quad \text{and} \quad \mathbf{E}(Y) = \left.\frac{\partial G}{\partial t}\right|_{s=t=1}.$$

Quite often we write $G_{st}(s, t)$ for $\partial^2 G/\partial s \partial t$, and so on, and in this form the covariance of X and Y is given by

(9) $$\operatorname{cov}(X, Y) = G_{st}(1, 1) - G_s(1, 1)G_t(1, 1).$$ ●

(10) Example 5.10 revisited: Golf Recall that you play n holes of golf, each of which you independently win, lose or tie, with repective probabilities p, q and r respectively. The numbers of wins, losses and ties are X, Y and Z respectively, with $X + Y + Z = n$.

(a) Find $\rho(X, Y)$. (b) Find $\operatorname{var}(X - Y)$.

Solution (a) By Example 3 above, we calculate

$$\mathbf{E}(x^X y^Y z^Z) = (px + qy + rz)^n = G(x, y, z) \quad \text{say.}$$

Hence

$$\mathbf{E}(X) = G_x(1, 1, 1) = np,$$

and

$$\operatorname{var}(X) = G_{xx}(1, 1, 1) + G_x(1) - (G_x(1))^2 = np(1 - p),$$

and

$$\mathbf{E}(XY) = G_{xy}(1, 1, 1) = n(n - 1)pq.$$

Therefore the correlation between X and Y is

(11) $$\rho(X, Y) = \frac{\operatorname{cov}(X, Y)}{(\operatorname{var}(X)\operatorname{var}(Y))^{\frac{1}{2}}} = \frac{n(n-1)pq - n^2pq}{(n^2p(1-p)q(1-q))^{\frac{1}{2}}} = -\left(\frac{pq}{(1-p)(1-q)}\right)^{\frac{1}{2}}.$$

You should compare the labour in this calculation with the more primitive techniques of Example 5.10.

(b) Using (6) with $a = 1$, $b = -1$, we have, on setting $W = X - Y$,

$$G_W(s) = \mathbf{E}(s^{X-Y}) = G(s, s^{-1}, 1) = (ps + qs^{-1} + r)^n.$$

Hence $dG_W/ds = n(p - qs^{-2})(ps + qs^{-1} + r)^{n-1}$, and

$$\frac{d^2G_W}{ds^2} = n(n-1)(p - qs^{-2})^2(ps + qs^{-1} + r)^{n-2} + 2nqs^{-3}(ps + qs^{-1} + r)^{n-1}.$$

Therefore

$$\operatorname{var}(W) = n(n-1)(p-q)^2 + 2nq + n(p-q) - n^2(p-q)^2$$
$$= n(p + q - (p-q)^2).$$ ●

Finally we record that joint generating functions provide a useful characterization of independence.

(12) Theorem Let X and Y have joint p.g.f. $G(s, t)$. Then X and Y are independent if and only if

(13) $$G(s, t) = G(s, 1)G(1, t).$$

Proof If (13) holds, then equating coefficients of $s^j t^k$ gives

$$\mathbf{P}(X = j, Y = k) = \mathbf{P}(X = j)\mathbf{P}(Y = k)$$

as required. The converse is immediate by Theorem 5.3.8. ■

(14) Example 5.5.8 revisited: Eggs Recall that the number X of eggs is Poisson with parameter λ, and eggs hatch independently with probability p. Let Y be the number which do hatch, and Z the number which do not. Show that Y and Z are independent, and also that $\rho(X, Y) = \sqrt{p}$.

Solution Conditional on $X = x$, the number Y of hatchings is binomial with p.g.f.

(15) $$\mathbf{E}(s^Y | X = x) = (ps + 1 - p)^x.$$

Hence, by conditional expectation,

$$\mathbf{E}(y^Y z^Z) = \mathbf{E}(y^Y z^{X-Y}) = \mathbf{E}\left(z^X \mathbf{E}\left(\left(\frac{y}{z}\right)^Y \middle| X\right)\right)$$
$$= \mathbf{E}\left(z^X\left(\frac{py}{z} + 1 - p\right)^X\right) \quad \text{by (15)}$$
$$= \exp(\lambda(py + (1-p)z - 1)) \quad \text{since } X \text{ is Poisson,}$$
$$= e^{\lambda p(y-1)} e^{\lambda(1-p)(z-1)}.$$

Hence Y and Z are independent by Theorem 12. Furthermore we see immediately that Y is Poisson with parameter λp.

To find $\rho(X, Y)$, we first find the joint p.g.f. of X and Y, again using conditional expectation. Thus

$$\mathbf{E}(s^X y^Y) = \mathbf{E}(s^X \mathbf{E}(y^Y | X)) = \mathbf{E}(s^X (py + 1 - p)^X) = \exp(\lambda(s(py + 1 - p) - 1)).$$

Hence, using (7), $\mathbf{E}(XY) = \lambda^2 p + \lambda p$, and so, using the first part, $\rho(X, Y) = \sqrt{p}$. You should compare this with the method of Example 5.5.8. ●

(16) Example: Pairwise independence Independent random variables X and Y each take the values $+1$ or -1 only, and $\mathbf{P}(X = 1) = a$, with $\mathbf{P}(Y = 1) = b$. Let $Z = XY$.

Show that there are values of a and b such that X, Y and Z are pairwise independent.

Solution Consider the joint probability generating function of X and Z.

$$
\begin{aligned}
G(s,t) &= \mathbf{E}(s^X t^Z) = \mathbf{E}(s^X t^{XY}) = \mathbf{E}(\mathbf{E}(s^X t^{XY} | X)) \\
&= \mathbf{E}\left(bs^X t^X + (1-b)\frac{s^X}{t^X}\right) \\
&= a\left(bst + (1-b)\frac{s}{t}\right) + (1-a)\left(\frac{b}{st} + (1-b)\frac{t}{s}\right) \\
&= \frac{abs^2t^2 + a(1-b)s^2 + (1-a)(1-b)t^2 + b(1-a)}{st} \\
&= \frac{1}{st}\left(as^2(bt^2 + 1 - b) + \frac{(1-a)(1-b)}{b}(bt^2 + 1 - b)\right. \\
&\qquad \left. + b(1-a) - (1-b)^2(1-a)b^{-1}\right)
\end{aligned}
$$

which factorizes into a product of a function of s and a function of t if

$$b^2 - (1-b)^2 = 0,$$

that is if $b = \frac{1}{2}$. In this case X and Z are independent. If $a = \frac{1}{2}$ then Y and Z are independent, and $a = b = \frac{1}{2}$ entails the pairwise independence of X, Y and Z. ●

6.6 Sequences

In Section 4.5 we defined the convergence of a sequence of mass functions. This can be usefully connected to the convergence of corresponding sequences of generating functions. For sequences of probability generating functions we have the following result, which we give without proof.

(1) **Theorem** Let $f(k)$ be a probability mass function with generating function

$$G(s) = \sum_0^\infty s^k f(k),$$

and suppose that for each $n \geqslant 1$, $f_n(k)$ is a probability mass function with generating function

$$G_n(s) = \sum_0^\infty s^k f_n(k).$$

Then, as $n \to \infty$, $f_n(k) \to f(k)$ for all k, if and only if $G_n(s) \to G(s)$ for all $0 < s < 1$.

We now use this to prove a result which we have already shown by more primitive methods.

(2) Example Let $(X_n;\ n \geqslant 1)$ be a sequence of random variables such that X_n has a binomial distribution with parameters n and λ/n, $\lambda > 0$. Then

$$\mathbf{E}(s^{X_n}) = \left(1 - \frac{\lambda}{n} + \frac{\lambda}{n}s\right)^n \to e^{\lambda(s-1)} \quad \text{as } n \to \infty.$$

This is the p.g.f. of a Poisson random variable, and so as $n \to \infty$,

$$\mathbf{P}(X_n = k) \to e^{-\lambda}\lambda^k/k! \qquad \bullet$$

It is often convenient to work with distributions and moment generating functions. In this case the following result (for which we give no proof) is useful.

(3) Theorem Let $\{F_n(x);\ n \geqslant 1\}$ be a sequence of distribution functions with corresponding moment generating functions $\{M_n(t):\ n \geqslant 1\}$. If $F(x)$ is a distribution having corresponding moment generating function $M(t)$, then, as $n \to \infty$, $M_n(t) \to M(t)$ for all t, if and only if $F_n(x) \to F(x)$, whenever $F(x)$ is continuous.

Additional conditions are required to link the convergence of a sequence of mass functions or distributions and the convergence of their moments. The following theorem (for which again we offer no proof) is for a sequence of distributions.

(4) Theorem Suppose that for each $n \geqslant 1$, the distribution $F_n(x)$ has moments $\{\mu_j(n);\ j \geqslant 1\}$, such that $|\mu_j(n)| < a_j < \infty$.

(i) Let $F_n(x) \to F(x)$, as $n \to \infty$, wherever $F(x)$ is continuous. Then as $n \to \infty$, for each j,

$$\mu_j(n) \to \mu_j < \infty,$$

and $(\mu_j;\ j \geqslant 1)$ are the moments of $F(x)$.

(ii) Conversely, for each $j \geqslant 1$, as $n \to \infty$, suppose that

$$\mu_j(n) \to \mu_j < \infty,$$

where $\{\mu_j;\ 1 \leqslant j\}$ are the moments of a unique distribution $F(x)$. Then, as $n \to \infty$, $F_n(x) \to F(x)$ wherever $F(x)$ is continuous.

There is a corresponding result for sequences of mass functions.

These theorems find applications (for example) in the theory of random graphs, and other combinatorial problems where moments are more tractable than distributions.

(5) Example Let X_n have the binomial distribution with parameters n and λ/n. Then by Example 6.2.11

$$\mu^{(k)} = \begin{cases} \dfrac{n!}{(n-k)!}\left(\dfrac{\lambda}{n}\right)^k & 1 \leqslant k \leqslant n \\ 0 & k > n \end{cases} \to \lambda^k$$

as $n \to \infty$. But by Example 6.2.12, these are the factorial moments of the Poisson distribution. Hence as $n \to \infty$, $\mathbf{P}(X_n = k) \to e^{-\lambda}\lambda^k/k!$, which we proved directly in Example 6.6.2, and earlier in Example 4.9. ●

(6) **Example: Matching again** Recall that we are assigning n distinct letters randomly to n matching envelopes, and X is the number of matched pairs (of letter and envelope) that result. Consider the kth factorial moment of X.

$$\mu^{(k)} = \mathbf{E}(X(X-1)\dots(X-k+1); \quad 1 \le k \le n.$$

Let I_j be the indicator of the event that the jth envelope contains the matching letter. Then the sum

(7) $$S = \sum_{j_1<j_2<\dots<j_k} I_{j_1}\dots I_{j_k}$$

is just the number of ways of choosing a set of size k from the set of matching pairs. But another way of writing this is as $\binom{X}{k}$. Hence

(8) $$\begin{aligned}\frac{\mu^{(k)}}{k!} &= \mathbf{E}\frac{(X(X-1)\dots(X-k+1))}{k!}\\ &= \mathbf{E}\left(\binom{X}{k}\right)\\ &= \mathbf{E}(S)\\ &= \binom{n}{k}\mathbf{E}(I_{j_1}\dots I_{j_k})\\ &= \binom{n}{k}\mathbf{P}\text{ (a given set of } k \text{ all match)}\\ &= \binom{n}{k}\left(\frac{1}{n}\cdot\frac{1}{n-1}\dots\frac{1}{n-k+1}\right) = \frac{1}{k!}.\end{aligned}$$

Hence

$$\mu^{(k)} = \begin{Bmatrix}1; & k \le n\\ 0; & k > n,\end{Bmatrix} \to 1 \quad \text{for all } k$$

as $n \to \infty$. But these are the factorial moments of the Poisson distribution with parameter 1, and so as $n \to \infty$

(9) $$\mathbf{P}(X = k) \to \frac{1}{ek!}.$$ ●

We conclude with an example which leads into the material of the next chapter.

(10) **Example 6.3.6 revisited** Recall that the p.g.f. of T_n (where T_n is the number of uniform random variables required to give a sum greater than n) is

$$G(s) = 1 + (s-1)\left(1 + \frac{s}{n}\right)^n.$$

What happens as $n \to \infty$? How do you interpret this?

Solution From (6.3.12), as $n \to \infty$

$$\mathbf{E}(s^{T_n}) \to 1 + (s - 1)e^s$$

or, equivalently

$$\mathbf{E}(e^{tT_n}) \to 1 + (e^t - 1)e^{e^t}.$$

It follows that

$$\mathbf{P}(T_n = k) \to \frac{1}{(k-1)!} - \frac{1}{k!}.$$

The limiting factorial moments have a simple form, for

$$\mathbf{E}((1 + t)^{T_n}) \to 1 + te^{1+t} = 1 + \sum_{k=1}^{\infty} \frac{e}{(k-1)!} t^k.$$

Hence in the limit $\mu^{(k)} = ek$, $k \geq 1$.

To interpret this we return to the original definition

$$T_n = \min\left\{k: \sum_1^n X_i > n\right\} = \min\left\{k: \sum_1^n \left(\frac{X_i}{n}\right) > 1\right\}$$

where each X_i/n is uniformly distributed on $\{1/n, 2/n, \ldots, 1\}$. However the limit of this sequence of uniform mass functions as $n \to \infty$ is *not* the mass function of a discrete random variable.

Intuitively you may feel that it is approaching the distribution of a variable which is uniformly distributed over the *interval* $[0, 1]$. This vague remark can in fact be given a meaning if we introduce new objects—namely, continuous random variables. This is the subject of the next two chapters; after much technical development it can be shown that the limit of T_n above is indeed the number of independent random variables, each uniform on $[0, 1]$, required to produce a sum greater than 1.

6.7 Regeneration

Many interesting and important sequences of random variables arise as some process evolves in time. Often a complete analysis of the process may be too difficult, and we seek simplifying ideas. One such concept, which recurs throughout probability, is the idea of regeneration. Here is an illustration:

(1) **Example: Maze** You are trying to traverse an unknown labyrinth. You set off at a constant speed from the clock by the portal, and each time a decision is required you choose at random from the alternatives. It is dark, and you have no pencil and paper, so a description of the process (that is, your route) is impossible.

However, each time you arrive back at the portal, you can look at the clock and so record T_r, the time at which you return to the clock for the rth time; $T_0 = 0$ say. Now it is clear from the set-up that, when you set off for a second time (at time T_1), your chance of following any given route around the maze is the same as when you set off for the first time. Thus the time until your second return, which is $T_2 - T_1$, has the same distribution as T_1 and is independent of T_1. The same is true of every subsequent interval between successive visits to the portal. These times $(T_n; n \geq 0)$

are regenerative in the sense that the distribution of your paths starting from T_n is the same as the distribution starting from T_m, for all m and n.

Of course, if you leave pebbles at junctions, or take a ball of string, or make a map, then this is no longer true. ●

Here is another archetypal illustration.

(2) **Example: Renewal** A machine started at $T_0 = 0$ uses a bit that wears out. As soon as it wears out, at time T_1, it is replaced by a similar bit, which in turn is replaced at T_2. Assuming that the machine performs much the same tasks as time passes, it seems reasonable to assume that the collection $(X_n; n \geqslant 1)$, where $X_n = T_n - T_{n-1}$, are independent and identically distributed. The replacement (or renewal) times are regenerative. ●

In fact we have already used this idea of re-starting from scratch; see (for example) Examples 4.4.9, 4.18 and 5.4.13.

Here is another elementary illustration.

Example Three players A, B and C take turns rolling a fair die in the order $ABCAB\ldots$ until one of them rolls a 5 or a 6. Let X_0 be the duration of the game (that is, the number of rolls). Let A_0 be the event that A wins, and let A_r be the event that A wins after the rth roll. Let W_i, $i \geqslant 1$, be the event that the game is won on the ith roll. Of course

$$A_0 = W_1 \cup \{W_1^c \cap W_2^c \cap W_3^c \cap A_3) \text{ where } \mathbf{P}(W_1) = \frac{1}{3}.$$

Now, as usual, we denote the indicator of any event E by $I(E)$, and so

$$\mathbf{E}(s^{X_0} I(W_1)) = \frac{1}{3}s.$$

Next we observe that if the first three rolls fail to yield 5 or 6, then the process regenerates (in the sense discussed above), so $X_0 = 3 + X_3$, where X_3 has the same distribution as X_0. Hence

$$\mathbf{E}(s^{X_0} I(A_3)) = \mathbf{E}(s^{3+X_3} I(A_3)) = s^3 \mathbf{E}(s^{X_0} I(A_0)).$$

Therefore, we can write

$$\mathbf{E}(s^{X_0} I(A_0)) = \mathbf{E}(s^{X_0}\{I(W_1) + I(W_1^c \cap W_2^c \cap W_3^c \cap A_3)\})$$
$$= \frac{1}{3}s + \frac{8}{27}s^3 \mathbf{E}(s^{X_0} I(A_0)).$$

Hence

$$\mathbf{E}(s^{X_0} I(A_0)) = \frac{\frac{1}{3}s}{1 - \frac{8}{27}s^3}.$$

Likewise, in an obvious notation,

$$\mathbf{E}(s^{X_0} I(B_0)) = \frac{\frac{2}{9}s^2}{1 - \frac{8}{27}s^3},$$

and

$$\mathbf{E}(s^{X_0} I(C_0)) = \frac{\frac{4}{27}s^3}{1 - \frac{8}{27}s^3}.$$

Hence

$$\mathbf{E}(s^{X_0}) = \mathbf{E}(s^{X_0}(I(A_0) + I(B_0) + I(C_0))) = \frac{4s^3 + 6s^2 + 9s}{27 - 8s^3}.$$

Now we consider a more general case.

Let H be some phenomenon (or happening) which may occur or not at any time $n = 1, 2, 3, \ldots$ Let H_n be the event that H occurs at time n, and define X_n to be the time interval between the $(n-1)$th and nth occurrences of H. Thus

$$X_1 = \min\{n > 0: H_n \text{ occurs}\}$$

$$X_1 + X_2 = \min\{n > X_1: H_n \text{ occurs}\}$$

and so on. We suppose that $(X_n;\ n \geq 2)$ are independent and identically distributed random variables with mass function $(f(k);\ k \geq 1)$ and p.g.f. $G(s)$. The first interval X_1 is independent of $(X_n;\ n \geq 2)$, but its mass function may or may not be the same as that of X_2. This gives rise to two cases:

Case (O) The *ordinary* case. The mass function of X_1 is $(f(k);\ k \geq 1)$, the same as X_2.

Case (D) The *delayed* case. The mass functions of X_1 is $(d(k);\ k \geq 1)$, and $\mathbf{E}(s^{X_1}) = D(s)$.

These two cases admit a conventional interpretation: in the ordinary case we suppose that H_0 occurred, so X_1 has the same mass function as the other intervals; in the delayed case H_0 did not occur, so X_1 may have a different mass function.

The mathematical structure described above is known as a *recurrent event process*, or alternatively as a *discrete renewal process*. The important point about such a process is that each time H occurs, the process regenerates itself, in the sense discussed above.

Now Examples 1 and 2 make it clear that there are two essentially different types of renewal process. In Example 1 there is always a chance that you do traverse the maze (or encounter the Minotaur), and so do not return to the entrance. That is

$$\mathbf{P}(X_2 < \infty) < 1.$$

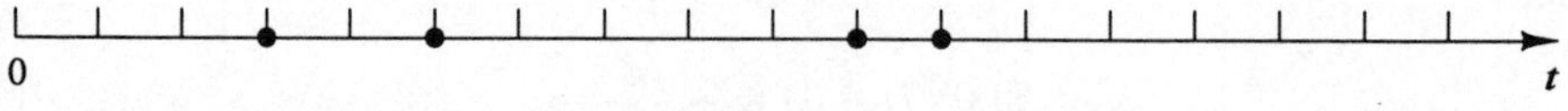

Figure 6.2 A delayed renewal process. Here $X_1 = T_1 = 3$; $X_2 = 2$, $X_3 = 5$, $X_4 = 1, \ldots$

Such a process is called *transient*. On the other hand, all machine bits wear out eventually, so in (2) we have

$$\mathbf{P}(X_2 < \infty) = 1.$$

Such a process is called *persistent* (or recurrent).

Now define the probabilities $u_n = \mathbf{P}(H_n)$, $n \geqslant 1$. A natural question is to ask whether this distinction between persistence and transience can also be observed in the properties of u_n. (The answer is yes, as we shall see.)

It is customary to make a further distinction between two different types of persistent renewal process.

Definition If $\mathbf{E}(X_2) = \infty$, then the process is said to be *null*; if $\mathbf{E}(X_2) < \infty$, then the process is said to be *non-null*.

Note that $\mathbf{E}(X_2)$ is sometimes known as the *mean recurrence time* of the process. ▲

Ordinary renewal is just a special case of delayed renewal of course, but it is convenient to keep them separate. We therefore define $u_0 = 1$, and $u_n = \mathbf{P}(H_n)$, in the ordinary case; and $v_0 = 0$, and $v_n = \mathbf{P}(H_n)$, in the delayed case. These have respective generating functions $U(s) = \sum_0^\infty u_n s^n$, and $V(s) = \sum_1^\infty v_n s^n$. (Remember that $U(s)$ and $V(s)$ are not *probability* generating functions in the sense in which we use that term.)

Now we have the following:

(3) **Theorem** (i)

$$U(s) = \frac{1}{1 - G(s)}$$

(ii)

$$V(s) = \frac{D(s)}{1 - G(s)}.$$

Proof By conditional probability, in Case (O),

$$u_n = \sum_{k=1}^{n} \mathbf{P}(H_n|X_1 = k)\mathbf{P}(X_1 = k).$$

However, given H_k, the probability of any later occurrence of H is as if the process started at k. That is to say

(4) $$\mathbf{P}(H_n|X_1 = k) = \mathbf{P}(H_n|H_k) = \mathbf{P}(H_{n-k}) = u_{n-k}, \quad n \geqslant k.$$

Hence

$$u_n = \sum_{k=1}^{n} u_{n-k} f(k).$$

Since the right hand sum is a convolution, its generating function is the product of the two generating functions $U(s)$ and $G(s)$, whence

(5) $$U(s) - 1 = U(s)G(s).$$

Likewise, in Case (D),

$$v_n = \sum_{k=1}^{n} \mathbf{P}(H_n|X_1 = k)\mathbf{P}(X_1 = k) = \sum_{k=1}^{n} \mathbf{P}(H_n|X_1 = k)d(k)$$

$$= \sum_{k=1}^{n} u_{n-k}d(k) \quad \text{by (4).}$$

Hence

(6) $$V(s) = U(s)D(s) = \frac{D(s)}{1 - G(s)} \quad \text{by (5).}$$ ■

Thus given $G(s)$ (and $D(s)$ in the delayed case) we can in principle find $\mathbf{P}(H_n)$, the probability that H occurs at time n, by expanding $U(s)$ in powers of s.

Conversely, given $V(s)$ (and $U(s)$ in the delayed case) we can find $G(s)$ and $D(s)$, and also decide whether the process is transient or persistent.

(7) **Corollary** If $U(1) < \infty$ then the process is transient. Otherwise it is persistent.

Proof This follows immediately from Theorem 3(i). ■

(8) **Example: Coincidences** Suppose that a number c of independent simple symmetric random walks are started simultaneously from the origin. Let H be the 'happening' that they are all at the origin, so H_{2n} is the event that all the c walks are at 0 on the $2n$th step. Show that H is persistent when $c = 2$, but H is transient for $c \geqslant 3$.

Solution For $c = 2$, we recall that

$$\mathbf{P}(H_{2n}) = (u_{2n})^2 = \left(\frac{1}{4^n}\binom{2n}{n}\right)^2$$

$$= \left(1 - \frac{1}{2}\right)^2\left(1 - \frac{1}{4}\right)^2 \cdots \left(1 - \frac{1}{2n}\right)^2$$

$$> \left(1 - \frac{1}{2}\right)^2\left(1 - \frac{1}{3}\right)\left(1 - \frac{1}{4}\right) \cdots \left(1 - \frac{1}{2n-1}\right)\left(1 - \frac{1}{2n}\right)$$

$$= \frac{1}{4n} \quad \text{on successive cancellation.}$$

Hence H is persistent as $\sum_n \mathbf{P}(H_{2n})$ diverges.

For $c \geqslant 3$, we have similarly that

$$\mathbf{P}(H_{2n}) = \left(\frac{1}{4^n}\binom{2n}{n}\right)^c = \left(\left(1 - \frac{1}{2}\right)^2\left(1 - \frac{1}{4}\right)^2 \cdots \left(1 - \frac{1}{2n}\right)^2\right)^{c/2}$$

$$< \left(\frac{1}{2}\cdot\frac{2}{3}\cdot\frac{3}{4}\cdot\frac{4}{5} \cdots \frac{2n-1}{2n}\cdot\frac{2n}{2n+1}\right)^{c/2} = \left(\frac{1}{2n+1}\right)^{c/2}.$$

Hence H is transient as $\sum_n (1/(2n+1))^{c/2} < \infty$. ●

(9) Example: Stationary renewal Let $X > 0$ have p.g.f. $G(s)$. Show that if $\mathbf{E}(X) < \infty$, then

$$H(s) = \frac{1}{\mathbf{E}(X)} \frac{1 - G(s)}{1 - s}$$

is the p.g.f. of a non-negative random variable.

Now consider the delayed case of a recurrent event process in which $\mathbf{E}(s^{X_2}) = G(s)$ and $\mathbf{E}(s^{X_1}) = H(s)$. Show that for all n

(10) $$\mathbf{P}(H_n) = \frac{1}{G'(1)}.$$

Solution From (6.1.6) we have that $H(s)$ is a power series with non-negative coefficients. Furthermore, by L'Hôpital's rule

$$H(1) = \lim_{s \uparrow 1} \frac{-G'(s)}{-\mathbf{E}(X)} = 1$$

Hence $H(s)$ is a p.g.f. Finally, if $D(s) = H(s)$ in (6), then

$$V(s) = \frac{1}{\mathbf{E}(X_2)(1 - s)},$$

and the result follows. ●

6.8 Random walks

Recall that if $(X_i; i \geqslant 1)$ are independent and identically distributed, then $S_n = S_0 + \sum_1^n X_i$ is a random walk. Since generating functions have been so useful in handling sums of random variables, we may expect them to be exceptionally useful in analysing random walks. If X has p.g.f. $G(z)$, then trivially we have, when $S_0 = 0$,

$$G_n(z) = \mathbf{E}(z^{S_n}) = (G(z))^n.$$

It follows that we can define the function H by

(1) $$H(z, w) = \sum_{n=0}^{\infty} w^n G_n(z) = (1 - wG(z))^{-1}.$$

This bivariate generating function tells us everything about S_n in principle, as $\mathbf{P}(S_n = r)$ is the coefficient of $z^r w^n$ in $H(z, w)$. However the analytical effort required to work at this level of generality is beyond our scope. We proceed by considering simple examples.

(2) Example: Simple symmetric random walk Let $(S_n = \sum_1^n X_i;\ n \geqslant 0)$ be a simple symmetric random walk, with $S_0 = 0$. Let H_n be the event that $S_n = 0$. Because steps of the walk are independent and identically distributed, it follows that visits to the origin form an ordinary renewal process. Here $u_n = \mathbf{P}(S_n = 0)$.

Define the first passage times

$$T_j = \min\{n > 0: S_n = j | S_0 = 0\},$$

and the generating functions $U(s) = \sum_0^\infty u_n s^n$, and $G_j(s) = \mathbf{E}(s^{T_j})$.

Find $U(s)$ and $G_0(s)$, and show that the simple symmetric random walk is persistent null.

Solution We give two methods of finding $U(s)$ and $G_j(s)$. For the first define

$$T_j^* = \min\{n: S_n = j - 1 | S_0 = -1\},$$

and let $\hat{T}_1$ be a random variable having the same distribution as T_1, but independent of T_1. Because the steps of the walk are independent, and symmetrically and identically distributed, it follows that

$$\mathbf{E}(s^{T_1}) = \mathbf{E}(s^{T_{-1}}), \tag{3}$$

$$T_2 = T_1 + \hat{T}_1, \tag{4}$$

and

$$\mathbf{E}(s^{T_2^*}) = \mathbf{E}(s^{T_2}) = \mathbf{E}(s^{T_1+\hat{T}_1}) \tag{5}$$

$$= (G_1(s))^2, \quad \text{by independence.} \tag{6}$$

Hence, by conditional expectation

$$G_1(s) = \mathbf{E}(\mathbf{E}(s^{T_1}|X_1)) = \frac{1}{2}\mathbf{E}(s^{T_1}|X_1 = 1) + \frac{1}{2}\mathbf{E}(s^{T_1}|X_1 = -1) \tag{7}$$

$$= \frac{1}{2}s + \frac{1}{2}\mathbf{E}(s^{1+T_2^*}) = \frac{1}{2}s + \frac{1}{2}s(G_1(s))^2 \quad \text{by (6).}$$

One root of (7) is a probability generating function, so this root is $G_1(s)$, namely

$$G_1(s) = (1 - (1 - s^2)^{\frac{1}{2}})/s.$$

Now using conditional expectation again

$$G_0(s) = \mathbf{E}(\mathbf{E}(s^{T_0}|X_1)) \tag{8}$$

$$= \frac{1}{2}\mathbf{E}(s^{T_0}|X_1 = 1) + \frac{1}{2}\mathbf{E}(s^{T_0}|X_1 = -1)$$

$$= \frac{1}{2}s\mathbf{E}(s^{T_1}) + \frac{1}{2}s\mathbf{E}(s^{T_{-1}}) = sG_1(s) \quad \text{by (3)}$$

$$= 1 - (1 - s^2)^{\frac{1}{2}}.$$

Hence

$$U(s) = (1 - s^2)^{-1/2} \quad \text{by Theorem 6.7.3.} \tag{9}$$

Alternatively we could observe that $S_{2n} = 0$ if and only if the walk has taken n steps in each direction. They may be taken in any order so

$$u_{2n} = \binom{2n}{n}\left(\frac{1}{2}\right)^{2n}. \tag{10}$$

Now recall that by the negative binomial theorem

$$\sum_0^\infty \binom{2n}{n}\left(\frac{1}{2}\right)^{2n} x^n = (1 - x)^{\frac{-1}{2}}, \tag{11}$$

and (9) and (8) follow.

Setting $s = 1$ shows that $G_0(1) = 1$ (and $U(1) = \infty$) so H is persistent. However

$$\frac{d}{ds}G_0(s) = \frac{s}{(1-s^2)^{\frac{1}{2}}},$$

and setting $s = 1$ shows that H is null; the expected number of steps to return to the origin is infinite as we know already, recall Example 5.6.27. ●

Now that we have the generating functions $U(s)$ and $G_0(s)$, we can provide slicker derivations of earlier results. For example

$$G_0(s) = 1 - (1-s^2)^{\frac{1}{2}} = 1 - (1-s^2)U(s).$$

Hence equating coefficients of s^{2k} gives (5.17.1)

(12) $$f_{2k} = u_{2k-2} - u_{2k}.$$

Also

$$s\frac{d}{ds}G_0(s) = s^2(1-s^2)^{-\frac{1}{2}} = s^2U(s)$$

and so equating coefficients again gives (5.17.6)

(13) $$2kf_{2k} = u_{2k-2}.$$

See Problem 41 for another simple application of this. Here is a trickier application.

(14) Example: Truncated walk Let $(S_n; n \geqslant 1)$ be a simple symmetric random walk with $S_0 = 0$, and let $T = \min\{n > 0: S_n = 0\}$. Let $T \wedge 2m = \min\{T, 2m\}$ and show that

(15) $$\mathbf{E}(T \wedge 2m) = 4mu_{2m} = 2\mathbf{E}(|S_{2m}|).$$

Solution We shall establish (15) by showing that all three terms have the same generating function. Equality then follows by the uniqueness theorem. Firstly

$$\sum_0^\infty 4mu_{2m}s^{2m} = 2s\sum_1^\infty 2ms^{2m-1}u_{2m} = 2s\frac{d}{ds}U(s) = \frac{2s^2}{(1-s^2)^{3/2}}.$$

Secondly, recalling (13) and (5.17.2),

$$\mathbf{E}(T \wedge 2m) = \sum_{k=1}^m 2kf_{2k} + 2m\mathbf{P}(T > 2m) = \sum_{k=1}^m u_{2k-2} + 2mu_{2m}.$$

Hence

$$\begin{aligned}\sum_m s^{2m}\mathbf{E}(T \wedge 2m) &= \sum_m s^{2m}\sum_{k=0}^{m-1} u_{2k} + s\sum_m 2ms^{2m-1}u_{2m}\\ &= \frac{s^2U(s)}{1-s^2} + sU'(s) \quad \text{using (6.1.10)}\\ &= \frac{2s^2}{(1-s^2)^{\frac{3}{2}}}.\end{aligned}$$

Finally, using the hitting time theorem (5.6.17),

$$
\begin{aligned}
\sum_m s^{2m}\mathbf{E}(|S_{2m}|) &= 2\sum_m s^{2m}\sum_{k=1}^{m} 2k\mathbf{P}(S_{2m}=2k)\\
&= 2\sum_m s^{2m}\sum_{k=1}^{m} 2mf_{2k}(2m)\\
&= 2s\frac{d}{ds}\sum_m s^{2m}\sum_{k=1}^{m} f_{2k}(2m)\\
&= 2s\frac{d}{ds}(G_2(s)+G_4(s)+G_6(s)+\ldots)\\
&= 2s\frac{d}{ds}\frac{(G_1(s))^2}{(1-(G_1(s))^2)} \quad \text{by Example (2)}\\
&= s\frac{d}{ds}((1-s^2)^{\frac{1}{2}}-1) = \frac{s^2}{(1-s^2)^{\frac{3}{2}}}.
\end{aligned}
$$

●

As a final example of the use of generating functions in random walks, we establish yet another arc-sine law.

(16) **Example: Arc-sine law for leads** Let $(S_n;\ n \geqslant 0)$ be a simple symmetric random walk with $S_0 = 0$. Of the first $2n$ steps, let L_{2n} denote the number which do not enter the negative half-line. Show that

(17) $$\mathbf{P}(L_{2n}=2k) = 4^{-n}\binom{2k}{k}\binom{2n-2k}{n-k}.$$

Solution Define the generating functions

(18) $$G_{2n}(s) = \mathbf{E}(s^{L_{2n}})$$

and

$$H(s,t) = \sum_{n=0}^{\infty} t^{2n} G_{2n}(s).$$

Let T be the number of steps until the walk first revisits zero, and recall that

(19) $$F(s) = \mathbf{E}(s^T) = \sum_1^{\infty} s^{2r} f(2r) = 1-(1-s^2)^{\frac{1}{2}}.$$

Now using conditional expectation

$$
\begin{aligned}
\mathbf{E}(s^{L_{2n}}) &= (\mathbf{E}(\mathbf{E}(s^{L_{2n}})|T)\\
&= \sum_{r=1}^{n}\mathbf{E}(s^{L_{2n}}|T=2r)f(2r) + \mathbf{E}(s^{L_{2n}}|T>2n)\sum_{r=n+1}^{\infty} f(2r).
\end{aligned}
$$

Now, depending on the first step,

$$\mathbf{P}(L_T=T) = \frac{1}{2} = \mathbf{P}(L_T=0),$$

and visits to zero constitute regeneration points for the process L_{2n}. Hence we may rewrite (18) as

$$G_{2n}(s) = \sum_{r=1}^{n} G_{2n-2r}\left(\frac{1}{2}(s^{2r}+1)\right)f(2r) + \sum_{r=n+1}^{\infty} f(2r)\left(\frac{1}{2}(s^{2n}+1)\right).$$

Multiplying by t^{2n}, and summing over n yields

$$H(s, t) = \frac{1}{2}H(s, t)(F(st) + F(t)) + \frac{1}{2}\cdot\frac{1-F(st)}{1-t^2s^2} + \frac{1}{2}\cdot\frac{1-F(t)}{1-t^2},$$

by the convolution theorem. Now substituting for $F(.)$ from (19) gives

$$H(s, t) = ((1-s^2t^2)(1-t^2))^{-\frac{1}{2}}.$$

The coefficient of $t^{2n}s^{2k}$ in this is (17). Now use Exercise 5.17.8 to produce the arc-sine distribution. ●

WORKED EXAMPLES AND EXERCISES

6.9 Example: Gambler's ruin and first passages

Let $S_n = a + \sum_{i=1}^{n} X_i$ be a simple random walk with $a \geqslant 0$.

(a) Suppose that $0 \leqslant S_n \leqslant K$, and that the walk stops as soon as either $S_n = 0$ or $S_n = K$. (In effect, this is the gambler's ruin problem.) Define $T_{a0} = \min\{n: S_n = 0\}$, and find $\mathbf{E}(s^{T_{a0}}) = F_a(s)$ (say).

(b) Now suppose that $K = \infty$, and define $T_{a0} = \min\{n: S_n = 0\}$. (In effect, this is the gambler's ruin with an infinitely rich opponent.) Find $\mathbf{E}(s^{T_{10}}) = F_{1,0}(s)$ (say).

Solution (a) Of course T_{a0} is defective in general, since the walk may stop at K. This makes no difference to the fact that, by conditional expectation, for $0 < a < K$,

(1)
$$\begin{aligned} F_a(s) &= \mathbf{E}(\mathbf{E}(s^{T_{a0}}|X_1)) = p\mathbf{E}(s^{T_{a0}}|X_1 = 1) + q\mathbf{E}(s^{T_{a0}}|X_1 = -1) \\ &= ps\mathbf{E}(s^{T_{a+1,0}}) + qs\mathbf{E}(s^{T_{a-1,0}}) = psF_{a+1}(s) + qsF_{a-1}(s). \end{aligned}$$

Since the walk stops at 0 or K,

(2)
$$F_K(s) = 0 \quad \text{and} \quad F_0(s) = 1.$$

The difference equation (1) has auxiliary equation $psx^2 - x + qs = 0$, with roots

$$\lambda_1(s) = (1 + (1-4pqs^2)^{\frac{1}{2}})/(2ps)$$
$$\lambda_2(s) = (1 - (1-4pqs^2)^{\frac{1}{2}})/(2ps).$$

Hence, in the by now familiar way, the solution of (1) which satisfies (2) is

(3)
$$F_a(s) = \frac{\lambda_1^K\lambda_2^a - \lambda_1^a\lambda_2^K}{\lambda_1^K - \lambda_2^K}.$$

(b) When the walk is unrestricted it is still the case, by conditional expectation, that

(4)
$$F_{1,0}(s) = \mathbf{E}(\mathbf{E}(s^{T_{10}}|X_1)) = ps\mathbf{E}(s^{T_{20}}) + qs.$$

However, using the same argument as we did above for the symmetric random walk yields $\mathbf{E}(s^{T_{20}}) = (\mathbf{E}(s^{T_{10}}))^2$. Substituting in (4) shows that $F_{1,0}(s)$ is either $\lambda_1(s)$ or $\lambda_2(s)$. However $\lambda_1(s)$ is not the p.g.f. of a non-negative random variable and so

(5)
$$F_{1,0}(s) = \lambda_2(s) = (1 - (1 - 4pqs^2)^{\frac{1}{2}})/(2ps).$$

(6) Exercise Let $T_{aK} = \min\{n: S_n = K\}$.
Show that

$$\mathbf{E}(s^{T_{aK}}) = \frac{\lambda_1^a - \lambda_2^a}{\lambda_1^K - \lambda_2^K}.$$

(7) Exercise What is the p.g.f. of the duration of the game in the gambler's ruin problem?

(8) Exercise What is $\mathbf{E}(s^{T_{a0}})$ for the unrestricted random walk?

(9) Exercise For the unrestricted random walk started at $a > 0$, find the probability that the walk ever visits 0, and $\mathbf{E}(T_{a0} | T_{a0} < \infty)$.

(10) Exercise Let S_n be a simple random walk with $S_0 = 0$. Let $T = \min\{n > 0: S_n = 0\}$. Show that $\mathbf{E}(s^T) = 1 - (1 - 4pqs^2)^{\frac{1}{2}}$. What is $\mathbf{E}(T | T < \infty)$?

6.10 Example: 'Fair' pairs of dice

You have the opportunity to play a game of craps with either 'Lucky' Luke or 'Fortunate' Fred. Whose dice shall you play with? Luke's two dice are perfectly regular cubes, but the faces bear unorthodox numbers:

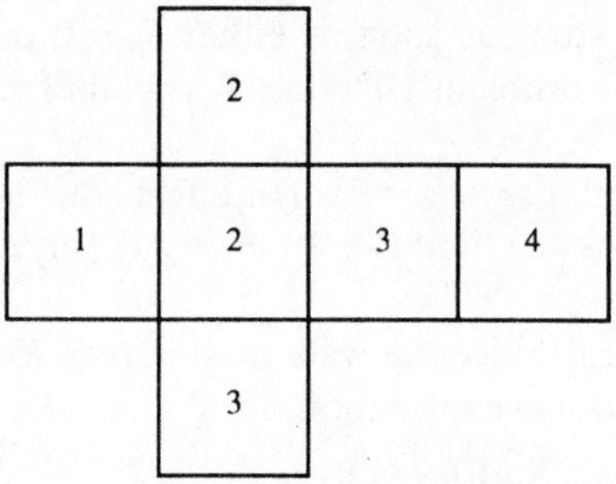

and

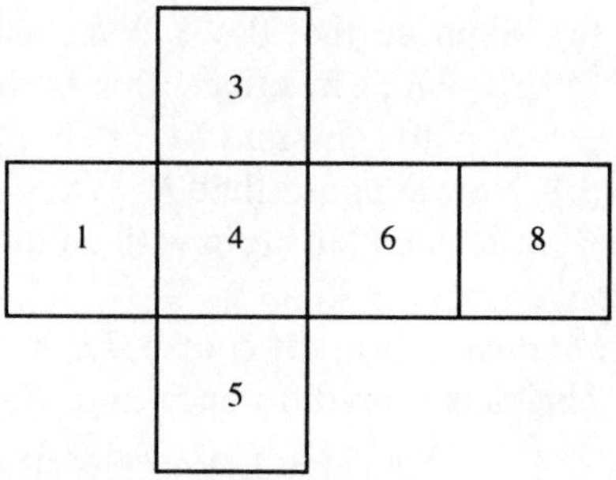

Luke explains that, when rolled, the sum of his two dice has the same mass function as the sum of two conventional fair dice; he uses these to ensure that no one can surreptitiously switch to unfair dice.

Fred's two dice are conventionally numbered, but are irregular cubes. Fred explains that these have been cleverly biased so that, when rolled, the sum has the same mass function as two fair dice; their irregular shape ensures that no-one can secretly switch to unfair dice.

Assuming you wish to play at the usual odds, whose dice should you use? (Sadly your own dice were confiscated by a casino last week.)

Solution Let X and Y be the scores of two fair dice. The p.g.f. of their sum is

$$\begin{aligned}\mathbf{E}(s^{X+Y}) &= \mathbf{E}(s^X)\mathbf{E}(s^Y) = (\tfrac{1}{6}(s + s^2 + s^3 + s^4 + s^5 + s^6))^2 \\ &= \frac{s^2}{36}(1 + 2s + 3s^2 + 4s^3 + 5s^4 + 6s^5 + 5s^6 + 4s^7 + 3s^8 + 2s^9 + s^{10}) \\ &= G(s) \quad \text{(say)}.\end{aligned}$$

Now the sum of Luke's dice $L_1 + L_2$ has p.g.f.

$$\mathbf{E}(s^{L_1+L_2}) = \frac{1}{6}(s + 2s^2 + 2s^3 + s^4)\frac{1}{6}(s + s^3 + s^4 + s^5 + s^6 + s^8) = G(s)$$

on multiplying out the brackets. So Luke's claim is correct.

On the other hand $G(s)$ can be factorized as

$$36G(s) = s^2(1 + s)^2(1 - s + s^2)^2(1 + s + s^2)^2,$$

where $1-s+s^2$ and $1+s+s^2$ are irreducible, having complex roots. Hence there are only two possibilities for the generating functions of Fred's dice:

either

(i) $$\mathbf{E}(s^{F_1}) = \frac{1}{2}s(1 + s)(1 - s + s^2)^2$$

and

$$\mathbf{E}(s^{F_2}) = \frac{s}{18}(1 + s)(1 + s + s^2)^2;$$

or (ii) the dice are fair:

$$\mathbf{E}(s^{F_1}) = \mathbf{E}(s^{F_2}) = \frac{1}{6}s(1 + s)(1 + s + s^2)(1 - s + s^2)$$
$$= \frac{1}{6}(s + s^2 + s^3 + s^4 + s^5 + s^6).$$

However, $(1 + s)(1 - s + s^2)^2 = 1 - s + s^2 + s^3 - s^4 + s^5$, and the negative coefficients ensure that $\frac{1}{2}s(1 + s)(1 - s + s^2)^2$ is not a p.g.f. The only remaining possibility is that the dice are fair, which palpably they are not.

This shows that the sum of two biased dice cannot have the same mass function as the sum of two fair dice. Thus Fred's claim is incorrect, his dice are as crooked as yours probably were.

You should play with Luke's dice.

(1) Exercise You have two fair tetrahedral dice whose faces are numbered conventionally 1, 2, 3, 4. Show how to re-number the faces so that the distribution of the sum is unchanged.

(2) Exercise Yet another regular Platonic solid is the dodecahedron with 12 pentagonal faces.

(a) Write down the generating function of the sum of two fair dodecahedra with faces numbered 1 to 12 inclusive

(b) Show that two such dodecahedral dice can be biased in such a way that their sum has the same distribution as the sum of the fair dice.

Hint: let

$$f(x) = x + x^{12} + (2 - \sqrt{3})(x^2 + x^{11}) + (5 - 2\sqrt{3})(x^3 + x^{10}) + (7 - 4\sqrt{3})(x^4 + x^9) + (10 - 5\sqrt{3})(x^5 + x^8) + (11 - 6\sqrt{3})(x^3 + x^7)$$

and

$$g(x) = (x + x^4 + x^7 + x^{10})(1 + \sqrt{3}x + x^2).$$

Consider $f(x)g(x)$.]

(3) Exercise Show that it is not possible to weight two conventional dice in such a way that the sum of the numbers shown is equally likely to take any value between 2 and 12 inclusive.

(4) Exercise Is it possible to re-number two fair dice so that their sum is equally likely to take any value between 2 and 12 inclusive?

(5) Exercise Can the sum of three biased dice have the same mass function as the sum of three fair dice?

Remark Some results of this example were recorded by SG Landry, LC Robertson and RM Shortt in the *American Mathematical Monthly*, 1988.

6.11 Example: Branching process

Let Z_n be the size at time n of the ordinary branching process defined in Example 6.3.16. Thus the rth individual in the nth generation (that is, at time n) is replaced by a family of size $X(r, n+1)$, where the $X(r, n+1)$ are independent and identically distributed, with mean μ, variance σ^2, and cumulant generating function

(1) $$\kappa(t) = \log(\mathbf{E}[\exp(tX(1, 1))]).$$

Show that

(2) $$\mathbf{E}(Z_n) = \mu^n.$$

Show also that

(3) $$\text{var}(Z_n) = \text{var}(Z_1)(\mathbf{E}(Z_1))^{n-1} + (\mathbf{E}(Z_1))^2\text{var}(Z_{n-1})$$

and hence find an expression for $\text{var}(Z_n)$ in terms of μ and σ, when $\mu \neq 1$.

Solution First recall the basic identity of branching processes, namely, given Z_{n-1}

(4) $$Z_n = \sum_{r=1}^{Z_{n-1}} X(r, n).$$

Now let the cumulant generating function of Z_n be $\kappa_n(t)$. Then by conditional expectation

$$\text{(5)} \quad \kappa_n(t) = \log(\mathbf{E}(e^{tZ_n})) = \log(\mathbf{E}(\mathbf{E}(e^{tZ_n}|Z_{n-1})))$$
$$= \log\left(\mathbf{E}\left(\mathbf{E}\exp\left[t\sum_{r=1}^{Z_{n-1}} X(r, n)\right]\middle| Z_{n-1}\right)\right) \quad \text{by (4)}$$
$$= \log(\mathbf{E}[(\mathbf{E}(e^{tX(1,1)}))^{Z_{n-1}}]) = \log(\mathbf{E}([e^{\kappa(t)}]^{Z_{n-1}}))$$
$$= \kappa_{n-1}(\kappa(t)).$$

Now expanding $\kappa_{n-1}(\kappa(t))$ using (6.4.12) gives

(6) $$\kappa_{n-1}(\kappa(t)) = \kappa(t)\mathbf{E}(Z_{n-1}) + \frac{1}{2}(\kappa(t))^2\text{var}(Z_{n-1}) + \dots$$
$$= \mu t\mathbf{E}(Z_{n-1}) + \frac{1}{2}\sigma^2\mathbf{E}(Z_{n-1})t^2 + \frac{1}{2}\mu^2\text{var}(Z_{n-1})t^2 + O(t^3)$$

on expanding $\kappa(t)$ using (6.4.12). Hence equating coefficients of t and t^2 in (5) and (6) now yields

(7) $$\mathbf{E}(Z_n) = \mu\mathbf{E}(Z_{n-1})$$

and

(8) $$\text{var}(Z_n) = \sigma^2\mu^{n-1} + \mu^2\text{var}(Z_{n-1}).$$

Iterating (7) gives (2), and equation (8) is just (3), as required. To solve the difference equation (8), we note first that $A\mu^{2n}$ is a solution of the reduced equation

$$\mathrm{var}(Z_n) = \mu^2 \mathrm{var}(Z_{n-1}).$$

By inspection a particular solution of (8), when $\mu \neq 1$, is given by $\sigma^2\mu^{n-1}/(1-\mu)$. Imposing the initial condition $\mathrm{var}(Z_1) = \sigma^2$ now shows that when $\mu \neq 1$, (8) has solution

(9) $$\mathrm{var}(Z_n) = \sigma^2\mu^{n-1}\frac{1-\mu^n}{1-\mu}.$$

(10) Exercise Find $\mathrm{var}(Z_n)$ when $\mu = 1$.

(11) Exercise Show that for $n > m$, $\mathbf{E}(Z_n Z_m) = \mu^{n-m}\mathbf{E}(Z_m^2)$.
Deduce that when $\mu = 1$,

$$\rho(Z_m, Z_n) = \left(\frac{m}{n}\right)^{\frac{1}{2}}.$$

(12) Exercise Find an expression for $\rho(Z_m, Z_n)$ when $\mu \neq 1$, and deduce that for $\mu > 1$ as $n, m \to \infty$, $\rho(Z_m, Z_n) \to 1$.

(13) Exercise If r is such that $r = \mathbf{E}(r^{Z_1})$, show that $\mathbf{E}(r^{Z_{n+1}} | Z_n) = r^{Z_n}$. What is r?

6.12 Example: Geometric branching

Let $(Z_n;\ n \geq 0)$ be an ordinary branching process with $Z_0 = 1$, and suppose that

(1) $$\mathbf{E}(s^{Z_1}) = \frac{1-p}{1-ps}.$$

(a) Show that for $p \neq \frac{1}{2}$

(2) $$\mathbf{E}(s^{Z_n}) = \frac{\rho^n - 1 - \rho s(\rho^{n-1} - 1)}{\rho^{n+1} - 1 - \rho s(\rho^n - 1)}$$

where $\rho = p/(1-p)$.

(b) Now let $(Z_n^*;\ n \geq 0)$ be an ordinary branching process with $Z_0 = 1$, family size distribution given by (1), and such that at time n, for all $n \geq 1$, one new particle is added to the population independently of Z_n.
Show that for $p < \frac{1}{2}$

(3) $$\lim_{n\to\infty} \mathbf{E}(s^{Z_n} | Z_n > 0) = \lim_{n\to\infty} \mathbf{E}(s^{Z_n^*}) = \frac{s(1-2p)}{1-p(1+s)}.$$

Solution (a) As usual let $\mathbf{E}(s^{Z_n}) = G_n(s)$. We establish (2) by induction. Assuming that (2) holds for n, it follows from Example 6..3.16 that

$$G_{n+1}(s) = G_n\left(\frac{1-p}{1-ps}\right)$$

$$= \frac{(\rho^n - 1)(1-ps) - p(\rho^{n-1} - 1)}{(\rho^{n+1} - 1)(1-ps) - p(\rho^n - 1)} \quad \text{by the induction hypothesis}$$

(4) $$= \frac{\rho^{n+1} - 1 - \rho s(\rho^n - 1)}{\rho^{n+2} - 1 - \rho s(\rho^{n+1} - 1)}.$$

Since (2) is true for $n = 1$, by (1), the result does follow by induction.

(b) Let $(\hat{Z}_n;\ n \geq 1)$ be a collection of independent random variables such that $\hat{Z}_n$

has the same distribution as Z_n. Now Z_n^* is the sum of the descendants of the initial individual, and the descendants of the fresh individual added at $n = 1$, and those of the next added at $n = 2$, and so on. That is to say Z_n^* has the same distribution as $1 + \hat{Z}_1 + \ldots + \hat{Z}_n$. Hence

(5)
$$\mathbf{E}(s^{Z_n^*}) = s\prod_{r=1}^{n} G_r(s) = \frac{s(\rho - 1)}{\rho^{n+1} - 1 - \rho s(\rho^n - 1)} \quad \text{by successive cancellation,}$$
$$\to \frac{s(\rho - 1)}{\rho s - 1} \quad \text{as } n \to \infty$$
$$= \frac{s(1 - 2p)}{1 - p(1 + s)}.$$

This is the generating function of a random variable with mass function

$$f(k + 1) = \left(1 - \frac{p}{1 - p}\right)\left(\frac{p}{1 - p}\right)^k.$$

For the other half of (3) we require the conditional generating function $\mathbf{E}(s^{Z_n} | Z_n > 0)$. Since $\mathbf{P}(Z_n = 0) = G_n(0)$, this is given by

(6)
$$\mathbf{E}(s^{Z_n} | Z_n > 0) = \frac{G_n(s) - G_n(0)}{1 - G_n(0)}.$$

Substituting for $G_n(.)$ from (2), we find

$$\mathbf{E}(s^{Z_n} | Z_n > 0) = \left(\frac{\rho^n - 1 - s\rho^n + \rho s}{\rho^{n+1} - 1 - s\rho^{n+1} + \rho s} - \frac{\rho^n - 1}{\rho^{n+1} - 1}\right)\Bigg/\left(1 - \frac{\rho^n - 1}{\rho^{n+1} - 1}\right)$$
$$\to \frac{s(1 - \rho)}{1 - \rho s} \quad \text{as } n \to \infty,$$

as required.

(7) **Exercise** Let Z_n be an ordinary branching process with family size mass function $\mathbf{P}(X = k) = 2^{-(k+1)}$; $k \geq 0$. Show that

$$G_n(s) = \frac{n - (n - 1)s}{n + 1 - ns}; \quad n \geq 0.$$

(8) **Exercise (7) continued** Show that in this case ($p = \frac{1}{2}$) we have

$$\mathbf{E}\left(\frac{Z_n}{n} \Big| Z_n > 0\right) \to 1, \quad \text{as } n \to \infty.$$

(9) **Exercise** (a) Let X be any non-negative random variable such that $\mathbf{E}(X) = 1$. Show that $\mathbf{E}(X | X > 0) \leq \mathbf{E}(X^2)$.
(b) Deduce that if $p > \frac{1}{2}$, $\mathbf{E}(Z_n \rho^{-n} | Z_n > 0) < 2p/(p - q)$.

(10) **Exercise** When $p < q$, find $\lim_{n\to\infty} \mathbf{E}(s^{Z_n^*} t^{Z_{n+m}^*})$.

6.13 Example: Waring's theorem; occupancy problems

Let $(A_i; 1 \leq i \leq n)$ be a collection of events (not necessarily independent). Let X be the number of these events which occur, and set $p_m = \mathbf{P}(X = m)$, $q_m = \mathbf{P}(X \geq m)$, and

$$s_m = \sum_{i_1<\ldots<i_m} \mathbf{P}(A_{i_1} \cap \ldots \cap A_{i_m}).$$

Show that

(a)
$$s_m = \mathbf{E}\binom{X}{m} = \sum_{i=m}^{n} \binom{i}{m} p_i;$$

(b)
$$p_m = \sum_{i=m}^{n} (-)^{i-m} \binom{i}{m} s_i;$$

and (c)
$$q_m = \sum_{i=m}^{n} (-)^{i-m} \binom{i-1}{m-1} s_i.$$

Solution (a) Recall that $I(A)$ is the indicator of the event A. Since

$$\mathbf{P}(A_{i_1} \cap \ldots \cap A_{i_m}) = \mathbf{E}(I(A_{i_1}) \ldots I(A_{i_m}))$$

it follows that the sum s_m is just the expected number of distinct sets of size m (of the A_i) that occur. But, given $X = i$, exactly $\binom{i}{m}$ such distinct sets of size m occur. Hence, by conditional expectation

(1)
$$s_m = \sum_{m}^{n} \binom{i}{m} \mathbf{P}(X = i) = \mathbf{E}\binom{X}{m}.$$

(b) Now define generating functions $G_s(z) = \sum_{m=0}^{n} z^m s_m$, and $G_q(z) = \sum_{m=0}^{n} z^m q_m$. By (1)

(2)
$$G_s(z-1) = \sum_{m=0}^{n} (z-1)^m s_m = \sum_{m=0}^{n} (z-1)^m \mathbf{E}\binom{X}{m} = \mathbf{E}\left(\sum_{m=0}^{X} (z-1)^m \binom{X}{m}\right)$$

$$= \mathbf{E}((1 + z - 1)^X) = \mathbf{E}(z^X) = \sum_{m=0}^{n} p_m z^m.$$

Now equating coefficients of z_m proves (b).

(c) By Theorem 3.6.7 or Theorem 6.1.5, we have

$$G_q(z) = \frac{1 - zG_X(z)}{1 - z}.$$

Hence by (2)

(3)
$$\frac{G_q(z) - 1}{z} = \frac{G_s(z-1) - 1}{z - 1}.$$

Equating coefficients of z^m yields (c).

(4) Exercise Show that

$$s_m = \sum_{i=m}^{n} \binom{i-1}{m-1} q_i.$$

(5) Exercise Let $t_m = \sum_{i_1<\ldots<i_m} \mathbf{P}(A_{i_1} \cup \ldots \cup A_{i_m})$. Find expressions for s_m, p_m and q_m in terms of $(t_i; 1 \leq i \leq n)$.

(6) **Exercise** If r balls are placed at random in n cells, so that each ball is independently equally likely to arrive in any cell, find the probability that exactly c cells are each occupied by exactly i balls.

6.14 Example: Bernoulli patterns and runs

A coin is tossed repeatedly, heads appearing with probability $p(= 1 - q)$ on each toss.

(a) Let X be the number of tosses required until the first occasion when successive tosses show HTH. Show that

(1) $$\mathbf{E}(s^X) = \frac{p^2qs^3}{1 - s + pqs^2 - pq^2s^3}.$$

(b) Let Y be the number of tosses required until the first occasion when three successive tosses show HTH or THT. Show that

(2) $$\mathbf{E}(s^Y) = \frac{pqs^3(1 - 2pqs + pqs^2)}{1 - s + pqs^2 - p^2q^2s^4}.$$

Solution (a) Consider the event that HTH does not appear in the first n tosses, and it does then appear as a result of the next three. The probability of this is $\mathbf{P}(X > n)p^2q$.

This event is the union of the following two disjoint events:

(i) the last two tosses of the n were HT, in which case $X = n + 1$;

and

(ii) the last two tosses of the n were not HT, in which case $X = n + 3$.

The probabilities of these two events are $\mathbf{P}(X = n + 1)pq$ and $\mathbf{P}(X = n + 3)$ respectively. Hence

(3) $$\mathbf{P}(X > n)p^2q = \mathbf{P}(X = n + 1)pq + \mathbf{P}(X = n + 3); \quad n \geq 0.$$

Multiplying (3) by s^{n+3}, summing over n, and recalling (6.1.7), yields

(4) $$\frac{1 - \mathbf{E}(s^X)}{1 - s}p^2qs^3 = pqs^2\mathbf{E}(s^X) + \mathbf{E}(s^X).$$

The required result (1) follows.

(b) First consider the event that neither HTH nor THT have appeared in the first n tosses and HTH then appears. This event is the union of three disjoint events:

(i) the last two tosses of the n were HT, so $Y = n + 1$, with the appearance of HTH;
(ii) the last two tosses of the n were TT, so $Y = n + 2$, with the appearance of THT;
(iii) otherwise $Y = n + 3$, with the appearance of HTH.

Let $f^1(n)$ denote the probability that $Y = n$ with the occurrence of HTH, and $f^2(n)$ the probability that $Y = n$ with the occurrence of THT. Then from the above we have

(5) $$\mathbf{P}(Y > n)p^2q = f^1(n+1)pq + f^2(n+2)p + f_1(n+3).$$

Hence, multiplying by s^{n+3} and summing over n,

(6) $$\frac{1-\mathbf{E}(s^Y)}{1-s}p^2qs^3 = pqs^2G^1(s) + spG^2(s) + G^1(s)$$

where $G^1(s) = \sum_n s^n f^i(n)$; $i = 1, 2$.

Second, consider the event that neither HTH or THT have appeared in the first n tosses and THT then occurs. Arguing as in the first case yields

(7) $$\mathbf{P}(Y > n)pq^2 = f^2(n+1)pq + f^1(n+2)q + f^1(n+3).$$

Hence

(8) $$\frac{1-\mathbf{E}(s^Y)}{1-s}pq^2s^3 = pqs^2G^2(s) + qsG^1(s) + G^2(s).$$

Now we also have

(9) $$\mathbf{E}(s^Y) = G_1(s) + G_2(s)$$

so solving (6), (8) and (9) for $\mathbf{E}(s^Y)$ yields (2) as required.

(10) Exercise When $p = \frac{1}{2}$ find $\mathbf{E}(X)$ for all possible triples of the form HHH, HHT, etc. Comment on your results.

(11) Exercise Show that $\mathbf{E}(X) = 1/p + 1/(qp^2)$.

(12) Exercise Show that $\mathbf{E}(Y) = (1 + pq + p^2q^2)/(pq(1-pq))$.

(13) Exercise Let Z be the number of tosses required for the first occurrence of HHH. Find $\mathbf{E}(s^Z)$ and show that

$$\mathbf{E}(Z) = \frac{1}{p} + \frac{1}{p^2} + \frac{1}{p^3}.$$

(14) Exercise Let W be the number of tosses required for the first appearance of either HHH or TTT. Show that

$$\mathbf{E}(s^W) = \frac{s^3(p^3 + q^3 + qp(p^2+q^2)s + p^2q^2s^2)}{1 - pqs^2 - pqs^3 - p^2q^2s^4}.$$

(15) Exercise Let V be the number of tosses required for the first appearance of either r consecutive heads or ρ consecutive tails. Find $\mathbf{E}(s^V)$, and show that

$$\mathbf{E}(V) = \left(\frac{p^r(1-p)}{1-p^r} + \frac{q^\rho(1-q)}{1-q^\rho}\right)^{-1}.$$

(16) Exercise Find the expected number of tosses required for the first appearance of $HTHTH$.

6.15 Example: Waiting for unusual light bulbs

The light bulbs in the sequence illuminating your room have independent and identically distributed lifetimes, so the replacement times form an ordinary renewal process (as defined in Example 6.7.2). Suppose the lifetimes are $(X_i;\ i \geq 1)$ with common mass function $f(k)$.

A bulb is called 'unusual' if its life is shorter than a or longer than b, where $a \leq b$. Let T be the time at which a bulb is first identified as being an unusual bulb. Show that (for integers a and b),

(1) $$\mathbf{E}(s^T) = \frac{\mathbf{E}(I_a s^{X_1}) + s^b \mathbf{E}(I_b)}{I - \mathbf{E}(s^{X_1} I_a^c I_b^c)}$$

where I_a and I_b are the indicators of the events $\{X_1 < a\}$ and $\{X_1 > b\}$ respectively. That is

$$I_a = I\{X_1 < a\}, \quad \text{with } I_a^c = 1 - I_a,$$

and

$$I_b = I\{X_1 > b\}, \quad \text{with } I_b^c = 1 - I_b.$$

Solution Since $I_a + I_b + I_a^c I_b^c = 1$, we can write

(2) $$\mathbf{E}(s^T) = \mathbf{E}(I_a s^T) + \mathbf{E}(I_b s^T) + \mathbf{E}(I_a^c I_b^c s^T).$$

Now on the event I_a, $X_1 = T$, because the first bulb failed before a and was identified as unusual at X_1. So

(3) $$\mathbf{E}(I_a s^T) = \mathbf{E}(I_a s^{X_1}).$$

On the event $I_a^c I_b^c$, the process regenerates at the first replacement $X_1 \in [a, b]$, and so $T = X_1 + T'$, where T' is independent of X_1 and has the same distribution as T. So

(4) $$\mathbf{E}(I_a^c I_b^c s^T) = \mathbf{E}(s^T)\mathbf{E}(I_a^c I_b^c s^{X_1}).$$

Finally on the event I_b, the first light bulb is identified as unusual when it survives beyond b, so

(5) $$\mathbf{E}(S^T I_b) = s^b \mathbf{E}(I_b).$$

Substituting (3), (4) and (5) into (2), gives (1).

(6) Exercise Find the expected time until a light bulb has a lifetime shorter than a. Evaluate this when $f_X(k) = qp^{k-1}$, $k \geq 1$.

(7) Exercise: Crossing the road Successive cars pass at instants X_i seconds apart ($i \geq 1$). You require b seconds in order to cross the road. If the random variables (X_i; $i \geq 1$) are independent and identically distributed, find your expected waiting time until you cross. Evaluate this when $f_X(k) = qp^{k-1}$; $k \geq 1$.

(8) Exercise Let L be the time until a light bulb has lasted longer than r. Show that

$$\mathbf{E}(s^L) = \frac{s^r \mathbf{P}(X > r)}{1 - \sum_1^r s^k \mathbf{P}(X = k)}.$$

(9) Exercise A biased coin is tossed repeatedly; on each toss it shows a head with probability $p(= 1 - q)$. Let W be the number of tosses until the first occasion when r consecutive tosses have shown heads. Show that

$$\mathbf{E}(s^W) = \frac{(1 - ps)p^r s^r}{1 - s + qp^r s^{r+1}}.$$

(10) Exercise In n tosses of a biased coin let L_n be the length of the longest run of heads, and set $\pi_{n,r} = \mathbf{P}(L_n < r)$. Show that

$$1 + \sum_{n=1}^{\infty} s^n \pi_{n,r} = \frac{1 - p^r s^r}{1 - s + qp^r s^{r+1}}.$$

PROBLEMS

1 Let $G(s) = \sum_0^\infty f_X(k)s^k$, where $f_X(k) = \mathbf{P}(X = k);\ k \geqslant 0$. Show that:

(a) $\sum_0^\infty \mathbf{P}(X < k)s^k = sG(s)/(1 - s)$;

(b) $\sum_0^\infty \mathbf{P}(X \geqslant k)s^k = (1 - sG(s))/(1 - s)$.

2 Find the probability generating function of each of the following distributions, and say where it exists.

(a) $f(k) = \dfrac{1}{n};\quad 1 \leqslant k \leqslant n.$

(b) $f(k) = \dfrac{1}{2n+1};\quad -n \leqslant k \leqslant +n.$

(c) $f(k) = \dfrac{1}{k(k+1)};\quad 1 \leqslant k.$

(d) $f(k) = \begin{cases} \dfrac{1}{2k(k+1)} & \text{for } k \geqslant 1 \\ \dfrac{1}{2k(k-1)} & \text{for } k \leqslant -1. \end{cases}$

(e) $f(k) = \dfrac{1-c}{1+c}c^{|k|};\quad k \in \mathbb{Z}, 0 < c < 1.$

3 Which of the following are probability generating functions, and when?

(a) $\exp(-\lambda(1 - G_X(s)))$; where $\lambda > 0$, and $G_X(s)$ is a p.g.f.

(b) $\sin\left(\dfrac{\pi s}{2}\right)$

(c) $\left(\dfrac{q}{1-ps}\right)^r$ $\Big\}\ p + q = 1;\ p > 0,\ q > 0$

(d) $(q + ps)^r$

(e) $1 - (1 - s^2)^{\frac{1}{2}}$

(f) $\alpha \log(1 + \beta s)$

4 If the random variable X has p.g.f. $G_X(s)$ show that for constants a and b the random variable $aX + b$ has p.g.f. $s^b G(s^a)$. For what values of s is this defined?

5 Let X have p.g.f. $G(s)$. Describe a random variable Y which has p.g.f. $G_Y(s) = G(s)(2 - G(s))^{-1}$. For what values of s is this defined?

6 A loaded die may show different faces with different probabilities. Show that it is not possible to load two traditional cubic dice in such a way that the sum of their scores is uniformly distributed on $\{2, 3, \ldots, 12\}$.

7 The three pairs of opposite faces of a fair die show 1, 2 and 3 respectively. The two faces of a fair coin show 1 and 2 respectively.

(a) Find the distribution of the sum of their scores when tossed together.

(b) Is it possible to weight the die in such a way that the sum of the scores is uniform on $\{2, 3, 4, 5\}$?

8 Let X have p.g.f. $G(s)$, and let E be the event that X is even. Show that

$$\mathbf{E}(s^X | E) = \frac{G(s) + G(-s)}{G(1) + G(-1)}.$$

9 Define the probability generating function of an integer valued random variable X, and show how it may be used to obtain the mean μ_X, variance σ_X^2, and third moment about the mean γ_X.
(a) Let $Y = \sum_{i=1}^{N} X_i$, where the X_i are independent integer valued random variables identically distributed as X. Let $\mu_X = 0$, and let N be an integer valued random variable distributed independently of the X_i. Show that $\sigma_Y^2 = \mu_N \sigma_X^2$, and $\gamma_Y = \mu_N \gamma_X$.
(b) Find σ_Y^2 when $\mu_X \neq 0$.

10 An unfair coin is tossed n times, each outcome is independent of all the others, and on each toss a head is shown with probability p. The total number of heads shown is X. Use the probability generating function of X to find:
(a) the mean and variance of X,
(b) the probability that X is even,
(c) the probability that X is divisible by 3.

11 Let the non-negative random variable X have p.g.f. $G_X(s)$. Show that

$$G(s) = \frac{1}{\mathbf{E}(X)} \cdot \frac{1 - G_X(s)}{1 - s}$$

is the p.g.f. of a non-negative random variable Y. When is $G(s) = G_X(s)$?

12 Let $G_1(s)$ and $G_2(s)$ be probability generating functions, and suppose that $0 \leq \lambda \leq 1$. Show that $\lambda G_1 + (1 - \lambda) G_2$ is a p.g.f. and interpret this result.

13 In a multiple choice examination a student chooses between one true and one false answer to each question. Assume the student answers at random, and let N be the number of such answers until she first answers two successive questions correctly. Show that $\mathbf{E}(s^N) = s^2(4 - 2s - s^2)^{-1}$. Hence find $\mathbf{E}(N)$ and $\mathbf{P}(N = k)$. Now find $\mathbf{E}(N)$ directly.

14 A number X of objects are ranked in order of beauty (with no ties). You pick one at random with equal probability of picking any.
(a) If $X - 1$ has a Poisson distribution with parameter λ, show that the p.g.f. of the rank of the object you pick is

$$s\frac{1 - e^{\lambda(s-1)}}{\lambda(1 - s)}.$$

What is the mean rank of your object?
(b) What if X has the logarithmic distribution, $f_X(k) = cp^k/(k + 1)$; $k \geq 1$?

15 A biased coin is tossed N times, where N is a Poisson random variable with parameter λ. Show that if H is the number of heads shown, and T the number of tails, then H and T are independent Poisson random variables. Find the mean and variance of $H - T$.

16 A biased coin is tossed N times where N is a random variable with finite mean. Show that if the numbers of heads and tails are independent, then N is Poisson.
[You may wish to use the fact that all continuous solutions of $f(x + y) = f(x)f(y)$ take the form $f(x) = e^{\lambda x}$ for some λ.]

17 Let X_n have a negative binomial distribution with parameters n and p $(= 1 - q)$. Show (using generating functions) that if $n \to \infty$ in such a way that $\lambda = nq$ remains constant, then $\lim_{n\to\infty} \mathbf{P}(X_n = k) = e^{-\lambda}\lambda^k/k!$. Show that $\mathbf{E}(X_n) = nqp^{-1}$ and $\operatorname{var}(X) = nqp^{-2}$.

18 The events $(A_n;\ n \geq 1)$ are independent and $\mathbf{P}(A_n) = 1 - \exp(-\lambda_n)$. Define $N = \min\{n\colon A_n \text{ occurs}\}$.
(a) Show that $\mathbf{E}(s^N) = s + (s - 1)\sum_{n=1}^{\infty} \exp\left(-\sum_1^n \lambda_k\right) s^n$.
(b) Find $\mathbf{E}(s^N)$ and $\mathbf{E}(N)$, when $\lambda_n = a + \log n$.

19 The probability of obtaining heads when a certain coin is tossed is p. The coin is tossed repeatedly until a sequence of three heads is obtained. If p_n is the probability that this event occurs in n throws show that

$$p_0 = p_1 = p_2 = 0,$$

$$p_3 = p^3,$$

and

$$p_n = p^3(1-p)\left(1 - \sum_{k=0}^{n-4} p_k\right) \quad \text{if } n \geq 4.$$

Show that the generating function $G(s) = \sum_{k=0}^{\infty} p_k s^k$ is given by

$$G(s) = \frac{p^3 s^3 (1 - ps)}{1 - s + p^3(1-p)s^4}.$$

Now find the expected number of throws of an unbiased coin needed to obtain three consecutive heads.

20 Each packet of a certain breakfast cereal contains one token, coloured either red, blue or green. The coloured tokens are distributed randomly among the packets, each colour being equally likely. Let X be the random variable which takes the value j when I find my first red token in the jth packet which I open. Obtain the probability generating function of X, and hence find its expectation.

More generally, suppose that there are tokens of m different colours, all equally likely. Yet Y be the random variable which takes the value j when I first obtain a full set, of at least one token of each colour, when I open my jth packet. Find the generating function of Y, and show that its expectation is $m(1 + \frac{1}{2} + \frac{1}{3} + \ldots + \frac{1}{m})$.

21 A gambler repeatedly plays the game of guessing whether a fair coin will fall heads or tails when tossed. For each correct prediction he wins £1, and for each wrong one he loses £1. At the start of play he holds £n (where n is a positive integer), and he has decided to stop play as soon as either (i) he has lost all his money, or (ii) he possesses £K, where K is a given integer greater than n. Let $p(n)$ denote for $1 \leq n \leq K-1$ the probability that he loses all his money, and let $p(0) = 1$, $p(K) = 0$. Show that $p(n) = \frac{1}{2}(p(n-1) + p(n+1))$; $(1 \leq n \leq K-1)$. Prove that if $G(s)$ is defined for all real s by

$$G(s) = \sum_{n=0}^{K-1} p(n) s^n$$

then, provided $s \neq 1$,

$$G(s) = \frac{1}{(1-s)^2}(1 - (2 - p(1))s + p(K-1)s^{K+1}).$$

Hence, or otherwise, show that $p(1) = 1 - 1/K$, $p(K-1) = 1/K$ and that in general, $p(n) = 1 - n/K$.

22 A class of particles behaves in the following way. Any particle in existence at time n is replaced at time $n+1$ by a random number of similar particles having probability mass function $f(k) = 2^{-k}$; $k > 0$, independently of all other particles. At time zero there is exactly one particle in existence and the set of all succeeding particles is called its descendants.

Let the total number of particles which have ever existed by time n be S_n. Show that the p.g.f. $G_n(z) = \mathbf{E}(z^{S_n})$ satisfies

$$G_n(z) = \frac{z}{2 - G_{n-1}(z)} \quad \text{for } 0 \leq z \leq 1 \text{ and } n \geq 1.$$

Deduce that with probability one, the number of particles that ever exist is finite, but that as $n \to \infty$, $\mathbf{E}(S_n) \to \infty$.

23 Let $G(s)$ be the generating function of the family size in an ordinary branching process. Let Z_n be the size of the population in the nth generation, and let T_n be the total number of individuals who have ever lived up to that time. Show that $H_n(s, t)$, the joint generating function of Z_n and T_n satisfies $H_n(s, t) = tG(H_{n-1}(s, t))$.

24 Show that for each integer n, $(s+n-1)(s+n-2)\dots s/n!$, is the probability generating function of some random variable X. Show that as $n \to \infty$, $\mathbf{E}(X)/\log n \to 1$.

25 Find the probability generating function of the distribution

$$\mathbf{P}(X = k) = \left(\frac{a}{1+a}\right)^{\lambda} \frac{\lambda(\lambda+1)\dots(\lambda+k-1)}{(1+a)^k k!}; \quad k > 0, \mathbf{P}(X = 0) = \left(\frac{a}{1+a}\right)^{\lambda}.$$

26 Let X and Y be independent Poisson random variables with parameters λ and μ respectively. Find the joint probability generating function of $X - Y$ and $X + Y$.

Find the factorial moments of $X + Y$, and the cumulants of $X - Y$.

27 Let X be a binomial random variable with parameters n and p, and let Y be a binomial random variable with parameters m and q $(= 1 - p)$. Find the distribution of $X - Y + m$ and explain why it takes the form it does.

28 A series of objects passes a checkpoint. Each object has (independently) probability p of being defective, and probability α of being subjected to a check which infallibly detects a defect if it is present. Let N be the number of objects passing the checkpoint before the first defective is detected, and let D be the number of these passed objects which were defective (but undetected). Find:
(a) the joint p.g.f. of D and N;
(b) $\mathbf{E}(D/N)$.
If the check is not infallible, but errs with probability δ, find the above two quantities in this case.

29 Let the sequence $(a_i;\ i \geq 0)$ be defined by

$$\frac{1}{(2-s)^{n+1}} = \sum_0^{\infty} a_i s^i.$$

Show that $\sum_{i=0}^{n} a_i = \frac{1}{2}$, and interpret this result in terms of random variables.
[*Hint*: $(1+x)^{n-r} = (1 - x/(1+x))^r (1+x)^n$.]

30 A two-dimensional random walk $(X_n, Y_n;\ n \geq 0)$ evolves in the following way. If $(X_n, Y_n) = (x, y)$, then the next step is to one of the four points $(x+1, y)$, $(x-1, y)$, $(x, y+1)$, $(x, y-1)$, with respective probabilities $\alpha_1, \beta_1, \alpha_2, \beta_2$, where $\alpha_1 + \beta_1 + \alpha_2 + \beta_2 = 1$. Initially $(X_0, Y_0) = (0, 0)$. Define $T = \min\{n;\ X_n + Y_n = m\}$.
Find the distribution of T.

31 In Problem 30, if $\alpha_1 = \beta_1$ and $\alpha_2 = \beta_2$, show that $\mathbf{E}(X_n^2 + Y_n^2) = n$.
Also in Problem 30, if $\alpha_1 = \beta_1 = \alpha_2 = \beta_2 = \frac{1}{4}$,
(a) show that $\mathbf{E}(T) = \infty$;
(b) show that the point at which the walk hits $x + y = m$ is a proper random variable;
(c) find its generating function $\mathbf{E}(s^{X_T - Y_T})$.

32 Use the identity $t(1+t)^{n-1} = (1+t)^n \sum_{i=0}^{\infty} (-t^{-1})^i$ to prove that

$$\binom{n}{i} - \binom{n}{i+1} + \dots + (-)^{n-i}\binom{n}{n} = \binom{n-1}{i-1}.$$

33 Let the generating function of the family size in an ordinary branching process be $G(s) = 1 - p(1-s)^{\beta};\ 0 < p, \beta < 1$. Show that if $Z_0 = 1$

$$\mathbf{E}(s^{Z_n}) = 1 - p^{1+\beta+\dots+\beta^{n-1}}(1-s)^{\beta^n}.$$

34 Let the generating function of the family size of an ordinary branching process be $G(s) = q + ps$, and let $\mathbf{E}(s^{Z_0}) = e^{\lambda(s-1)}$. Let $T = \min\{n;\ Z_n = 0\}$. Show that

$$\mathbf{P}(T = n) = e^{-\lambda p^{n+1}} - e^{-\lambda p^n}.$$

35 Let the number of tosses required for a fair coin to show a head be T. An integer X is picked at random from $\{1, \ldots, T\}$ with equal probability $\frac{1}{T}$ of picking any one. Find $G_X(s)$.

36 Show that for $\alpha > 0$, $\beta > 0$, $\alpha + \beta < 1$,

$$G(s, t) = \frac{\log(1 - \alpha s - \beta t)}{\log(1 - \alpha - \beta)}$$

is a bivariate p.g.f. Find the marginal p.g.f.s and the covariance.

37 Let X and Y be independent with rth cumulants $\kappa_r^{(X)}$ and $\kappa_r^{(Y)}$. Show that $X + Y$ has rth cumulant $\kappa_r^{(X)} + \kappa_r^{(Y)}$.

38 Let X have cumulants κ_r; $r \geqslant 1$ and moments μ_r; $r \geqslant 1$. Show that $\kappa_1 = \mathbf{E}(X)$, $\kappa_2 = \operatorname{var}(X)$, and $\kappa_3 = \mu_3 - 3\mu_1\mu_2 + 2\mu_1^3$.

39 Show that the joint probability mass function

$$f(x, y) = \frac{1 - \lambda - \mu}{\mu}\binom{x + y - 1}{x}\lambda^x\mu^y; \quad x \geqslant 0, y \geqslant 1$$

has joint p.g.f.

$$G(s, t) = \frac{(1 - \lambda - \mu)t}{1 - \lambda s - \mu t}.$$

What is $\operatorname{cov}(X, Y)$?

40 Let X_m have generating function $(p/(1 - qs))^m$ where $p = 1 - q > 0$. Show that as $m \to 0$

$$\mathbf{E}(s^{X_m} | X_m > 0) \to \frac{\log(1 - qs)}{\log(1 - q)}.$$

41 Prove the identity

$$\sum_{k=0}^{n}\binom{2k}{k}4^{-k} = (2n + 1)\binom{2n}{n}4^{-n}.$$

Now let S_n be a simple symmetric random walk with $S_0 = 0$. Let v_n be the expected number of visits of the walk to zero, up to and including time n. Show that (including the initial visit)

$$v_{2n} = v_{2n+1} = (2n + 1)\binom{2n}{n}2^{-2n}.$$

42 Let $(S_n;\ n \geqslant 0)$ be a simple random walk with $S_0 = 0$. Let R_r be the number of steps until the walk first revisits the origin for the rth time, and let T_{0r} be the number of steps until the walk first visits r. Show that

$$\mathbf{E}(s^{T_{0r}}) = \left(\frac{1}{2qs}\right)^r \mathbf{E}(s^{R_r}).$$

Deduce that $\mathbf{P}(R_r = n) = (2q)^r\mathbf{P}(T_r = n - r)$, and hence that, as $n \to \infty$

$$\frac{\mathbf{P}(R_r = n)}{\mathbf{P}(R_1 = n)} \to r.$$

[H Kesten and F Spitzer have shown that this remains true for a wider class of random walks. (*J. d'Anal. Math.* Vol. 11, 1963).]

7
Continuous random variables

7.1 Density and distribution

Hitherto we have assumed that a random variable can take any one of only a countable set of values. However, suppose your height is 5 feet or 6 feet (or somewhere in between). Then at previous times (however briefly) your height in feet has taken every value in $[1, 5]$, including $\sqrt{2}$, e, π and so on. (And each value can be taken more than once, since you are taller in the morning than in the evening.) Thus, if X is the height of a randomly selected member of the population, the state space of X is not countable. There are many other simple examples of variables which may take any one of an uncountable number of values; for example: the brightness of a randomly chosen star; the time until some cell divides; the velocity of a comet; the direction of the wind, and so on. Think of some yourself.

In view of these remarks, we are about to introduce a new class of random variables such that the state space Ω is uncountable, and $X(\omega)$ may take any one of an uncountable number of real values. However, before we embark on this task, it is as well to reassure you that, despite their separate presentation, these new random variables share most of the useful properties of discrete random variables. Also, many of these properties are proved in exactly the same way as in the discrete case and (even better) we are able to use much of the same notation.

Thus, as in the discrete case, we start with a probability function $\mathbf{P}(.)$ defined on a collection $\mathcal{F}$ (the event space) of subsets of Ω (the sample space). Then we think of a random variable X as a real valued function $X(\omega)$ defined for each $\omega \in \Omega$.

Our first requirement (as in the discrete case) is a function that tells us about the relative likelihoods of possible values of X. Happily we already have such a function; recall the following:

(1) Definition The distribution function F of the random variable X is the function

(2)
$$F(x) = \mathbf{P}(A_x)$$

where A_x is the event $A_x = \{\omega: X(\omega) \leq x\}$, $x \in \mathbb{R}$. ▲

We usually write (2) as

(3)
$$F(x) = \mathbf{P}(X \leq x),$$

and denote F by $F_X(x)$ when we wish to stress the role of X.

Notice that for Definition 1 to be meaningful, the event A_x must be in $\mathcal{F}$, so that we know $\mathbf{P}(A_x)$. Thus $X(.)$ only qualifies to appear in Definition 1 if $A_x \in \mathcal{F}$ for all x. This is true throughout this book; the implications of this so-called *measurability* condition are explored in more advanced books.

(4) **Example: Uniform distribution** You devise an experiment in which the outcome is equally likely to be any point Q in the interval $[0, 1]$. Thus the sample space is the set of points $(Q: Q \in [0, 1])$. The event space $\mathcal{F}$ will include all intervals in $[0, 1]$; we omit the proof that such an $\mathcal{F}$ exists, but you should rest assured that it does exist. Define the random variable $X(Q)$ to be the distance from the origin O to Q. From the nature of the experiment, if

$$A_{ab} = \{Q: Q \in (a, b)\}, \qquad 0 \leq a \leq b \leq 1$$

then $\mathbf{P}(A_{ab}) = b - a$. Hence X has distribution function

$$F_X(x) = \mathbf{P}(A_{0x}) = \begin{cases} 0 & \text{if} \quad x < 0 \\ x & \text{if} \quad 0 \leq x \leq 1 \\ 1 & \text{if} \quad x > 1. \end{cases}$$ ●

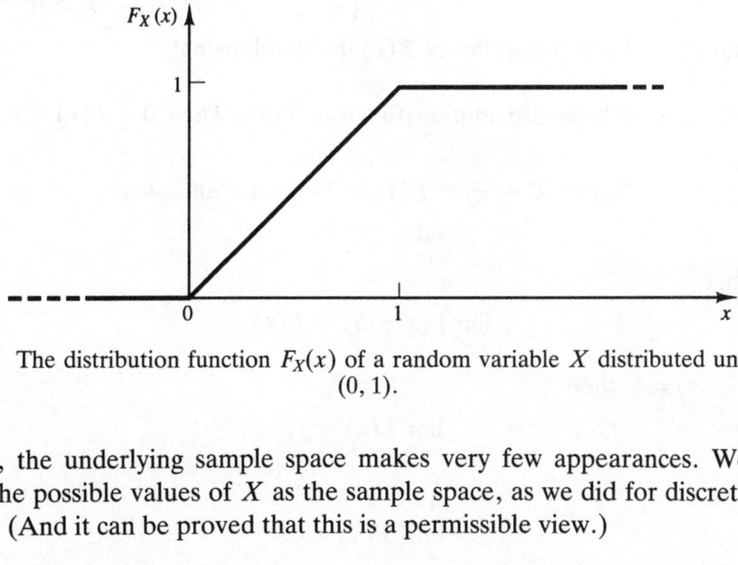

Figure 7.1 The distribution function $F_X(x)$ of a random variable X distributed uniformly on $(0, 1)$.

In future, the underlying sample space makes very few appearances. We tend to think of the possible values of X as the sample space, as we did for discrete random variables. (And it can be proved that this is a permissible view.)

(5) **Example** A point Q is picked at random in a triangle of area a, with base of length b. Let X be the perpendicular distance from Q to the base. What is the distribution function $F_X(x)$?

Solution Let the height of the triangle AP be h. The event $X > x$ occurs when Q lies inside the triangle ABC. For reasons of symmetry

$$\mathbf{P}(Q \in \text{ABC}) = (\text{area of ABC})/a = \left(\frac{h - x}{h}\right)^2.$$

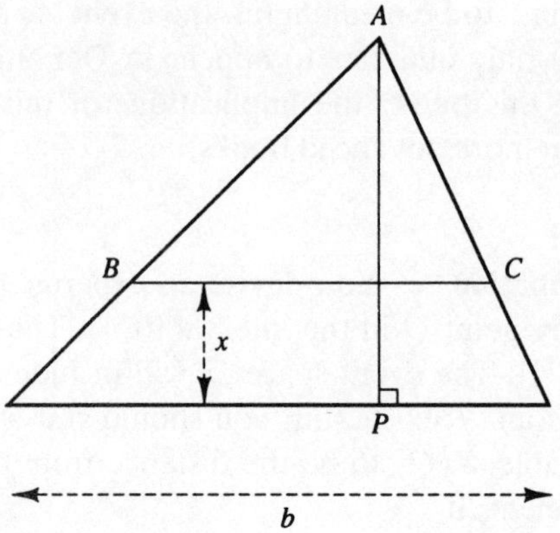

Figure 7.2 The altitude AP is of length h.

Hence

$$F_X(x) = \mathbf{P}(X \leqslant x) = 1 - \mathbf{P}(Q \in \text{ABC}) = \begin{cases} 0 & x < 0 \\ 1 - \left(\dfrac{h-x}{h}\right)^2 & 0 \leqslant x \leqslant h \\ 1 & x > h. \end{cases} \qquad \bullet$$

We summarize the basic properties of $F(x)$ in the following:

(6) **Theorem** Let X have distribution function $F(x)$. Then $0 \leqslant F(x) \leqslant 1$ for all x, and

(7) $$\mathbf{P}(x < X \leqslant y) = F(y) - F(x) \quad \text{for all } x \leqslant y, \\ \geqslant 0.$$

(8) If $h > 0$ then

$$\lim_{h \to 0} F(x + h) = F(x).$$

(9) If $\mathbf{P}(|X| < \infty) = 1$, then

$$\lim_{x \to \infty} F(x) = 1,$$

and

$$\lim_{x \to -\infty} F(x) = 0.$$

In plain words, this theorem says that as x increases, $F(x)$ is non-decreasing, continuous on the right, and lies between 0 and 1. It can be shown conversely that any function with these properties is the distribution function of some random variable.

Proof The first result follows from (2) because $0 \leqslant \mathbf{P}(A_x) \leqslant 1$. To show (7) note that

$$\{\omega\colon x < X \leqslant y\} = A_y \cap A_x^c = A_y \backslash A_x.$$

Because $A_x \subseteq A_y$ for $x \leqslant y$, we have

$$\mathbf{P}(x < X \leqslant y) = \mathbf{P}(A_y) - \mathbf{P}(A_x) = F(y) - F(x) \geqslant 0$$

by the non-negativity of $\mathbf{P}(.)$. To prove (8) we use the continuity of $\mathbf{P}(.)$, see (1.5.4). Let $(h_k;\, k \geqslant 1)$ be any sequence decreasing to zero, and let $A_{(k)}$ be the event that $X \leqslant x + h_k$. Then

$$\begin{aligned}\lim_{h\to 0} F(x+h) &= \lim_{k\to\infty} F(x+h_k) = \lim_{k\to\infty} \mathbf{P}(A_{(k)}) \\ &= \mathbf{P}(\lim_{k\to\infty} A_{(k)}) = \mathbf{P}(A_x) = F(x)\end{aligned}$$

as required. Finally, for (9),

$$\lim_{x\to\infty} F(x) = \lim_{n\to\infty} \mathbf{P}(A_n) = \mathbf{P}(\Omega) \quad \text{by (1.5.4).}$$

The last part is proved similarly. ■

Although the distribution function has not played a very active role so far in this book, it now assumes a greater importance. One reason for this is Theorem 6 above, which shows that $F(x)$ really is a function that can tell us how likely X is to be in some simple subset of the real line. Another reason is the following simple corollary of Example 5.

(10) Corollary If $F(x)$ is continuous then for all x $\mathbf{P}(X = x) = 0$.

Proof

$$\begin{aligned}\mathbf{P}(X = x) &= \lim_{n\to\infty} \mathbf{P}\left(x - \frac{1}{n} < X \leqslant x\right) &&\text{by (1.5.4)} \\ &= \lim_{n\to\infty}\left(F(x) - F\left(x - \frac{1}{n}\right)\right) &&\text{by Theorem 6} \\ &= 0 \quad \text{since } F \text{ is continuous.}\end{aligned}$$

■

(11) Example Let X be uniformly distributed on (0, 1). Then

$$F(x) = \begin{cases} 0 & x \leqslant 0 \\ x & 0 < x < 1 \\ 1 & x \geqslant 1 \end{cases}$$

is clearly continuous, so that $\mathbf{P}(X = x) = 0$ for all x. ■

If F_X is continuous, then X is known as a continuous random variable. We now define a particularly important class of continuous random variables.

(12) Definition Let X have distribution function F. If the derivative

(13) $$\frac{dF}{dx} = F'(x)$$

exists at all but a finite number of points, and the function f defined by

(14)
$$f(x) = \begin{cases} F'(x) & \text{where } F'(x) \text{ exists} \\ 0 & \text{elsewhere,} \end{cases}$$

satisfies

(15)
$$F(x) = \int_{-\infty}^{x} f(v)\,dv,$$

then X is said to be a continuous random variable with density $f(x)$. ▲

It follows from (15) that if X has density $f(x)$ then for $C \subseteq \mathbb{R}$

$$\mathbf{P}(X \in C) = \int_C f(x)\,dx,$$

when both sides exist. (In line with the remark following (3), they exist if $\{\omega\colon X(\omega) \in C\} \in \mathcal{F}$.)

Example 4 revisited: Uniform density In Example 4 we found that if X was chosen uniformly at random in $[0, 1]$ then

$$F_X(x) = \begin{cases} 0 & x < 0 \\ x & 0 \leqslant x \leqslant 1 \\ 1 & x > 1. \end{cases}$$

It follows that X has a density

$$f_X(x) = \begin{cases} 1 & 0 < x < 1 \\ 0 & \text{otherwise.} \end{cases}$$ ●

Figures 7.3 to 7.6 illustrate the density and distribution of random variables Y and Z, which are uniformly distributed on (a, b) and $(a, b) \cup (c, d)$ respectively.

Example (5) revisited For the point Q picked at random in a triangle, where X is the perpendicular distance from Q to the base, we found

$$F_X(x) = \begin{cases} 0 & x < 0 \\ 1 - \left(\dfrac{h - x}{h}\right)^2 & 0 \leqslant x \leqslant h \\ 1 & x > h. \end{cases}$$

It follows that X has density

$$f_X(x) = \begin{cases} 2(h - x)/h^2 & 0 < x < h \\ 0 & \text{otherwise.} \end{cases}$$ ●

There are continuous random variables which do not have a density, but none appear in this text. Therefore whenever X is said to be continuous here, it always follows that X has a density f.

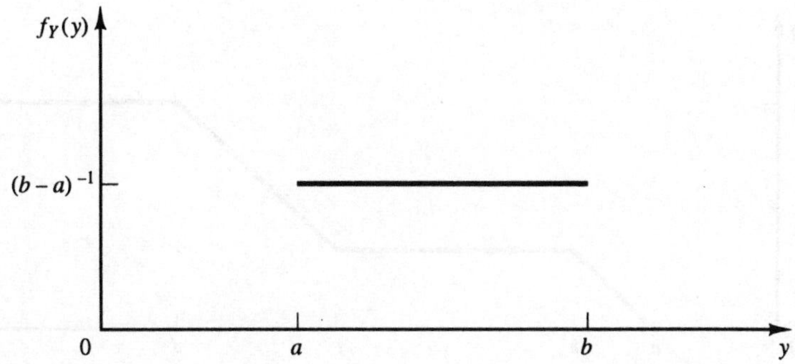

Figure 7.3 The density function $f(y)$ of a random variable distributed uniformly on (a, b).

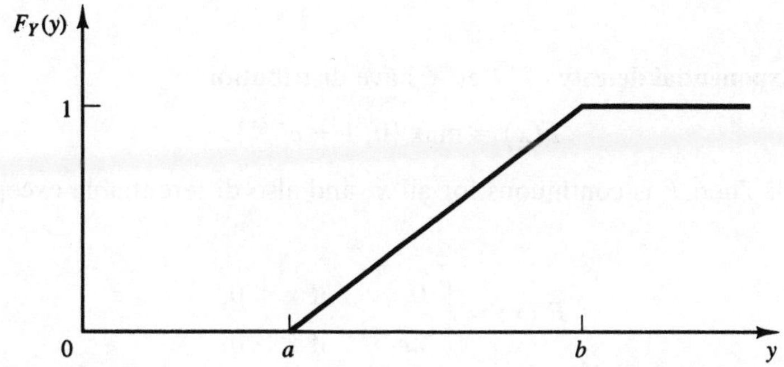

Figure 7.4 The distribution function $F(y)$ of a random variable distributed uniformly on (a, b).

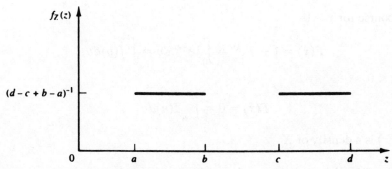

Figure 7.5 The density function of a random variable distributed uniformly on $(a, b) \cup (c, d)$, where $a < b < c < d$.

Let us consider a basic example. In the discrete case the geometric distribution is of the form $F(x) = 1 - q^x$, if $x = 1, 2, \ldots$ It is natural to consider the analogous distribution in the continuous case, which turns out to be equally (if not more) important.

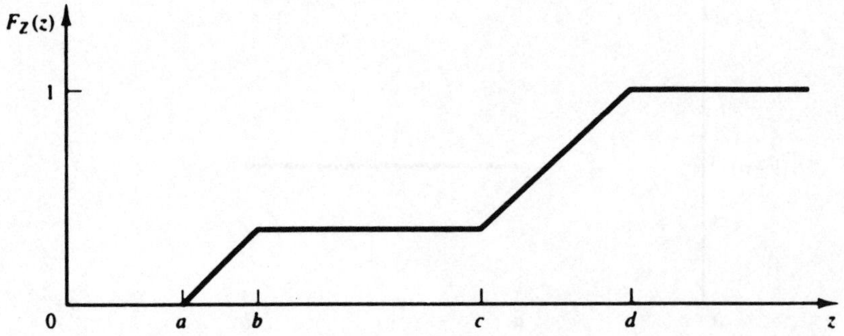

Figure 7.6 The distribution function of a random variable distributed uniformly on $(a, b) \cup (c, d)$, where $a < b < c < d$.

(16) Example: Exponential density Let X have distribution

$$F(x) = \max\{0, 1 - e^{-\lambda x}\},$$

where $\lambda > 0$. Then F is continuous for all x, and also differentiable except at $x = 0$, so

$$F'(x) = \begin{cases} 0 & \text{if } x < 0, \\ \lambda e^{-\lambda x} & \text{if } x > 0. \end{cases}$$

Now let

$$f(x) = \begin{cases} \lambda e^{-\lambda x} & \text{if } x > 0, \\ 0 & x \leqslant 0. \end{cases}$$

Then of course for $x > 0$

$$F(x) = 1 - e^{-\lambda x} = \int_0^x \lambda e^{-\lambda v} dv = \int_{\infty}^x f(v) dv,$$

and for $x < 0$

$$F(x) = 0 = \int_{-\infty}^x f(v) dv.$$

Hence $f(x)$ is a density of X. ●

Notice that $F'(0)$ does not exist, and also that the function

$$f(x) = \begin{cases} \lambda e^{-\lambda x} & x \geqslant 0 \\ 0 & x < 0 \end{cases}$$

would also satisfy $F(x) = \sum_{-\infty}^{x} f(v) dv$. This illustrates the fact that (15) does not uniquely determine $f(x)$ given $F(x)$. However, this problem can only arise at a finite number of points and (to some extent) it does not matter what value we give $f(x)$ at any of these exceptional points. Usually we make it zero, but consider the following:

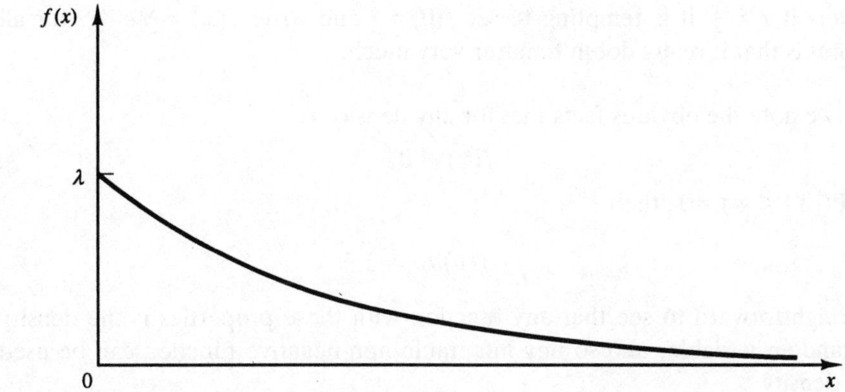

Figure 7.7 The density of an exponential random variable with parameter λ; $f(x) = \lambda e^{-\lambda x}$ for $x \geq 0$.

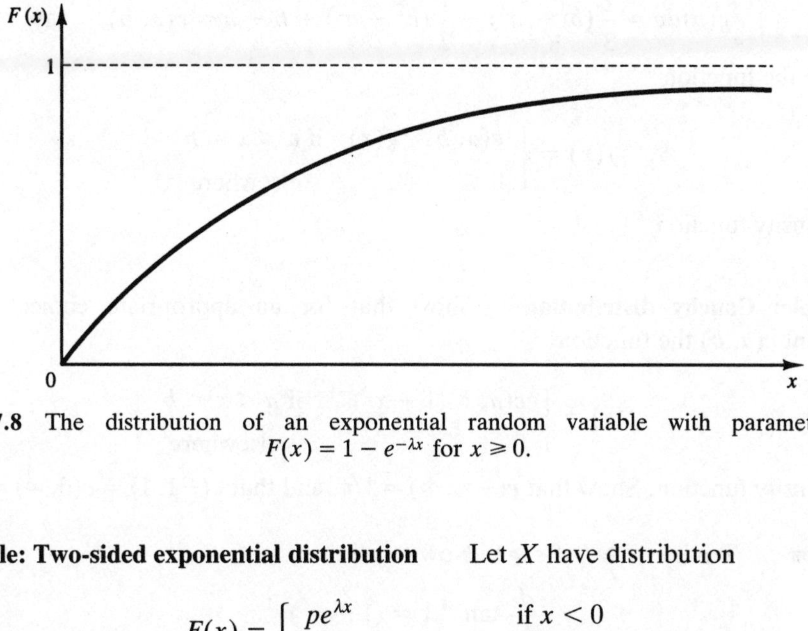

Figure 7.8 The distribution of an exponential random variable with parameter λ; $F(x) = 1 - e^{-\lambda x}$ for $x \geq 0$.

(17) Example: Two-sided exponential distribution Let X have distribution

$$F(x) = \begin{cases} pe^{\lambda x} & \text{if } x < 0 \\ 1 - (1-p)e^{-\lambda x} & \text{if } x \geq 0, \end{cases}$$

where $0 < p < 1$ and $\lambda > 0$. Then $F(x)$ is continuous and

$$F'(x) = \begin{cases} \lambda p e^{\lambda x} & \text{if } x < 0 \\ \lambda(1-p)e^{-\lambda x} & \text{if } x > 0. \end{cases}$$

A suitable density is

$$f(x) = \begin{cases} \lambda p e^{\lambda x} & \text{if } x < 0 \\ 0 & \text{if } x = 0 \\ \lambda(1-p)e^{-\lambda x} & \text{if } x > 0. \end{cases}$$

However, if $p = \frac{1}{2}$, it is tempting to set $f(0) = \frac{\lambda}{2}$ and write $f(x) = \frac{1}{2}\lambda e^{-\lambda|x|}$ for all x. The point is that it really doesn't matter very much. ●

Finally we note the obvious facts that for any density f,

(18) $$f(x) \geqslant 0,$$

and if $\mathbf{P}(|X| < \infty) = 1$, then

(19) $$\int_{-\infty}^{\infty} f(v)dv = 1.$$

It is straightforward to see that any function with these properties is the density of some random variable, and so any integrable non-negative function can be used to form a density.

Example The function $g(x) = x^2 - x + 1$ is easily seen to be non-negative for all x, and

$$\int_{+a}^{+b} g(v)dv = \frac{1}{3}(b^3 - a^3) - \frac{1}{2}(b^2 - a^2) + b - a = c(a, b), \quad \text{say.}$$

Hence the function

$$f(x) = \begin{cases} c(a, b)^{-1}g(x) & \text{if } a < x < b \\ 0 & \text{elsewhere} \end{cases}$$

is a density function. ●

(20) Example: Cauchy distribution Show that for an appropriate choice of the constant $c(a, b)$ the function

$$f(x) = \begin{cases} c(a, b)(1 + x^2)^{-1} & \text{if } a < x < b \\ 0 & \text{elsewhere} \end{cases}$$

is a density function. Show that $c(-\infty, \infty) = 1/\pi$, and that $c(-1, 1) = c(0, \infty) = 2/\pi$.

Solution Trivially $f \geqslant 0$ if $c \geqslant 0$. Now recall that

$$\frac{d}{dx}\tan^{-1} x = (1 + x^2)^{-1}.$$

Thus $\int_a^b f(x)dx = 1$ if and only if $c(a, b)^{-1} = \tan^{-1} b - \tan^{-1} a$. In particular

$$c(-\infty, \infty)^{-1} = \frac{\pi}{2} + \frac{\pi}{2} = \pi$$

as required, and

$$c(-1, 1)^{-1} = \frac{\pi}{4} + \frac{\pi}{4} = c(0, \infty)^{-1}.$$ ●

In general then, given a non-negative function $g(x)$, the function

$$f(x) = g(x)\left(\int_{-\infty}^{\infty} g(v)dv\right)^{-1}$$

is a density, if the integral exists. To discover whether it does, the following technique is useful. If we can find a constant b such that for all $n > 0$, $\int_0^n g(v)dv < b < \infty$, then

$$\int_0^\infty g(v)dv = \lim_{n\to\infty} \int_0^n g(v)dv$$

exists by monotone convergence.

(21) Example: Normal density Show that

$$f = c\exp\left(-\frac{1}{2}x^2\right) \quad \text{for all } x \in \mathbb{R}$$

can be a density.

Solution For any $n > 1$

$$\int_{-n}^n \exp\left(-\frac{1}{2}v^2\right)dv < 2\left(\int_0^1 dv + \int_1^n e^{-v}dv\right) < 2(1 + e^{-1}).$$

Hence $c^{-1} = \int_{-\infty}^\infty \exp(-\frac{1}{2}v^2)dv$ exists, and f is a density for this c. ●

Remark In fact it can be shown that

(22) $$c^{-1} = (2\pi)^{1/2}.$$

The proof of this is not quite trivial; we give it in Example 8.3.8. Also note that there are other normal densities; the one in Example 21 is called the *standard* normal density denoted by $N(0, 1)$, and by

(23) $$\phi(x) = (2\pi)^{-1/2}\exp(-\tfrac{1}{2}x^2).$$

Its distribution is $\Phi(x)$, given by

(24) $$\Phi(x) = \int_{-\infty}^x \phi(v)dv.$$

(25) Example: Gamma distribution Show that for α, λ, $x > 0$, the function $f(x) = c\lambda^\alpha x^{\alpha-1}e^{-\lambda x}$ can be a density. When α is a positive integer, show that $c^{-1} = (\alpha - 1)!$

Solution First we show that the integral $\int_0^\infty x^{\alpha-1}e^{-\lambda x}dx$ exists. Since $x^{\alpha-1}e^{-\frac{1}{2}\lambda x} \to 0$ as $x \to \infty$, there is some $m < \infty$ such that for $x > m$, $x^{\alpha-1}e^{-\frac{1}{2}\lambda x} < 1$. Hence for $n > m$,

$$\int_0^n v^{\alpha-1}e^{-\lambda v}dv < \int_0^m v^{\alpha-1}e^{-\lambda v}dv + \int_m^n e^{-\frac{1}{2}\lambda v}dv < \int_0^m v^{\alpha-1}e^{-\lambda v}dv + 2\lambda^{-1}e^{-\frac{1}{2}\lambda m} = b \text{ (say)}.$$

Hence $c^{-1} = \int_0^\infty \lambda^\alpha x^{\alpha-1}e^{-\lambda x}dx$ exists, and if α is a positive integer then repeated integration by parts gives $c^{-1} = (\alpha - 1)!\int_0^\infty \lambda e^{-\lambda v}dv = (\alpha - 1)!$ ●

This density is known as the gamma density with parameters α and λ. When α is not an integer, the integral above defines a function of α known as the gamma function, and denoted by

(26) $$\Gamma(\alpha) = \int_0^\infty \lambda^\alpha v^{\alpha-1} e^{-\lambda v} dv = \int_0^\infty u^{\alpha-1} e^{-u} du.$$

The density function in this case is

(27) $$f(x) = \frac{1}{\Gamma(\alpha)} \lambda^\alpha x^{\alpha-1} e^{-\lambda x} \quad \text{for } x > 0.$$

In particular, when $\lambda = \frac{1}{2}$ and $\alpha = \frac{n}{2}$, the density

(28) $$f(x) = \frac{1}{\Gamma(\frac{n}{2})} \lambda^{\frac{n}{2}} x^{\frac{n}{2}-1} e^{\frac{-1}{2}x}$$

is known as the $\chi^2(n)$ density, and it is referred to as chi-squared with n degrees of freedom.

We conclude with another way of producing densities.

(29) Example: Mixtures Let $f_1(x)$ and $f_2(x)$ be density functions, and let $f_3(x) = \gamma f_1(x) + (1-\gamma) f_2(x)$ where $0 \leqslant \gamma \leqslant 1$. Then $f_3(x) \geqslant 0$, and

$$\int f_3 = \gamma \int f_1 + (1-\gamma) \int f_2 = 1.$$

Hence f_3 is a density, and is said to be a *mixture* of f_1 and f_2. ●

For example, the two-sided exponential density of Example 17 may now be seen as a mixture of $f_1 = \lambda e^{\lambda x} (x < 0)$ and $f_2 = \lambda e^{-\lambda x}$ $(x > 0)$ with $\gamma = p$.

7.2 Functions of random variables

Suppose that X and Y are random variables such that $Y = g(X)$, where $g(.)$ is some given function. If we know the density of X, can we find the distribution of Y? In general terms the answer is straightforward because, by the properties of densities and distributions,

(1) $$F(y) = \mathbf{P}(Y \leqslant y) = \mathbf{P}(g(X) \leqslant y) = \int_C f_X(v) dv$$

where $C = \{v: g(v) \leqslant y\}$. Then, if $F(y)$ is continuous and differentiable, we can go on to find the density of Y, if it exists. Here are some simple examples of this idea in practice.

(2) Example Let X be uniformly distributed on $(0, 1)$ with density

$$f(x) = \begin{cases} 1 & \text{if } 0 < x < 1 \\ 0 & \text{otherwise.} \end{cases}$$

If $Y = -\lambda^{-1} \log X$, where $\lambda > 0$, what is the density of Y?

Solution First we seek the distribution of Y:

$$F_Y(y) = \mathbf{P}(-\lambda^{-1}\log X \leqslant y) = \mathbf{P}(\log X \geqslant -\lambda y)$$

$$= \mathbf{P}(X \geqslant \exp(-\lambda y)) = \begin{cases} 1 - e^{-\lambda y} & \text{for } y \geqslant 0 \\ 0 & \text{otherwise.} \end{cases}$$

Hence, the derivative exists except at $y = 0$, and

$$f_Y(y) = \begin{cases} \lambda e^{-\lambda y} & \text{if } y > 0 \\ 0 & \text{if } y \leqslant 0 \end{cases}$$

This is the exponential density with parameter λ. ●

Some care is required if $g(.)$ is not one–one.

(3) **Example** Let X be uniformly distributed on $[-1, 1]$. Find the density of $Y = X^r$ for non-negative integers r.

Solution First note that X has distribution function $F(x) = \frac{1}{2}(1 + x)$ for $-1 \leqslant x \leqslant 1$. Now, if r is odd then the function $g(x) = x^r$ maps the interval $[-1, 1]$ onto itself in one–one correspondence. Hence, routinely:

$$\mathbf{P}(Y \leqslant y) = \mathbf{P}(X^r \leqslant y) = \mathbf{P}(X \leqslant y^{1/r}) = \frac{1}{2}(1 + y^{1/r}) \quad \text{for } -1 \leqslant y \leqslant 1,$$

and Y has density

$$f(y) = \frac{1}{2r} y^{\frac{1}{r}-1}, \quad -1 \leqslant y \leqslant 1.$$

If r is even then $g(x) = x^r$ takes values in $[0, 1]$ for $x \in [-1, 1]$. Therefore

$$\mathbf{P}(Y \leqslant y) = \mathbf{P}(0 \leqslant X^r \leqslant y) = \mathbf{P}(-y^{1/r} \leqslant X \leqslant y^{1/r}) = y^{1/r} \quad \text{for } 0 \leqslant y \leqslant 1.$$

Hence Y has density

$$f(y) = \frac{1}{r} y^{\frac{1}{r}-1}, \quad 0 \leqslant y \leqslant 1.$$

Finally, if $r = 0$, then $X^r = 1$, $F_Y(y)$ is not continuous (having a jump from 0 to 1 at $y = 1$) and so Y does not have a density in this case. Obviously Y is discrete, with $\mathbf{P}(Y = 1) = 1$. ●

(4) **Example** Let X have the standard normal distribution with density

$$f(x) = (2\pi)^{-\frac{1}{2}} \exp\left(-\frac{1}{2}x^2\right).$$

Find the density of $Y = \sigma X + \mu$ for given constants μ and $\sigma \neq 0$. Also find the density of $Z = X^2$.

Solution Adopting the by now familiar technique:

(5)
$$\mathbf{P}(\sigma X + \mu \leqslant y) = \mathbf{P}(\sigma X \leqslant y - \mu) = \begin{cases} \mathbf{P}\left(X \leqslant \dfrac{y-\mu}{\sigma}\right) & \text{if } \sigma > 0 \\ \mathbf{P}\left(X \geqslant \dfrac{y-\mu}{\sigma}\right) & \text{if } \sigma < 0 \end{cases}$$

$$= \begin{cases} F_X\left(\dfrac{y-\mu}{\sigma}\right) & \text{if } \sigma > 0 \\ 1 - F_X\left(\dfrac{y-\mu}{\sigma}\right) & \text{if } \sigma < 0 \end{cases}$$

Hence, differentiating (5) with respect to y,

(6)
$$f_Y(y) = \frac{1}{|\sigma|} f_X\left(\frac{y-\mu}{\sigma}\right) = \frac{1}{(2\pi\sigma^2)^{\frac{1}{2}}} \exp\left(-\frac{1}{2}\left(\frac{y-\mu}{\sigma}\right)^2\right).$$

Secondly

$$\mathbf{P}(X^2 \leqslant z) = \mathbf{P}(X \leqslant \sqrt{z}) - \mathbf{P}(X \leqslant -\sqrt{z}) = F_X(\sqrt{z}) - F_X(-\sqrt{z}).$$

Differentiating now gives

(7)
$$f_Z(z) = \frac{1}{2\sqrt{z}} f_X(\sqrt{z}) + \frac{1}{2\sqrt{z}} f_X(-\sqrt{z}) = \frac{1}{\sqrt{2\pi z}} \exp\left(-\frac{1}{2} z\right). \quad \bullet$$

Remark The density given by (6) is known as the *normal* density with parameters μ and σ^2, sometimes denoted by $N(\mu, \sigma^2)$. The standard normal density of Example 7.1.21 was $N(0, 1)$ because $\phi(x)$ has $\mu = 0$ and $\sigma = 1$.

The density given by (7) is the gamma density of (7.1.23) with parameters $\frac{1}{2}$ and $\frac{1}{2}$. This is known as the chi-squared density with parameter 1, sometimes denoted by $\chi^2(1)$. This is a special case of (7.1.28).

(8) Example: Inverse functions Let X have distribution function $F(x)$, where $F(x)$ is continuous and strictly increasing. Let $g(x)$ be a function satisfying $F(g) = x$. Because $F(x)$ is continuous and strictly increasing, this defines $g(x)$ uniquely for every x in $(0, 1)$. The function $g(.)$ is called the inverse function of $F(.)$ and is often denoted by

$$g(x) = F^{-1}(x).$$

Clearly F is the inverse function of g, that is

(9)
$$g(F(x)) = F(g(x)) = x,$$

and $g(x)$ is an increasing function.

(a) Use this function to show that $Y = F(X)$ is uniformly distributed on $(0, 1)$.
(b) Show that if U is uniform on $(0, 1)$, then $Z = F^{-1}(U)$ has distribution $F(z)$.

Solution (a) As usual we seek the distribution function

$$\mathbf{P}(Y \leqslant y) = \mathbf{P}(F(X) \leqslant y) = \mathbf{P}(g(F(X)) \leqslant g(y)) = \mathbf{P}(X \leqslant g(y)) \quad \text{by (9)}$$
$$= F(g(y)) = y \quad \text{by (9)}.$$

(b) Again

$$\mathbf{P}(F^{-1}(U) \leqslant z) = \mathbf{P}(F(g(U)) \leqslant F(z)) = \mathbf{P}(U \leqslant F(z)) \quad \text{by (9)}$$
$$= F(z).$$ ●

Though we have introduced them separately, discrete and continuous variables do have close links. Here are some examples to show this.

(10) Example: Step functions Let X have distribution function $F(x)$ and density f. Define the function $S: \mathbb{R} \to \mathbb{Z}$ by

(11) $$S(X) = k, \text{ if } k \leqslant X < k + 1,$$

where k is any integer. Then $S(X)$ is an integer valued discrete random variable with mass function

(12) $$f_S(k) = \mathbf{P}(k \leqslant X < k + 1) = \int_k^{k+1} f(v)dv.$$

Obviously $\mathbf{P}(S(X) \leqslant X) = 1$, and $F_S(x) \geqslant F_X(x)$, and

(13) $$|S(X) - X| \leqslant 1.$$ ●

Now equation (13) shows that the integer valued $S(X)$ is (in some sense) a rough approximation to the continuous random variable X. It is easy to get much better approximations as follows.

(14) Example: Discrete approximation As usual X has density $f(x)$; suppose also that $X > 0$. For fixed n, with $0 \leqslant r \leqslant 2^n - 1$, and $k \geqslant 0$, define

$$S_n(X) = k + r2^{-n} \text{ if } k + r2^{-n} \leqslant X < k + (r + 1)2^{-n}.$$

Then $S_n(X)$ is a discrete random variable taking values in $(k + r2^{-n};\ k \geqslant 0, 0 \leqslant r \leqslant 2^n - 1)$, with

$$\mathbf{P}(S_n(X) = k + r2^{-n}) = \int_{k+r2^{-n}}^{k+(r+1)2^{-n}} f(v)dv.$$

Again we have $S_n(X) \leqslant X$, but this time, by the construction,

(15) $$|S_n(X) - X| \leqslant 2^{-n}.$$

Thus, by choosing n large enough, we can find a discrete random variable $S_n(X)$ such that $|X - S_n(X)|$ is as small as we please. ●

In fact it can be shown that we can find a simple random variable (taking only a finite number of values) which is arbitrarily close to X, but in a weaker sense than (15). (See Problem 12.)

7.3 Simulation of random variables

A random variable is a mathematical concept (having no other existence) which is suggested by the outcomes of real experiments. Thus, tossing a coin leads us to define an $X(.)$ such that $X(H) = 1$, $X(T) = 0$, and X is the number of heads. The coin exists, X is a concept. A natural next step, having developed theorems about mathematical coins (such as the arc-sine laws) is to test them against reality. However the prospect of actually tossing a large enough number of coins to check the arc-sine laws is rather forbidding.

Luckily we have machines to do large numbers of boring and trivial tasks quickly, namely computers. These can be persuaded to produce a very large number of numbers $(u_i;\ i \geq 1)$ which are sprinkled evenly and 'randomly' over the interval $(0, 1)$. The word randomly appeared in inverted commas because each u_i is not really random. Since the machine was programmed to produce it, the outcome is known in advance, but such numbers behave for many practical purposes as though they were random. They are called pseudo-random numbers.

Now if we have a pseudo-random number u from a collection sprinkled uniformly in $(0, 1)$, we can look to see if $u < \frac{1}{2}$, in which case we call it 'heads', or $u > \frac{1}{2}$ in which case we call it 'tails'. This process is called *simulation*; we have simulated tossing a coin.

Different problems produce different random variables, but computers find it easiest to produce uniform pseudo-random numbers. We are thus forced to consider appropriate transformations of uniform random variables, and therefore many of the results of Section 7.2 find concrete applications when we seek to simulate random variables. A natural first question (before 'how') is *why* might we want to simulate such random variables? Some examples should suffice to answer this question.

(1) **Example: Epidemic** An infection is introduced into a population. For each individual the incubation period is a random variable X, the infectious period is a random variable Y, and the number of further individuals infected is a random variable N depending on Y and the behaviour of the infected individual. What happens? Unfortunately exact solutions to such problems are very rare, and for many diseases (such as the so-called 'slow viruses') X and Y are measured in decades, so experiments are impractical. However, if we could simulate X and Y and the infection process N, then we could produce one simulated realization (not a real realization) of the epidemic. With a fast computer we could do this many times and gain a pretty accurate idea of how the epidemic would progress (if our assumptions were correct). ●

(2) **Example: Toll booths** Motorists are required to pay a fee before entering a toll road. How many toll booths should be provided to avoid substantial queues? Once again an experiment is impractical. However, very simple apparatus can provide us with the rates and properties of traffic on equivalent roads. If we then simulate the workings of the booth and test it with the actual traffic flows, we should obtain reasonable estimates of the chances of congestion. ●

Because of the ready availability of large numbers of uniform pseudo-random numbers, interest is concentrated on finding transformations which then yield random variables of arbitrary type. We have seen several in Section 7.2. Here is another idea.

Example: Composition The pseudo-random variable U is uniformly distributed on $(0, 1)$. Show how to simulate a random variable with density

$$f_X = \frac{1}{4}(x^{-\frac{1}{2}} + (1-x)^{-\frac{1}{2}}), \quad 0 < x < 1.$$

Solution Recall that if U is uniform on $(0, 1)$ then U^2 has density $f_1(x) = \frac{1}{2}x^{-\frac{1}{2}}$. Next consider

$$\mathbf{P}(1 - U^2 \leqslant x) = \mathbf{P}(U \geqslant (1-x)^{\frac{1}{2}}) = 1 - (1-x)^{\frac{1}{2}}.$$

Hence $1 - U^2$ has density $f_2(x) = \frac{1}{2}(1-x)^{-\frac{1}{2}}$. Now toss a coin (real or simulated), and write

$$X = \begin{cases} U^2 & \text{if it's heads} \\ 1 - U^2 & \text{if it's tails.} \end{cases}$$

Then

$$f_X(x) = \frac{1}{2}f_1(x) + \frac{1}{2}f_2(x) = \frac{1}{4}(x^{-\frac{1}{2}} + (1-x)^{-\frac{1}{2}})$$

as required. ●

We shall describe other methods of simulation as the necessary ideas are developed.

7.4 Expectation

Random variables with a density may have an expected value, similarly to random variables with a mass function.

(1) **Definition** Let X have density $f(x)$. If $\int_{-\infty}^{\infty}|v|f(v)dv < \infty$ then X has an expected value, which is given by

(2) $$\mathbf{E}(X) = \int_{-\infty}^{\infty} vf(v)dv.$$ ▲

(3) **Example: Uniform density** Let X be uniformly distributed on (a, b). Then

$$\mathbf{E}(X) = \int_a^b \frac{v}{b-a}dv = \frac{1}{2}(b-a).$$ ●

(4) **Example: Exponential density** Let X have density $f(x) = \lambda e^{-\lambda x}$ for $x \geqslant 0$. Then

$$\mathbf{E}(X) = \int_0^{\infty} v\lambda e^{-\lambda v}dv = \lambda^{-1}.$$ ●

(5) **Example: Normal density** Let X have the $N(\mu, \sigma^2)$ density. Then

$$\mathbf{E}(X) = \frac{1}{\sigma(2\pi)^{\frac{1}{2}}}\int_{-\infty}^{\infty} v \exp(-(v-\mu)^2/(2\sigma^2))\,dv$$

$$= \frac{1}{\sigma(2\pi)^{\frac{1}{2}}}\int_{-\infty}^{\infty}(v-\mu)\exp\left(-\frac{1}{2}\left(\frac{v-\mu}{\sigma}\right)^2\right)dv$$

$$+\frac{\mu}{\sigma(2\pi)^{\frac{1}{2}}}\int_{-\infty}^{\infty}\exp\left(-\frac{1}{2}\left(\frac{v-\mu}{\sigma}\right)^2\right)dv$$

$$= \frac{1}{(2\pi)^{\frac{1}{2}}}\int_{-\infty}^{\infty} u\exp\left(-\frac{1}{2}u^2\right)du + \frac{\mu}{(2\pi)^{\frac{1}{2}}}\int_{-\infty}^{\infty}\exp\left(-\frac{1}{2}u^2\right)du$$

on making the substitution $u = (v-\mu)/\sigma$ in both integrands. The first integrand is an odd function, so the integral over $\mathbb{R}$ is zero. The second term is μ by Example 7.1.21 and 7.1.22. Hence $\mathbf{E}(X) = \mu$. ●

Expectation may be infinite, as the next example shows.

(6) **Example: Pareto density** Let X have density $f(x) = (\alpha - 1)x^{-\alpha}$ for $x \geqslant 1$, and $\alpha > 1$. Then if $\alpha \leqslant 2$, the expected value of X is infinite because

$$\mathbf{E}(X) = \lim_{n\to\infty}\int_1^n \frac{(\alpha-1)v}{v^\alpha}\,dv = (\alpha-1)\lim_{n\to\infty}\int_1^n \frac{1}{v^{\alpha-1}}\,dv,$$

which diverges to ∞ for $\alpha - 1 \leqslant 1$. However, for $\alpha > 2$

$$\mathbf{E}(X) = \int_1^\infty \frac{(\alpha-1)}{v^{\alpha-1}}\,dv = \frac{(\alpha-1)}{(\alpha-2)}.$$ ●

Then again, the expectation of X may not exist, as the next example shows.

(7) **Example: Cauchy density** Let X have density

$$f(x) = \frac{1}{\pi(1+x^2)}, \qquad -\infty < x < \infty.$$

Since $\int_0^a v(\pi(1+v^2))^{-1}dv$ diverges as $a \to -\infty$ and as $a \to +\infty$, X does not have an expected value. ●

It is appropriate to give a moment to considering why we define $\mathbf{E}(X)$ by Definition 1. This definition is at least plausible, by analogy with the definition

$$\mathbf{E}(X) = \sum_{v=-\infty}^{\infty} vf(v),$$

in the discrete case. Of course Definition 1 is much more than just a plausible analogy, but a complete account of expectation is well beyond our scope. However we can use Example 7.2.14 to give a little more justification for Definition 1. Let $k + r2^{-n} = a(k, r, n)$. Recall from (7.2.15) that $|S_n - X| < 2^{-n}$. Now by definition, since $S_n(X)$ is discrete,

$$\mathbf{E}(S_n(X)) = \sum_{k,r} a(k,r,n) \int_{a(k,r,n)}^{a(k,r+1,n)} f(v)dv$$

$$= \sum_{k,r} \int_{a(k,r,n)}^{a(k,r+1,n)} (vf(v) + (a(k,r,n) - v)f(v))dv.$$

Since $|a(k,r,n) - v| < 2^{-n}$, it can be shown (with more work, which we omit) that

$$\mathbf{E}(S_n(X)) = \int_{-\infty}^{\infty} vf(v)dv + \epsilon_n = \mathbf{E}(X) + \epsilon_n$$

where $\epsilon_n \to 0$ as $n \to \infty$. An explicit demonstration may be helpful here.

(8) Example Let X be uniform on $(0, 1)$ with mean value $\mathbf{E}(X) = \frac{1}{2}$. Then $S_n(X)$ is uniformly distributed on $\{0, 2^{-n}, 2.2^{-n}, 3.2^{-n}, \ldots, 1 - 2^{-n}\}$. Therefore

$$\mathbf{E}(S_n(X)) = \sum_{r=0}^{2^n-1} r2^{-n}.2^{-n} = \frac{1}{2}(2^n - 1)2^{-n}$$

$$= \mathbf{E}(X) - 2^{-(n+1)} \to \mathbf{E}(X), \quad \text{as } n \to \infty.$$ ●

Thus our definitions of expectation for discrete and continuous variables are at least consistent in some way. In more advanced books a single definition of $\mathbf{E}(.)$ is given, which is shown to yield our definitions as special cases.

Next we return to considering functions of random variables. Suppose we are given random variables Y and X related by $Y = g(X)$. What is $\mathbf{E}(Y)$?

If we know the density of X, then we may be able to find $\mathbf{E}(Y)$ by first discovering $f_Y(y)$, if it exists. This is often an unattractive procedure. We may do much better to use the following theorem, which we state without proof.

(9) Theorem Let random variables X and Y satisfy $Y = g(X)$, where X has density $f(x)$. Then Y has an expected value if $\int_{-\infty}^{\infty} |g(v)| f(v)dv < \infty$, and in this case

(10) $$\mathbf{E}(Y) = \int_{-\infty}^{\infty} g(v)f(v)dv.$$

The proof of this is straightforward but long. An heuristic discussion of the type above shows that if we represent the distribution of X as a limit of discrete distributions, and then formally proceed to this limit in Theorem 4.3.4, equation (10) is the result. Again, this only makes (10) plausible, it doesn't provide the proof, which is beyond our scope.

This important result implies that the useful consequences of Theorem 4.3.4 remain true for random variables with a density. In particular Theorem 4.3.6 remains true; the proofs of most parts are just typographical variants of the proofs in the discrete case; just replace $\sum$ by $\int$.

We describe one important and less trivial case in detail, namely the analogy of Theorem 4.3.11.

(11) Theorem Let the non-negative random variable X have density f, distribution F, and finite expected value $\mathbf{E}(X)$. Then

$$\mathbf{E}(X) = \int_0^\infty (1 - F(x))dx. \tag{12}$$

Proof For any finite y we may integrate by parts to obtain

$$\int_0^y xf(x)dx = -x(1 - F(x))|_0^y + \int_0^y (1 - F(x))dx. \tag{13}$$

But we have

$$y(1 - F(y)) = y\int_y^\infty f(x)dx \leqslant \int_y^\infty xf(x)dx \to 0$$

as $y \to \infty$, because $\mathbf{E}(X) < \infty$. Hence we can let $y \to \infty$ in (13) to prove the theorem. ■

We can use this to prove a useful special case of Theorem 9.

(14) Example Let the non-negative random variable X have density f, and let $g(X) \geqslant 0$. Show that $\mathbf{E}(g(X)) = \int_0^\infty g(v)f(v)dv$.

Solution

$$\begin{aligned}\mathbf{E}(g(X)) &= \int_0^\infty \mathbf{P}(g(X) \geqslant v)dv \quad \text{by (12)}\\ &= \int_0^\infty \int_{x:g(x)\geqslant v} f(x)dxdv = \int_0^\infty f(x)\int_0^{g(x)} dvdx = \int_0^\infty f(x)g(x)dx.\end{aligned}$$

as required. The interchange in the order of integration is justified by a theorem on double integrals, which we omit. ●

The various moments of a random variable with a density are defined just as they were for discrete random variables, that is to say: $\mu_k = \mathbf{E}(X^k)$, and

$$\sigma_k = \mathbf{E}((X - \mathbf{E}(X))^k).$$

(15) Example: Normal density Let X have the density $N(0, \sigma^2)$. Find μ_k for all k.

Solution If k is odd, then $x^k \exp(-x^2/(2\sigma^2))$ is an odd function. Hence $\mu_k = 0$, if k is odd. If $k = 2n$, then integrating by parts gives

$$\begin{aligned}\mu_{2n} &= \frac{1}{\sigma(2\pi)^{\frac{1}{2}}}\int_{-\infty}^\infty v^{2n}\exp(-v^2/(2\sigma^2))dv\\ &= \frac{1}{\sigma(2\pi)^{\frac{1}{2}}}\left(-v^{2n-1}\sigma^2\exp(-v^2/(2\sigma^2))|_{-\infty}^\infty + \int_{-\infty}^\infty (2n-1)\sigma^2 v^{2n-2}\exp(-v^2/(2\sigma^2))dv\right)\\ &= (2n - 1)\sigma^2\mu_{2n-2} = \sigma^{2n}\frac{(2n)!}{2^n n!}\end{aligned}$$

on iterating and observing that $\mu_0 = 1$. Hence in particular $\mu_2 = \sigma^2$. ●

Finally, and thankfully, we are pleased to record that the expectation $\mathbf{E}(X)$ of a continuous random variable X has the same useful basic properties that we established for the discrete case in Section 4.6. For convenience we recall them here.

(16) **Theorem** Let a and b be constants, and let g and h be functions. Then:

(i) if $g(X)$ and $h(X)$ have finite mean then

$$\mathbf{E}(g(X) + h(X)) = \mathbf{E}(g(X)) + \mathbf{E}(h(X));$$

(ii) if $\mathbf{P}(a \leqslant X \leqslant b) = 1$ then $a \leqslant \mathbf{E}(X) \leqslant b$;
(iii) if h is non-negative then for $a > 0$, $\mathbf{P}(h(X) \geqslant a) \leqslant \mathbf{E}(h(X)/a)$;
(iv) if g is convex then $\mathbf{E}(g(X)) \geqslant g(\mathbf{E}(X))$.

Proof The proof is an exercise for you. ■

7.5 Moment generating functions

In dealing with integer valued discrete random variables we found the probability generating function exceptionally useful (see Chapter 6). It would be welcome to have such a useful workhorse available for random variables with densities. Of course, if X has a density then $\mathbf{P}(X = x) = 0$, so we cannot expect the probability generating function to be of much use. Fortunately another function will do the job.

(1) **Definition** If X has density f, then X has *moment generating function* $M_X(t)$ given by

(2) $$M_X(t) = \mathbf{E}(e^{tX}) = \int_{-\infty}^{\infty} e^{tv} f(v)dv. \quad ▲$$

We shall only be interested in $M_X(t)$ for those values of t for which it is finite; this includes $t = 0$, of course. It is particularly pleasant when $M_X(t)$ exists in a neighbourhood of zero, but it is beyond our scope to explain all the reasons for this.

(3) **Example: Uniform density** Let X be uniform on $[0, a]$. Find $\mathbf{E}(e^{tX})$. Where does it exist?

Solution

$$\mathbf{E}(e^{tX}) = \int_0^a \frac{1}{a} e^{tv} dv = \left[\frac{1}{at} e^{tv}\right]_0^a = \frac{e^{at} - 1}{at}.$$

This exists for all t, including $t = 0$, where it takes the value 1. ●

(4) **Example: Gamma density** Recall from (7.1.24) that the gamma function $\Gamma(\alpha)$ is defined for any $\alpha > 0$ and $\lambda > 0$ by

(5) $$\Gamma(\alpha) = \int_0^{\infty} x^{\alpha-1} \lambda^{\alpha} e^{-\lambda x} dx.$$

Hence

$$f(x) = \frac{\lambda^\alpha}{\Gamma(\alpha)} x^{\alpha-1} e^{-\lambda x}, \quad x \geqslant 0,$$

is the density of a random varaible x. Find $\mathbf{E}(e^{tX})$. Where does it exist?

Solution

$$\mathbf{E}(e^{tX}) = \int_0^\infty e^{tv} \frac{\lambda^\alpha}{\Gamma(\alpha)} v^{\alpha-1} e^{-\lambda v} dv = \frac{\lambda^\alpha}{\Gamma(\alpha)} \int_0^\infty v^{\alpha-1} e^{-(\lambda-t)v} dv.$$

The integral exists if $\lambda > t$, and then making the substitution $(\lambda - t)v = u$ gives

(6) $$M_X(t) = \left(\frac{\lambda}{\lambda - t}\right)^\alpha \int_0^\infty \frac{u^{\alpha-1}}{\Gamma(\alpha)} e^{-u} du = \left(\frac{\lambda}{\lambda - t}\right)^\alpha \quad \text{by (5)},$$

for $-\infty < t < \lambda$. ●

(7) **Example: Normal density** Let X be a standard normal random variable. Then

$$\sqrt{2\pi}\, M_X(t) = \int_{-\infty}^\infty \exp\left(-\frac{1}{2}x^2 + tx\right) dx = \int_{-\infty}^\infty \exp\left(-\frac{1}{2}(x-t)^2 + \frac{1}{2}t^2\right) dx$$

$$= e^{\frac{1}{2}t^2} \int_{-\infty}^\infty \exp\left(-\frac{1}{2}v^2\right) dv, \quad \text{setting } x - t = v,$$

$$= e^{\frac{1}{2}t^2} \sqrt{2\pi}.$$

So $M_X(t) = e^{\frac{1}{2}t^2}$. Now by (7.2.4) if Y is $N(\mu, \sigma^2)$, $M_Y(t) = e^{\mu t + \frac{1}{2}\sigma^2 t^2}$. ●

You may ask, why is $M_X(t)$ called the moment generating function? The answer lies in the following formal expansion.

(8) $$\mathbf{E}(e^{tX}) = \mathbf{E}\left(\sum_{k=0}^\infty \frac{X^k t^k}{k!}\right) = \sum_{k=0}^\infty \mathbf{E}\frac{(X^k)t^k}{k!} = \sum_{k=0}^\infty \frac{\mu_k t^k}{k!}.$$

Thus, provided the interchange of expectation and summation at (8) is justified, we see that $M_X(t)$ is the (exponential) generating function of the moments μ_k.

Note that the word 'exponential' is always omitted in this context, and that the required interchange at (8) is permissible if $M_X(t)$ exists in an interval which includes the origin.

You may also ask, do we always know the density $f_X(x)$, if we know $M_X(t)$? After all, the probability generating function does uniquely determine the corresponding mass function. Unfortunately the answer is no in general, for there do exist densities not uniquely determined by their moments. However, none appear here; every density in this book is uniquely determined by its moment generating function (if it has one). We state the following *inversion theorem* without proof.

(9) **Theorem** If X has moment generating function $M(t)$, where for some $a > 0$, $M(t) < \infty$ for $|t| < a$, then the distribution of X is determined uniquely.

Furthermore, $M(t) = \sum_{k=0}^\infty \frac{1}{k!} t^k \mathbf{E}(X^k)$.

The moment generating function is especially useful in dealing with sequences of random variables; the following theorem is the basis of this assertion. We state it without proof.

(10) **Theorem: Continuity theorem** Let $(F_n(x); n \geqslant 1)$ be a sequence of distribution functions with corresponding moment generating functions $(M_n(t); n \geqslant 1)$ which exist for $|t| < b$. Suppose that as $n \to \infty$ $M_n(t) \to M(t)$ for $|t| \leqslant a < b$, where $M(t)$ is the m.g.f. of the distribution $F(x)$. Then as $n \to \infty$, $F_n(x) \to F(x)$ at each point x where $F(x)$ is continuous.

The main application of this theorem arises when $M(t) = e^{\frac{1}{2}t^2}$ and $F(x) = \Phi(x)$, as we shall see in the next chapter when we come to the celebrated central limit theorem. Here is a preliminary note.

Note: The *O*–*o* notation In considering limits of sequences of functions, we shall quite often produce large and unwieldly expressions of which only one or two terms remain in the limit. Rather than keep a precise record of the essentially irrelevant terms, it is convenient to have a special compact notation for them.

Definition If $g(n)$ and $h(n)$ are two functions of n, then we write

$$h(n) = O(g(n)) \quad \text{as } n \to \infty$$

if $|h(n)/g(n)| < c$ for all large enough n and some finite constant c. ▲

For example, as $n \to \infty$,

$$n^2 + \log n = O(n^2) \quad \text{with } c = 2$$

and

$$n^2 + n^{\frac{3}{2}} = O(n^2) \quad \text{with } c = 2.$$

Observe that this is an abuse of notation (= being the abused symbol) because it does *not* follow from these two examples that $\log n = n^{\frac{3}{2}}$. Also if $h(n) = O(g(n))$ and $k(n) = O(g(n))$ then $h(n) + k(n) = O(g(n))$.

A similar definition holds for small values of the argument.

Definition If $g(x)$ and $h(x)$ are two functions of x, then we write $h(x) = O(g(x))$ as $x \to 0$, if $|h(x)/g(x)| < c$ for all small enough x and some constant c. ▲

Often an even cruder representation will suffice.

Definition If $g(x)$ and $h(x)$ are two functions of x then we write $h(x) = o(g(x))$ as $x \to \infty$ if $\lim_{x\to\infty} (h(x)/g(x)) = 0$. Likewise $h(x) = o(g(x))$ as $x \to 0$ if $\lim_{x\to 0} (h(x)/g(x)) = 0$. ▲

For example, $x^2 = o(x)$ as $x \to 0$, and $x = o(x^2)$ as $x \to \infty$. For another example, $x + x \log x + x^2 = o(1)$ as $x \to 0$.

We use this new notation in the following famous result.

(11) **Example: de Moivre–Laplace theorem** For each $n \geq 1$ let X_n be a binomial random variable with parameters n and p. Let $q = 1 - p$, and define

$$Y_n = \frac{X_n - np}{(npq)^{\frac{1}{2}}}.$$

Show that as $n \to \infty$

$$\mathbf{P}(Y_n \leq x) \to \Phi(x) = \int_{-\infty}^{x} (2\pi)^{-\frac{1}{2}} e^{-y^2/2} dy.$$

Solution We use Theorem 10. First calculate the moment generating function

(12) $$\mathbf{E}(e^{tY_n}) = \mathbf{E}\left(\exp\left(\frac{t(X_n - np)}{(npq)^{\frac{1}{2}}}\right)\right) = \left[\mathbf{E}\exp\left(\frac{t(X_1 - p)}{(npq)^{\frac{1}{2}}}\right)\right]^n$$

$$= \left[p\exp\left(\frac{qt}{(npq)^{\frac{1}{2}}}\right) + q\exp\left(\frac{-pt}{(npq)^{\frac{1}{2}}}\right)\right]^n.$$

Next we expand the two exponential terms in (12) to give

(13) $$\mathbf{E}(e^{tY_n}) = \left(1 + \frac{t^2}{2n} + O(n^{-\frac{3}{2}})\right)^n.$$

Now we recall the useful result which says that for constant a

(14) $$\lim_{n\to\infty}\left(1 + \frac{a}{n} + o(n^{-1})\right)^n = e^a.$$

Applying this to (13) shows that

$$\lim_{n\to\infty} \mathbf{E}(e^{tY_n}) = e^{\frac{1}{2}t^2},$$

which is the m.g.f. of the standard normal distribution as required. (More demanding readers should note that they can prove (14) by first taking logarithms.) ●

The appearance of the normal distribution in these circumstances is one of the most remarkable results in the theory of probability. The first proof, due to de Moivre, was greatly improved by Laplace. Their methods were very different from those used here, relying on fairly precise direct estimates of the binomial probabilities. We outline a modern version of their proof in Example 7.19.

7.6 Conditional distributions

Just as in the discrete case, it is often necessary to consider the distribution of a random variable X conditional upon the occurrence of some event A. By definition of conditional probability, we have

(1) $$F_{X|A}(x) = \mathbf{P}(X \leq x | A) = \mathbf{P}(\{\omega: X \leq x\} \cap A)/\mathbf{P}(A) = \mathbf{P}(X \leq x; A)/\mathbf{P}(A), \text{ say.}$$

(Obviously A has to be in $\mathcal{F}$, the event space.) The case that arises most commonly is when A is an event of the form

(2) $$A = \{\omega: a < X \leq b\},$$

that is, we seek the distribution of X conditional on its lying in some subset of its range.

(3) **Example** Let $a < b < c < d$. Let X be uniform on (a, d) and let $A = \{\omega: b < X(\omega) \le c\}$.

Then by (1)

$$\mathbf{P}(X \le x | A) = \frac{\mathbf{P}(X \le x; b < X \le c)}{\mathbf{P}(b < X \le c)} = \begin{cases} \dfrac{x-b}{d-a} \Big/ \left(\dfrac{c-b}{d-a}\right) & \text{if } b < x \le c \\ 0 & \text{otherwise} \end{cases}$$

$$= \frac{x-b}{c-b} \quad \text{for } b < x \le c.$$

Thus the distribution of X given A is just uniform on (b, c). More generally it is easy to see that a uniform random variable, constrained to lie in any subset A of its range, is uniformly distributed over the subset A. ●

Since $\mathbf{P}(X \le x|A)$ is a distribution, it may have an expectation. For example, suppose that X has density f, and A is given by (2). Then by (1)

(4) $$\mathbf{P}(X \le x|A) = \frac{F(x) - F(a)}{F(b) - F(a)} \quad \text{for } a < x \le b,$$

and differentiating yields the conditional density

(5) $$f_{X|A}(x) = \begin{cases} \dfrac{f(x)}{F(b) - F(a)} & \text{if } a < x \le b \\ 0 & \text{otherwise.} \end{cases}$$

Notice that $\int_a^b f_{X|A}(v)dv = 1$, as it must. Then we may define the conditional expectation

(6) $$\begin{aligned} \mathbf{E}(X|A) &= \int_a^b \frac{vf(v)}{F(b) - F(a)} dv \\ &= a + \int_a^b \frac{F(b) - F(v)}{F(b) - F(a)} dv, \quad \text{on integrating by parts,} \\ &= \int_0^\infty (1 - F_{X|A}(v))dv \end{aligned}$$

on using (4). Notice that this is in agreement with Theorem 7.4.11, as of course it must be.

(7) **Example: Exponential density and lack-of-memory** Let X be exponentially distributed with parameter λ. Show that

(8) $$\mathbf{P}(X > s + t|X > s) = e^{-\lambda t} = \mathbf{P}(X > t).$$

Find $\mathbf{E}(X|X > s)$ and $\mathbf{E}(X|X \le s)$.

Solution Trivially

$$\mathbf{P}(X > s + t | X > s) = \mathbf{P}(X > s + t)/\mathbf{P}(X > s) = e^{-\lambda(s+t)}/e^{-\lambda s} = e^{-\lambda t}.$$

Hence

(9) $$\mathbf{E}(X | X > s) = s + \int_0^\infty e^{-\lambda t} dt = s + \mathbf{E}(X).$$

We remark that the remarkable identity (8) is known as the *lack-of-memory property* of the exponential distribution. Finally

$$\mathbf{E}(X | X \leq s) = \int_0^s \frac{\mathbf{P}(s \geq X > v)}{\mathbf{P}(s \geq X)} dv = \frac{1}{\lambda} - \frac{s}{e^{\lambda s} - 1}. \quad \bullet$$

7.7 Ageing and survival

Many classic examples of continuous random variables arise as waiting times or survival times. For instance, the time until the cathode-ray tube in your television fails, the time until you are bitten by a mosquito after disembarking in the tropics, the time until a stressed metal component fails due to fatigue. For definiteness we consider the *lifetime* T of some *device* or component. The device is said to *fail* at time T.

It is often useful to quantify the ageing process of a device, in particular we may wish to compare a device of given age with a new one. (We are all familiar with the fact that it is not necessarily always a good thing to replace a working component with a new one. This fact is embodied in the popular saying: 'If it works, don't fix it'.) Let T have distribution F and density f. The following quantities turn out to be of paramount importance in comparing devices of different ages.

The *survival* function

(1) $$\bar{F}(t) = 1 - F(t) = \mathbf{P}(T > t).$$

The *hazard* function

(2) $$H(t) = -\log(1 - F(t)).$$

The *hazard rate* function

(3) $$r(t) = \frac{f(t)}{\bar{F}(t)} = \frac{f(t)}{1 - F(t)} = \frac{dH(t)}{dt}.$$

The last equality explains why $r(t)$ is called the hazard rate. Integrating (3) yields

(4) $$\exp\left(-\int_0^t r(s) ds\right) = \bar{F}(t).$$

Before we explain the significance of these quantities, you are warned that terminology in the literature of ageing is quite chaotic. Note that:

(i) the survival function is also known as the survivor function, reliability function, or hazard function;
(ii) the hazard function is also known as the log-survivor function;
(iii) the hazard rate function is also known as the failure rate function, mortality function, or hazard function.

Beware!

Now let A_t denote event that $T > t$. Then

$$F_{T|A_t}(s+t) = \mathbf{P}(T \leqslant s+t | T > t) = \frac{F(s+t) - F(t)}{1 - F(t)}.$$

This is the probability that the device fails during $(t, t+s)$, given that it has not failed by time t. Now

(5) $$\lim_{s \to 0} \frac{1}{s} F_{T|A_t}(s+t) = (1 - F(t))^{-1} \lim_{s \to 0} \frac{F(t+s) - F(t)}{s} = \frac{f(t)}{1 - F(t)} = r(t).$$

Thus $r(t)$ may be thought of as the 'intensity' of the probability that a device aged t will fail.

(6) **Example: Exponential life** If T has an exponential density, then $\bar{F}(t) = e^{-\lambda t}$, $H(t) = \lambda t$, and

$$r(t) = \lambda.$$

This constant hazard rate is consonant with the lack-of-memory property mentioned in Example 7.6.7. Very roughly speaking, the device cannot remember how old it is, and so the failure intensity remains constant. ●

We see that intuitively there is a distinction between devices for which $r(t)$ increases, essentially they are 'wearing out', and those for which $r(t)$ decreases, they are 'bedding in'.

A simple and popular density in this context is the Weibull density, which can exhibit both types of behaviour.

(7) **Example: Weibull life** If T has density

$$f(t) = \alpha t^{\alpha - 1} \exp(-t^\alpha), \quad t > 0, \alpha > 0,$$

then it has distribution $F(t) = 1 - \exp(-t^\alpha)$. Hence

$$\bar{F}(t) = \exp(-t^\alpha)$$

and so

$$\frac{\mathbf{P}(T > t + s | T > s)}{\mathbf{P}(T > t)} = \exp(-(t+s)^\alpha + s^\alpha + t^\alpha)$$

which is >1 or <1 according as $\alpha < 1$ or $\alpha > 1$. (To see this, just consider the stationary value of $x^\alpha + (1-x)^\alpha - 1$ at $x = \frac{1}{2}$.) Hence if $\alpha < 1$, the chance of lasting a further time t (conditional on $T > s$) increases with s. On the other hand if $\alpha > 1$, this chance decreases with s. ●

The behaviour of $r(t)$ is not the only measure of comparison between new and old devices. There is a large hierarchy of measures of comparison, which we display formally as follows (in the notation of (1)–(3)).

(8) Definition

(i) If $r(t)$ increases then T is (or has) *increasing failure rate*, denoted by IFR.

(ii) If $\dfrac{H(t)}{t}$ increases then T is (or has) *increasing failure rate average*, denoted by IFRA.

(iii) If for all $s \geqslant 0$, $t \geqslant 0$,

$$H(s+t) \geqslant H(s) + H(t)$$

then T is *new better than used*, denoted by NBU.

(iv) If for all $t \geqslant 0$

$$\mathbf{E}(T) \geqslant \mathbf{E}(T - t \mid A_t)$$

then T is *new better than used in expectation*, denoted by NBUE.

(v) If for all $0 \leqslant s < t < \infty$

$$\mathbf{E}(T - s \mid A_s) \geqslant \mathbf{E}(T - t \mid A_t)$$

then T has (or is) *decreasing mean residual life*, denoted by DMRL.

The random variable T may also be decreasing failure rate: DFR; decreasing failure rate on average: DFRA; new worse than used: NWU; new worse than used in expectation: NWUE; or increasing mean residual life: IMRL. All these are defined in the obvious way, analogous to (i)–(v). ▲

It can be shown that the following relationships hold between these concepts:

$$\begin{array}{ccccc} & & \text{IFRA} \;\Rightarrow\; \text{NBU} & & \\ & \nearrow & & \searrow & \\ \text{IFR} & & & & \text{NBUE} \\ & \searrow & & \nearrow & \\ & & \text{DMRL} & & \end{array}$$

Some of these implications are trivial, and some are established in Example 7.16 below. These ideas are linked to another concept, that of stochastic ordering.

7.8 Stochastic ordering

As in Section 7.7, let T be a non-negative random variable. In general, let $R(s)$ be a random variable whose distribution is that of $T - s$ given that $T > s$, namely

(1) $$F_R(x) = \mathbf{P}(T - s \leqslant x \mid T > s).$$

We refer to $R(s)$ as the *residual life* (of T at s).

The above example shows that if T has the exponential density, then its residual life is also exponentially distributed with constant mean.

More generally $F_{R(s)}$ may depend on s, and more significantly it may do so in a systematic way; the following definition is relevant here.

(2) **Definition** Let X and Y be random variables. If

(3) $$\bar{F}_X(x) \geqslant \bar{F}_Y(x) \quad \text{for all } x$$

then X is said to be stochastically larger than Y. ▲

Now we can supply a connection with the ideas of the preceding section (7.7).

(4) **Example** If T is a random variable with residual life $R(s)$, $s > 0$, show that T has increasing failure rate if and only if $R(s)$ is stochastically larger than $R(t)$ for all $s < t$.

Solution First we find

(5) $$\begin{aligned}\mathbf{P}(R(t) > x) &= \mathbf{P}(T - t > x | T > t) = \bar{F}(t + x)/\bar{F}(t)\\ &= \exp\left(-\int_0^{t+x} r(s)ds\right)\exp\left(\int_0^t r(s)ds\right) \quad \text{by (7.7.4)}\\ &= \exp\left(-\int_t^{t+x} r(s)ds\right).\end{aligned}$$

Differentiating (5) with respect to t we have

(6) $$\frac{\partial}{\partial t}\mathbf{P}(R(t) > x) = (r(t) - r(t + x))\exp\left(-\int_t^{t+x} r(s)ds\right).$$

Since $\exp(-\int rds)$ is positive, and $r(t) - r(t + x)$ is positive or negative according as T is DFR or IFR, the result follows. ●

Finally we have the useful:

(7) **Theorem** If X is stochastically larger than Y, then $\mathbf{E}(X) \geqslant \mathbf{E}(Y)$.

Proof We prove this when $X \geqslant 0$ and $Y \geqslant 0$. (The general result is left as an exercise.) From Theorem 7.4.11

$$\begin{aligned}\mathbf{E}(X) &= \int_0^\infty \bar{F}_X(x)dx \geqslant \int_0^\infty \bar{F}_Y(x)dx \quad \text{by hypothesis,}\\ &= \mathbf{E}(Y).\end{aligned}$$ ■

7.9 Random points

Picking a point Q at random in the interval (0, 1) yielded the uniform density (of the length OQ). It is intuitively attractive to consider problems that involve picking one or more points at random in other nice geometrical figures, such as discs, squares, triangles, spheres, and so on. Indeed this idea is so natural that mathematicians had already started doing this kind of thing in the eighteenth century, and one of the most celebrated articles on the subject is that of M W Crofton in the 1885 edition of the *Encyclopaedia Britannica*. Such questions also have applications in statistics.

Confining ourselves to two dimensions for definiteness, suppose a point Q is picked at random in a region R of area $|R|$. Then it is natural to let the probability $\mathbf{P}(S)$, that Q lies in a set $S \subseteq R$, be given by

(1) $$\mathbf{P}(S) = \frac{|S|}{|R|},$$

where, now, $|S|$ denotes the area of S. It follows from the properties of area that $\mathbf{P}(.)$ has the required properties of a probability function, and we can proceed to solve a variety of simple problems using elementary geometry. The following is typical.

(2) **Example** A point Q is picked at random in the unit square. What is the probability v that it is nearer to the centre O of the square than to its perimeter?

Solution By symmetry we need consider only the sector $0 \leq y \leq x \leq \frac{1}{2}$. Then the point (x, y) is nearer to O than the perimeter if $\sqrt{x^2 + y^2} < \frac{1}{2} - x$, that is, if in this sector,

$$x < \frac{1}{4} - y^2, \text{ for } 0 \leq y \leq \frac{1}{2}(\sqrt{2} - 1).$$

Hence the area is given by an integral and

$$v = 8\int_0^{\frac{1}{2}(\sqrt{2}-1)} \left(\frac{1}{4} - y^2 - y\right) dy = \frac{4}{3}\sqrt{2} - \frac{5}{3}. \qquad \bullet$$

An equally trivial but much more important example is the following.

(3) **Example** Let $f = f(x)$ be an integrable function with $0 \leq f(x) \leq 1$ for $0 \leq x \leq 1$. Let Q be picked at random in the unit square, and let A_v be the set of points such that $0 \leq x \leq v$, and $0 \leq y \leq f(x)$. Then from (1),

(4) $$\mathbf{P}(Q \in A_v) = \int_0^v f(x)dx. \qquad \bullet$$

This trivial result has at least two important applications. The first we have met already in Example 5.8.9.

(5) **Example: Hit-or-miss Monte Carlo integration** Let $f(x)$ and Q be as defined in Example 3, and declare Q a *hit* if Q lies below $f(x)$, $0 \leq x \leq 1$. Then the probability of a hit is $\mathbf{P}(A_1) = \int_0^1 f(x)dx$. Now we pick a sequence of such points $Q_1, Q_2, \ldots$ and let X_n be the number of hits. If points are picked independently, then X_n is a binomial random variable with parameters n and $\mathbf{P}(A_1)$, and we have shown that as $n \to \infty$, for $\epsilon > 0$,

$$\mathbf{P}\left(\left|n^{-1}X_n - \int_0^1 f(x)dx\right| > \epsilon\right) \to 0.$$

This therefore offers a method for evaluating the integral $\int_0^1 f(x)dx$. In practice one would be unlikely to use this method in one dimension, but you might well use the analogous method to evaluate $\int f(\mathbf{x})d\mathbf{x}$ where $\mathbf{x}$ is a vector in (say) eleven dimensions. $\bullet$

(6) **Example: Simulation** With Q and $f(x)$ defined as above, consider the probability that Q lies in A_v given that it is a hit. By definition this has probability

$$\mathbf{P}(A_v|A_1) = \frac{\int_0^v f(x)dx}{\int_0^1 f(x)dx}.$$

By inspection the function $F(v) = \mathbf{P}(A_v|A_1)$ is the distribution function of the x-coordinate of Q given that it is a hit. This procedure therefore offers a method of simulating a random variable X with density function

(7) $$f_X(x) = \frac{f(x)}{\int_0^1 f(x)dx}.$$

You can just pick a point Q and, if it is a hit, let its x-coordinate be X. ●

A natural next step is to consider events defined jointly by a number of points picked independently in a region R. One famous example is Sylvester's problem: for four points picked at random in R, what is the probability that one of them lies in the triangle formed by the other three? This is too difficult for us, but we can consider an amusing simpler problem to illustrate a few of the basic ideas.

(8) **Example: Two points in a disc** Let $\lambda(r)$ be the expected value of the distance $L(r)$ between two points Q_1 and Q_2, each distributed uniformly (and independently) over a disc of radius r. Show that

(9) $$\lambda(r) = \frac{128r}{45\pi}.$$

Solution This can be done by a brutal integration; here is a better way, discovered by M V Crofton in 1885.

Consider a disc of radius $x + h$, which we may think of as a disc D of radius x, surrounded by an annulus A of width h. Then if Q_1 and Q_2 are dropped at random on to the disc of radius $x + h$ we have (using independence and the properties of the uniform density) that

(10) $$\mathbf{P}(Q_1 \in D \cap Q_2 \in D) = \left(\frac{\pi x^2}{\pi(x+h)^2}\right)^2 = 1 - \frac{4h}{x} + o(h).$$

Also

$$\mathbf{P}(Q_1 \in D \cap Q_2 \in A) = \frac{\pi x^2}{\pi(x+h)^2}\left(1 - \frac{\pi x^2}{\pi(x+h)^2}\right) = \frac{2h}{x} + o(h)$$

and $\mathbf{P}(Q_1 \in A \cap Q_2 \in A) = o(h)$. Hence, by conditional expectation

(11) $$\lambda(x+h) = \mathbf{E}(L(x+h)|Q_1 \in D; Q_2 \in D)\left(1 - \frac{4h}{x} + o(h)\right) + 2\mathbf{E}(L(x+h)|Q_1 \in D; Q_2 \in A)\left(\frac{2h}{x} + o(h)\right) + o(h).$$

Now $\mathbf{E}(L(x+h)|Q_1 \in D; Q_2 \in A)$ is just the mean distance of a random point Q_1 in a disc of radius x, from a point Q_2 on its circumference (plus a quantity which is $o(h)$). Hence taking plane polar coordinates with Q_2 as origin:

$$\mathbf{E}(L(x+h)|Q_1 \in D; Q_2 \in A) = \frac{1}{\pi x^2}\int_{-\pi/2}^{\pi/2}\int_0^{2x\cos\theta} v^2\,dv\,d\theta + o(h) = \frac{32x}{9\pi} + o(h).$$

Returning to (11), note that $\mathbf{E}(L(x+h)|Q_1 \in D;\ Q_2 \in D) = \lambda(x)$; hence rearranging (11) and letting $h \to 0$ gives

$$\frac{d\lambda(x)}{dx} = \lim_{h\to 0}\frac{1}{h}\left(\lambda(x+h) - \lambda(x)\right) = \frac{-4}{x}\lambda(x) + \frac{128}{9\pi}.$$

Integrating this, and observing that $\lambda(0) = 0$, we have

$$\lambda(x) = \frac{128x}{45\pi}$$

as required. ●

Using the same idea, and with a lot more toil, we can find the density of L.

The next natural step is to pick lines (or other objects) at random and ask how they divide up the region R in random tessellations or coverings. This is well beyond our scope, but the trivial Example 7.17 illustrates some of the problems.

WORKED EXAMPLES AND EXERCISES

7.10 Example: Using a uniform random variable

The random variable U is uniformly distributed on $(0, 1)$.

(a) Can you use U to get a random variable with density

(1) $$f_0(y) = 12\left(y - \frac{1}{2}\right)^2 \quad \text{for } 0 < y < 1?$$

(b) Actually you really want a random variable with density

(2) $$f(x) = 3\left(\left(x - \frac{1}{2}\right)^2 + \frac{1}{8}|1 - 2x|^{\frac{1}{2}}\right) \quad \text{for } 0 < x < 1,$$

and in your pocket is a fair coin. Explain how the coin is useful.

Solution If $g(U)$ is a continuous increasing function, and $Y = g(U)$, then

$$F_Y(y) = \mathbf{P}(g(U) \leq y) = \mathbf{P}(U \leq g^{-1}(y)) = g^{-1}(y) \quad \text{since } U \text{ is uniform.}$$

From (1) we have the distribution of interest

$$F_Y(y) = \int_0^y 12\left(y - \frac{1}{2}\right)^2 dy = 4\left(y - \frac{1}{2}\right)^3 + \frac{1}{2}.$$

Hence, if we find a function $g(.)$ such that

$$g^{-1}(y) = 4\left(y - \frac{1}{2}\right)^3 + \frac{1}{2},$$

then $g(U)$ has the density (1) as required. Setting $y = g(u)$, and solving

$$u = 4\left(g(u) - \frac{1}{2}\right)^3 + \frac{1}{2},$$

we find immediately that

$$g(u) = \left(\frac{u}{4} - \frac{1}{8}\right)^{\frac{1}{3}} + \frac{1}{2}$$

is the required function $g(.)$.

For the second part, we notice that

$$\frac{1}{|1-2x|^{\frac{1}{2}}} = \frac{1}{(1-2x)^{\frac{1}{2}}} \quad \text{if } 0 < x < \frac{1}{2},$$

and that

$$f_1(x) = \begin{cases} (1-2x)^{-\frac{1}{2}} & \text{if } 0 < x < \frac{1}{2} \\ 0 & \text{elsewhere} \end{cases}$$

is a density function. By the method of the first part, or by inspection, we see that

$$g_1(U) = \frac{1}{2}(1 - U^2)$$

is a random variable with density $f_1(x)$. (To see this just make the simple calculation

$$\mathbf{P}(g_1(U) \leqslant x) = \mathbf{P}\left(\frac{1}{2}(1-U^2) \leqslant x\right) = \mathbf{P}(U \geqslant (1-2x)^{\frac{1}{2}}) = 1 - (1-2x)^{\frac{1}{2}},$$

and differentiate to get the density f_1.)

Likewise

$$f_2(x) = \begin{cases} (2x-1)^{-\frac{1}{2}} = |1-2x|^{\frac{1}{2}} & \text{if } \frac{1}{2} < x < 1 \\ 0 & \text{elsewhere} \end{cases}$$

is a density function, and

$$g_2(U) = \frac{1}{2}(1 + U^2)$$

is a random variable with density f_2. (You can check this, as we did for g_1 and f_1.)

Now you take the coin and toss it three times. Let A be the event that you get either three heads or three tails, B the event that you get two heads and a tail (in any order) and C the event that you get two tails and a head (in any order). Define the random variable

(3)
$$X = \begin{cases} \left(\frac{U}{4} - \frac{1}{8}\right)^{\frac{1}{3}} + \frac{1}{2} & \text{if } A \text{ occurs} \\ \frac{1}{2}(1 + U^2) & \text{if } B \text{ occurs} \\ \frac{1}{2}(1 - U^2) & \text{if } C \text{ occurs} \end{cases}$$

Then the density of X is just a mixture of the densities of $g(U)$, $g_1(U)$ and $g_2(U)$, namely

$$f_X(x) = \frac{1}{4} f_0(x) + \frac{3}{8} f_1(x) + \frac{3}{8} f_2(x) = 3\left(x - \frac{1}{2}\right)^2 + \frac{3}{8}|1 - 2x|^{-\frac{1}{2}}$$

as required.

(4) Exercise Explain how you would use U to get a random variable with density

$$f(x) = \frac{3}{4}\left(1 + (2x - 1)^2\right) \quad \text{if } 0 < x < 1.$$

(5) Exercise Show that $Y = \gamma(-\log U)^{\frac{1}{\beta}}$ has a *Weibull* distribution.

(6) Exercise Let the random variable X be defined by

$$X = \begin{cases} (2U)^{\frac{1}{2}} & \text{if } U < \frac{1}{2} \\ 2 - (2 - 2U)^{\frac{1}{2}} & \text{if } U \geq \frac{1}{2}. \end{cases}$$

Show that X has a triangular density on $[0, 2]$.

(7) Exercise Find the densities of:
(a) $\tan(\pi U)$;
(b) $\tan(\frac{\pi}{2}U)$.

7.11 Example: Normal distribution

Let

(1)
$$\phi(x) = (2\pi)^{-\frac{1}{2}} e^{-x^2/2}; \ \Phi(x) = \int_{-\infty}^{x} \phi(u)\,du.$$

(a) Define the sequence of functions $H_n(x)$; $n \geq 0$, by

(2)
$$(-)^n \frac{d^n \phi(x)}{dx^n} = H_n(x)\phi(x); \ H_0 = 1.$$

Show that $H_n(x)$ is a polynomial in x of degree n. What is $H_1(x)$?

(b) Define Mills' ratio $r(x)$ by $r(x)\phi(x) = 1 - \Phi(x)$. Show that for $x > 0$

(3)
$$\frac{1}{x} - \frac{1}{x^3} < r(x) < \frac{1}{x}.$$

Solution First make the important observation that

(4)
$$\frac{d\phi}{dx} = \frac{d}{dx}\left((2\pi)^{-\frac{1}{2}} e^{-x^2/2}\right) = -x\phi.$$

(a) We use induction. First, by (2)

$$-H_{n+1}(x)\phi(x) = +\frac{d}{dx}(H_n(x)\phi(x)) = H'_n(x)\phi(x) - H_n(x)x\phi(x), \text{ by (4)}.$$

Hence

(5)
$$H_{n+1}(x) = xH_n(x) - H'_n(x)$$

and by (2) and (4), $H_1(x) = x$. The result follows by induction as claimed.

(b) For the right hand inequality we consider

$$1 - \Phi(x) = \int_x^\infty \phi(u)du \leqslant \int_x^\infty \frac{u}{x}\phi(u)du \quad \text{for } x > 0,$$

$$= -\frac{1}{x}\int_x^\infty \phi'(u)du \quad \text{by (4)},$$

$$= \frac{1}{x}\phi(x).$$

For the left hand inequality we consider

$$1 - \Phi(x) = \int_x^\infty \phi(u)du = -\int_x^\infty \frac{\phi'(u)}{u}du \quad \text{by (4)}$$

$$= -\left[\frac{\phi(u)}{u}\right]_x^\infty - \int_x^\infty \frac{\phi(u)}{u^2}du \quad \text{on integrating by parts,}$$

$$= \frac{\phi(x)}{x} + \int_x^\infty \frac{\phi'(u)}{u^3}du \quad \text{by (4),}$$

(6)
$$= \frac{\phi(x)}{x} - \frac{\phi(x)}{x^3} + \int_x^\infty \frac{3\phi(u)}{u^4}du \quad \text{on integrating by parts,}$$

$$\geqslant \phi(x)\left\{\frac{1}{x} - \frac{1}{x^3}\right\}$$

as required.

Remark For large x these bounds are clearly very tight.

(7) Exercise The polynomials $H_n(x)$ are known as Hermite (or Chebyshov–Hermite) polynomials. Show that they are orthogonal with respect to $\phi(x)$ over $\mathbb{R}$, which is to say that

$$\int_{-\infty}^{\infty} H_n(x)H_m(x)\phi(x)dx = \begin{cases} 0 & m \neq n \\ n! & m = n. \end{cases}$$

(8) Exercise Show that the exponential generating function of the H_n is

$$\sum_{n=0}^{\infty} H_n(x)\frac{t^n}{n!} = e^{tx - \frac{1}{2}t^2}.$$

(9) Exercise Show that for $x > 0$

$$\frac{1}{x} - \frac{1}{x^3} < r(x) < \frac{1}{x} - \frac{1}{x^3} + \frac{3}{x^5}.$$

(10) Exercise Let X have the Weibull distribution $F(x) = 1 - \exp(-(\lambda t)^2)$. Show that

$$\frac{1}{2}\lambda^{-2}t^{-1} - \frac{1}{4}\lambda^{-4}t^{-3} < \mathbf{E}(X - t|X > t) < \frac{1}{2}\lambda^{-2}t^{-1}.$$

7.12 Example: Bertrand's paradox

(a) A point P is chosen at random inside a circular disc of radius a. What is the probability that its distance from O, the centre of the disc, is less than d? Let X be the length of the chord of the disc of which P is the midpoint. Show that $\mathbf{P}(X > \sqrt{3}a) = \frac{1}{4}$.

(b) Now choose another chord as follows. A point Q is fixed on the circumference of the disc and a point P is chosen at random on the circumference. Let the length of PQ be Y. Show that $\mathbf{P}(Y > \sqrt{3}a) = \frac{1}{3}$.

Solution (a) If P is less than d from the centre, then it lies inside the disc of radius d with area πd^2. Therefore the required probability is

(1) $$\pi d^2/(\pi a^2) = d^2/a^2.$$

Now $X > \sqrt{3}a$ if and only if the chord subtends an angle greater than $\frac{2\pi}{3}$ at the centre of the disc. This occurs (see Figure 7.9) if and only if OP has length less than $\frac{1}{2}a$.

Hence, by (1)

$$\mathbf{P}(X > \sqrt{3}a) = \frac{(a/2)^2}{a^2} = \frac{1}{4}.$$

(b) As in (a) we observe that $Y > \sqrt{3}a$ if and only if PQ subtends an angle greater than $2\pi/3$ at O. This occurs if and only if P lies on the dashed interval of the circumference of the disc in Figure 7.10. Since this interval is one third of the circumference, $\mathbf{P}(Y > \sqrt{3}a) = \frac{1}{3}$.

(2) **Exercise** In part (b) suppose that Q is picked at random as well as P. What is $\mathbf{P}(Y > \sqrt{3}a)$?

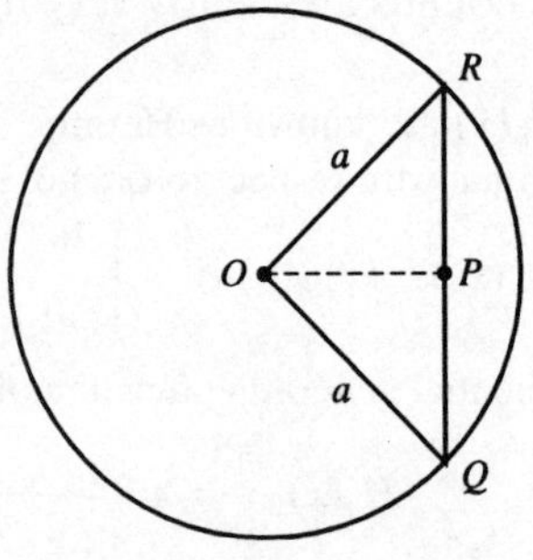

Figure 7.9 Bertrand's paradox. In this case $X < \sqrt{3}a$ because $OP > \frac{1}{2}a$.

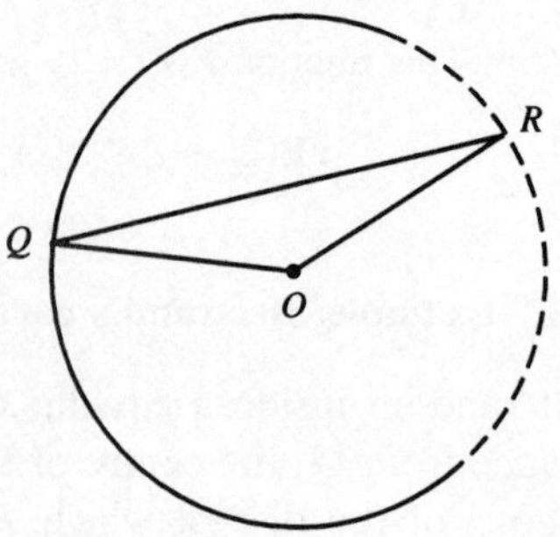

Figure 7.10 Bertrand's paradox. In this case $X > \sqrt{3}a$.

(3) **Exercise** A point P is picked at random on an arbitrarily chosen radius of the disc. Let Z be the length of the chord of which P is the midpoint. Show that $\mathbf{P}(Z > \sqrt{3}a) = \frac{1}{2}$.

(4) **Exercise** A point Q is fixed on the circumference. The chord is drawn which makes an angle Θ with the tangent at Q, where Θ is uniform on $(0, \pi)$. If the length of this chord is W show that $\mathbf{P}(W > \sqrt{3}a) = \frac{1}{3}$.
Is it just a coincidence that this answer is the same as (b) above?

7.13 Example: Stock control

A manufacturer of bits and bobs has a shop. Each week it is necessary to decide how many bits to deliver to the shop on Monday, in the light of the following information.

(i) Delivering y bits costs c per bit, plus a fixed delivery charge k.
(ii) Any bit unsold at the weekend has to be packed, stored, insured and discounted over the weekend, at a total cost of h per bit.
(iii) If the shop sells every bit before the weekend, then further customers that week are supplied by post at the end of the week; this costs p per bit, due to postage, packing, paperwork and other penalties, and $p > c$.
(iv) The demand Z for bits each week is a random variable with density $f(z)$, and distribution $F(z)$, where $F(0) = 0$.

If the manager seeks to minimize the expected costs of her decision, and she has x bits in the shop over the weekend, approximately how many bits should she order on Monday morning?

Note that the customer pays the same whether the bit comes from the shop or factory. Note also that the problem implicitly assumes that we are content with a continuous approximation to what is actually a discrete problem.

Solution If nothing is delivered, then costs are

$$p(Z - x) \quad \text{if} \quad Z > x;$$
$$h(x - Z) \quad \text{if} \quad Z < x.$$

Hence expected costs are

(1) $$\lambda(x) = p\int_x^\infty (z - x)f(z)dz + h\int_0^x (x - z)f(z)dz.$$

If $y - x$ bits are delivered, to bring the stock of bits to y, then expected costs are

(2) $$\mu(x, y) = k + c(y - x) + \lambda(y).$$

Now

$$\frac{\partial\mu}{\partial y} = c + \lambda'(y) = c + hF(y) - p(1 - F(y))$$

and

$$\frac{\partial^2\mu}{\partial y^2} = (h + p)f(y) \geq 0.$$

Since $\mu'(0) < 0$, and $\mu'(y) > 0$ for large y, it follows that $\mu(x, y)$ has a unique minimum at the value $\hat{y}$ such that

(3) $$F(\hat{y}) = \frac{p-c}{p+h}.$$

Thus, if any delivery is made, the expected total costs are minimized by choosing $y = \hat{y}$, and the minimum is

$$\mu(x, \hat{y}) = k + c(\hat{y} - x) + \lambda(\hat{y}).$$

The only alternative is to have no delivery, with expected total cost $\lambda(x)$. Hence the optimal policy is to have no delivery when $x > \hat{y}$ or

$$\lambda(x) \leqslant k + c\hat{y} + \lambda(\hat{y}) - cx,$$

and to deliver $\hat{y} - x$ when $x < \hat{y}$ and

$$\lambda(x) > k + c\hat{y} + \lambda(\hat{y}) - cx.$$

Now if we set $g(x) = \lambda(x) + cx$, we have

$$g'(x) = c - p + (h + p)F(x)$$

and

$$g''(x) = (h + p)F(x) \geqslant 0.$$

Since $g'(0) < 0$ and $g'(\hat{y}) = 0$, it follows that there is a unique point $\hat{x}$ such that

(4) $$g(\hat{x}) = \lambda(\hat{x}) + c\hat{x} = k + c\hat{y} + \lambda(\hat{y}).$$

Hence the optimal policy takes the simple form:

deliver no bits if $x \geqslant \hat{x}$; or
deliver $\hat{y} - x$ bits if $x < \hat{x}$;

where $\hat{y}$ satisfies (3) and $\hat{x}$ satisfies (4).

(5) Exercise What is the optimal policy if the fixed delivery cost k is zero?

(6) Exercise Suppose that the postal deliveries also have a set-up cost, so that posting y bits costs $m + py$. If demand is exponentially distributed with distribution

$$F(x) = \begin{cases} 1 - e^{-\lambda(x-a)}; & x \geqslant a \\ 0 \quad ; & x < a, \end{cases}$$

find the optimal delivery policy.

7.14 Example: Obtaining your visa

A certain consular clerk will answer the telephone only on weekdays at about 10.00 a.m. On any such morning it is an evens chance whether he is at his desk or not; if he is absent no one answers, and days are independent. The line is never engaged.

If he is at his desk, the time T that he takes to answer the telephone is a random variable such that

(1) $$\mathbf{P}(T \leqslant t) = \begin{cases} 0 & t \leqslant 1 \\ 1 - t^{-1} & t > 1. \end{cases}$$

(a) If you telephone this clerk one morning, and do not hang up, what is the probability that the telephone rings for at least a time s?

(b) You adopt the following procedure. Each day until you are successful you telephone the clerk, and hang up at time s if he has not answered by then.

Show that in order to minimize the expected time you spend listening to the ringing tone, you should choose s to be the unique positive root s_0 of $\log s = (s+1)(s-2)$.

Solution (a) Let R be the ringing time. Conditioning on whether the clerk is there or not we have

(2) $$\mathbf{P}(R > s) = \frac{1}{2}\mathbf{P}(R > s \mid \text{absent}) + \frac{1}{2}\mathbf{P}(R > s \mid \text{present})$$

$$= \begin{cases} 1 & \text{for } s < 1 \\ \frac{1}{2} + \frac{1}{2}\cdot\frac{1}{s} & \text{for } s \geq 1, \text{ by (1).} \end{cases}$$

(b) If your call is successful, then the expected time for which the telephone rings is

(3) $$\mathbf{E}(R \mid R < s) = \int_0^s \mathbf{P}(R > x \mid R < s)\,dx \quad \text{by (7.4.12)}$$

$$= \int_0^s \frac{\mathbf{P}(x < R < s)\,dx}{\mathbf{P}(R < s)} = 1 + \int_1^s \frac{s}{s-1}(x^{-1} - s^{-1})\,dx = \frac{s\log s}{s-1}, \; s > 1.$$

The number of unsuccessful calls has a geometric mass function with parameter $\rho = \frac{1}{2}(1 - \frac{1}{s})$, and expectation

(4) $$\frac{1-\rho}{\rho} = \frac{\frac{1}{2}(1+\frac{1}{s})}{\frac{1}{2}(1-\frac{1}{s})} = \frac{s+1}{s-1}.$$

Hence the expected time spent listening to the ringing tone is

(5) $$\omega(s) = \frac{s(s+1)}{s-1} + \frac{s\log s}{s-1}.$$

Differentiating with respect to s gives $\omega'(s) = (s-1)^{-2}(s^2 - s - 2 - \log s)$. Thus a stationary value in $(1, \infty)$ occurs at a zero of $s^2 - s - 2 - \log s$. That is where $(s-2)(s+1) = \log s$.

(6) Exercise Show that there is just one such zero, and by inspection of (5) this stationary value is a minimum $\omega_{\min}$.

(7) Exercise Show that $\omega(s) \leq 2s^2/(s-1)$ and deduce that $\omega_{\min} \leq 8$.

(8) Exercise More generally, suppose that the clerk is in his office with probability p, and that $\mathbf{P}(T \leq x) = F(x)$.

Show that

$$\mathbf{E}(R) = \frac{s}{pF(s)} - \int_0^s \frac{F(x)}{F(s)}\,dx.$$

When $F(x) = \frac{x}{1+x}$, show that

$$\mathbf{E}(R) = (1+s)((1-p)p^{-1} + s^{-1}\log(1+s)).$$

7.15 Example: Pirates

Expensive patented (or trade marked) manufactures are often copied, and the copies sold as genuine. You are replacing part of your car; with probability p you buy a pirate part, with probability $1-p$ a genuine part. In each case lifetimes are exponential, pirate parts with parameter μ, genuine parts with parameter λ, where $\lambda < \mu$. The life of the part you install is T. Is T IFR or DFR? Does it make any difference if $\lambda > \mu$?

Solution By conditional probability

$$\mathbf{P}(T > t) = \bar{F}(t) = pe^{-\mu t} + (1-p)e^{-\lambda t}.$$

Hence, setting $q = 1 - p$, we have

(1) $$r(t) = f(t)/\bar{F}(t) = \frac{\mu p + \lambda q e^{(\mu-\lambda)t}}{p + qe^{(\mu-\lambda)t}} = \lambda + \frac{p(\mu-\lambda)}{p + qe^{(\mu-\lambda)t}}.$$

This decreases as t increases. Hence your part has DFR.

It makes no difference if $\lambda > \mu$. This is obvious anyway by symmetry, but also $r(t)$ given by (1) decreases as t increases if $\lambda > \mu$.

(2) Exercise What happens if $\lambda = \mu$?

(3) Exercise Suppose the part has survived for a time t after you install it.

(a) Show that the probability π that it is a pirate part is given by

$$\pi(t) = \frac{p}{p + (1-p)e^{(\mu-\lambda)t}}.$$

(b) Find the limit of $\pi(t)$ as $t \to \infty$, and explain why the answer depends on whether $\lambda > \mu$ or $\lambda < \mu$.

(4) Exercise Let X have density f and m.g.f. $M_X(\theta) = \mathbf{E}(e^{\theta X})$. Show that

$$\frac{d^2}{d\theta^2}\log(M_X(\theta)) > 0.$$

[You have shown that $M_X(\theta)$ is log–convex, if you are interested.]

(5) Exercise Due to variations in the manufacturing process, the lifetime T is exponential with parameter Λ where Λ has density $f(\lambda)$. Use the preceding exercise to show that T is DFR.

(6) Exercise Let T_Λ be a family of random variables indexed by a parameter Λ where Λ is a random variable with density $f(\lambda)$. Let $M(t)$ be the continuous mixture

$$M(t) = \mathbf{P}(T_\Lambda \leq t) = \int_0^\infty F_{T_\lambda}(t) f(\lambda)\, d\lambda.$$

Show that if $F_{T_\lambda}(t)$ is DFR for all λ, then $M(t)$ is DFR. [*Hint*: the Cauchy–Schwarz inequality says that $\mathbf{E}(XY) \leq (\mathbf{E}(X^2)\mathbf{E}(Y^2))^{\frac{1}{2}}$.]

7.16 Example: Failure rates†

Let T have distribution $F(t)$.

(a) Show that T is IFRA if and only if, for all $0 \leq \alpha \leq 1$,

(1) $$(\bar{F}(t))^\alpha \leq \bar{F}(\alpha t).$$

(b) Show also that if T is IFRA, then it is NBU.

†See Section 7.7 for expansions of the acronyms.

Solution (a) By definition, T is IFRA if $H(t)/t = \frac{1}{t}\int_0^t r(v)dv$ is increasing in t. This is the same as saying that for all $0 \le \alpha \le 1$

$$\frac{1}{\alpha t}\int_0^{\alpha t} r(v)dv \le \frac{1}{t}\int_0^t r(v)dv.$$

But, by (7.7.4), this is equivalent to

(2) $$\frac{-1}{\alpha}\log \bar{F}(\alpha t) \le -\log \bar{F}(t).$$

Now (1) follows as required because e^x is a monotone increasing function of x.

(b) Since $H(t)/t$ is increasing in t, for all $0 \le \alpha \le 1$ we have $H(\alpha t) \le \alpha H(t)$, and $H((1-\alpha)t) \le (1-\alpha)H(t)$. Hence

(3) $$H(\alpha t) + H(t - \alpha t) \le H(t).$$

Setting $\alpha t = s$ gives condition (iii) in Definition 7.7.8 for NBU.

(4) Exercise Show that if T is IFR then it is IFRA and DMRL.

(5) Exercise Show that if T is NBU or DMRL then it is NBUE.

(6) Exercise Let T have a gamma density with parameters 2 and λ. Find $H(t)$ and $r(t)$. Is T IFR?

7.17 Example: Triangles

A point P is chosen at random along a rod of length l.

(a) The rod is bent at P to form a right angle, thus forming the two shorter sides of a right-angled triangle. Let Θ be the smallest angle in this triangle. Find $\mathbf{E}(\tan\Theta)$ and $\mathbf{E}(\cot\Theta)$.

(b) The rod is now cut into two pieces at P. A piece is picked at random and cut in half.

What is the probability that the three pieces of the rod can form a triangle of any kind? Show that, conditional on the event that a triangle can be formed, the probability that it has no obtuse angle is $2(\sqrt{2}-1)$.

Solution Without loss of generality we can suppose the rod to be the unit interval, so that the length OP is a random variable X uniformly distributed on $[0, 1]$.

(a) Since Θ is the smallest angle

$$\tan\Theta = \begin{cases} \dfrac{X}{1-X}; & 0 \le X < \frac{1}{2} \\[2ex] \dfrac{1-X}{X}; & \frac{1}{2} \le X \le 1. \end{cases}$$

Hence

(1) $$\mathbf{E}(\tan\Theta) = \int_0^{\frac{1}{2}} \frac{x}{1-x}\,dx + \int_{\frac{1}{2}}^{1} \frac{1-x}{x}\,dx = 2\log 2 - 1 \simeq 0.39.$$

For variety and instruction we choose a different method of finding $\mathbf{E}(\cot\Theta)$. Let $Y = \cot\Theta$. Then for $y \geqslant 1$,

$$\begin{aligned} F(y) = \mathbf{P}(Y \leqslant y) &= \mathbf{P}\left(\frac{X}{1-X} \leqslant y\right) + \mathbf{P}\left(\frac{1-X}{X} \leqslant y\right) \\ &= \mathbf{P}\left(X \leqslant \frac{y}{1+y}\right) + \mathbf{P}\left(X \geqslant \frac{1}{1+y}\right) \\ &= F_X\left(\frac{y}{1+y}\right) + 1 - F_X\left(\frac{1}{1+y}\right) \\ &= 1 + \frac{y-1}{y+1} = \frac{2y}{y+1}. \end{aligned}$$

Hence, differentiating

(2) $$f_Y(y) = \frac{2}{y+1} - \frac{2y}{(y+1)^2} = \frac{2}{(y+1)^2}.$$

Thus

$$\mathbf{E}(\cot\Theta) = \mathbf{E}(Y) = \int_1^{\infty} \frac{2y}{(y+1)^2}\,dy = \infty.$$

(b) Suppose (without loss of generality) that the piece cut in half has length $1 - X$. Then, if it exists, the triangle is isosceles with sides $\frac{1}{2}(1-X)$, $\frac{1}{2}(1-X)$, X. This is possible if and only if $\frac{1}{2}(1-X) > \frac{1}{2}X$, which occurs if and only if $X < \frac{1}{2}$, which has probability $\frac{1}{2}$.

There is an obtuse angle (between the two sides of equal length) if and only if

$$\frac{\frac{1}{2}X}{\frac{1}{2}(1-X)} > \frac{1}{\sqrt{2}}$$

which occurs if and only if $X > \sqrt{2} - 1$.

Hence

$$\begin{aligned} \mathbf{P}\,(\text{no obtuse angle} \mid \text{the triangle exists}) &= \frac{\mathbf{P}(\{X < \sqrt{2}-1\} \cap \{X < \frac{1}{2}\})}{\mathbf{P}(X < \frac{1}{2})} \\ &= \frac{\mathbf{P}(X < \sqrt{2}-1)}{\mathbf{P}(X < \frac{1}{2})} = 2(\sqrt{2}-1). \end{aligned}$$

(3) Exercise What is the distribution of the length $X \wedge (1 - X)$ of the shortest side of the triangle?

(4) Exercise The longest side is $X \vee (1 - X)$. Show that

$$\frac{\mathbf{E}\{X \wedge (1-X)\}}{\mathbf{E}\{X \vee (1-X)\}} = \frac{1}{3},$$

where $x \wedge y = \min\{x, y\}$ and $x \vee y = \max\{x, y\}$.

(5) Exercise Find $\dfrac{\mathbf{E}(\sin\Theta)}{\mathbf{E}(\cos\Theta)}$.

(6) Exercise Show that the hypotenuse of the triangle has density

$$f(y) = \frac{2y}{(2y^2 - 1)^{\frac{1}{2}}}, \quad \frac{1}{\sqrt{2}} \leq y \leq 1.$$

(7) Exercise Let X have density

$$f_X(x) = \frac{1}{B(a, b)} x^{a-1}(1 - x)^{b-1} \quad ; 0 < x < 1.$$

Show that $\mathbf{E}(\cot \Theta)$ is finite if and only if $a > 1$ and $b > 1$.

7.18 Example: Stirling's formula

Let $\Gamma(x)$ be the gamma function defined by

(1) $$\Gamma(x) = \int_0^\infty t^{x-1} e^{-t} dt.$$

(a) Show that

(2) $$\Gamma(x) = x^{x-\frac{1}{2}} e^{-x} \int_{-x^{\frac{1}{2}}}^{\infty} (1 + ux^{-\frac{1}{2}})^{x-1} e^{-ux^{\frac{1}{2}}} du.$$

(b) Show that for fixed u, the integrand converges to $\exp(-\frac{1}{2}u^2)$ as $x \to \infty$, and deduce that as $x \to \infty$

(3) $$\Gamma(x) e^x x^{-x+\frac{1}{2}} \to \int_{-\infty}^{\infty} \exp\left(-\frac{1}{2}u^2\right) du.$$

You may assume that $\log(1 + x) = x - \frac{1}{2}x^2 + O(x^3)$ for $|x| < 1$.

Proof (a) Making the substitution $t = x + ux^{\frac{1}{2}}$ in (1) gives (2).
(b) Let the integrand in (2) be $f(x, u)$. Then for $u < x^{\frac{1}{2}}$,

(4) $$\log f(x, u) = -ux^{\frac{1}{2}} + (x - 1)\log(1 + ux^{\frac{1}{2}}) = -\frac{1}{2}u^2 - ux^{-\frac{1}{2}} + O(x^{-1})$$

$$\to -\frac{1}{2}u^2 \quad \text{as } x \to \infty.$$

Now if we were justified in saying that

(5) $$\lim_{x\to\infty} \int f(x, u) du = \int \lim_{x\to\infty} f(x, u) du,$$

then (3) would follow from (2), (4) and (5). However, it is a basic result in calculus that if $0 \leq f(x, u) \leq g(u)$, where $\int_{-\infty}^{\infty} g(u) du < \infty$, then (5) is justified. All we need to do is find a suitable $g(u)$.

First, for $x^{\frac{1}{2}} > 1$ and $u \geq 0$,

$$e^u f(x, u) = e^{-u(x^{1/2}-1)} (1 + ux^{\frac{1}{2}})^{x-1} \to 0 \quad \text{as } u \to \infty.$$

Hence $f(x, u) < M_1 e^{-u}$, $u \geq 0$, for some constant M_1. Second, for $u < 0$, the function $e^{-u} f(x, u)$ has a maximum where

$$-(1 + x^{\frac{1}{2}}) + (x - 1)(u + x^{\frac{1}{2}})^{-1} = 0,$$

that is at $u = -1$. Hence

$$f(x, u) \leq e^u \max_{u<0} \{e^{-u} f(x, u)\} = e^u(1 - x^{-\frac{1}{2}})^{x-1} \exp(1 + x^{\frac{1}{2}})$$

$$\to e^u e^{\frac{1}{2}} \quad \text{as } x \to \infty.$$

Hence, for some constant M_2, $f(x, u) < M_2 e^u$, $u < 0$. Therefore $f(x, u) < Me^{-|u|}$ for some M, and we have our $g(u)$.

(6) Exercise Show that $\lim_{n\to\infty} n! e^n n^{-n-\frac{1}{2}} = (2\pi)^{\frac{1}{2}}$.

(7) Exercise Let $(S_n;\ n \geq 0)$ be a simple random walk with $S_0 = 0$. Given that $S_{2n} = 0$, find the probability P_b that $S_r = b$ for some r such that $0 \leq r \leq 2n$. Show that if $n \to \infty$ and $b \to \infty$ in such a way that $b = yn^{\frac{1}{2}}$, then $P_b \to e^{-y^2}$.

(8) Exercise: de Moivre–Laplace theorem Let S_n be binomial with parameters n and p. Define

$$Y_n = \frac{S_n - np}{(npq)^{\frac{1}{2}}}, \quad q = 1 - p, \quad \text{and} \quad y_k = \frac{k - np}{(npq)^{\frac{1}{2}}}.$$

Show that as $n \to \infty$

$$\mathbf{P}(S_n = k) = \left(\frac{n}{2\pi k(n-k)}\right)^{\frac{1}{2}} \left(\frac{np}{k}\right)^k \left(\frac{nq}{n-k}\right)^{n-k} (1 + o(1)).$$

Deduce that as $n \to \infty$

$$\mathbf{P}(Y_n = y_k) = \left(\frac{1}{(2\pi npq)^{\frac{1}{2}}} \exp\left(-\frac{1}{2} y_k^2\right)\right) (1 + O(n^{-\frac{1}{2}})).$$

Conclude that for fixed finite a and b, as $n \to \infty$

$$\mathbf{P}(a < Y_n < b) \to \frac{1}{(2\pi)^{\frac{1}{2}}} \int_a^b e^{-\frac{1}{2}y^2} dy = \Phi(b) - \Phi(a).$$

(9) Exercise Let $(S_n;\ n \geq 0)$ be a simple random walk with $S_0 = 0$. Given that $S_{2n} = 2j$, show that the probability that the last visit to the origin was at the $2r$th step is

$$f_r = \frac{j}{n - r} \binom{n+j}{r} \binom{n-j}{r} \bigg/ \binom{2n}{2r}, \quad 0 \leq r < n.$$

Show that if r, j and n all increase in such a way that $\dfrac{r}{n} \to x$, $j/\sqrt{n} \to y$, then

$$\lim_{r\to\infty} f_r = \frac{y}{\sqrt{\pi x}} \frac{1}{(1 - x)^{\frac{3}{2}}} \exp\left(\frac{-xy^2}{1 - x}\right), \quad 0 < x < 1.$$

Remark The result of Exercise 6, which is known as Stirling's formula, was established by de Moivre in 1730. The formula actually proved by Stirling in 1730 was

(10) $$n!\left(n + \frac{1}{2}\right)^{-(n+\frac{1}{2})} e^{n+\frac{1}{2}} \to (2\pi)^{\frac{1}{2}}, \quad \text{as } n \to \infty.$$

(11) Exercise Prove (10).

PROBLEMS

1 Let $f(x) = c(\alpha, \beta)(x - \alpha)(\beta - x)$.
For what values of x and $c(\alpha, \beta)$ can f be a density function?

2 Let X have distribution $F(x)$. Show that $\mathbf{P}(X = x) > 0$ if and only if $F(x)$ is discontinuous at x.

3 The beta function $B(a, b)$ is given by $B(a, b) = \int_0^1 v^{a-1}(1-v)^{b-1}dv$; $a > 0$, $b > 0$. The beta distribution has density

$$f(x) = \frac{1}{B(a, b)} x^{a-1}(1-x)^{b-1} \quad \text{for } 0 < x < 1.$$

If X has the beta distribution, show that $\mathbf{E}(X) = B(a+1, b)/B(a, b)$. What is $\text{var}(X)$?

4 For what value of c is $f = c(\sin x)^\alpha(\cos x)^\beta$; $0 < x < \pi/2$, a density function?

5 What is the distribution function of the random variable having the beta density with $a = b = \frac{1}{2}$?

6 Let X have the density $f = \exp(-x - \exp(-x))$ for $x \in \mathbb{R}$. What is the distribution function of X?

7 Let X be exponentially distributed with parameter λ. What is the density of $Y = e^{aX}$? For what values of λ does $\mathbf{E}(Y)$ exist?

8 Let X have the gamma density with parameters α and λ. Show that

$$\mu_k = \alpha(\alpha+1)\ldots(\alpha+k-1)\lambda^{-k}, \quad \text{and}$$

$$\text{var}(X) = \frac{\alpha}{\lambda^2}.$$

9 Let X have the standard normal density, and $a > 0$. Show that

$$\mathbf{P}(X > x + ax^{-1} | X > x) \to e^{-a} \text{ as } x \to \infty.$$

10 (a) Let X have the standard normal density. Show that $|X|$ has distribution function $F = 2\Phi(x) - 1$, for $x > 0$.

(b) Let X have distribution $F(x)$. What is the distribution of $|X|$? What is the density of $|X|$ if it exists?

11 An elastic string has modulus of elasticity λ and natural length l_0. A mass m is attached to one end, the other being fixed. The period of oscillation of the mass when slightly displaced is $2\pi\sqrt{ml_0/\lambda}$. Suppose that the modulus of elasticity is uniformly distributed on $[a, b]$. What is the density and expectation of the period?

12 Let X have density $f(x)$. Construct a simple random variable $S_n(X)$ such that given $\epsilon > 0$, $\mathbf{P}(|S_n(X) - X| > \epsilon) < 2^{-n}$. (Assume X is proper.)

13 If X is exponentially distributed find the m.g.f. of X, $\mathbf{E}(e^{tX})$.

14 A point Q is chosen at random inside an equilateral triangle of unit side. Find the density of the perpendicular distance X to the nearest side of the triangle.

15 For what value of c is $\mathbf{E}((X-c)^2)$ least?

16 Suppose a machine's lifetime T has hazard rate $\lambda\sqrt{t}$, where $\lambda > 0$. Find $\mathbf{P}(T > t)$.

17 Suppose that X has distribution function

$$F(x) = 1 - \exp\left(-\int_0^x g(u)du\right)$$

for some function $g(.)$. Show that this is possible if and only if $g(u) \geq 0$, and $\int_0^\infty g(u)du = \infty$.

18 What are the cumulants of the normal density?

19 What are the cumulants of the exponential density?

20 You have two independent random variables, each uniform on $(0, 1)$. Explain how you would use them to obtain a random variable X with density

$$f(x) = \frac{3}{5}\left(1 + x + \frac{1}{2}x^2\right) \quad \text{for } 0 \leq x \leq 1.$$

21 Define $I(a, b) = \int_0^\infty \exp(-a^2u^2 - b^2u^{-2})du$ for $a, b > 0$. Show that:

(a) $I(a, b) = a^{-1}I(1, ab)$;

(b) $\dfrac{\partial I}{\partial b} = -2I(1, ab)$;

(c) $I(a, b) = \dfrac{\sqrt{\pi}}{2a}e^{-2ab}$.

22 Use the result of Problem 21 to find the m.g.f. of the following densities:

(a) $f(x) = \alpha x^{-\frac{1}{2}}\exp(-\beta/x - \gamma x)$ for $x > 0$; $\beta, \gamma > 0$. What is α?

(b) $f(x) = (2\pi x^3)^{-\frac{1}{2}}\exp(-(2x)^{-1})$ for $x > 0$.

23 Let X be a standard normal random variable. What is the density of X^{-2}?

24 Let X be a standard normal random variable. Find the m.g.f. and density of X^2.

25 What is the moment generating function of the two-sided exponential density? Where is it defined?

26 Let U be uniform on $(0, 1)$. Show that, if $[a]$ denotes the integer part of a,

$$X = 1 + \left[\frac{\log U}{\log(1 - p)}\right]$$

has a geometric distribution.

27 Let U be uniform on $(0, 1)$. Show how to use U to simulate a random variable with density

$$f(x) = \frac{24}{25}\left(1 + \frac{1}{2}\left(x - \frac{1}{2}\right)^2\right); \quad 0 \leqslant x \leqslant 1.$$

28 Let P and Q be two points chosen independently and uniformly in $(0, a)$. Show that the distance between P and Q has density $2(a - x)a^{-2}$ for $0 < x < a$.

29 **Continuous mixture** Let $f(\theta, x)$ be the exponential density with parameter θ for $0 \leqslant x$. Let $g(\theta) = \nu e^{-\nu\theta}$ for $\theta \geqslant 0$, $\nu > 0$. Show that $\int_0^\infty f(\theta, x)g(\theta)d\theta$ is a density.

30 Let X_n be a Poisson random variable with parameter n. Show that as $n \to \infty$, $\mathbf{P}(X_n \leqslant n + \sqrt{n}x) \to \Phi(x)$.

31 Let U be uniform on $(0, 1)$. Show how to simulate a random variable X with the Pareto distribution given by $F(x) = 1 - x^{-d}$; $x > 1$, $d > 1$.

32 Let X_α have gamma density with parameters α and 1, so $f_\alpha(x) = \dfrac{x^{\alpha-1}e^{-x}}{(\alpha-1)!}$. Let $\phi_\alpha(x)$ be the density of $Y_\alpha = (X_\alpha - \alpha)\alpha^{-\frac{1}{2}}$. Show that as $\alpha \to \infty$, $\phi_\alpha(x) \to \phi(x)$ where $\phi(x)$ is the standard normal density.

33 Let X have mean μ and variance σ^2. Show that $\mathbf{P}(|X - \mu| \leqslant a\sigma) \geqslant 1 - a^{-2}$, for any $a > 0$. Let Z be a standard normal random variable, and define $Y = \alpha + \beta Z + \gamma Z^2$. Find the mean and variance of Y, and show that

$$\mathbf{P}\left(|Y - \alpha - \gamma| \leqslant \frac{1}{2}\alpha\right) \geqslant 1 - 4\alpha^{-2}(\beta^2 + 2\gamma^2).$$

8

Jointly continuous random variables

8.1 Joint density and distribution

Often it is necessary to consider the joint behaviour of several random variables which may each take an uncountable number of possible values. Just as for discrete random vectors, we need to define a variety of useful functions, and develop the appropriate machinery to set them to work. For simplicity in definitions and theorems, we start by considering a pair of random variables (X, Y) taking values in $\mathbb{R}^2$. This theoretical outline can be easily extended to larger collections of random variables $(X_1, X_2, \ldots, X_n)$ taking values in $\mathbb{R}^n$, with a correspondingly greater expenditure of notation and space.

As usual, we should start with a sample space Ω, an event space $\mathcal{F}$, and a probability function $\mathbf{P}$, such that for all x and y

$$A_{xy} = \{\omega: X \leqslant x, Y \leqslant y\} \in \mathcal{F}.$$

Then

(1) $$F(x, y) = \mathbf{P}(A_{xy}) = \mathbf{P}(X \leqslant x, Y \leqslant y)$$

is the joint distribution function of X and Y. In fact we suppress this underlying structure, and begin with random variables X and Y having joint distribution $F(x, y)$ given by (1). A special class of such jointly distributed random variables is of great importance.

(2) Definition Let $F(x, y)$ be a joint distribution. Suppose that $\frac{\partial^2 F}{\partial x \partial y}$ exists, and is non-negative, except possibly on a finite collection of lines in $\mathbb{R}^2$. Suppose further that the function $f(x, y)$ defined by

$$f(x, y) = \begin{cases} \dfrac{\partial^2 F}{\partial x \partial y} & \text{where this exists} \\ 0 & \text{elsewhere,} \end{cases}$$

satisfies

(3) $$F(x, y) = \int_{-\infty}^{x}\int_{-\infty}^{y} f(u, v)\, du\, dv.$$

Then X and Y, being random variables having the (joint) distribution F, are said to be (jointly) continuous with (joint) density function $f(x, y)$. ▲

The words 'joint' and 'jointly' are often omitted to save time and trees. Sometimes we write $f_{X,Y}(x, y)$ and $F_{X,Y}(x, y)$ to stress the role of X and Y, or to avoid ambiguity.

(4) **Example: Uniform distribution** Suppose you pick a point Q at random in the rectangle $R = (x, y: 0 < x < a, 0 < y < b)$. Then from the properties of the uniform distribution (see (7.8.1)) we have

(5)
$$F(x, y) = \begin{cases} 1 & \text{if } x \geqslant a, y \geqslant b \\ \dfrac{xy}{ab} & \text{if } 0 \leqslant x \leqslant a, 0 \leqslant y \leqslant b \\ \dfrac{y}{b} & \text{if } x \geqslant a, 0 \leqslant y \leqslant b \\ \dfrac{x}{a} & \text{if } 0 \leqslant x \leqslant a, y \geqslant b \\ 0 & \text{elsewhere.} \end{cases}$$

Differentiating wherever possible gives

(6)
$$\frac{\partial^2 F}{\partial x \partial y} = \begin{cases} \dfrac{1}{ab} & \text{if } 0 < x < a, 0 < y < b \\ 0 & \text{if } x < 0 \text{ or } x > a \\ & \text{or } y < 0 \text{ or } y > b. \end{cases}$$

Hence, the function

(7)
$$f(x, y) = \begin{cases} \dfrac{1}{ab} & \text{if } 0 < x < a, 0 < y < b \\ 0 & \text{otherwise} \end{cases}$$

satisfies (3), and is the density of X and Y. It is uniformly distributed over the rectangle R. Furthermore, if A is a subset of R with area $|A|$, then using (7.8.1) (and a theorem about double integrals), we have

(8)
$$\mathbf{P}((X, Y) \in A) = \frac{|A|}{ab} = \iint_{(x,y)\in A} f(x, y)\, dxdy. \qquad \bullet$$

In fact, a version of the useful relationship (8) holds true for all densities $f(x, y)$. This is important enough to state formally as a theorem, which we do not prove.

(9) **Theorem** If X and Y have density $f(x, y)$, and $\mathbf{P}((X, Y) \in A)$ exists, then it is given by

$$\mathbf{P}((X, Y) \in A) = \iint_{(x,y)\in A} f(x, y)\, dxdy.$$

Note that the condition that the probability exists is equivalent to saying that $\{\omega: (X(\omega), Y(\omega)) \in A\} \in \mathcal{F}$. This is another demonstration of the fact that, while we can just about suppress $(\Omega, \mathcal{F}, \mathbf{P})$ at this elementary level, further rigorous progress is not possible without bringing the underlying probability space into play. The attractive result (9) may then be proved.

Here is a simple example of Theorem 9 in use.

(10) Example Let X and Y have density

$$f(x, y) = \begin{cases} 8xy & \text{if } 0 < y < x < 1 \\ 0 & \text{elsewhere.} \end{cases}$$

What are $\mathbf{P}(2X > 1, 2Y < 1)$ and $\mathbf{P}(X + Y > 1)$? Find $F(x, y)$.

Solution Notice that the constraints $2X > 1$, $2Y < 1$ require that $(X, Y) \in S$, where S is the square with vertices $(\frac{1}{2}, 0)$, $(1, 0)$, $(\frac{1}{2}, \frac{1}{2})$, $(1, \frac{1}{2})$. Hence

$$\mathbf{P}(2X > 1, 2Y < 1) = 8\int_{\frac{1}{2}}^{1}\int_{0}^{\frac{1}{2}} f(x, y)\, dydx = 8\int_{\frac{1}{2}}^{1} x\, dx \int_{0}^{\frac{1}{2}} y\, dy = \frac{3}{8}.$$

Likewise $X + Y > 1$ if $(X, Y) \in T$, where T is the triangle with vertices $(\frac{1}{2}, \frac{1}{2})$, $(1, 0)$, $(1, 1)$. Hence

$$\mathbf{P}(X + Y > 1) = 8\int_{\frac{1}{2}}^{1}\int_{1-x}^{x} xy\, dy\, dx = \frac{5}{6}.$$

Finally

$$F(x, y) = \int_{0}^{y}\int_{v}^{x} 8uv\, du\, dv = 2x^2y^2 - y^4. \qquad \bullet$$

The geometrical problems of Section 7.9 can now be reformulated and generalized in this new framework. Obviously "picking a point Q at random in some region R", is what we would now describe as picking (X, Y) such that X and Y are jointly uniform in R. More generally we can allow (X, Y) to have joint density $f(x, y)$ in R.

Example: More triangles The random variables X and Y have joint density

$$f(x, y) = \begin{cases} cx^a & \text{if } x < 1,\ y < 1,\ x + y > 1,\ a > -1, \\ 0 & \text{otherwise.} \end{cases}$$

(a) What is c?
(b) What is $F(x, y)$?
(c) Show that it is possible to construct a triangle with sides $X, Y, 2 - X - Y$, with probability one.
(d) Show that the angle opposite to the side of length Y is obtuse with probability

$$p_0 = c\int_{0}^{1} \frac{x^{a+1} - x^{a+2}}{2-x}\, dx.$$

(e) When $a = 0$, show that $p_0 = 3 - 4\log 2$.

Solution (a) Since $\iint f(x, y)\, dx\, dy = 1$, this entails

$$c^{-1} = \int_{0}^{1} x^a \int_{1-x}^{1} dy\, dx = (a + 2)^{-1}.$$

(b) Using (3) gives

$$F(x, y) = \int_{1-y}^{x}\int_{1-u}^{y} cu^a \, dv \, du = \frac{a+2}{a+1} y x^{a+1} + x^{a+2} - \frac{a+2}{a+1} x^{a+1} + \frac{1}{a+1}(1-y)^{a+2}.$$

(c) Three such lengths form a triangle if $X + Y > 2 - X - Y$, $X + 2 - X - Y > Y$, and $Y + 2 - X - Y > X$. But these constraints are just those which define the region in which $f(x, y)$ is non-zero, and $\iint f \, dx \, dy = 1$.

(d) If θ is the angle opposite Y, then

$$\cos\theta = \frac{X^2 + (2 - X - Y)^2 - Y^2}{2X(2 - X - Y)} < 0$$

if θ is an obtuse angle. Hence in this case

$$Y > \frac{X^2 - 2X + 2}{2 - X} = g(X), \quad \text{say.}$$

Now $g(x) \geqslant 1 - x$, (with equality only at $x = 0$). Hence p_0 is given by

$$\mathbf{P}(\theta \text{ is obtuse}) = \mathbf{P}(Y > g(X)) = \int_0^1 \int_{g(x)}^1 f(x, y) \, dy dx$$

$$= c\int_0^1 (1 - g(x))x^a \, dx = c\int_0^1 \frac{x - x^2}{2 - x} x^a \, dx.$$

(e) When $a = 0$

$$p_0 = 2\int_0^1 \frac{x - x^2}{2 - x} dx = 2\int_0^1 \frac{2x - x^2}{2 - x} + \frac{2 - x}{2 - x} - \frac{2}{2 - x} dx = 3 - 4\log 2. \quad \bullet$$

Next we record that as a result of (1) and Theorem 9, $f(x, y)$ and $F(x, y)$ have the following elementary properties, analogous to those of f and F in the discrete case.

First, $F(x, y)$ is obviously non-decreasing in x and y. More strongly we have

(11)
$$\begin{aligned} 0 &\leqslant \mathbf{P}(a < X \leqslant b, c \leqslant Y \leqslant d) \\ &= \mathbf{P}(a < X \leqslant b, Y \leqslant d) - \mathbf{P}(a < X \leqslant b, Y \leqslant c) \\ &= F(b, d) - F(a, d) - F(b, c) + F(a, c). \end{aligned}$$

Secondly, if X and Y are finite with probability 1, then

(12)
$$1 = \int_{-\infty}^{\infty}\int_{-\infty}^{\infty} f(u, v) \, du \, dv = \lim_{x,y\to\infty} F(x, y).$$

Thirdly, knowledge of $F(x, y)$ and $f(x, y)$ will also provide us with the separate distributions and densities of X and Y. Thus

(13)
$$F_X(x) = \mathbf{P}(X \leqslant x) = \lim_{y\to\infty} \mathbf{P}(X \leqslant x, Y \leqslant y) = \int_{-\infty}^{\infty}\int_{-\infty}^{x} f(u, v) \, du dv,$$

and

(14)
$$f_X(x) = \frac{d}{dx} F_X(x) = \int_{-\infty}^{\infty} f(x, v) \, dv.$$

Likewise

(15)
$$f_Y(y) = \int_{-\infty}^{\infty} f(u, y) \, du$$

and

(16) $$F_Y(y) = \lim_{x\to\infty} F(x, y).$$

Here are some examples to illustrate these properties. Note that in future we shall specify $f(x, y)$ only where it is non-zero.

(17) Example Verify that the function $f(x, y) = 8xy$ for $0 < y < x < 1$ is a density. For what value of c is $f(x, y) = cxy$ for $0 < x < y < 1$, a density? Find the density of X in the second case.

Solution Since $f > 0$ and

$$\int_0^1\int_0^x 8xy\,dydx = \int_0^1 4x^3\,dx = 1,$$

f is indeed a density. By symmetry $c = 8$ in the second case also, and we have

$$f_X(x) = \int_x^1 8xy\,dy = 4x(1 - x^2).$$

●

(18) Example The function $H(x, y) = 1 - e^{-(x+y)}$ for $x > 0$, $y > 0$, is non-decreasing in x and y, and $0 \leqslant H \leqslant 1$. Is it a distribution?

Solution No, because $\frac{\partial^2 H}{\partial x \partial y}$ exists and is negative in $x > 0$, $y > 0$. Alternatively note that

$$H(1, 1) - H(1, 0) - H(0, 1) + H(0, 0) = 2e^{-1} - 1 - e^{-2} < 0$$

which cannot (as it should) be the value of $\mathbf{P}(0 < X \leqslant 1, 0 < Y \leqslant 1)$. ●

(19) Example: Bivariate normal density Verify that when $\sigma, \tau > 0$

$$f(x, y) = \frac{1}{2\pi\sigma\tau(1-\rho^2)^{\frac{1}{2}}}\exp\left[-\frac{1}{2(1-\rho^2)}\left(\frac{x^2}{\sigma^2} - \frac{2\rho xy}{\sigma\tau} + \frac{y^2}{\tau^2}\right)\right]$$

is a density for $|\rho| < 1$, and find the marginal densities $f_X(x)$ and $f_Y(y)$.

Solution From (14), if $f(x, y)$ is a density we have

$$\begin{aligned} f_X(x) &= \int_{-\infty}^{\infty} f(x, y)\,dy \\ &= \frac{1}{2\pi\sigma\tau(1-\rho^2)^{\frac{1}{2}}}\int_{-\infty}^{\infty}\exp\left[-\frac{1}{2(1-\rho^2)}\left(\left(\frac{y}{\tau} - \frac{\rho x}{\sigma}\right)^2 + \frac{x^2}{\sigma^2} - \frac{\rho^2 x^2}{\sigma^2}\right)\right]dy \end{aligned}$$

Now setting $\dfrac{y}{\tau} - \dfrac{\rho x}{\sigma} = u$, and recalling that

$$\int_{-\infty}^{\infty}\exp\left(-\frac{u^2}{2(1-\rho^2)}\right)\tau\,du = (2\pi(1-\rho^2))^{\frac{1}{2}}\tau,$$

yields

$$f_X(x) = \frac{1}{(2\pi)^{\frac{1}{2}}\sigma}\exp\left(-\frac{x^2}{2\sigma^2}\right).$$

This is the $N(0, \sigma^2)$ density, and so f satisfies (12) and is non-negative. It is therefore a density. Interchanging the roles of x and y in the above integrals shows that $f_Y(y)$ is the $N(0, \tau^2)$ density. ●

8.2 Change of variables

We have interpreted the random vector (X, Y) as a random point Q picked in $\mathbb{R}^2$ according to some density $f(x, y)$, where (x, y) are the Cartesian coordinates of Q. Of course the choice of coordinate system is arbitrary; we may for some very good reasons choose to represent Q in another system of coordinates (u, v), where (x, y) and (u, v) are related by $u = u(x, y)$ and $v = v(x, y)$. What now is the joint density of $U = u(X, Y)$ and $V = v(X, Y)$?

Equally, given a pair of random variables X and Y, our real interest may well lie in some function or functions of X and Y. What is their (joint) distribution?

As we have remarked above, at a symbolic or formal level, the answer is straightforward. For U and V above, and $A = \{x, y: u(x, y) \leqslant w, v(x, y) \leqslant z\}$ then, by Theorem 8.1.9,

$$F_{U,V}(w, z) = \int_A f_{X,Y}(x, y)\, dx\, dy.$$

The problem is to turn this into a more tractable form.

Fortunately there are well known results about changing variables within a multiple integral which provide the answer. We state without proof a theorem for a transformation T satisfying the following conditions. Let C and D be subsets of $\mathbb{R}^2$. Suppose that T given by

$$T(x, y) = (u(x, y), v(x, y))$$

maps C one–one onto D, with inverse T^{-1} given by

$$T^{-1}(u, v) = (x(u, v), y(u, v))$$

which maps D one–one onto C. We define

$$J(u, v) = \frac{\partial x}{\partial u}\frac{\partial y}{\partial v} - \frac{\partial x}{\partial v}\frac{\partial y}{\partial u}$$

where the derivatives are required to exist and be continuous in D. Then we have the following result.

(1) **Theorem** Let X and Y have density $f(x, y)$ which is zero outside C. Then $U = u(X, Y)$ and $V = v(X, Y)$ have joint density

$$f_{U,V}(u, v) = f_{X,Y}(x(u, v), y(u, v))|J(u, v)| \quad \text{for } (u, v) \in D.$$

Here are some examples of this theorem in use.

(2) **Example** Suppose $Q = (X, Y)$ is uniformly distributed over the circular disc of radius 1. Then X and Y have joint density

(3) $$f(x, y) = \frac{1}{\pi} \quad \text{for } x^2 + y^2 \leqslant 1.$$

However, it seems more natural to use polar rather than Cartesian coordinates in this case. These are given by $r = (x^2 + y^2)^{\frac{1}{2}}$, and $\theta = \tan^{-1}(y/x)$, with inverse $x = r\cos\theta$ and $y = r\sin\theta$. They map $C = \{x, y: x^2 + y^2 \leqslant 1\}$ one–one onto $D = \{r, \theta: 0 \leqslant r \leqslant 1, 0 < \theta \leqslant 2\pi\}$.

In this case

$$J(r, \theta) = \frac{\partial x}{\partial r}\frac{\partial y}{\partial \theta} - \frac{\partial x}{\partial \theta}\frac{\partial y}{\partial r} = r\cos^2\theta + r\sin^2\theta = r.$$

Hence the random variables $R = r(X, Y)$ and $\Theta = \theta(X, Y)$ have joint density given by

(4) $$f_{R,\Theta}(r, \theta) = \frac{r}{\pi} \quad \text{for } 0 \leqslant r \leqslant 1, 0 < \theta \leqslant 2\pi.$$

Notice that $f(r, \theta)$ is not uniform, as was $f(x, y)$. ●

(5) **Example** Let $Q = (X, Y)$ be uniformly distributed over the ellipse C with boundary satisfying

$$\frac{x^2}{a^2} + \frac{y^2}{b^2} = 1,$$

of area $|C|$.

What is $\mathbf{P}(X > Y, X > -Y)$?

Solution Here the transformation $x = ar\cos\theta$ and $y = br\sin\theta$ maps the ellipse one–one onto the circular disc with radius 1. Furthermore $J = abr$. Now X and Y have density

$$f(x, y) = \frac{1}{|C|}, \quad \text{for } (x, y) \in C,$$

so R and Θ have joint density

$$f(r, \theta) = \frac{abr}{|C|}, \quad \text{for } 0 \leqslant r < 1, 0 < \theta \leqslant 2\pi.$$

Hence $|C| = \pi ab$, and

$$\mathbf{P}(X > Y, X > -Y) = \mathbf{P}\left(-1 < \frac{Y}{X} < 1\right) = \mathbf{P}\left(-1 < \frac{b}{a}\tan\Theta < 1\right) = \frac{1}{\pi}\tan^{-1}\frac{a}{b},$$

since Θ is uniform on $(0, 2\pi)$. ●

8.3 Independence

As usual, independence is an extremely important property; its definition is by now familiar.

(1) **Definition** Jointly distributed random variables are independent if, for all x and y,

$$\mathbf{P}(X \leqslant x, Y \leqslant y) = \mathbf{P}(X \leqslant x)\mathbf{P}(Y \leqslant y).$$

In terms of distributions, this equivalent to the statement that

(2) $$F(x, y) = F_X(x)F_Y(y).$$ ▲

For random variables with a density it follows immediately by differentiating, that

(3) $$f(x, y) = f_X(x)f_Y(y)$$

if X and Y are independent. Using the basic property of densities (Theorem 8.1.9), now further shows that if $C = (x, y: x \in A, y \in B)$ and X and Y are independent, then

(4) $$\iint_C f(x, y)\,dxdy = \int_A f_X(x)\,dx \int_B f_Y(y)\,dy.$$

(Assuming of course that the integrals exist.)

Finally, if the random variables U and V satisfy $U = g(X)$, $V = h(Y)$, and X and Y are independent, then U and V are independent. To see this just let

(5) $$A = (x: g(x) \leqslant u) \quad \text{and} \quad B = (g: h(y) \leqslant v),$$

and the independence follows from (4) and (2). An important and useful converse is the following.

(6) Theorem If X and Y have density $f(x, y)$, and for all x and y it is true that

$$f(x, y) = f_X(x)f_Y(y),$$

then X and Y are independent.

The proof follows immediately from a standard theorem on multiple integrals (just consider $\int_{-\infty}^{x}\int_{-\infty}^{y} f(u, v)\,du\,dv$) and we omit it.

(7) Example: Uniform distribution Let X and Y have the uniform density over the unit circular disc C, namely

$$f(x, y) = \pi^{-1} \quad \text{for } (x, y) \in C.$$

(a) Are X and Y independent?
(b) Find $f_X(x)$ and $f_Y(y)$.
(c) If $X = R\cos\Theta$, and $Y = R\sin\Theta$, are R and Θ independent?

Solution (a) The set $\{x, y: x \leqslant -1/\sqrt{2}, y \leqslant -1/\sqrt{2}\}$ lies outside C, so

$$F\left(-\frac{1}{\sqrt{2}}, -\frac{1}{\sqrt{2}}\right) = 0.$$

However the intersection of the set $\{x: x \leqslant -1/\sqrt{2}\}$ with C has non-zero area, so

$$F_X\left(-\frac{1}{\sqrt{2}}\right)F_Y\left(-\frac{1}{\sqrt{2}}\right) > 0.$$

Therefore X and Y are not independent.

(b) By (8.1.14)

$$f_X(x) = \int_{-1}^{1} f(x, y)\,dy = \frac{1}{\pi}\int_{-(1-x^2)^{\frac{1}{2}}}^{(1-x^2)^{\frac{1}{2}}} dy = \frac{2}{\pi}(1 - x^2)^{\frac{1}{2}}.$$

Likewise

$$f_Y(y) = \frac{2}{\pi}(1 - y^2)^{\frac{1}{2}}.$$

(c) By Example 8.2.4, R and Θ have joint density

$$f_{R,\Theta}(r, \theta) = \frac{r}{\pi}, \quad \text{for } 0 \leq r < 1, 0 < \theta \leq 2\pi.$$

Hence

$$f_\Theta(\theta) = \int_0^1 f(r, \theta)\, dr = \frac{1}{2\pi}; \quad 0 < \theta \leq 2\pi,$$

and

$$f_R(r) = \int_0^{2\pi} f(r, \theta)\, d\theta = 2r; \quad 0 \leq r < 1.$$

Hence $f(r, \theta) = f_\Theta(\theta) f_R(r)$, and so R and Θ are independent. ●

Example: Bertrand's paradox again Suppose we choose a random chord of a circle C radius a, as follows. A point P is picked at random (uniformly) inside C. Then a line through P is picked independently of P at random (that is, its direction Θ is uniform on $(0, 2\pi)$). Let X be the length of the chord formed by the intersection of this line with the circle. Show that

$$\mathbf{P}(X > a\sqrt{3}) = \frac{1}{3} + \frac{\sqrt{3}}{2\pi}.$$

Solution Let R be the distance from the centre of the circle to P; by the above, R has distribution given by $\mathbf{P}(R/a \leq r) = r^2$; $0 \leq r \leq 1$. Now $X > a\sqrt{3}$, if and only if $2R \sin\Theta < a$, as you can see by inspecting Figure 8.1. Hence

$$\mathbf{P}(X > a\sqrt{3}) = \frac{2}{\pi}\int_0^{\pi/2} \mathbf{P}\left(R < \frac{a}{2\sin\theta}\right) d\theta$$
$$= \frac{2}{\pi}\int_0^{\pi/6} d\theta + \frac{2}{\pi}\int_{\pi/6}^{\pi/2} \frac{1}{4}\operatorname{cosec}^2\theta\, d\theta = \frac{1}{3} + \frac{\sqrt{3}}{2\pi}.$$

Compare this with the results of Example 7.18. ●

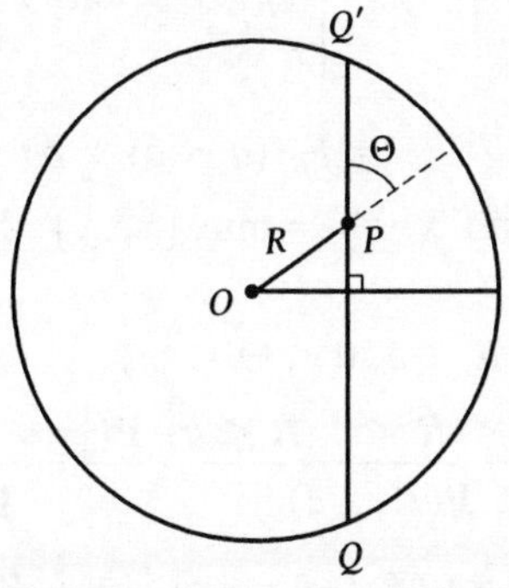

Figure 8.1 Bertrand's paradox.

(8) Example: Normal densities Let X and Y be independent with common density $f(x) = k\exp(-\frac{1}{2}x^2)$ for all x.

(a) Show that $k = (2\pi)^{-\frac{1}{2}}$.
(b) Show that $X^2 + Y^2$ and $\tan^{-1}(Y/X)$ are independent random variables.
(c) If $a > 0 < b < c$ and $0 < \alpha < \frac{1}{2}\pi$, find the probability that $b < (X^2 + Y^2)^{\frac{1}{2}} < c$ and $\frac{1}{4}\pi < \tan^{-1}(Y/X) < \frac{1}{2}\pi$, given that $(X^2 + Y^2)^{\frac{1}{2}} < a$, $Y > 0$, and $\tan^{-1}(Y/X) < \frac{1}{3}\pi$.

Solution Since X and Y are independent they have joint density

$$f(x, y) = k^2 \exp\left(-\frac{1}{2}(x^2 + y^2)\right).$$

Make the change of variables to polar coordinates, so that by Theorem 8.2.1 the random variables $R = (X^2 + Y^2)^{\frac{1}{2}}$ and $\Theta = \tan^{-1}(Y/X)$ have joint density

$$f(r, \theta) = k^2 r \exp\left(-\frac{1}{2}r^2\right) \quad \text{for } 0 \leqslant r < \infty, 0 < \theta \leqslant 2\pi.$$

Hence R has density

$$f_R(r) = r\exp\left(-\frac{1}{2}r^2\right) \quad \text{for } 0 \leqslant r < \infty,$$

and Θ has density

$$f_\Theta(\theta) = k^2 \quad \text{for } 0 < \theta \leqslant 2\pi.$$

It follows immediately that

(a) $k^2 = (2\pi)^{-1}$, and
(b) $f(r, \theta) = f_R(r)f_\Theta(\theta)$, so that Θ and R are independent by Theorem 6. Hence Θ and R^2 are independent.
(c) Finally, note that

(9) $$\mathbf{P}\left(\frac{1}{4}\pi < \Theta < \frac{1}{2}\pi, 0 < \Theta < \frac{1}{3}\pi\right) = \mathbf{P}\left(\frac{1}{4}\pi < \Theta < \frac{1}{3}\pi\right) = \frac{\pi}{12},$$

and

$$\mathbf{P}(b < R < c, R < a) = \begin{cases} F_R(c) - F_R(b) & \text{if } c < a \\ F_R(a) - F_R(b) & \text{if } b < a \leqslant c \\ 0 & \text{otherwise} \end{cases}$$

(10) $$= F_R((a \wedge c) \vee b) - F_R(b)$$

where $x \wedge y = \min\{x, y\}$ and $x \vee y = \max\{x, y\}$. Now, because R and Θ are independent

$$\mathbf{P}(b < R < c, \tfrac{1}{4}\pi < \Theta < \tfrac{1}{2}\pi | R < a, 0 < \Theta < \tfrac{1}{3}\pi)$$

$$= \frac{\mathbf{P}(b < R < c, R < a)}{\mathbf{P}(R < a)} \frac{\mathbf{P}(\frac{1}{4}\pi < \Theta < \frac{1}{2}\pi, 0 < \Theta < \frac{1}{3}\pi)}{\mathbf{P}(0 < \Theta < \frac{1}{3}\pi)}$$

$$= \frac{1}{4F_R(a)}(F_R((a \wedge c) \vee b) - F_R(b)), \quad \text{by (9) and (10).}$$ ●

8.4 Sums, products and quotients

We now return to the question of the distribution of functions of random vectors, and take a brief look at some particularly important special cases. Of these the most important is the sum of two random variables.

Theorem Let X and Y have joint density $f(x, y)$. Show that if $Z = X + Y$ then

(1) $$f_Z(z) = \int_{-\infty}^{\infty} f(u, z-u)\, du$$

and that if X and Y are independent then

(2) $$f_Z(z) = \int_{-\infty}^{\infty} f_X(u) f_Y(z-u)\, du.$$

Proof First notice that by (8.3.3) the result (2) follows immediately from (1) when X and Y are independent. Turning to the proof of (1), we give two methods of solution.

Method I Let A be the region in which $u + v \leq z$. Then

$$\mathbf{P}(Z \leq z) = \iint_{(u,v)\in A} f(u, v)\, du\, dv = \int_{-\infty}^{\infty} \int_{-\infty}^{z-u} f(u, v)\, dv du$$
$$= \int_{-\infty}^{\infty} \int_{-\infty}^{z} f(u, w-u)\, dw du \quad \text{on setting } v = w - u.$$

Now differentiating with respect to z gives $f_Z(z) = \int_{-\infty}^{\infty} f(u, z-u)\, du$.

Method II This time we use the change of variable technique of Section 8.2. Consider the transformation $z = x + y$, and $u = x$, with inverse $x = u$, and $y = z - u$. Here $J = 1$. This satisfies the conditions of Theorem 8.2.1 and so $U = u(X, Y)$ and $Z = z(X, Y)$ have joint density $f(u, z-u)$. We require the marginal density of Z, which is of course just (1). ∎

(3) Example Let X and Y have the bivariate normal distribution of Example 8.1.19,

$$f(x, y) = \frac{1}{2\pi\sigma\tau(1-\rho^2)^{1/2}} \exp\left(-\frac{1}{2(1-\rho^2)}\left(\frac{x^2}{\sigma^2} - \frac{2\rho xy}{\sigma\tau} + \frac{y^2}{\tau^2}\right)\right).$$

Find the density of $aX + bY$ for constants a and b.

Solution The joint density of $U = aX$ and $V = bY$ is

$$g(u, v) = \frac{1}{ab} f\left(\frac{u}{a}, \frac{v}{b}\right).$$

Hence by the above theorem, the density of $Z = U + V = aX + bY$ is

(4) $$f_Z(z) = \int_{-\infty}^{\infty} \frac{1}{ab} f\left(\frac{u}{a}, \frac{z-u}{b}\right) du.$$

Rearranging the exponent in the integrand we have, after a little manipulation

$$\frac{-1}{2(1-\rho^2)}\left(\frac{u^2}{a^2\sigma^2}-\frac{2\rho u(z-u)}{ab\sigma\tau}+\frac{(z-u)^2}{b^2\tau^2}\right)$$
$$=\frac{-1}{2(1-\rho^2)}\left(\alpha\left(u-\frac{\beta}{\alpha}z\right)^2+\frac{z^2}{\alpha}\frac{(1-\rho^2)}{a^2b^2\sigma^2\tau^2}\right)$$

where

$$\alpha=\frac{1}{a^2\sigma^2}+\frac{2\rho}{ab\sigma\tau}+\frac{1}{b^2\tau^2},$$

and

$$\beta=\frac{\rho}{ab\sigma\tau}+\frac{1}{b^2\tau^2}.$$

Setting $u=v+\frac{\beta}{\alpha}z$ in the integrand, we evaluate

$$\int_{-\infty}^{\infty}\exp\left(-\frac{\alpha v^2}{2(1-\rho^2)}\right)dv=\left(\frac{2\pi(1-\rho^2)}{\alpha}\right)^{\frac{1}{2}}.$$

Hence, after a little more manipulation, we find that

$$f_Z(z)=\frac{1}{(2\pi\xi^2)^{\frac{1}{2}}}\exp\left(-\frac{z^2}{2\xi^2}\right)$$

where $\xi^2=a^2\sigma^2+2\rho ab\sigma\tau+b^2\tau^2$. That is to say Z is $N(0,\xi^2)$. ●

One very important special case arises when $\rho=0$. The above result then shows we have proved the following.

(5) **Theorem** Let X and Y be independent normal random variables having the densities $N(0,\sigma^2)$ and $N(0,\tau^2)$. Then the sum $Z=aX+bY$ has the density $N(0,a^2\sigma^2+b^2\tau^2)$. ■

Next we turn to products and quotients.

(6) **Theorem** Let X and Y have joint density $f(x,y)$. Then the density of $Z=XY$ is

(7) $$f(z)=\int_{-\infty}^{\infty}\frac{1}{|u|}f\left(u,\frac{z}{u}\right)du$$

and the density of $W=\dfrac{X}{Y}$ is

$$f(w)=\int_{-\infty}^{\infty}|u|f(uw,u)\,du.$$

Proof We use Theorem 8.2.1 again. Consider the transformation $u=x$, and $z=xy$, with inverse $x=u$, and $y=z/u$. Here

$$J(u, z) = \begin{vmatrix} 1 & 0 \\ \dfrac{-z}{u^2} & \dfrac{1}{u} \end{vmatrix} = u^{-1}.$$

This satisfies the conditions of Theorem 8.2.1 and so $U = X$ and $Z = XY$ have joint density

(8) $$f(u, z) = \frac{1}{|u|} f\left(u, \frac{z}{u}\right).$$

The result (7) follows immediately as it is the marginal density of Z obtained from $f(u, z)$.

Alternatively it is possible to derive the result directly by the usual plod, as follows.

$$\begin{aligned}
\mathbf{P}(XY \leqslant z) &= \mathbf{P}\left(X > 0, Y \leqslant \frac{z}{X}\right) + \mathbf{P}\left(X < 0, Y \geqslant \frac{z}{X}\right) \\
&= \int_{-\infty}^{0} \int_{z/u}^{\infty} f(u, v)\, dv\, du + \int_{0}^{\infty} \int_{-\infty}^{z/u} f(u, v)\, dv\, du \\
&= \int_{-\infty}^{0} \int_{-\infty}^{z} f\left(u, \frac{t}{u}\right) \frac{dt}{(-u)}\, du + \int_{0}^{\infty} \int_{-\infty}^{z} f\left(u, \frac{t}{u}\right) \frac{dt}{u}\, du \\
&= \int_{-\infty}^{z} \int_{-\infty}^{\infty} f\left(u, \frac{t}{u}\right) \frac{du}{|u|}\, dt.
\end{aligned}$$

The required density is obtained by comparison of this expression with (7.1.15).

Now we turn to the quotient $W = X/Y$. First, let $V = 1/Y$. Then, by definition

$$F_{X,V}(x, v) = \mathbf{P}(X \leqslant x, V \leqslant v) = \mathbf{P}\left(X \leqslant x, Y \geqslant \frac{1}{v}\right) = \int_{-\infty}^{x} \int_{1/v}^{\infty} f(s, t)\, ds dt.$$

Hence, on differentiating, the joint density of X and Y^{-1} is given by

$$f_{X,V}(x, v) = \frac{1}{v^2} f\left(x, \frac{1}{v}\right).$$

Now $W = XV$, so by the first part

$$f_W(w) = \int_{-\infty}^{\infty} \frac{1}{|u|} \frac{u^2}{w^2} f\left(u, \frac{u}{w}\right) du = \int_{-\infty}^{\infty} |v| f(vw, v)\, dv$$

on setting $u = vw$ in the integrand. Alternatively of course, you can obtain this by using Theorem 8.2.1 directly via the transformation $w = x/y$ and $u = y$; or you can proceed via the routine plod. ■

As usual, here are some illustrative examples.

(9) Example Let X and Y be independent with respective density functions $f_X(x) = xe^{-\frac{x^2}{2}}$ for $x > 0$, and $f_Y(y) = \pi^{-1}(1 - y^2)^{-\frac{1}{2}}$ for $|y| < 1$. Show that XY has a normal distribution.

Solution When X and Y are independent, we have $f(x, y) = f_X(x) f_Y(y)$, and Theorem 6 takes the special form

$$f(z) = \int_{-\infty}^{\infty} \frac{1}{|u|} f_X(u) f_Y\left(\frac{z}{u}\right) du = \int_{u>z} \frac{1}{|u|} u e^{-\frac{u^2}{2}} \pi^{-1} \left(1 - \frac{z^2}{u^2}\right)^{-\frac{1}{2}} du$$

$$= \frac{1}{\pi} \int_z^{\infty} \frac{e^{-\frac{u^2}{2}}}{(u^2 - z^2)^{\frac{1}{2}}} u \, du.$$

Now we make the substitution $u^2 = z^2 + v^2$, to find that

$$f(z) = \frac{1}{\pi} e^{-\frac{z^2}{2}} \int_0^{\infty} e^{-\frac{v^2}{2}} dv = \frac{e^{-\frac{z^2}{2}}}{(2\pi)^{\frac{1}{2}}}$$

which is the $N(0, 1)$ density. ●

(10) Example Let X and Y have density $f(x, y) = e^{-x-y}$, for $x > 0$, $y > 0$. Show that $U = X/(X + Y)$ has the uniform density on $(0, 1)$.

Solution To use Theorem 6 we need to know the joint density of X and $V = X + Y$. A trivial application of Theorem 8.2.1 shows that X and V have density $f(x, v) = e^{-v}$, for $0 < x < v < \infty$. Hence by Theorem 6,

$$\begin{aligned} f(u) &= \int_0^{\infty} v e^{-v} \, dv, \quad \text{for } 0 < uv < v \\ &= 1 \qquad\qquad\quad, \quad \text{for } 0 < u < 1. \end{aligned}$$

Alternatively we may use Theorem 8.2.1 directly by considering the transformation

$$u = \frac{x}{x + y}, \qquad v = x + y,$$

with

$$x = uv, \qquad y = v(1 - u),$$

and $|J| = v$. Hence $U = X/(X + Y)$ and $V = X + Y$ have density $f(u, v) = ve^{-v}$, for $v > 0$ and $0 < u < 1$. The marginal density of U is 1, as required. ●

8.5 Expectation

Suppose that the random variable $Z = g(X, Y)$ has density $f(z)$. Then by definition

$$\mathbf{E}(Z) = \int_{-\infty}^{\infty} z f(z) \, dz$$

provided that $\mathbf{E}(|Z|) < \infty$. However, suppose we know only the joint density $f(x, y)$ of X and Y. As we have discovered above, finding the density of $g(X, Y)$ may not be a trivial matter. Fortunately this task is rendered unnecessary by the following result, which we state without proof.

(1) Theorem If X and Y have joint density $f(x, y)$ and

$$\int_{-\infty}^{\infty} \int_{-\infty}^{\infty} |g(u, v)| f(u, v) \, du dv < \infty,$$

then

(2) $$\mathbf{E}(g(X, Y)) = \int_{-\infty}^{\infty} \int_{-\infty}^{\infty} g(u, v) f(u, v) \, du dv.$$

This useful result has the same pleasing consequences as did the corresponding result (Theorem 5.3.1) for discrete random variables.

(3) **Corollary** Let X and Y have finite expectations. Then

(i) $\mathbf{E}(aX + bY) = a\mathbf{E}(X) + b\mathbf{E}(Y)$, for any constants a and b.
(ii) If $\mathbf{P}(X \leq Y) = 1$, then $\mathbf{E}(X) \leq \mathbf{E}(Y)$.

Suppose further that $\mathbf{E}(X^2)$ and $\mathbf{E}(Y^2)$ are finite. Then

(iii) $\mathbf{E}(X) \leq \mathbf{E}(|X|) \leq (\mathbf{E}(X^2))^{\frac{1}{2}}$,

and

(iv) $\mathbf{E}(XY) \leq (\mathbf{E}(X^2)\mathbf{E}(Y^2))^{\frac{1}{2}}$.

Recall that this last result is the Cauchy–Schwarz inequality.

Finally suppose that $\mathbf{E}(g(X))$ and $\mathbf{E}(h(Y))$ are finite, and that X and Y are independent. Then

(v) $\mathbf{E}(g(X)h(Y)) = \mathbf{E}(g(X))\mathbf{E}(h(Y))$.

We omit the proofs of these results. Generally speaking, the proofs follow the same line of argument as in the discrete case, with the difference that those proofs used results about rearrangement of sums, whereas these proofs use standard results about multiple integrals.

Some important expectations deserve special mention. Just as we did for discrete random variables, we define the covariance as

$$\operatorname{cov}(X, Y) = \mathbf{E}((X - \mathbf{E}(X))(Y - \mathbf{E}(Y))),$$

and the correlation as

$$\rho(X, Y) = \frac{\operatorname{cov}(X, Y)}{(\operatorname{var}(X)\operatorname{var}(Y))^{\frac{1}{2}}}.$$

When X and Y are independent, then it follows from Corollary 3(v) that $\operatorname{cov}(X, Y) = \rho(X, Y) = 0$, but not conversely.

(4) **Example** Let Θ be uniformly distributed over $(0, \alpha)$. Find $\operatorname{cov}(\sin\Theta, \cos\Theta)$, and show that for $\alpha = k\pi$, $(k \neq 0)$, $\sin\Theta$ and $\cos\Theta$ are uncorrelated and not independent.

Solution Routine calculations proceed thus:

$$\mathbf{E}(\sin\Theta\cos\Theta) = \frac{1}{\alpha}\int_0^\alpha \frac{1}{2}\sin 2\theta\, d\theta = \frac{1}{4\alpha}(1 - \cos 2\alpha),$$

and $\mathbf{E}(\sin\Theta) = (1/\alpha)(1 - \cos\alpha)$, and $\mathbf{E}(\cos\Theta) = (1/\alpha)\sin\alpha$, Hence

$$\operatorname{cov}(\sin\Theta, \cos\Theta) = \frac{1}{4\alpha}(1 - \cos 2\alpha) - \frac{1}{\alpha^2}\sin\alpha(1 - \cos\alpha).$$

This covariance is zero whenever $\alpha = k\pi$, $(k \neq 0)$ and so for these values of α, $\rho = 0$. However $\sin\Theta$ and $\cos\Theta$ are not independent. This is obvious because

$\sin\Theta = \cos(\frac{\pi}{2} - \Theta)$; but we can verify it formally by noting that

$$\mathbf{P}\left(\sin\Theta > \frac{3}{4}, \cos\Theta > \frac{3}{4}\right) = 0 \neq \mathbf{P}\left(\sin\Theta > \frac{3}{4}\right)\mathbf{P}\left(\cos\Theta > \frac{3}{4}\right).$$ ●

Generating functions have been so useful above that it is natural to introduce them again now. Since we are dealing with jointly continuous random variables, the obvious candidate for our attention is a joint moment generating function.

(5) Definition Let X and Y be jointly distributed. The *joint moment generating function* of X and Y is

$$M_{X,Y}(s, t) = \mathbf{E}(e^{sX+tY}).$$ ▲

If this exists in a neighbourhood of the origin, then it has the same attractive properties as the ordinary m.g.f. That is, it determines the joint distribution of X and Y uniquely, and also it does yield the moments, in that

(6) $$\frac{\partial^{m+n}}{\partial s^m \partial t^n} M(s, t)\bigg|_{s=t=0} = \mathbf{E}(X^m Y^n).$$

Furthermore, just as joint p.g.f.s factorize for independent discrete random variables, it is the case that joint m.g.f.s factorize for independent continuous random variables. That is to say:

(7) $$M_{X,Y}(s, t) = M_X(s) M_Y(t)$$

if and only if X and Y are independent.

We offer no proofs for the above statements, as a proper account would require a wealth of analytical details. Nevertheless, we shall use them freely as required.

(8) Example Let X and Y have density $f(x, y) = e^{-y}, 0 < x < y < \infty$. Then

$$M_{X,Y}(s, t) = \int_0^\infty \int_x^\infty e^{sx+ty-y}\, dy dx = ((1-t)(1-s-t))^{-1}.$$

Hence, differentiating, we obtain

$$\frac{\partial M}{\partial s}\bigg|_{s=t=0} = 1, \quad \frac{\partial M}{\partial t}\bigg|_{s=t=0} = 2, \quad \frac{\partial^2 M}{\partial s \partial t}\bigg|_{s=t=0} = 3.$$

Thus $\operatorname{cov}(X, Y) = 1$. ●

(9) Example Let X and Y be independent and identically distributed normal random variables. Let $M = \frac{1}{2}(X + Y)$ and $V = (X - M)^2 + (Y - M)^2$. Show that M and V are independent.

Solution We will use an obvious extension of (7). Let $\mathbf{E}(X) = \mu$, and $\operatorname{var}(X) = \sigma^2$. Then consider the joint moment generating function of M, $X - M$, and $Y - M$:

$$\mathbf{E}(\exp(sM + t(X - M) + u(Y - M)))$$
$$= \mathbf{E}\left(\exp\left(\frac{1}{2}(s + t + u)X + \frac{1}{2}(s - t - u)Y\right)\right)$$

$$= \exp\left(\frac{\mu}{2}(s+t+u) + \frac{\sigma^2}{8}(s+t+u)^2\right)\exp\left(\frac{\mu}{2}(s-t-u) + \frac{\sigma^2}{8}(s-t-u)^2\right),$$

since X and Y are independent,

(10) $$= \exp\left(\mu s + \frac{\sigma^2}{4}s^2\right)\exp\left(\frac{\sigma^2}{4}(t+u)^2\right).$$

Hence M is independent of the random vector $(X - M,\ Y - M)$, and so M is independent of V. ●

This remarkable property of the normal distribution extends to any independent collection $(X_i;\ 1 \leq i \leq n)$ of $N(\mu, \sigma^2)$ random variables, and is known as the independence of sample mean and sample variance property.

(11) Example Let X and Y be independent and identically distributed with mean zero, variance 1 and moment generating function $M(t)$, which is thrice differentiable at 0. Show that if $X + Y$ and $X - Y$ are independent, then X and Y are normally distributed.

Solution By the independence of X and Y

(12) $$\begin{aligned} M(s+t)M(s-t) &= \mathbf{E}(e^{(s+t)X})\mathbf{E}(e^{(s-t)Y}) = \mathbf{E}(e^{s(X+Y)+t(X-Y)}) \\ &= \mathbf{E}(e^{s(X+Y)})\mathbf{E}(e^{t(X-Y)}), \end{aligned}$$

by the independence of $X + Y$ and $X - Y$

$$= (M(s))^2 M(t)M(-t),$$

using the independence of X and Y again. Next we note that by the conditions of the problem $M'(0) = \mathbf{E}(X) = 0$, and $M''(0) = \mathbf{E}(X^2) = 1$. Now differentiating (12) twice with respect to t, and then setting $t = 0$ gives

$$\begin{aligned} M(s)M''(s) - (M'(s))^2 &= (M(s))^2(M(0)M''(0) - (M'(0))^2) \\ &= (M(s))^2(\mathbf{E}(X^2) - (\mathbf{E}(X))^2) = (M(s))^2. \end{aligned}$$

Integrating this differential equation once gives

$$\frac{M'(s)}{M(s)} = s$$

and integrating again yields $M(s) = \exp(\frac{1}{2}s^2)$. Since this is the moment generating function of the $N(0, 1)$ density, it follows that X and Y have this density. ●

We are of course already aware from Chapter 6 that generating functions are of considerable value in handling sums of independent random variables.

(13) Example Let $(X_1, \ldots, X_n)$ be independent having the $N(0, 1)$ density. Show that $Y = \sum_{i=1}^n X_i^2$ has a $\chi^2(n)$ density.

Solution With a view to using moment generating functions, we first find

$$\mathbf{E}(e^{tX_1^2}) = \int_{-\infty}^{\infty} \frac{1}{(2\pi)^{\frac{1}{2}}}\exp\left(tx^2 - \frac{1}{2}x^2\right)dx = \frac{1}{(1-2t)^{\frac{1}{2}}}.$$

Hence

(14)
$$\mathbf{E}(e^{tY}) = (\mathbf{E}(e^{tX_1^2}))^n \quad \text{by independence}$$
$$= \frac{1}{(1-2t)^{\frac{n}{2}}}$$

and by (7.1.28) and (7.5.6), this is the m.g.f. of the $\chi^2(n)$ density. Hence by Theorem 7.5.9 Y has a $\chi^2(n)$ density. ●

Many results about sums of random variables can now be established by methods which, if not trivial, are at least straightforward.

(15) Example Let $(X_k;\ k \geqslant 1)$ be independent and identically distributed exponential random variables, with parameter λ. Then for $S_n = \sum_{k=1}^{n} X_k$,

$$\mathbf{E}(e^{tS_n}) = (\mathbf{E}(e^{tX_1}))^n \quad \text{by independence}$$
$$= \left(\frac{\lambda}{\lambda - t}\right)^n \quad \text{by Example 7.5.4}$$

and so S_n has a gamma distribution by (7.5.6). ●

8.6 Conditional density and expectation

Suppose that X and Y have joint density $f(x, y)$, and we are given the value of Y. By analogy with the conditional mass function which arises when X and Y are discrete, we make the following definition.

(1) Definition If X and Y have joint density $f(x, y)$, then the conditional density of X given $Y = y$ is given by

(2)
$$f_{X|Y}(x|y) = \begin{cases} \dfrac{f(x, y)}{f_Y(y)} & \text{if } 0 < f_Y(y) < \infty \\ 0 & \text{elsewhere.} \end{cases}$$ ▲

We observe immediately that $f_{X|Y}(x|y)$ is indeed a density, because it is non-negative and

$$\int_{-\infty}^{\infty} f_{X|Y}(x|y)\,dx = \int_{-\infty}^{\infty} \frac{f(x, y)}{f_Y(y)}\,dx = \frac{1}{f_Y(y)} \cdot f_Y(y) = 1.$$

The corresponding conditional distribution function is

$$F_{X|Y}(x, y) = \int_{-\infty}^{x} f_{X|Y}(u|y)\,du = \mathbf{P}(X \leqslant x | Y = y),$$

and we have

(3)
$$\mathbf{P}(X \in A | Y = y) = \int_{x \in A} f_{X|Y}(x|y)\,dx.$$

(4) Example Let (X, Y) be the coordinates of the point Q uniformly distributed on a circular disc of unit radius. What is $f_{Y|X}(y|x)$?

Solution Recall that for the marginal density $f_X(x) = (2/\pi)(1 - x^2)^{\frac{1}{2}}$. Hence by definition

$$f_{Y|X}(y|x) = \frac{f(x, y)}{f_X(x)} = \frac{1}{\pi} \frac{\pi}{2(1 - x^2)^{\frac{1}{2}}}$$
$$= \frac{1}{2}(1 - x^2)^{-\frac{1}{2}} \quad \text{for } |y| < (1 - x^2)^{\frac{1}{2}}.$$

This conditional density is uniform on $(-(1 - x^2)^{\frac{1}{2}}, (1 - x^2)^{\frac{1}{2}})$, which is consistent with our earlier observations about conditioning of uniform densities. ●

(5) **Example** Let X and Y be independent and exponential with parameter λ. Show that the density of X conditional on $X + Y = v$ is uniform on $(0, v)$.

Solution To use (1) we need to take some preliminary steps. First note that the joint density of X and Y is $f(x, y) = \lambda^2 e^{-\lambda(x+y)}$ for $x > 0, y > 0$.

Next we need the joint density of X and $X + Y$ so we consider the transformation $u = x$ and $v = x + y$, with inverse $x = u$ and $y = v - u$, so that $J = 1$. Hence by Theorem 8.2.1

$$f_{U,V}(u, v) = \lambda^2 e^{-\lambda v} \quad \text{for } 0 < u < v < \infty.$$

It follows that

$$f_V(v) = \int_0^v \lambda^2 e^{-\lambda v}\, du = \lambda^2 v e^{-\lambda v},$$

and so by definition

$$f_{U|V}(u|v) = \frac{f(u, v)}{f_V(v)} = \frac{1}{v} \quad \text{for } 0 < u < v.$$

This is the required uniform density. ●

This striking result is related to the lack-of-memory property of the exponential density.

Now, since $f_{X|Y}(x|y)$ is a density it may have an expected value, which naturally enough is called *conditional expectation*.

(6) **Definition** If $\int_{\mathbb{R}} |x| f_{X|Y}(x|y)\, dx < \infty$ then the *conditional expectation* of X given $Y = y$ is given by

$$\mathbf{E}(X|Y = y) = \int_{\mathbb{R}} x f_{X|Y}(x|y)\, dx.$$ ▲

(7) **Example (5) revisited** If X and Y are independent and exponential, then we showed that the density of X given $X + Y = v$ is uniform on $(0, v)$. Hence

$$\mathbf{E}(X|X + Y = v) = \frac{1}{2} v.$$ ●

Actually this is otherwise obvious because, for reasons of symmetry,

$$\mathbf{E}(X|X + Y = v) = \mathbf{E}(Y|X + Y = v),$$

and trivially $\mathbf{E}(X+Y|X+Y=v)=v$. Hence the result follows, provided it is true that for random variables X, Y and V, we have

(8) $$\mathbf{E}(X+Y|V=v)=\mathbf{E}(X|V=v)+\mathbf{E}(Y|V=v).$$

In fact this is true, as we now show.

(9) Theorem Let X, Y and V have joint density $f(x,y,v)$. Then (8) holds.

Proof The joint density of $W=X+Y$ and V is $\int_{-\infty}^{\infty} f(w-u,u,v)\,du$. Then by definition

$$\mathbf{E}(X+Y|V=v)=\frac{1}{f_V(v)}\int_{-\infty}^{\infty}\int_{-\infty}^{\infty} wf(w-u,u,v)\,du dw.$$

Now consider the transformation $x=w-u$ and $y=u$, with inverse $w=x+y$ and $u=y$, so that $J=1$. Changing the variables in the double integral accordingly, and using standard results about such double integrals shows that

$$\begin{aligned}\mathbf{E}(X+Y|V=v)&=\frac{1}{f_V(v)}\int_{-\infty}^{\infty}\int_{-\infty}^{\infty}(x+y)f(x,y,v)\,dx\,dy\\&=\frac{1}{f_V(v)}\int_{-\infty}^{\infty} xf_{X,V}(x,v)\,dx+\frac{1}{f_V(v)}\int_{-\infty}^{\infty} yf_{Y,V}(y,v)\,dy\\&=\mathbf{E}(X|V=v)+\mathbf{E}(Y|V=v).\end{aligned}$$

■

Next we make the important observation that by writing

(10) $$\psi(y)=\mathbf{E}(X|Y=y)$$

we emphasize the fact that the conditional expectation of X given Y is a function of Y. If the value of Y is left unspecified, we write

$$\psi(Y)=\mathbf{E}(X|Y)$$

on the understanding that when $Y=y$, $\psi(Y)$ takes the value $\mathbf{E}(X|Y=y)$ defined above.

It is therefore natural to think of $\mathbf{E}(X|Y)$ as a random variable which is a function of Y. (And a more rigorous analysis can indeed justify this assumption.) Just as in the discrete case, its expected value is $\mathbf{E}(X)$.

(11) Theorem The expected value of $\psi(Y)$ is $\mathbf{E}(X)$.

Proof Since $\psi(Y)$ is a function of Y, we can calculate its expected value in the usual way as

$$\begin{aligned}\mathbf{E}(\psi(Y))&=\int_{-\infty}^{\infty}\psi(y)f_Y(y)\,dy\\&=\int_{-\infty}^{\infty}\int_{-\infty}^{\infty} xf_{X|Y}(x|y)f_Y(y)\,dx\,dy \quad \text{by Definition 6}\\&=\int_{-\infty}^{\infty}\int_{-\infty}^{\infty} xf(x,y)\,dx\,dy=\mathbf{E}(X).\end{aligned}$$

■

We recall an earlier example.

(12) Example Let X and Y have density $f(x, y) = 8xy$ for $0 < y < x < 1$. Find $\mathbf{E}(X|Y)$ and $\mathbf{E}(Y|X)$.

Solution Since

$$f_X(x) = \int_0^x 8xy\, dy = 4x^3,$$

and

$$f_Y(y) = \int_y^1 8xy\, dx = 4y(1 - y^2),$$

we have $f_{X|Y}(x|y) = 2x/(1 - y^2), \quad y < x < 1$. Hence

$$\mathbf{E}(X|Y) = \int_Y^1 \frac{2x^2}{(1 - Y^2)}\, dx = \frac{2}{3}\left(\frac{1 - Y^3}{1 - Y^2}\right).$$

Likewise $f_{Y|X}(y|x) = 2y/x^2 \quad 0 < y < x$, and therefore

$$\mathbf{E}(Y|X) = \int_0^X \frac{2y^2}{X^2}\, dy = \frac{2}{3}X.$$

●

The identity in Theorem 11 can also be used to calculate probabilities by the simple device of letting X be the indicator of the event of interest.

(13) Example Let U and Y have density $f(u, y)$. What is $\mathbf{P}(U < Y)$?

Solution Let X be the indicator of $U < Y$. Then

$$\mathbf{P}(U < Y) = \mathbf{E}(X) = \mathbf{E}(\mathbf{E}(X|Y)) \quad \text{by Theorem 11}$$

$$= \mathbf{E}\left(\int_{-\infty}^{Y} f(u|Y)\, du\right) = \int_{-\infty}^{\infty}\int_{-\infty}^{y} f(u, y)\, du dy \quad \text{by (2)}$$

Of course we could have written this down immediately by Theorem 8.1.9. ●

(14) Example: Bivariate normal Let X and Y have the bivariate normal density

(15) $$f(x, y) = \frac{1}{2\pi\sigma\tau(1 - \rho^2)^{\frac{1}{2}}} \exp\left(-\frac{1}{2(1 - \rho^2)}\left(\frac{x^2}{\sigma^2} - \frac{2\rho xy}{\sigma\tau} + \frac{y^2}{\tau^2}\right)\right).$$

(a) Find the conditional density of X given $Y = y$.
(b) Find $\mathbf{E}(e^{tXY})$, and hence find the density of $Z = X_1Y_1 + X_2Y_2$, where (X_1, Y_1) is independent of (X_2, Y_2) and each has the density (15).

Solution (a) From Example 8.1.19 we know that Y has the $N(0, \tau^2)$ density. Hence

$$f_{X|Y}(x|y) = \frac{f(x, y)}{f_Y(y)} = \frac{1}{\sigma(2\pi(1 - \rho^2)^{1/2})} \exp\left(-\frac{1}{2(1 - \rho^2)}\left(\frac{x}{\sigma} - \frac{\rho y}{\tau}\right)^2\right).$$

Hence the conditional density of X given $Y = y$ is $N(\rho\sigma y/\tau, \sigma^2(1 - \rho^2))$.

(b) By part (a), the conditional moment generating function of X given Y is

$$M_{X|Y}(t) = \exp\left(\frac{\rho\sigma Y}{\tau}t + \frac{1}{2}\sigma^2(1-\rho^2)t^2\right).$$

Hence by conditional expectation

$$\begin{aligned}\mathbf{E}(e^{tXY}) &= \mathbf{E}(\mathbf{E}(e^{tXY}|Y)) = \mathbf{E}(M_{X|Y}(tY)) \\ &= \mathbf{E}\left(\exp\left[\left(\frac{\rho\sigma}{\tau}t + \frac{1}{2}\sigma^2(1-\rho^2)t^2\right)Y^2\right]\right) \\ &= \frac{1}{(1-2\rho\sigma\tau t - \sigma^2\tau^2(1-\rho^2)t^2)^{\frac{1}{2}}} \quad \text{using Example 8.5.13}\end{aligned}$$

and so $X_1Y_1 + X_2Y_2$ has moment generating function

$$\begin{aligned}M(t) &= (1-2\rho\sigma\tau t - \sigma^2\tau^2(1-\rho^2)t^2)^{-1} \\ &= \frac{1-\rho}{2}\frac{1}{1+\sigma\tau(1-\rho)t} + \frac{1+\rho}{2}\frac{1}{1-\sigma\tau(1+\rho)t}.\end{aligned}$$

Hence $Z = X_1Y_1 + X_2Y_2$ has an asymmetric bilateral exponential density

(16) $$f(z) = \begin{cases}\dfrac{1+\rho}{2}\exp(-\sigma\tau(1+\rho)z) & \text{if } z > 0 \\ \dfrac{1-\rho}{2}\exp(\sigma\tau(1-\rho)z) & \text{if } z < 0.\end{cases}$$ ●

We note without proof that $\psi(Y)$ has the useful properties which we recorded in the discrete case. Among the most important is that

(17) $$\mathbf{E}(Xg(Y)|Y) = g(Y)\psi(Y)$$

for any function $g(Y)$ of Y.

Finally we stress that conditional expectation is important in its own right, it should not be regarded merely as a stage on the way to calculating something else.

For example, suppose that X and Y are random variables, and we wish to record the value of X. Unfortunately X is inaccessible to measurement, we can record only the value of Y. Can this help us to make a good guess at X?

First we have to decide what a 'good' guess $g(Y)$ at X is. We decide that $g_1(Y)$ is a better guess than $g_2(Y)$ if

(18) $$\mathbf{E}[(g_1(Y)-X)^2] < \mathbf{E}[(g_2(Y)-X)^2].$$

According to this (somewhat arbitrary) rating, it turns out that the best guess at X given Y is $\psi(Y) = \mathbf{E}(X|Y)$.

(19) Theorem For any function $g(Y)$ of Y

$$\mathbf{E}[(X-g(Y))^2] \geqslant \mathbf{E}[(X-\psi(Y))^2].$$

Proof Using (17), we have

(20) $$\mathbf{E}[(X-\psi)(\psi-g)] = \mathbf{E}[(\psi-g)\mathbf{E}(X-\psi|Y)] = 0.$$

Hence

$$\mathbf{E}[(X-g)^2] = \mathbf{E}[(X-\psi+\psi-g)^2] = \mathbf{E}[(X-\psi)^2] + \mathbf{E}[(\psi-g)^2] \text{ by (20)}$$
$$\geqslant \mathbf{E}[(X-\psi)^2].$$

■

8.7 Transformations; order statistics

We introduced the change of variable technique in Section 8.2. We return to this topic to consider a particularly important class of transformations, namely those which are linear. Thus, let the random vector $(X_1, \ldots, X_n)$ have joint density $f(x_1, \ldots, x_n)$. Suppose that for $1 \leqslant i, j \leqslant n$, and some constants a_{ij},

(1) $$Y_i = \sum_{j=1}^{n} a_{ij} X_j.$$

What can be said about the joint density of $(Y_1, \ldots, Y_n)$?

In Section 8.2 we required such transformations to be invertible, and we make the same restriction here; therefore we suppose that the matrix $A = (a_{ij})$ has an inverse $A^{-1} = (b_{ij}) = B$. A sufficient condition for this is that the determinant $\det A$ is not zero. Then we have the following useful result, which we state without proof.

(2) Theorem Suppose that $(X_1, \ldots, X_n)$ has density $f_X(x_1, \ldots, x_n)$, and that $(Y_1, \ldots, Y_n)$ is related to $(X_1, \ldots, X_n)$ by

$$Y_i = \sum_{j=1}^{n} a_{ij} X_j$$

and

$$X_i = \sum_{j=1}^{n} b_{ij} Y_j,$$

where $BA = I$, the identity matrix, and $\det A \neq 0$. Then the density of $(Y_1, \ldots, Y_n)$ is given by

(3) $$f_Y(y_1, \ldots, y_n) = \frac{1}{|\det A|} f_X(x_1(y_1, \ldots, y_n) \ldots x_n(y_1, \ldots, y_n))$$
$$= |\det B| f_X(x_1, \ldots, x_n).$$

(4) Example: Normal sample Let $(X_1, \ldots, X_n)$ be independent $N(0, 1)$ random variables, and define

(5) $$Y_j = \sum_{j=1}^{n} x_j a_{ij} \quad \text{for } 1 \leqslant j \leqslant n,$$

where the matrix $A = (a_{ij})$ is an orthogonal rotation with $\det A = 1$, and, denoting the transpose of A by A^T,

$$AA^T = I = A^T A.$$

(a) Show that $(Y_1, \ldots, Y_n)$ are independent $N(0, 1)$ random variables.

(b) Deduce that the sample mean

$$\bar{X} = \frac{1}{n}\sum_1^n X_i$$

and the sample variance

$$s^2 = \frac{1}{n-1}\sum_{i=1}^n (X_i - \bar{X})^2$$

and independent, and that $(n-1)s^2$ has a χ^2 density.

Solution (a) It is convenient to use the standard notation for vectors and matrices in problems of this type. Thus we write $\boldsymbol{x}^T$ for the transpose of $\boldsymbol{x}$, where

(6) $$\boldsymbol{x} = (x_1, \ldots, x_n) = \boldsymbol{y}\boldsymbol{A}^T \quad \text{from (5).}$$

Furthermore

$$\sum_{i=1}^n x_i^2 = \boldsymbol{x}\boldsymbol{x}^T = \boldsymbol{y}\boldsymbol{A}^T\boldsymbol{A}\boldsymbol{y}^T = \boldsymbol{y}\boldsymbol{y}^T = \sum_{i=1}^n y_i^2.$$

Hence by (3), $(Y_1, \ldots, Y_n)$ have density

(7) $$f_Y = \frac{1}{(2\pi)^{\frac{n}{2}}}\exp\left(-\frac{1}{2}\sum_{i=1}^n y_i^2\right).$$

Since this factorizes, $(Y_1, \ldots, Y_n)$ are independent with the $N(0, 1)$ density.

(b) Now let (a_{ij}) be any rotation such that $a_{1j} = n^{-\frac{1}{2}}$, giving

$$Y_1 = \sum_{j=1}^n \frac{1}{\sqrt{n}} X_j = \sqrt{n}\,\bar{X}.$$

Then

$$\begin{aligned}(n-1)s^2 &= \sum_{i=1}^n X_i^2 - 2\bar{X}\sum_{i=1}^n X_i + n\bar{X}^2 \\ &= \sum_{i=1}^n X_i^2 - n\bar{X}^2 \\ &= \sum_{i=1}^n Y_i^2 - Y_1^2 = \sum_2^n Y_i^2.\end{aligned}$$

Hence s^2 is independent of $\bar{X}$, by the independence of Y_1 and $(Y_2, \ldots, Y_n)$. Finally, since each Y_i is $N(0, 1)$, $(n-1)s^2$ has a $\chi^2(n-1)$ density by Example 8.5.13. ●

A particularly important linear transformation is the one which places $(X_1, \ldots, X_n)$ in non-decreasing order. Thus

$$\begin{aligned} Y_1 &= \text{smallest of } X_1, \ldots, X_n \\ Y_2 &= \text{second smallest of } X_1, \ldots, X_n \\ &\vdots \\ Y_n &= \text{largest of } X_1, \ldots, X_n. \end{aligned}$$

We assume that each X_k has a density $f(x_k)$, so that the chance of ties is zero. It is customary to use the special notation

$$Y_k = X_{(k)},$$

and then $X_{(1)}, X_{(2)}, \ldots, X_{(n)}$ are known as the *order statistics* of $X_1, \ldots, X_n$. Now the above transformation is linear, but not one–one. It is in fact many–one; to see this, suppose that $y_1 < y_2 < \ldots < y_n$. Then the outcomes

$$X_1 = y_1, X_2 = y_2, \ldots, X_n = y_n$$

and

$$X_2 = y_1, X_1 = y_2, \ldots, X_n = y_n$$

both yield the same set of order statistics, namely

$$X_{(1)} = y_1, X_{(2)} = y_2, \ldots, X_{(n)} = y_n.$$

However, if $(\pi(1), \ldots, \pi(n))$ is any one of the $n!$ distinct permutations of the first n integers, and R_π is the region $x_{\pi(1)} < x_{\pi(2)} < \ldots < x_{\pi(n)}$, then the transformation

$$x_{(k)} = x_{\pi(k)}; \quad 1 \leqslant k \leqslant n$$

is one–one and linear. In the notation of (2), we have

$$a_{ij} = \begin{cases} 1 & \text{if } i = \pi(j) \\ 0 & \text{otherwise,} \end{cases}$$

and $|\det A| = 1$. Therefore the density of $X_{(1)}, \ldots, X_{(n)}$ is $\prod_{i=1}^n f(y_i)$.

Now we observe that $X_1, X_2, \ldots, X_n$ lies in just one of the $n!$ regions R_π, and hence the order statistics have joint density

(8) $$n!\prod_{i=1}^n f(y_i) \quad \text{for } y_1 < y_2 < \ldots < y_n.$$

Here are some applications of this useful result.

(9) Example Let $(X_1, \ldots, X_n)$ be independently and uniformly distributed on $(0, a)$. Then by (8) their order statistics have the density

(10) $$f = \frac{n!}{a^n} \quad \text{for } y_1 < y_2 < \ldots < y_n.$$ ●

It follows from (8) that we may in principle obtain the marginal density of any subset of the order statistics by performing appropriate integrations. For small subsets this is actually unnecessary.

(11) Example Show that $X_{(k)}$ has density

$$f_{(k)}(y) = k\binom{n}{k} f(y)(1 - F(y))^{n-k}[F(y)]^{k-1}.$$

Solution The event $X_{(k)} \leqslant y$ occurs if and only if at least k of the X_i lie in $(-\infty, y]$. Hence

(12) $$F_{(k)}(y) = \sum_{j=k}^{n} \binom{n}{j} [F(y)]^j (1 - F(y))^{n-j}.$$

Now, differentiating to obtain the density,

$$f_{(k)}(y) = f(y) \sum_{j=k}^{n} \left[j\binom{n}{j} F^{j-1}(1-F)^{n-j} - (j+1)\binom{n}{j+1} F^j (1-F)^{n-(j+1)} \right]$$

(13) $$= f(y) k \binom{n}{k} F^{k-1}(1-F)^{n-k}$$

by successive cancellation in the sum. ●

8.8 The Poisson process

A recurring idea in previous chapters has been that of a series of events or happenings which may occur repeatedly at random times denoted by $T_1, T_2, \ldots$ For example we have considered light bulbs which may fail and be replaced at $(T_n; n \geqslant 1)$, or machine bits which may wear out and be renewed at $(T_n; n \geqslant 1)$ and so on. Other practical problems may have this structure also; for example the T_n may be the times at which my telephone rings, or cars arrive at the toll booth, or meteorites fall from the sky, or you get stung by a wasp.

You can think of many more such examples yourself, and it is clear that it would be desirable to have a general theory of such processes. This is beyond our scope, but we can now consider one exceptionally important special case of such processes.

The basic requirement is that the times between events should be independent and identically distributed random variables $(X_k; k \geqslant 1)$; we assume further that they have an exponential distribution.

(1) Definition Let $(X_k; k \geqslant 1)$ be independent identically distributed exponential random variables with parameter λ. Let $T_0 = 0$, and set

$$T_n = \sum_{1}^{n} X_k; \quad n \geqslant 1.$$

Define

(2) $$N(t) = \max\{n: T_n \leqslant t\}; \quad t \geqslant 0.$$

Then $N(t)$ is a *Poisson process with parameter* λ. ▲

A couple of remarks are in order here. First note that $N(t)$ is just the number of happenings or events by time t; $N(t)$ is constant until an event occurs, when it increases by 1. Second, the collection $(N(t); t \geqslant 0)$ is an uncountable collection of random variables. We have said nothing about such collections up to now, and so our analysis of $N(t)$ must of necessity be rather informal.

Our first result explains why $N(t)$ is called a Poisson process.

(3) **Theorem** $N(t)$ has mass function

$$f_N(k) = \frac{e^{-\lambda t}(\lambda t)^k}{k!}; \quad k \geqslant 0.$$

(4) ***Proof*** First we note that from Definition 2 of $N(t)$, the event $N(t) \geqslant k$ occurs if and only if $T_k \leqslant t$. It follows that

(5) $$\mathbf{P}(N(t) \geqslant k) = \mathbf{P}(T_k \leqslant t),$$

and since T_k has a gamma density by (8.5.15) we have

$$f_N(k) = \mathbf{P}(N(t) \geqslant k) - \mathbf{P}(N(t) \geqslant k+1) = \mathbf{P}(T_k \leqslant t) - \mathbf{P}(T_{k+1} \leqslant t)$$
$$= \int_0^t \left(\frac{\lambda^k v^{k-1}}{(k-1)!} - \frac{\lambda^{k+1} v^k}{k!} \right) e^{-\lambda v}\, dv = \frac{e^{-\lambda t}(\lambda t)^k}{k!}$$

after an integration by parts. As an alternative, we could argue straight from (5) and (6.1.7) that

$$\frac{1 - s\mathbf{E}(s^{N(t)})}{1-s} = \sum_0^\infty s^k \mathbf{P}(N(t) \geqslant k) = 1 + \sum_{k=1}^\infty s^k \int_0^t \frac{\lambda^k v^{k-1}}{(k-1)!} e^{-\lambda v}\, dv$$
$$= 1 + s\lambda \int_0^t e^{\lambda vs - \lambda v}\, dv = 1 - \frac{s}{1-s}[e^{\lambda v(s-1)}]_0^t$$
$$= 1 + \frac{s}{1-s} - \frac{se^{\lambda t(s-1)}}{1-s} = \frac{1 - se^{\lambda t(s-1)}}{1-s}$$

and the Poisson mass function of $N(t)$ follows. ∎

Our next result is one of the most strking and important properties of $N(t)$, from which many other results flow.

(6) **Theorem: Conditional property of the Poisson process** Let $N(t)$ be a Poisson process as defined in Definition 1. Conditional on the event $N(t) = k$, the k random variables $T_1, \ldots, T_k$ have conditional density

(7) $$f_{T|N=k}(t_1, \ldots, t_k) = \frac{k!}{t^k}; \quad 0 < t_1 < t_2 < \ldots < t_k \leqslant t.$$

Before proving (7) let us interpret it. From Example 8.7.9 we recognize that the density $k!/t^k$ is the density of the order statistics of k independent random variables, each uniform on $(0, t)$. Thus Theorem 6 can be more dramatically expressed thus: given $N(t) = k$, the k events of the process are independently and uniformly distributed on $(0, t)$.

Proof of (6) Since $X_1, \ldots, X_k$ are independent and exponential, they have joint density

(8) $$f(x_1, \ldots, x_k) = \lambda^k \exp(-\lambda(x_1 + \ldots + x_k)).$$

Next observe that the transformation

$$t_n = \sum_{i=1}^{n} x_i; \quad 1 \le n \le k+1$$

is linear and invertible with $|J| = 1$. Hence by Theorem 8.7.2 the random variables $T_n = \sum_1^n X_i$; $1 \le n \le k+1$ have joint density

(9) $$f(t_1, \ldots, t_{k+1}) = \lambda^{k+1} e^{-\lambda t_{k+1}}; \quad 0 < t_1 < \ldots < t_{k+1}.$$

Now

$$\begin{aligned} \mathbf{P}(0 < T_1 < t_1 < T_2 < \ldots < T_k < t_k; N(t) = k) & \\ &= \mathbf{P}(0 < T_1 < t_1 < \ldots < T_k < t_k < t < T_{k+1}) \\ &= \lambda^k t_1(t_2 - t_1) \ldots (t_k - t_{k-1}) e^{-\lambda t} \end{aligned}$$

on integrating the density (9). Hence the conditional distribution of $T_1, \ldots, T_k$ given $N(t) = k$ is

(10) $$\begin{aligned} \mathbf{P}(T_1 \le t_1, \ldots, T_k \le t_k | N(t) = k) &= \mathbf{P}(T_1 \le t_1 < \ldots < T_k; N(t) = k)/\mathbf{P}(N(t) = k) \\ &= t_1(t_2 - t_1) \ldots (t_k - t_{k-1}) \frac{k!}{t^k}. \end{aligned}$$

Now differentiating (10) with respect to all of $t_1, \ldots, t_k$ gives (7) as required. ■

As we have remarked, this result finds many applications, see Example 8.16 for some of them. For the moment we content ourselves with showing that $N(t)$ has the so-called *independent increments* property.

(11) **Theorem: The Poisson process has independent increments** Let $N(t)$ be a Poisson process, as usual, and let $s < t \le u < v$. Then $N(t) - N(s)$ is independent of $N(v) - N(u)$.

Proof Let $W = N(t) - N(s)$ and $Z = N(v) - N(u)$. Then by conditional expectation

(12) $$\mathbf{E}(w^W z^Z) = \mathbf{E}[\mathbf{E}(w^W z^Z | N(v))].$$

However, conditional on $N(v) = k$, these events are independently and uniformly distributed in $(0, v)$, whence $(W, Z, k - W - Z)$ has a trinomial distribution with

(13) $$\mathbf{E}(w^W z^Z | N(v) = k) = \left(\left(\frac{t-s}{v}\right) w + \left(\frac{v-u}{v}\right) z + \left(\frac{s+u-t}{v}\right)\right)^k.$$

Hence combining (12) and (13) and Theorem (3) gives

(14) $$\begin{aligned} \mathbf{E}(w^W z^Z) &= \exp(\lambda w(t-s) + \lambda z(v-u) + \lambda(s+u-t) - \lambda v) \\ &= \exp(\lambda(t-s)(w-1) + \lambda(v-u)(z-1)) = \mathbf{E}(w^W)\mathbf{E}(z^Z) \end{aligned}$$

as required. ■

We may also observe from (14) that since $\mathbf{E}(w^W) = \exp(\lambda(t-s)(w-1))$, it follows that W is Poisson with parameter $\lambda(t-s)$. That is to say $N(t)-N(s)$ has the same mass function as $N(t-s)$. This property may be called *homogeneity* or the property of *stationary increments*.

8.9 Two limit theorems

Perhaps surprisingly (as we are towards the end of the book) this is an appropriate moment to reconsider our basic ideas about chance. Suppose we are given a number n of similar observations or measurements, denoted by $x_1, \ldots, x_n$. For example, the x_i may be the height of each of n men, or they may be the lifetimes of n light bulbs, or they may be the weight of potatoes yielded by each of n plants. By 'similar' in this context we mean that no measurement has any generic reason to be larger or smaller than the others; the potatoes are of the same variety and grown in the same circumstances; the light bulbs are of the same make and type, the men are of the same age and race.

It is convenient to have one number which gives an idea of the size of a typical x_i, and a popular candidate for this number is the average $\bar{x}$ given by

$$\bar{x}_n = \frac{1}{n}\sum_{i=1}^{n} x_i.$$

One reason for the popularity of $\bar{x}$ is that it is empirically observed that, as n increases, the sequence $\bar{x}_n$ undergoes smaller and smaller fluctuations, and indeed exhibits behaviour of the kind we call convergent.

A special case of such measurements arises when each x_i takes the value 1 or 0 according as to whether some event A occurs or not. Then $\bar{x}_n$ is the proportion of times that A occurs in n trials, and the fact that $\bar{x}_n$ fluctuates less and less as n increases is sometimes used as a basis to justify the axioms of probability.

Of course in mathematical terms, we think of x_i as the outcome of some random variable X_i. It follows that, if our theory of probability is as relevant as we have claimed, then the sequence

$$\bar{X}_n = \frac{1}{n}\sum_{i=1}^{n} X_i$$

ought also to exhibit a similar kind of regularity in the long run as $n \to \infty$. What kind might there be?

To gain some insight into the problem, consider the case when each x_i is 1 if A occurs, and 0 otherwise. In the mathematical formulation of this, X_i is the indicator of the event A, and we assume the X_i are independent. Then $\sum_1^n X_i$ is a binomial random variable with parameters n and p, and we have shown in Examples 4.17 and 7.5.11 that as $n \to \infty$

(1) $$\mathbf{P}\left(\left|\frac{1}{n}\sum_{i=1}^{n} X_i - p\right| > \varepsilon\right) \to 0,$$

and

(2) $$\mathbf{P}\left(\frac{1}{(npq)^{\frac{1}{2}}}\sum_{i=1}^{n}(X_i - p) \le x\right) \to \Phi(x)$$

where $\Phi(x)$ is the standard normal distribution. (Indeed we proved something even stronger than (1) in Example 4.17 and Theorem 5.8.8.)

It seems that $n^{-1}\sum_1^n(X_i - \mathbf{E}(X_1))$ is settling down around $\mathbf{E}(X_i) = p$, and that the distribution of $(n \operatorname{var}(X_1))^{-\frac{1}{2}}\sum_1^n(X_i - \mathbf{E}(X_1))$ is getting closer to the standard normal distribution $\Phi(x)$. More generally we showed in Theorem 5.8.6 that (1) holds for any collection of independent discrete random variables with the same mean and variance. We called this the weak law of large numbers.

This is deliberately vague and informal, but it should now seem at least plausible that the following results might be true.

(3) **Theorem** Let $(X_k; k \geq 1)$ be independent and identically distributed random variables with mean μ, variance $\sigma^2 < \infty$, and moment generating function $M_X(t)$, $|t| < a$. Then we have:

(i) **Weak law of large numbers** For $\varepsilon > 0$, as $n \to \infty$

(4)
$$\mathbf{P}\left(\left|\frac{1}{n}\sum_{i=1}^{n}(X_i - \mu)\right| > \varepsilon\right) \to 0.$$

(ii) **Central limit theorem** As $n \to \infty$

(5)
$$\mathbf{P}\left(\frac{1}{\sigma\sqrt{n}}\sum_{i=1}^{n}(X_i - \mu) \leq x\right) \to \Phi(x) = \int_{-\infty}^{x}(2\pi)^{-\frac{1}{2}}e^{-\frac{1}{2}y^2}\,dy.$$

It is a remarkable fact that both of these are indeed true, and we now prove them.

Proof of (4) The essential step here is to recall Chebyshov's inequality, for then we may write

$$\begin{aligned}\mathbf{P}\left(\frac{1}{n}\left|\sum_{1}^{n}(X_i - \mu)\right| > \varepsilon\right) &\leq \varepsilon^{-2}\mathbf{E}\left(n^{-2}\left(\sum_{1}^{n}(X_i - \mu)\right)^2\right) \\ &= \varepsilon^{-2}n^{-2}\mathbf{E}\left(\sum_{1}^{n}(X_i - \mu)^2\right) \quad \text{by independence,} \\ &= \varepsilon^{-2}n^{-1}\sigma^2 \to 0 \quad \text{as } n \to \infty\end{aligned}$$
■

Note that the proof here is the same as that of Theorem 5.8.6. So it is not necessary for (4) that the X_i be identically distributed or have an m.g.f., it is sufficient that they have the same mean and variance.

Proof of (5) The essential step here is to recall the continuity theorem (7.5.10), for then we may write

$$\begin{aligned}\mathbf{E}\left(\exp\left[\frac{t}{\sigma\sqrt{n}}\sum_{1}^{n}(X_i - \mu)\right]\right) &= \left(\mathbf{E}\left(\exp\left[\frac{t}{\sigma\sqrt{n}}(X_1 - \mu)\right]\right)\right)^n, \quad \text{by independence} \\ &= (M_Y(t/(\sigma\sqrt{n})))^n \quad \text{where } Y = X - \mu,\end{aligned}$$

$$= \left(1 + \mathbf{E}(Y^2)\frac{t^2}{2\sigma^2 n} + o\left(\frac{t^2}{n}\right)\right)^n \quad \text{by Theorem 7.5.9}$$

$$\to e^{\frac{1}{2}t^2}, \text{ since } \mathbf{E}(Y^2) = \sigma^2.$$

Now we recall that $e^{\frac{1}{2}t^2}$ is the moment generating function of the standard normal density $\phi(x)$, and (5) follows by the continuity theorem (7.5.10). ■

WORKED EXAMPLES AND EXERCISES

8.10 Example: Bivariate normal density

Let X and Y have the standard bivariate normal joint density

$$f(x, y) = \frac{1}{2\pi(1-\rho^2)^{\frac{1}{2}}} \exp\left(-\frac{x^2 - 2\rho xy + y^2}{2(1-\rho^2)}\right).$$

Show that the joint moment generating function of X and Y is

$$\exp\left(\frac{1}{2}(s^2 + 2\rho st + t^2)\right).$$

Solution We are asked to find

(1) $$M = \mathbf{E}(e^{sX+tY}) = \int_{-\infty}^{\infty}\int_{-\infty}^{\infty} f(x, y)e^{sx+ty}\, dxdy.$$

Now after a little thought we observe that the terms in the exponents in the integrand can be rearranged to give

$$M = \frac{1}{2\pi(1-\rho^2)^{\frac{1}{2}}} \iint \exp\left(-\frac{x^2}{2} + x(s + t\rho) - \frac{1}{2}\left(\frac{y - \rho x}{(1-\rho^2)^{\frac{1}{2}}}\right)^2 + \frac{y - \rho x}{(1-\rho^2)^{\frac{1}{2}}} t(1-\rho^2)^{\frac{1}{2}}\right) dxdy.$$

This suggests that we make the change of variables $u = x$, $v = (y - \rho x)/(1-\rho^2)^{\frac{1}{2}}$ in the integral. This map is one–one, and $J = (1-\rho^2)^{\frac{1}{2}}$. Hence

$$M = \frac{1}{2\pi} \int_{-\infty}^{-\infty}\int_{-\infty}^{-\infty} \exp\left(-\frac{1}{2}u^2 + (s + t\rho)u - \frac{1}{2}v^2 + t(1-\rho^2)^{\frac{1}{2}}v\right) dudv.$$

Since the integrand factorizes, we now recognize the right hand side as being equal to

$$\mathbf{E}(e^{(s+t\rho)U})\mathbf{E}(e^{t(1-\rho^2)^{\frac{1}{2}}V}),$$

where U and V are standard normal random variables. But we know the m.g.f. $\mathbf{E}(e^{tV})$ of a standard normal random variable V to be $e^{\frac{1}{2}t^2}$. Hence

(2) $$M = e^{\frac{1}{2}(s+t\rho)^2} e^{\frac{1}{2}t^2(1-\rho^2)}$$

as required.

(3) Exercise Find the conditional m.g.f. of Y given X.

(4) Exercise Use (3) to find $\mathbf{E}(e^{sX+tY})$. Show that $\rho(X, Y) = \operatorname{cov}(X, Y) = \rho$.

(5) Exercise Let $X_1, X_2, \ldots, X_n$ be independent standard normal variables. Let $W = \sum_{i=1}^n \alpha_i X_i$ and $Z = \sum_{i=1}^n \beta_i X_i$. When are Y and Z independent?

(6) **Exercise** Find the distribution of $aX + bY$, where X and Y have the bivariate normal distribution.

8.11 Example: Partitions

(a) The random variables X and Y are independently and uniformly distributed on $(0, a)$. Find the density of U, V and W, where $U = \min\{X, Y\}$, $V = |X - Y|$ and $W = a - \max\{X, Y\}$.

(b) Use this to show that if three points are picked independently and uniformly on the perimeter of a circle of radius r, then the expected area of the resulting triangle is $3r^2/(2\pi)$.

Solution (a) We give three methods of solution.
Method I: *Basic plod* (i) By independence

$$\mathbf{P}(U \leqslant u) = 1 - \mathbf{P}(X > u; Y > u) = 1 - \left(\frac{a-u}{a}\right)^2.$$

(ii) By the basic property of densities, if we let C be the set $\{x, y: |x - y| \leqslant v\}$, then

$$\mathbf{P}(V \leqslant v) = a^{-2}\int_C dx\, dy = 1 - \left(\frac{a-v}{a}\right)^2.$$

(iii) By independence

$$\mathbf{P}(W \leqslant w) = \mathbf{P}(\max\{X, Y\} \geqslant a - w) = 1 - \left(\frac{a-w}{a}\right)^2.$$

Hence U, V and W have the same density:

(1) $$f(z) = \frac{2}{a^2}(a - z), \quad \text{for } 0 < z < a.$$

Method II: *Crofton's route* Let $F(a, v)$ be the distribution of V, and consider $F(a + h, v)$. By conditioning on the three events {both X and Y lie in $(0, a)$}, {one of X, Y lies in $(0, a)$} and {neither of X, Y lie in $(0, a)$} we find that

$$F(a + h, v) = F(a, v)\left(\frac{a}{a+h}\right)^2 + \left(\frac{v}{a} + o(h)\right)\frac{2ha}{(a+h)^2} + o(h).$$

Hence, rearranging and taking the limit as $h \to 0$, we have

(2) $$\frac{\partial F}{\partial a}(a, v) = -\frac{2}{a}F(a, v) + \frac{2v}{a^2}.$$

Integrating (2) using the condition $F(a, a) = 1$, gives

(3) $$F(a, v) = \frac{2av - v^2}{a^2}.$$

The densities of U and W may be found by the same method (exercise).

Method III*: Symmetry* Suppose we pick three points independently at random on the perimeter of a circle with perimeter of length a. Then choose any of the three as origin and 'unwrap' the perimeter onto $(0, a)$. The other two points are distributed as X and Y. However, by the symmetry of the original problem the three lengths U, V and W have the same density. By method I, part (i), it is $2(a-z)/a^2$.

(b) Let θ, ϕ, ψ be the angles subtended at the centre by the three sides of the triangle. The area of the triangle is $A = \frac{1}{2}r^2(\sin\theta + \sin\phi + \sin\psi)$; note that this expression is still valid when an angle is obtuse. However by part (a), each of the arc lengths $r\theta$, $r\phi$, and $r\psi$ has the same density $2(2\pi r - z)/(2\pi r)^2$. Hence θ has density $2(2\pi - \theta)/(2\pi)^2$, and

$$\mathbf{E}(A) = \frac{3}{2}r^2\mathbf{E}(\sin\theta) = \frac{3}{2}r^2\int_0^{2\pi}\frac{2(2\pi-\theta)\sin\theta}{(2\pi)^2}\,d\theta = \frac{3r^2}{2\pi}.$$

(4) Exercise Show that the probability that U, V and W can form a triangle is $\frac{1}{4}$.

(5) Exercise Find the densities of U and W by method II.

(6) Exercise Suppose that $X_1, X_2, \ldots, X_n$ are independently and uniformly distributed on $(0, 1)$ with order statistics $X_{(1)}, \ldots, X_{(n)}$. What is the density of $X_{(k+1)} - X_{(k)}$; $1 \le k \le n-1$?

(7) Exercise (6) continued What is the joint density of $X_{(k+1)} - X_{(k)}$ and $X_{(j+1)} - X_{(j)}$ for $j \neq k$?

(8) Exercise Two points are picked at random on the perimeter (including its diameter) of a semicircle with radius 1. Show that the expected area of the resulting triangle they make with the midpoint of the diameter is $1/(2+\pi)$.

(9) Exercise Write down the joint density of U and W; then integrate to derive (1) by a fourth method.

8.12 Example: Buffon's needle

An infinite horizontal table is marked with a rectangular grid comprising two families of distinct lines A and B. The lines of A are parallel, and the distance between neighbouring lines is $2a$. All the lines of B are perpendicular to every line of A, and are distance $2b$ apart.

A thin symmetrical needle of length $2l$, where $l < \min\{a, b\}$, is thrown at random on to the table.

(a) Show that the probability that the needle intersects both an A-line and a B-line is

(1) $$\mathbf{P}(A \cap B) = \frac{l^2}{\pi ab}.$$

(b) Show that the probability that the needle intersects an A-line and does not intersect a B-line is

(2) $$\mathbf{P}(A \cap B^c) = \frac{2bl - l^2}{\pi ab}.$$

Solution The centre C of the needle must fall in some $2a \times 2b$ rectangle R, whose sides are A-lines and B-lines. The words 'at random' mean that the centre is

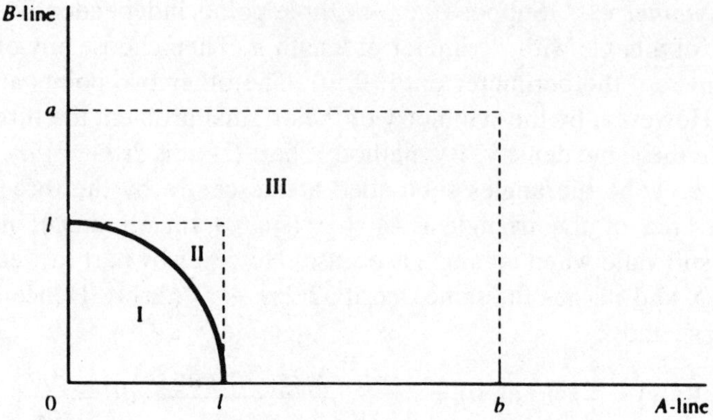

Figure 8.2 Buffon's needle.

uniformly distributed over R, and the angle Θ that the needle makes with any fixed line is uniformly distributed. By symmetry we can suppose that C lies in one quarter of R, namely the $a \times b$ rectangle Q, and also that $0 \leq \theta < \pi$. That is to say, we assume that $C = (X, Y)$ and Θ are jointly uniform on $\{0 \leq x < a\} \times \{0 \leq y < b\} \times \{0 \leq \theta < \pi\}$ with joint density $(\pi ab)^{-1}$.

(i) Now consider the figure. The needle can intersect both A and B only when $C = (x, y)$ lies in the positive quadrant of the circle, radius l, centred at the origin (region I).

If the angle it makes with OB lies between $\pm\sin^{-1}(x/l)$, then it cuts only OA. Likewise the needle cuts only OB if it lies within the angle $-\sin^{-1}(y/l) < \Theta < \sin^{-1}(y/l)$.

Therefore, when $X = x > 0$, $Y = y > 0$ and $x^2 + y^2 \leq l^2$, the probability of two intersections is $\pi - 2\sin^{-1}(x/l) - 2\sin^{-1}(y/l)$. Hence

(3) $$\mathbf{P}(A \cap B) = \frac{1}{\pi ab}\int \pi - 2\sin^{-1}\left(\frac{x}{l}\right) - 2\sin^{-1}\left(\frac{y}{l}\right) dx\, dy$$

where the integral is over $x > 0$, $y > 0$, $x^2 + y^2 \leq l^2$. Now

$$\int_0^l \int_0^{(l^2-x^2)^{\frac{1}{2}}} \sin^{-1}\left(\frac{x}{l}\right) dx\, dy = \int_0^l (l^2 - x^2)^{\frac{1}{2}} \sin^{-1}\left(\frac{x}{l}\right) dx$$

$$= \int_0^{\pi/2} l^2 \theta \cos^2\theta\, d\theta, \quad \text{with the obvious substitution,}$$

$$= \left(\frac{\pi^2}{16} - \frac{1}{4}\right) l^2.$$

Hence substituting into (3) gives

$$\mathbf{P}(A \cap B) = \frac{1}{\pi ab}\left(\frac{l^2\pi^2}{4} - 4l^2\left(\frac{\pi^2}{16} - \frac{1}{4}\right)\right) = \frac{l^2}{\pi ab}.$$

(ii) For $\mathbf{P}(A \cap B^c)$, we examine Figure 8.2 again. Firstly, if C is in region I, then the needle cuts A and not B if Θ lies in an angle $2\sin^{-1}(x/l)$, as we remarked above.

Secondly if C lies in region II (that is $0 \leq y < l$, but $x^2 + y^2 > l^2$) then the needle cuts A and not B if it lies in an angle of size $2\cos^{-1}(y/l)$. Hence

$$\pi ab\mathbf{P}(A \cap B^c) = \int_{\mathrm{I}} 2\sin^{-1}\left(\frac{x}{l}\right) dx\, dy + \int_{\mathrm{II}} 2\cos^{-1}\left(\frac{y}{l}\right) dxdy$$

$$= \int_0^l 2(l^2 - x^2)^{\frac{1}{2}} \sin^{-1}\left(\frac{x}{l}\right) dx + \int_0^l 2(l - (l^2 - y^2)^{\frac{1}{2}}) \cos^{-1}\left(\frac{y}{l}\right) dy$$

$$+ \int_0^l 2(a - l)\cos^{-1}\left(\frac{y}{l}\right) dy$$

$$= l^2 \int_0^{\frac{\pi}{2}} 2\theta(\cos^2\theta - \sin^2\theta)\, d\theta + l\int_0^{\frac{\pi}{2}} (2l + 2(a - l))\theta \sin\theta\, d\theta$$

$$= 2al - l^2, \quad \text{as required.}$$

(4) **Exercise** Show that the probability that the needle intersects no line of the grid is

$$1 - \frac{2l}{\pi b} - \frac{2l}{\pi a} + \frac{l^2}{\pi ab}.$$

(Do this in two ways, one of which is an integral.)

(5) **Exercise** Suppose the table is marked with only one set of parallel lines, each distance $2a$ from its next neighbour. Show that the probability that a needle of length $2l < 2a$ intersects a line is $2l/\pi a$. (Do this two ways also.)

(6) **Exercise** Consider the problem of Exercise 5 when $2l > 2a$. Show that the probability of an intersection is

$$\frac{2}{\pi}\cos^{-1}\left(\frac{a}{l}\right) + \frac{2l}{\pi a}\left(1 - \left(1 - \frac{a^2}{l^2}\right)^{\frac{1}{2}}\right).$$

(7) **Exercise** Suppose (instead of a needle) you roll a penny of radius l on to the grid of A-lines and B-lines. What is the probability that when it topples over it intersects a line? When is this an evens chance?

8.13 Example: Targets

(a) Let $(X_i;\ 1 \leq i \leq 2n + 1)$ be independently and uniformly distributed over $(-1, 1)$, and let $Y_n = X_{(n+1)}$ so that Y_n is the sample median of the X_i. Find the density of Y_n, and hence evaluate the integral $\int_0^1 (1 - x^2)^n\, dx$.

(b) Now n shots hit a circular target. The points of impact are independently and uniformly distributed over the circle. Let Z_n be the radius of the largest circle concentric with the target which includes no hit. Find $\mathbf{E}(Z_n)$.

Solution (a) First note that the uniform distribution on $(-1, 1)$ is $F(x) = \frac{1}{2}(1 + x)$. Now let A_k be the event that $X_{(n+1)} = X_k$; this occurs of course if n of the X_i are greater than X_k, and the remaining n are less than X_k. Then

$$\mathbf{P}(Y_n \leq y) = \sum_{k=1}^{2n+1} \mathbf{P}(Y_n \leq y \cap A_k) = (2n + 1)\mathbf{P}(Y_n \leq y \cap A_1), \quad \text{by symmetry,}$$

$$= (2n + 1)\int_{-1}^{y} f_{X_1}(y)\binom{2n}{n}\left(\frac{1 + y}{2}\right)^n \left(1 - \frac{(1 + y)}{2}\right)^n dy,$$

by conditional probability.

Hence Y_n has density $f_Y(y) = ((2n+1)!/(n!)^2)((1-y^2)^n/2^{2n+1})$. Since this is a density, its integral over $(-1, 1)$ is unity, so

$$\int_0^1 (1-y^2)^n \, dy = \frac{2^{2n}(n!)^2}{(2n+1)!}.$$

(Alternatively you could write down the density for $f_Y(y)$ using the known density of order statistics.)

(b) Let R_i be the distance of the ith hit from the centre of the target. Since hits are uniform, $\mathbf{P}(R_i \leq x) = x^2$ for $0 \leq x \leq 1$. Obviously

$$\mathbf{P}(Z_n > x) = \mathbf{P}(R_i > x \text{ for all } i) = \prod_{i=1}^{n} \mathbf{P}(R_i > x), \quad \text{by independence,}$$
$$= (1-x^2)^n.$$

Hence

$$\mathbf{E}(Z_n) = \int_0^1 \mathbf{P}(Z_n > x) \, dx = \int_0^1 (1-x^2)^n \, dx = \frac{2^{2n}(n!)^2}{(2n+1)!}.$$

(1) Exercise Find $\operatorname{var}(Y_n)$.

(2) Exercise Let A_n be the area of the smallest circle concentric with the target which includes all the hits. Find $\mathbf{E}(A_n)$.

(3) Exercise The hit furthest from the centre of the target is deleted. What now is the expected area of the smallest circle concentric with the target which includes all the remaining hits?

(4) Exercise Let R_n be the distance of the furthest hit from the centre of the target. Show that as $n \to \infty$, $\mathbf{P}(n(1 - R_n) \leq x) \to 1 - e^{-2x}$.

8.14 Example: Gamma densities

Let X and Y be independent, having gamma distributions with parameters $\{\alpha, \lambda\}$ and $\{\beta, \lambda\}$ respectively.

(a) Find the joint density of $U = X + Y$ and $V = \dfrac{X}{X+Y}$.

(b) Deduce that $\mathbf{E}\left(\dfrac{X}{X+Y}\right) = \dfrac{\mathbf{E}(X)}{\mathbf{E}(X) + \mathbf{E}(Y)}$.

(c) What is the density of V?

Solution (a) We use the change of variables technique. The transformation $u = x + y$, $v = x/(x+y)$ for $x, y > 0$, is a one–one map of the positive quadrant onto the strip $0 < v < 1$, $u > 0$, with inverse $x = uv$, and $y = u(1-v)$. Hence $J = u$, and by Theorem 8.2.1, U and V have joint density

$$f(u, v) = \frac{\lambda^\alpha \lambda^\beta}{\Gamma(\alpha)\Gamma(\beta)} (uv)^{\alpha-1}(u(1-v))^{\beta-1} e^{-\lambda uv} e^{-\lambda u(1-v)} u \tag{1}$$
$$= c_1 u^{\alpha+\beta-1} e^{-\lambda u} c_2 v^{\alpha-1}(1-v)^{\beta-1}, \tag{2}$$

where c_1 and c_2 are constants.

Hence U and V are independent, as $f(u, v)$ has factorized.

(b) Using the independence of U and V gives

(3) $$\mathbf{E}(X) = \mathbf{E}(UV) = \mathbf{E}(U)\mathbf{E}(V) = (\mathbf{E}(X) + \mathbf{E}(Y))\mathbf{E}\left(\frac{X}{X+Y}\right),$$

as required.

(c) A glance at (2) shows that V has the beta density with parameters α and β.

(4) Exercise Show that $\dfrac{\Gamma(\alpha)\Gamma(\beta)}{\Gamma(\alpha+\beta)} = \int_0^1 x^{\alpha-1}(1-x)^{\beta-1}\,dx$.

(5) Exercise Show that $\Gamma(\frac{1}{2}) = \pi^{\frac{1}{2}}$.

(6) Exercise Let the random variable Z have density $c(1+x^2)^{-m}$, $m > \frac{1}{2}$, $-\infty < x < \infty$. Show that $c^{-1} = \pi^{\frac{1}{2}}\Gamma(m-\frac{1}{2})/\Gamma(m)$.

8.15 Example: Simulation – the rejection method

(a) Let U and X be independent random variables such that U is uniform on $(0, 1)$ and X has density $f_X(x)$; suppose that there exists a constant a that for all x the function $f_S(x)$ satisfies

(1) $$0 \leqslant f_S(x) \leqslant af_X(x)$$

and $\int_{-\infty}^{\infty} f_S(x)\,dx = 1$. Show that

(2) $$\mathbf{P}(X \leqslant x | aUf_X(X) \leqslant f_S(X)) = \int_{-\infty}^{x} f_S(y)\,dy.$$

(b) Explain how this result may be used to produce realizations of a random variable Z with density $f_S(z)$.

Solution (a) By conditional probability

(3) $$\begin{aligned}\mathbf{P}(X \leqslant x | aUf_X(X) \leqslant f_S(X)) &= \frac{\mathbf{P}(X \leqslant x, aUf_X(X) \leqslant f_S(X))}{\mathbf{P}(aUf_X(X) \leqslant f_S(X))} \\ &= \frac{\int_{-\infty}^{x} \mathbf{P}(aUf_X(x) \leqslant f_S(x)) f_X(x)\,dx}{\int_{-\infty}^{\infty} \mathbf{P}(aUf_X(x) \leqslant f_S(x)) f_X(x)\,dx} \\ &= \int_{-\infty}^{x} \frac{f_S(x)}{af_X(x)} f_X(x)\,dx \Big/ \int_{-\infty}^{\infty} \frac{f_S(x)}{af_X(x)} f_X(x)\,dx, \qquad \text{by (1)}, \\ &= \int_{-\infty}^{x} f_S(x)\,dx.\end{aligned}$$

(b) Suppose we have independent realizations of U and X. Then the above equation says that conditional on the event $A = \{aUf_X(X) \leqslant f_S(X)\}$, X has density $f_S(x)$. In familiar notation we have

$$f_{X|A}(x) = f_S(x).$$

Now suppose we have a sequence $(U_k, X_k;\ k \geqslant 1)$ of random variables which have the same distributions as (U, X). For every pair (U_k, X_k) for which A occurs, the random variable X_k has density $f_S(x)$, and we can write $Z = X_k$. Then Z has density $f_S(z)$.

Remark It is implicit in the question that we want a random variable with density $f_S(x)$, and so any pair (U_k, X_k) for which A^c occurs is rejected. This explains the title of the example (although in the circumstances you might think a better title would be the *conditional* method). Obviously this offers a method for simulating random variables with an arbitrary density $f_S(x)$, subject only to the constraint that we have to be able to simulate X with density $f_X(x)$ which satisfies (1).

(4) **Exercise** Find the mass function and mean of the number N of pairs (U_k, X_k) that are rejected before the first occasion on which A occurs. What does this imply about a?

(5) **Exercise** If X is exponential with parameter 1, show that (2) takes the form $\mathbf{P}(X \leq x | aU_1U_2 \leq f_S(X)) = F_S(x)$, where U_1 and U_2 are independent and uniform on $(0, 1)$. Hence describe how you would simulate a random variable with density $f_S(x) = (2/\pi)^{\frac{1}{2}} e^{-x^2/2}$, $x > 0$.

(6) **Exercise** Let U_1 and U_2 be independent and uniform on $(0, 1)$. Let $X = -\log U_1$, $Y = -\log U_2$. What is the density of X conditional on $Y > \frac{1}{2}(X - 1)^2$?

8.16 Example: The inspection paradox

Let $N(t)$ be a Poisson process, and at each time $t > 0$, define $C(t)$ to be the time since the most recent event. (This is called the *current life* or *age*.) Further, define $B(t)$ to be the time until the next event (this is called the *balance of life* or *excess life*). Show that $B(t)$ and $C(t)$ are independent, and find the distribution of $C(t)$. What is $\mathbf{E}(B + C)$?
[*Note*: by convention, if $N(t) = 0$ we set $C(t) = t$.]

Solution Recall that we used the conditional property of the Poisson process to show that $N(t)$ has independent increments. Now

$$\begin{aligned}\mathbf{P}(B(t) > y, C(t) > z) &= \mathbf{P}(N(t+y) - N(t) = 0, N(t) - N(t-z) = 0)\\ &= \mathbf{P}(N(t+y) - N(t) = 0)\mathbf{P}(N(t) - N(t-z) = 0),\\ &\qquad \text{by the independence of increments,}\\ &= \mathbf{P}(B(t) > y)\mathbf{P}(C(t) > z).\end{aligned}$$

Furthermore, we showed that $N(t) - N(t - z)$ has the same distribution as $N(z)$, for t and $t - z$ both non-negative. Hence

(1) $$\mathbf{P}(C(t) > z) = \begin{cases} 1 & z < 0 \\ e^{-\lambda z} & 0 \leq z \leq t \\ 0 & z > t. \end{cases}$$

Likewise

$$\mathbf{P}(B(t) > y) = \begin{cases} 1 & y < 0 \\ e^{-\lambda y} & y > 0. \end{cases}$$

Hence

(2) $$\mathbf{E}(B+C) = \frac{1}{\lambda} + \int_0^t \lambda t e^{-\lambda t}\,dt + te^{-\lambda t} = \frac{2}{\lambda} - \frac{1}{\lambda}e^{-\lambda t}.$$

Remark If we suppose $N(t)$ is the number of renewals of (say) light bulbs, then (2) says that the expected life of the light bulb inspected at time t is $2/\lambda - 1/\lambda e^{-\lambda t}$ which is greater than the expected life of a randomly selected light bulb, which is $1/\lambda$.

It may seem as though we make light bulbs last longer by inspecting them, this is the 'paradox'. Of course this is not so, it is just that if you only look once, you are more likely to see a longer-lived light bulb. This is related to other sampling paradoxes mentioned previously, see for example 'congregations'.

(3) **Exercise: The Markov property** Show that for any $t_1 < t_2 < \ldots < t_n$, the process $N(t)$ has the so-called Markov property:

$$\mathbf{P}(N(t_n) = j_n | N(t_{n-1}) = j_{n-1}, \ldots, N(t_1) = j_1) = \mathbf{P}(N(t_n) = j_n | N(t_{n-1}) = j_{n-1}).$$

(4) **Exercise: The shower problem** Your telephone is called at the instants of a Poisson process with parameter λ. Each day you take a shower of duration Y starting at time X, where X and Y are jointly distributed in hours (and not independent). Show that the number of times that the telephone is called while you are in the shower has a Poisson distribution with parameter $\lambda\mathbf{E}(Y)$. (Asssume $0 \leq X \leq X + Y \leq 24$.)

(5) **Exercise** Aesthetes arrive at a small art gallery at the instants of a Poisson process of parameter λ. The kth arrival spends a time X_k in the first room and Y_k in the second room, and then leaves. The random variables X_k and Y_k are not independent, but (X_k, Y_k) is independent of (X_j, Y_j) for $j \neq k$. At time t, let R_1 and R_2 be the number of aesthetes in the respective rooms. Show that R_1 and R_2 are independent Poisson random variables.

(6) **Exercise** Find $\operatorname{cov}(N(s), N(t))$.

8.17 Example: von Neumann's exponential variable

Let the sequence of random variables $X_1, X_2, X_3, \ldots$ be independent and identically distributed with density f and distribution F. Define the random variable R by

(1) $$R = \min\{n: X_1 \geq X_2 \geq \ldots \geq X_{n-1} < X_n\}.$$

(a) Show that $\mathbf{P}(R = r) = (r-1)/r!$, and that

$$\mathbf{P}(X_R \leq x) = \exp(1 - F(x)) - e(1 - F(x)).$$

(b) Now let X_n be uniformly distributed on $(0, 1)$ for all n. Show that

(2) $$\mathbf{P}(X_1 \leq x; R = r) = \frac{x^{r-1}}{(r-1)!} - \frac{x^r}{r!}.$$

Deduce that

(3) $$\mathbf{P}(X_1 \leq x | R \text{ is even}) = \frac{1 - e^{-x}}{1 - e^{-1}}.$$

Finally, define a random variable V as follows. A sequence $X_1, X_2, \ldots, X_R$ is a 'run'; it is 'odd' if R is odd, otherwise it is 'even'. Generate runs until the first even run, and then let V equal the number N of odd runs plus X_1 in the even run. Show that V has density e^{-v} for $v > 0$.

Solution (a) Let $X_{(1)} \leqslant X_{(2)} \leqslant \ldots \leqslant X_{(r)}$ be the order statistics of $(X_1, \ldots, X_r)$. By symmetry $(X_1, \ldots, X_r)$ is equally likely to be any one of the $r!$ permutations of $(X_{(1)}, \ldots, X_{(r)})$. For the $r - 1$ permutations of the form:

$$(X_{(r)}, \ldots, X_{(k+1)}, X_{(k-1)}, \ldots, X_{(1)}, X_{(k)}), \quad 2 \leqslant k \leqslant r,$$

the event $R = r$ occurs, and for no others. Hence

$$\mathbf{P}(R = r) = \frac{r-1}{r!}.$$

The above remarks also show that

$$\begin{aligned}
\mathbf{P}(X_R \leqslant x) &= \sum_{r=2}^{\infty} \frac{1}{r!} \sum_{k=2}^{r} \mathbf{P}(X_{(k)} \leqslant x) \\
&= \sum_{r=2}^{\infty} \frac{1}{r!} \sum_{k=2}^{r} \sum_{j=k}^{r} \binom{r}{j} (F(x))^j (1 - F(x))^{r-j} \quad \text{by (8.7.12)} \\
&= \sum_{r=2}^{\infty} \frac{1}{r!} (rF(x) - 1 + (1 - F(x))^r) = e(F(x) - 1) + \exp(1 - F(x)),
\end{aligned}$$

on summing the series. It is easy to check that this is continuous and non-decreasing as x increases, and differentiation gives the density of X_R:

$$f_{X_R}(x) = ef(x)(1 - \exp(1 - F(x))).$$

(b) Now observe that the event $\{R > r\} \cap \{X_1 \leqslant x\}$ occurs if and only if $X_k \leqslant x$ for $1 \leqslant k \leqslant r$ (which has probability $(F(x))^r$, and $X_1 \geqslant X_2 \geqslant X_3 \geqslant \ldots \geqslant X_r$ (which has probability $1/r!$). Hence, when X_k is uniform on $(0, 1)$,

$$\mathbf{P}(R > r; X_1 \leqslant x) = \frac{x^r}{r!}$$

and so

$$\mathbf{P}(R = r; X_1 \leqslant x) = \frac{x^{r-1}}{(r-1)!} - \frac{x^r}{r!}.$$

Now summing over even values of R, we have

$$\mathbf{P}(X_1 \leqslant x, R \text{ is even}) = 1 - e^{-x},$$

and hence

$$\mathbf{P}(X_1 \leqslant x | R \text{ is even}) = (1 - e^{-x})/(1 - e^{-1}).$$

This shows that $\mathbf{P}(R \text{ is even}) = 1 - e^{-1}$, and so by independence of runs, N is a geometric random variable, with mass function

$$\mathbf{P}(N = n) = e^{-n}(1 - e^{-1}), \quad \text{for } n > 0.$$

Finally, let us denote X_1 in the even run by X_0. Then $\mathbf{P}(X_0 > x) =$

$1-(1-e^{-x})/(1-e^{-1})$, from (3). Hence

(4)
$$\begin{aligned}\mathbf{P}(V>v) &= \mathbf{P}(N \geqslant [v]+1) + \mathbf{P}(N=[v];\, X_0 > v-[v]) \\ &= e^{-[v]-1} + ((1-e^{-1})e^{-[v]})\left(1-\frac{1-e^{-v+[v]}}{1-e^{-1}}\right) \\ &= e^{-v} \quad \text{for } 0<v<\infty.\end{aligned}$$

Thus V is exponentially distributed.

Remark This method of generating exponential random variables from uniform ones was devised by von Neumann in 1951. Notice that it is computationally economical, in that it is necessary to store only the number of odd runs to date, and the first X_1 in the run in progress. Also the expected number of uniform random variables used for each exponential random variable is small. Since the original result, the method has been extended to generate other continuous random variables from uniform r.v.s.

(5) **Exercise** What is the density of X_{R-1}?

(6) **Exercise: Bad luck** As above $X_1, X_2, \ldots$ are independent and identically distributed with density f and distribution F. Define $T = \min\{n: X_n > X_1\}$. Find $\mathbf{P}(T=n)$ and show that T has infinite expectation. Show that X_T has distribution $F_{X_T}(x) = F(x) + (1-F(x))\log(1-F(x))$.

(7) **Exercise** Explain why the above exercise is entitled 'Bad luck'.

(8) **Exercise** Use the result of Exercise 6 to show that when X_1 has an exponential distribution, X_T has a gamma distribution. Why is this obvious without going through the analysis of Exercise 6?

PROBLEMS

1 When is $f(x, y) = xy + ax + by + 1$ a joint density function on $0 \leqslant x, y \leqslant 1$? Can it be the joint density of random variables X and Y which are independent?

2 Find $\text{cov}(X, Y)$ for the joint density of Problem 1.

3 Let X and Y have joint density $f(x, y) = c\exp(-x-y)$ for $x>0,\ y>0$. Find: (a) c; (b) $\mathbf{P}(X+Y>1)$; (c) $\mathbf{P}(X<Y)$.

4 Let X and Y have joint density $f(x, y) = g(x+y)$ for $x \geqslant 0,\ y \geqslant 0$. Find the density of $Z = X + Y$.

5 Let X and Y have joint density $f = c(1+x^2+y^2)^{-\frac{3}{2}}$ for all x and y. (a) What is c? (b) Find the marginal density of X.

6 Let X and Y have the joint density of Problem 5, and define $W = X^2 + Y^2$, $Z = Y/X$. Show that W and Z are independent.

7 Let U and V be independently and uniformly distributed on $(0, 1)$. Find the joint density of $X = U^{\frac{1}{2}}/(U^{\frac{1}{2}} + V^{\frac{1}{2}})$ and $Y = U^{\frac{1}{2}} + V^{\frac{1}{2}}$. By considering $\mathbf{P}(X \leqslant x | Y \leqslant 1)$, devise a rejection sampling procedure for simulating a random variable with density $6x(1-x)$; $0<x<1$.

8 Let U_1, U_2 and U_3 be independently and uniformly distributed on $(0, 1)$. Show that the density of $U_{(2)}$ is $6x(1-x)$; $0<x<1$.

9 Let U_1, U_2, U_3 and U_4 be independently and uniformly distributed on (0, 1). What is the density of $X = \log(U_1U_1)/\log(U_1U_2U_3U_4)$?

10 Let X and Y be independent normal random variables, and set $U = X + Y$, $V = X - Y$. Show that U and V are independent if and only if $\text{var}(X) = \text{var}(Y)$.

11 **Simulation using bivariate rejection** Let U and V be independent and uniform on (0, 1). Define the random variables

$$Z = (2U - 1)^2 + (2V - 1)^2,$$
$$X = (2U - 1)(2Z^{-1}\log Z^{-1})^{\frac{1}{2}}$$

and

$$Y = (2V - 1)(2Z^{-1}\log Z^{-1})^{\frac{1}{2}}.$$

Show that the conditional joint density of X and Y given $Z < 1$, is $(2\pi)^{-1}\exp(-\frac{1}{2}(x^2 + y^2))$. Explain how this provides a method for simulating normal random variables.

12 Let X and Y be independent exponential random variables with respective parameters λ and μ. Find $\mathbf{P}(\max\{X, Y\} \leq aX)$ for $a > 0$.

13 Let X and Y have joint density $f = ye^{-y(x+1)}$, $0 \leq x, y < \infty$.
What is the conditional density of X given Y?

14 A spherical melon has radius 1. Three gravid insects alight independently (for oviposition) at A, B and C, where A, B and C are uniformly distributed on the surface of the melon. For any two insects, if the distance between them (along a great circle of the melon) is less than $\pi/2$, then they detect each other's presence and will both fly off to seek an unoccupied melon. Show that the probability that exactly one insect is left in possession of the melon is $3(\pi - 1)/4\pi$, and that the probability that all three remain on the melon is $1/4\pi$.

15 Let X and Y have joint density $f(x, y) = c\sin(x + y)$, $0 < x, y < \pi/2$.
Show that $c = \frac{1}{2}$; $\text{cov}(X, Y) = \frac{1}{2} - \frac{1}{16}\pi^2$, and $\rho(X, Y) = (8 - \pi^2)/(\pi^2 + 8\pi - 32)$.

16 Let X and Y be exponential with parameters λ and μ respectively. Now define $U = X \wedge Y$, and $V = X \vee Y$. Find $\mathbf{P}(U = X)$, and show that U and $V - U$ are independent.

17 Let $(X_i;\ i \geq 1)$ be independent with the uniform density on $(-1, 1)$. Let the density of $\sum_{i=1}^n X_i$ be $f_n(x)$. Show that

$$f_n(x) = \frac{1}{2}\int_{x-1}^{x+1} f_{n-1}(u)\,du \quad \text{for } n \geq 2,$$

and deduce that for any integer k, the density $f_n(x)$ is a polynomial in x for $x \in [k, k + 1)$.

18 Let X and Y have the bivariate normal density of Example 8.4.3.

(a) If $\sigma = \tau$, what is $\mathbf{E}(X|X + Y)$?
(b) If $\sigma \neq \tau$, what are $\mathbf{E}(X|X + Y)$ and $\mathbf{E}(Y|X + Y)$?

19 Let $(X_n;\ n \geq 1)$ be independent and uniformly distributed on (0, 1). Define

$$T = \min\left\{n\colon \sum_{i=1}^{n} X_i > 1\right\}.$$

Show that $\mathbf{P}(T \geq j + 1) = 1/j!$ for $j \geq 1$. Deduce that $\mathbf{E}(T) = e$.

20 Let (X_1, X_2, X_3) be independent and uniformly distributed on (0, 1). What is the probability that the lengths X_1, X_2, X_3 can form a triangle?

21 Let $(U_i;\ i \geq 1)$ be independently and uniformly distributed on (0, 1), and define $M_n = \max\{U_1, \ldots, U_n\}$. Show that, as $n \to \infty$, the distribution of $Z_n = n(1 - M_n)$ converges to an exponential distribution.

22 Let $(X_i;\ i \geq 1)$ be independent exponential random variables each with parameter μ. Let N be independent of the X_i having mass function $f_N(n) = (1-p)p^{n-1};\ n \geq 1$. What is the density of $Y = \sum_{i=1}^{N} X_i$?

23 Let $N(t)$ be a Poisson process, $C(t)$ its current life at t, and X_1 the time of the first event. Show that $\operatorname{cov}(X_1, C(t)) = \frac{1}{2}t^2 e^{-\lambda t}$.

24 **Simulating gamma** Let U and X be independent where U is uniform on $(0, 1)$ and X is exponential with parameter $\alpha^{-1} \leq 1$. Show that the density of X conditional on

$$\left(\frac{eX}{\alpha}\right)^{\alpha-1} \exp\left(-\left(\frac{\alpha-1}{\alpha}\right)X\right) \geq U$$

is $x^{\alpha-1}e^{-x}/\Gamma(\alpha)$. Why is this of value?

25 (a) Let X be exponential with parameter 1. Show that X/λ is exponential with parameter λ.
(b) Let $(X_i;\ 1 \leq i \leq n)$ be independent and exponential with parameter 1. Use the lack-of-memory property of the exponential density to show that $\max\{X_1, \ldots, X_n\}$ has the same distribution as $X_1 + X_2/2 + \ldots + X_n/n$.

26 Let X_1, X_2, X_3 and X_4 be independent standard normal random variables. Show that $W = X_1/X_2$ has the Cauchy density, and $Z = |X_1X_2 + X_3X_4|$ has an exponential density.

27 Let X and Y be independent Poisson random variables each with parameter n. Show that as $n \to \infty$, $\mathbf{P}(X - Y \leq \sqrt{2n}\,x) \to \Phi(x)$.

28 Let $(U_i;\ i \geq 1)$ be a collection of independent random variables each uniform on $(0, 1)$. Let X have mass function $f_X(x) = (e-1)e^{-x};\ x \geq 1$ and let Y have mass function $f_Y(y) = 1/(e-1)y!,\ y \geq 1$. Show that $Z = X - \max\{U_1, \ldots, U_Y\}$ is exponential. (Assume X and Y are independent of each other and of the U_i).

29 Let X and Y be gamma with parameters $(\alpha, 1)$ and $(\beta, 1)$ respectively. Find the conditional density of X given $X + Y = z$.

30 Let X and Y be independent standard normal random variables. Show that the pair X and Z, where $Z = \rho X + (1-\rho^2)^{\frac{1}{2}}Y$, $|\rho| \leq 1$, has a standard bivariate normal density.

31 Let X and Y have joint moment generating function $M(s, t)$, and define $K(s, t) = \log M(s,t)$. Show that $K_s(0, 0) = \mathbf{E}(X)$, $K_{ss}(0, 0) = \operatorname{var}(X)$ and $K_{st}(0, 0) = \operatorname{cov}(X, Y)$.

32 A sorcerer has hidden a ring in one of an infinite number of boxes numbered $\ldots, -2, -1, 0, 1, 2, 3, \ldots$ You only have time to look in eleven boxes. The sorcerer gives you a hint. He tosses 100 fair coins and counts the number of heads. He does not tell you this number, nor does he tell you the number of the box with the ring in it, but he tells you the sum of these two numbers.
(a) If the sum is 75, which eleven boxes should you look in?
(b) Give an approximation to the probability of finding the ring.

$$\left(\frac{1}{\sqrt{2\pi}}\int_0^{\frac{11}{10}} e^{-\frac{u^2}{2}}\,du = 0.36.\right)$$

33 **Multivariate normal density** Let $(Y_1, \ldots, Y_n)$ be independent, each having the $N(0, 1)$ density. If

$$X_i = \sum_j a_{ij}Y_j + b_i \quad \text{for } 1 \leq i, j \leq n,$$

then $(X_1, \ldots, X_n)$ are said to have multivariate normal density. Find the joint m.g.f. $M_n(t_1, \ldots, t_n) = \mathbf{E}(\exp(\sum_{i=1}^{n} t_iX_i))$. Deduce that the following three statements are equivalent:
(a) the random variables $(X_1, \ldots, X_n)$ are independent.
(b) $(X_1, \ldots, X_n)$ are pairwise independent.
(c) $\operatorname{cov}(X_i, X_j) = 0 \quad$ for $1 \leq i \neq j \leq n$.

34 A sequence of random variables $X_1, X_2, \ldots$ is said to obey the Central Limit Theorem (CLT) if and only if the distribution of $(S_n - \mathbf{E}(S_n))/\sqrt{\text{var}(S_n)}$ tends to the standard normal distribution, where $S_n = \sum_{i=1}^{n} X_i$.

State sufficient conditions on (X_n) for the sequence to obey the CLT and say which of your conditions are necessary.

Let $(U_n(\lambda_n))$ be a sequence of independent random variables having the Poisson distribution with non-zero means (λ_n). In each of the following cases determine whether the sequence (X_n) obeys the CLT:

(i) $X_n = U_n(1)$;
(ii) $X_n = U_n(1) + n$
(iii) $X_n = U_n(\frac{1}{2})!$;
(iv) $X_n = U_{2n}(1)/(1 + U_{2n-1}(1))$;
(v) $X_n = U_n(n)$.

35 Let X_1 and X_2 be independent with the same density $f(x)$. Let U be independent of both, and uniformly distributed on $(0, 1)$. Let $Y = U(X_1 + X_2)$. Find $f(x)$ such that Y can also have density $f(x)$.

36 **Molecules** A molecule M has velocity $\boldsymbol{v} = (v_1, v_2, v_3)$ in Cartesian corrdinates. Suppose that v_1, v_2 and v_3 have joint density

$$f(x, y, z) = (2\pi\sigma^2)^{-\frac{3}{2}} \exp\left(-\frac{1}{2\sigma^2}(x^2 + y^2 + z^2)\right).$$

Show that the density of the magnitude $|\boldsymbol{v}|$ of $\boldsymbol{v}$ is

$$f(w) = \left(\frac{2}{\pi}\right)^{\frac{1}{2}} \sigma^{-3} w^2 \exp\left(-\frac{1}{2\sigma^2} w^2\right), \quad w > 0.$$

37 Let C be a circle radius r with centre O. Choose two points P and Q independently at random in C. Show that the probability that the triangle OPQ contains an obtuse angle is $\frac{3}{4}$.
[*Note*: no integration is required.]

38 Given a fixed line AB, a point C is picked at random such that $\max\{AC, BC\} \leqslant AB$. Show that the probability that the triangle ABC contains an obtuse angle is

$$\frac{1}{8}\pi\left(\frac{1}{3}\pi - \frac{1}{4}\sqrt{3}\right)^{-1}.$$

[*Note*: no integration is required. This is a version of a problem given by Lewis Carroll. To combat insomnia, he solved mathematical problems in his head; this one was solved on the night of 20 January 1884. He collected a number of these mental exercises in a book entitled *Pillow problems* (Macmillan, 1895).]

9
Markov chains

9.1 The Markov property

In previous chapters we have found it useful and interesting to consider sequences of independent random variables. However, many observed sequences in the natural world are patently not independent. Consider for example the air temperature outside your window on successive days, or the sequence of morning fixes of the price of gold. It is desirable and necessary to consider more general types of sequences of random variables.

After some thought, you may agree that for many such systems it is reasonable to suppose that, if we know exactly the state of the system today, then its state tomorrow should not further depend on its state yesterday (or on any previous state). This informal (and vague) preamble leads to the following formal (and precise) statement of the Markov property for a sequence of random variables.

(1) ***Definition*** Let $X = (X_n;\ n \geqslant 0)$ be a sequence of random variables taking values in a countable set S, called the *state space*. If, for all $n \geqslant 0$ and all possible values of $i, k, k_0, \ldots, k_{n-1}$, we have

(2) $$\mathbf{P}(X_{n+1} = k | X_0 = k_0, \ldots, X_n = i) = \mathbf{P}(X_{n+1} = k | X_n = i) = \mathbf{P}(X_1 = k | X_0 = i)$$

then X is said to be a *Markov chain*, or to have the *Markov property*. We write

$$p_{ik} = \mathbf{P}(X_1 = k | X_0 = i)$$

where $(p_{ik};\ i \in S,\ k \in S)$ are known as the transition probabilities of the chain. ▲

Sometimes we write $p_{i,k}$ for p_{ik}, and you are warned that some books use p_{ki} to denote p_{ik}.

Another popular rough and ready way of interpreting the formal condition (2), is to say that for a Markov chain the future is conditionally independent of the past, given the present.

Notice that in some applications it is more natural to start the clock at $n = 1$, so the chain is $X = (X_n;\ n \geqslant 1)$. Very occasionally it is convenient to suppose the chain extends in both directions so that $X = (X_n;\ -\infty < n < \infty)$.

The state space S is often a subset of the integers $\mathbb{Z}$, or a subset of the set of ordered pairs of integers $\mathbb{Z}^2$. Markov chains may take values in some countable set which happens not to be a subset of the integers. However, this set can immediately be placed in one–one correspondence with some appropriate subset of the integers, and the states relabelled accordingly.

(3) **Example: Simple random walk** Let $(S_n;\ n \geq 0)$ be a simple random walk. Since the steps $(S_{n+1} - S_n;\ n \geq 0)$ are independent, the sequence S_n clearly has the Markov property, and the transition probabilities are given by

$$p_{ik} = \mathbf{P}(S_{n+1} = k | S_n = i) = \begin{cases} p & \text{if } k = i + 1 \\ q & \text{if } k = i - 1 \\ 0 & \text{otherwise.} \end{cases}$$

The state space S is the set of integers $\mathbb{Z}$. ●

(4) **Example: Branching process** Let Z_n be the size of the nth generation in an ordinary branching process. Since family sizes are independent, $Z = (Z_n;\ n \geq 0)$ is a Markov chain. The transition probabilities are given by

$$p_{ik} = \mathbf{P}(Z_{n+1} = k | Z_n = i) = \mathbf{P}\left[\sum_{r=1}^{i} Y_r = k\right]$$

where $Y_1, \ldots, Y_i$ are the i families of the nth generation given that $Z_n = i$. The state space is the set of non-negative integers $\mathbb{Z}^+$. ●

When S is a finite set, X is known as a finite Markov chain. Until further notice we consider finite chains (unless it is specifically stated otherwise) and write $|S| = d$.

(5) **Example: Information source** A basic concern of telecommunications engineers is the transmission of signals along a channel. Signals arise at a *source*, and to devise efficient methods of communication it is necessary to have models for such sources. In general it is supposed that the source produces a sequence of symbols randomly drawn from a finite alphabet A. By numbering the symbols from 1 to $|A|$, the output becomes a sequence of random variables $(X_n;\ n \geq 1)$, called a message. Various assumptions can be made about the output of sources, but a common and profitable assumption is that they have the Markov property. In this case the output is a finite Markov chain, and the source is called a simple Markov source. ●

Having formally defined a Markov chain X, we emphasize that there are many ways of presenting the idea of a Markov chain to the mind's eye. You should choose the one which best suits the context of the problem, and your own psyche. For example:

(i) a particle performs a random walk on the vertices of a graph. The distribution of its next step depends on where it is, but not on how it got there.

(ii) a system may be in any one of d states. The distribution of its next state depends on its current state, but not on its previous states.

Because of this imagery, we talk equivalently of chains visiting k, being at k, taking the value k, and so on. Whatever the choice of concept, the notation is always essentially that of Definition 1, but (to avoid repetitive strain injury) some abbreviations of notation are widespread. Thus we commonly write

$$\mathbf{P}(X_{n+1} = k|X_0 = k_0, \ldots, X_{n-1} = k_{n-1}, X_n = i) = \mathbf{P}(X_{n+1} = k|X_0, \ldots, X_n).$$

If we wish to stress or specify the initial value of X, then we write $\mathbf{P}(X_{n+1} = k|X_0 = k_0, \ldots, X_n)$, and so on.

Note that the Markov property as defined in (1) is equivalent to each of the following properties, which it is occasionally convenient to take as definitive. First:

(6) $$\mathbf{P}(X_{n+m} = k|X_0, \ldots, X_n) = \mathbf{P}(X_{n+m} = k|X_n)$$

for any positive m and n. Second:

(7) $$\mathbf{P}(X_{n_r} = k|X_{n_1}, \ldots, X_{n_{r-1}}) = \mathbf{P}(X_{n_r} = k|X_{n_{r-1}})$$

for any $n_1 < n_2 < \ldots < n_r$. And third:

(8) $$\mathbf{P}(X_1 = k_1, \ldots, X_{r-1} = k_{r-1}, X_{r+1} = k_{r+1}, \ldots, X_n = k_n|X_r = k_r)$$
$$= \mathbf{P}(X_1 = k_1, \ldots, X_{r-1} = k_{r-1}|X_r = k_r)\mathbf{P}(X_{r+1} = k_{r+1}, \ldots, X_n = k_n|X_r = k_r).$$

You are asked to prove the equivalence of Definition 1 and (6), (7) and (8) in Problem 6. Notice that (8) expresses in a precise form our previously expressed rough idea that given the present state of a Markov chain, its future is independent of its past. Finally it should be noted that the Markov property is preserved by some operations, but not by others, as the following examples show.

(9) Example: Sampling Let X be a Markov chain. Show that the sequence $Y_n = X_{2n}$; $n \geqslant 0$, is a Markov chain.

Solution Since X is a Markov chain we can argue as follows

$$\begin{aligned}\mathbf{P}(Y_{n+1} = k|Y_0, \ldots, Y_n = i) &= \mathbf{P}(X_{2n+2} = k|X_0, \ldots, X_{2n} = i)\\ &= \mathbf{P}(X_{2n+2} = k|X_{2n} = i) \quad \text{by (7)}\\ &= \mathbf{P}(Y_{n+1} = k|Y_n = i).\end{aligned}$$

So Y is a Markov chain. It is said to be *imbedded* in X. ●

(10) Example If X is a Markov chain with state space S_X, show that

$$Y_n = (X_n, X_{n+1}); \quad n \geqslant 0,$$

is a Markov chain. What are its transition probabilities?

Solution The state space of Y is a collection of ordered pairs of the states of X, that is to say

$$S_Y = \{(s_1, s_2): s_1 \in S_X, s_2 \in S_X\}.$$

Now

$$\begin{aligned}\mathbf{P}(Y_{n+1} = (j, k)|Y_0, \ldots, Y_n) &= \mathbf{P}(X_{n+2} = k, X_{n+1} = j|X_0, \ldots, X_{n+1})\\ &= \mathbf{P}(X_{n+2} = k, X_{n+1} = j|X_{n+1}, X_n) \quad \text{since } X \text{ is Markov,}\\ &= \mathbf{P}(Y_{n+1} = (j, k)|Y_n).\end{aligned}$$

So Y is Markov. Also

(11) $$\mathbf{P}(Y_{n+1} = (k, l)|Y_n = (i, j)) = \mathbf{P}(X_{n+2} = l|X_n = k)\delta_{jk} = p_{kl}\delta_{kj}$$

where

$$\delta_{kj} = \begin{cases} 1 & \text{if } k = j \\ 0 & \text{otherwise} \end{cases}$$

is the usual Kronecker delta. ●

(12) Example Let X be a Markov chain. Show that $Y_n = |X_n|$; $n \geq 0$, is not necessarily a Markov chain.

Solution Let X have state space $S = \{-1, 0, 1\}$, and transition probabilities zero except for

$$p_{-1,0} = \frac{1}{2}, \; p_{-1,1} = \frac{1}{2}, \; p_{0,-1} = 1, \; p_{1,0} = 1.$$

Then

$$\mathbf{P}(Y_{n+1} = 1|Y_n = 1, Y_{n-1} = 1) = \mathbf{P}(X_{n+1} = 1|X_n = 1, X_{n-1} = -1) = 0.$$

But $\mathbf{P}(Y_{n+1} = 1|Y_n = 1) = \mathbf{P}(Y_{n+1} = 1|X_n \in \{-1, 1\})$, which is not necessarily zero. So Y is not Markov. ●

Notice that the states -1 and $+1$ for X_n produce one state $+1$ for Y_n; they are said to be *lumped* together. The example shows that lumping states together can destroy the Markov property. Conversely, given a sequence Y_n which is *not* a Markov chain, it is sometimes possible to construct a Markov chain involving Y_n by enlarging the state space.

(13) Example A machine can be in one of two states, working (denoted by s_0), or repair (denoted by s_1). Each day, if working, it may break down with probability α independently of other days. It takes r days to repair, where $r > 1$. Now if X_n is the state of the machine on the nth day, this is not a Markov chain. To see this note that

$$\mathbf{P}(X_{n+1} = s_0|X_n = s_1, X_{n-1} = s_0) = 0$$

but

$$\mathbf{P}(X_{n+1} = s_0|X_n = X_{n-1} = \ldots = X_{n-r+1} = s_1) = 1.$$

However, suppose we now let the state space be $S = \{s_0, s_1, \ldots, s_r\}$ where $X_n = s_i$ if the machine has been in repair for i days. Then

$$\mathbf{P}(X_{n+1} = s_{i+1}|X_n = s_i, X_{n-1}, \ldots) = 1, \quad \text{if } 1 \leq i \leq r - 1,$$
$$\mathbf{P}(X_{n+1} = s_0|X_n = s_r, \ldots) = 1,$$

and so on. It is easy to see that X_n now is a Markov chain. ●

9.2 Transition probabilities

Recall that X is a Markov chain with state space S, where $|S| = d$. The transition probabilities p_{ik} are given by

(1) $$p_{ik} = \mathbf{P}(X_{n+1} = k | X_n = i) \quad \text{for } n \geqslant 0.$$

The $d \times d$ matrix (p_{ij}) of transition probabilities is called the transition matrix and is denoted by $\boldsymbol{P}$. Let us first record two simple but important facts about $\boldsymbol{P}$. Because $(p_{ik};\ k \in S)$ is a conditional mass function we have

(2) $$p_{ik} \geqslant 0 \quad \text{for all} \quad i \quad \text{and} \quad k;$$

and

(3) $$\sum_{k \in S} p_{ik} = 1.$$

Any matrix $\boldsymbol{P}$ satisfying (2) and (3) is called *stochastic*. We remark that if in addition

(4) $$\sum_{i \in S} p_{ik} = 1$$

then $\boldsymbol{P}$ is *doubly stochastic*. Also, if (3) is replaced by the condition

(5) $$\sum_{k \in S} p_{ik} \leqslant 1$$

then a matrix satisfying (2) and (5) is called *substochastic*. For example, the simple random walk of Example 9.1.3 is doubly stochastic. If $p_{ij} > 0$ for all i and j, then $\boldsymbol{P}$ is called *positive*.

Now, given that $X_0 = i$, the distribution of X_n is denoted by

$$p_{ik}(n) = \mathbf{P}(X_n = k | X_0 = i) = \mathbf{P}(X_{n+m} = k | X_m = i)$$

because of (9.1.2). Trivially, of course,

(6) $$\sum_{k \in S} p_{ik}(n) = 1.$$

These probabilities are called the n-step transition probabilities, and they describe the random evolution of the chain.

Note that $p_{ij}(n)$ is a function of three variables, the two states i, j and the time n. In more complicated expressions involving several such probabilities you should use the symbols i, j, k, l to denote states and the symbols m, n, r, t to denote time (possibly with suffices).

Some simple special cases illustrate these notions.

(7) **Example (9.1.3) continued: Simple random walk** Recall that $(S_n;\ n \geqslant 0)$ are the successive values of a simple random walk. If $S_n = k$ and $S_0 = i$, then from Theorem 5.6.4 we have

(8) $$p_{ik}(n) = \begin{cases} \dbinom{n}{\frac{1}{2}(n+k-i)} p^{\frac{1}{2}(n+k-i)}(1-p)^{\frac{1}{2}(n-k+i)} & \text{if } n+k-i \text{ is even} \\ 0 & \text{otherwise.} \end{cases}$$

[Note that this chain has infinite state space.] ■

(9) **Example: Survival** A traffic sign stands in a vulnerable position. Each day, independently of other days, it may be demolished by a careless motorist with probability q. In this case the city engineer replaces it with a new one at the end of the day. At the end of day n, let X_n denote the number of days since the sign in position was newly installed. Show that X_n is a Markov chain, and find p_{ik} and $p_{ik}(n)$.

[Note that this chain has infinite state space.]

Solution By construction

$$X_{n+1} = \begin{cases} X_n + 1 & \text{with probability } 1 - q = p \\ 0 & \text{with probability } q. \end{cases}$$

Since the choice of outcomes is independent of previous days, X_n is a Markov chain, and

$$p_{ik} = \begin{cases} p & \text{if } k = i + 1 \\ q & \text{if } k = 0 \\ 0 & \text{otherwise.} \end{cases}$$

For the n-step transition probabilities we note that either the sign survives for all n days, or has been struck in the meantime. Hence

(10) $$p_{ik}(n) = p^n \quad \text{if } k = i + n,$$

and

(11) $$p_{ik}(n) = qp^k \quad \text{if } 0 \leqslant k \leqslant n - 1.$$ ●

Returning to the general case, we examine the relationship between p_{ik} and $p_{ik}(n)$. It is a remarkable and important consequence of the Markov property (9.1.1) that the random evolution of the chain is completely determined by p_{ik}, as the following theorem shows.

(12) **Theorem: Chapman–Kolmogorov equations** Let X have transition matrix $\mathbf{P}$. For any i and k in S, and any positive m and n, we have

(13) $$p_{ik}(m + n) = \sum_{j \in S} p_{ij}(m) p_{jk}(n)$$

and also

(14) $$p_{ik}(n + 1) = \sum_{j_1 \in S} \ldots \sum_{j_n \in S} p_{ij_1} p_{j_1 j_2} \ldots p_{j_n k}.$$

Proof Recall that if $(A_j;\ j \leqslant d)$ is a collection of disjoint events such that $\bigcup_1^d A_j = \Omega$, then for any events B and C

$$\mathbf{P}(B|C) = \sum_{j=1}^{d} \mathbf{P}(B \cap A_j|C).$$

Hence, setting $A_j = \{X_m = j\}$, we have

$$p_{ik}(m+n) = \sum_{j\in S} \mathbf{P}(X_{m+n} = k, X_m = j | X_0 = i)$$

$$= \sum_{j\in S} \mathbf{P}(X_{m+n} = k | X_m = j, X_0 = i)\mathbf{P}(X_m = j | X_0 = i),$$

by conditional probability,

$$= \sum_{j\in S} \mathbf{P}(X_{m+n} = k | X_m = j)\mathbf{P}(X_m = j | X_0 = i),$$

by the Markov property,

$$= \sum_{j\in S} p_{ij}(m) p_{jk}(n).$$

Hence in particular

(15) $$p_{ik}(n+1) = \sum_{j_1\in S} p_{ij_1} p_{j_1k}(n)$$

$$= \sum_{j_1\in S}\sum_{j_2\in S} p_{ij_1} p_{j_1j_2} p_{j_2k}(n-1)$$

$$\vdots$$

$$= \sum_{j_1\in S} \cdots \sum_{j_n\in S} p_{ij_1} \cdots p_{j_nk}$$

by repeated application of (15).

An alternative proof of (14) is provided by the observation that the summation on the right hand side is the sum of the probabilities of all the distinct paths of n steps which lead from i to k. Since these are mutually exclusive, and one of them must be used to make the trip from i to k, the result follows. ■

The n-step transition probabilities $p_{ik}(n)$ tell us how the mass function of X_n depends on X_0. If X_0 itself has mass function

(16) $$\alpha_i = \mathbf{P}(X_0 = i)$$

and X_n has mass function

(17) $$\alpha_i^{(n)} = \mathbf{P}(X_n = i)$$

then, by conditional probability, they are related by

(18) $$\alpha_k^{(n)} = \sum_{i\in S} \alpha_i p_{ik}(n).$$

The probabilities $\alpha_i^{(n)}$ are sometimes called the *absolute probabilities* of X_n.

Now we notice that the d^2 n-step transition probabilities ($p_{ik}(n)$; $1 \le i \le d$, $1 \le k \le d$) can be regarded as a matrix $\boldsymbol{P}_n$, and the absolute probabilities ($\alpha_i^{(n)}$;

$1 \leqslant i \leqslant d$) as a row vector $\boldsymbol{\alpha}_n$. It follows from Theorem 12 and (18) that

(19)
$$\boldsymbol{P}_{m+n} = \boldsymbol{P}_m \boldsymbol{P}_n = \boldsymbol{P}^{m+n}$$

and $\boldsymbol{\alpha}_n = \boldsymbol{\alpha} \boldsymbol{P}^n$, where $\boldsymbol{\alpha} = (\alpha_1, \ldots, \alpha_d)$.

(20) Example: Two state chain The following simple but important example is very helpful in illustrating these and other ideas about Markov chains. Let X have state space $S = \{1, 2\}$ and transition matrix

$$\boldsymbol{P} = \begin{pmatrix} 1-\alpha & \alpha \\ \beta & 1-\beta \end{pmatrix}.$$

You can verify by induction that

$$(\alpha + \beta)\boldsymbol{P}^n = \begin{pmatrix} \beta & \alpha \\ \beta & \alpha \end{pmatrix} + (1 - \alpha - \beta)^n \begin{pmatrix} \alpha & -\alpha \\ -\beta & \beta \end{pmatrix}.$$

Hence, for example,

$$p_{12}(n) = \frac{\alpha}{\alpha + \beta} - \frac{\alpha}{\alpha + \beta}(1 - \alpha - \beta)^n.$$ ●

Descending once again from the general to the particular, we identify some special varieties of chain which have attractive properties that we will find very useful later on.

(21) Definition If for some $n_0 < \infty$ we have

$$p_{ij}(n_0) > 0 \quad \text{for all} \quad i \quad \text{and} \quad j,$$

then the chain is said to be *regular*. ▲

(22) Example Let X have transition probabilities

$$\boldsymbol{P} = \begin{pmatrix} 0 & 1 \\ \frac{1}{2} & \frac{1}{2} \end{pmatrix}.$$

Then $\boldsymbol{P}$ is not positive, but

$$\boldsymbol{P}^2 = \begin{pmatrix} \frac{1}{2} & \frac{1}{2} \\ \frac{1}{4} & \frac{3}{4} \end{pmatrix} > 0,$$

so $\boldsymbol{P}$ is regular. ●

Roughly speaking a chain is regular if there is a time such that, no matter where it started, the chain could be anywhere in S. Some chains satisfy the weaker condition that every state can be reached from every other state with non-zero probability. This is called *irreducibility*:

(23) Definition A chain X is *irreducible* if for each i and k in S there exists an $n_0 < \infty$ such that $p_{ik}(n_0) > 0$. ▲

(24) Example Let X have transition matrix

$$\boldsymbol{P} = \begin{pmatrix} 0 & 1 \\ 1 & 0 \end{pmatrix}.$$

Then

$$\boldsymbol{P}^{2n} = \begin{pmatrix} 1 & 0 \\ 0 & 1 \end{pmatrix}; \quad n \geqslant 0$$

and

$$\boldsymbol{P}^{2n+1} = \begin{pmatrix} 0 & 1 \\ 1 & 0 \end{pmatrix}; \quad n \geqslant 0.$$

Hence X is neither positive nor regular but it is irreducible. In fact it is said to be *periodic* with period 2, because $p_{ii}(n) > 0$ iff n is even. A state with no period greater than 1 is *aperiodic*. ●

(25) Example Let X and Y be independent regular Markov chains with transition matrices $\boldsymbol{P} = (p_{ik})$ and $\boldsymbol{Q} = (q_{ik})$ respectively. Show that $Z_n = (X_n, Y_n)$; $n \geqslant 0$, is a regular Markov chain.

Solution Using the independence of X and Y

$$\begin{aligned}\mathbf{P}(Z_n = (k, l) | Z_{n-1} = (i, j), Z_{n-2}, \ldots, Z_0) &= \mathbf{P}(X_n = k | X_{n-1} = i, X_{n-2}, \ldots, X_0) \\ &\quad \times \mathbf{P}(Y_n = l | Y_{n-1} = j, \ldots, Y_0) \\ &= p_{ik} q_{jl}\end{aligned}$$

since X and Y are Markov chains. Therefore Z is a Markov chain. Likewise Z has n-step transition probabilities $p_{ik}(n)q_{jl}(n)$.

Finally, since $\boldsymbol{P}$ and $\boldsymbol{Q}$ are regular, there exists n_0 and m_0 (both finite) such that $p_{ik}(m_0)$ and $q_{jl}(n_0)$ are both positive for all i, k and all j, l respectively. Hence $p_{ik}(m_0 n_0) q_{jl}(m_0 n_0) > 0$, for all i, j, k, l, and so Z is regular. ●

Note two further bits of jargon. A set C of states is called *closed* if $p_{ik} = 0$ for all $i \in C$, $k \notin C$. Furthermore, if C is closed and $|C| = 1$, then this state is called *absorbing*.

We conclude this section with two examples drawn from communication theory.

(26) Example: Entropy Let the random vector $\boldsymbol{X}_n = (X_0, \ldots, X_n)$ have joint mass function $f(x_0, \ldots, x_n)$. Then the *entropy* (also called *uncertainty*) of $\boldsymbol{X}_n$ is defined as

$$H(\boldsymbol{X}_n) = -\mathbf{E}[\log(f(X_0, \ldots, X_n))]$$

(with the convention that $0 \log 0 = 0$). Let $X_0, \ldots, X_n$ be the first $n+1$ values of a Markov chain with transition matrix $\boldsymbol{P}$ and initial mass function $\boldsymbol{\alpha}$. Show that in this case

$$H(\boldsymbol{X}_n) = -\mathbf{E}[\log(\alpha_{X_0})] - \sum_{r=1}^{n} \mathbf{E}[\log(p_{X_{r-1}X_r})]. \tag{27}$$

Solution Since X is a Markov chain $f(x_0, x_1, \ldots, x_n) = \alpha_{x_0}p_{x_0x_1} \ldots p_{x_{n-1}x_n}$. Hence

$$\mathbf{E}[\log(f(X_0, \ldots, X_n))]$$

$$= \sum_{x_0 \in S} \ldots \sum_{x_n \in S} \alpha_{x_0}p_{x_0x_1} \ldots p_{x_{n-1}x_n}(\log \alpha_{x_0} + \log p_{x_0x_1} + \ldots + \log p_{x_{n-1}x_n})$$

$$= \sum_{x_0} \alpha_{x_0} \log \alpha_{x_0} + \sum_{x_0, x_1} \alpha_{x_0}p_{x_0x_1} \log p_{x_0x_1} + \ldots + \sum_{x_{n-1}, x_n} \alpha_{x_{n-1}}p_{x_{n-1}x_n} \log p_{x_{n-1}x_n}$$

$$= \mathbf{E}[\log(\alpha_{X_0})] + \sum_{r=1}^{n} \mathbf{E}[\log p_{X_{r-1}X_r}]$$

as required, yielding (27). ●

(28) Example: Simple Markov source Let the random variable X and the random vector $\boldsymbol{Y}$ be jointly distributed, and denote the conditional mass function of X given $\boldsymbol{Y}$ by $f(x|\boldsymbol{y})$. Then the *conditional entropy* of X with respect to $\boldsymbol{Y}$ is defined to be

$$H(X|\boldsymbol{Y}) = -\mathbf{E}[\mathbf{E}(\log f(X|\boldsymbol{Y})|\boldsymbol{Y})]$$

$$= -\sum_{y}\sum_{x} f(x|\boldsymbol{y}) \log f(x|\boldsymbol{y}) \mathbf{P}(\boldsymbol{Y} = \boldsymbol{y}).$$

Let $X_0, \ldots, X_{n+1}$ be the output from the Markov source defined in Example 9.1.5. Show that

$$H(X_{n+1}|X_0, \ldots, X_n) = H(X_{n+1}|X_n).$$

Solution Let $\boldsymbol{Y}$ be $(X_0, \ldots, X_n)$. Then by the Markov property

$$f(x|\boldsymbol{y}) = \mathbf{P}(X_{n+1} = x|X_0 = y_0, \ldots, X_n = y_n) = \mathbf{P}(X_{n+1} = x|X_n = y_n)$$

$$= p_{y_nx}.$$

Hence

$$H(X_{n+1}|\boldsymbol{Y}) = -\sum_{y}\sum_{x} p_{y_nx} \log p_{y_nx} \mathbf{P}(X_0, = y_0, \ldots, X_n = y_n)$$

$$= -\sum_{y_n}\sum_{x} p_{y_nx} \log p_{y_nx} \mathbf{P}(X_n = y_n) = \mathbf{H}(X_{n+1}|X_n).$$ ●

9.3 First passage times

For any two states i and k of X, we shall often be interested in the time it takes for the chain to travel from i to k. This is not merely a natural interest, these quantities are also of theoretical and practical importance. For example, in the simple gambler's ruin problem the state 0 entails ruin, and in the simple branching process $X = 0$ entails extinction.

(1) **Definition** For a Markov chain X with $X_0 = i$:

(a) when $i \neq k$, the *first passage time* to k from i is defined to be

$$T_{ik} = \min\{n \geqslant 0: X_n = k | X_0 = i\};$$

the *mean first passage time* is

(2)
$$\mu_{ik} = \mathbf{E}(T_{ik}).$$

(b) when $i = k$, the *recurrence time* of i is defined to be

$$T_i = \min\{n > 0: X_n = i | X_0 = i\};$$

the *mean recurrence time* is

(3)
$$\mu_i = \mathbf{E}(T_i). \qquad \blacktriangle$$

Note the simple but important fact that the chain has not entered k by time n if and only if

(4)
$$T_{ik} > n.$$

(5) **Example** Let X have transition matrix

$$\boldsymbol{P} = \begin{pmatrix} \frac{1}{3} & \frac{2}{3} \\ \frac{1}{4} & \frac{3}{4} \end{pmatrix}.$$

Then given $X_0 = 1$, the chain enters 2 as soon as it leaves 1. Hence

$$\mathbf{P}(T_{12} = r) = \frac{2}{3}.\left(\frac{1}{3}\right)^{r-1}; \quad r \geqslant 1,$$

and

$$\mu_{12} = \sum_{r=1}^{\infty} r\frac{2}{3}\left(\frac{1}{3}\right)^{r-1} = \frac{3}{2}.$$

Likewise, first return to 1 at the rth step occurs after $r - 2$ consecutive visits to 2, so

$$\mathbf{P}(T_1 = r) = \begin{cases} \frac{1}{3} & ; \quad r = 1 \\ \frac{2}{3}\left(\frac{3}{4}\right)^{r-2}\frac{1}{4}; & r \geqslant 2. \end{cases}$$

Hence

$$\mu_1 = \frac{1}{3} + \sum_{r=2}^{\infty} r\frac{1}{6}\left(\frac{3}{4}\right)^{r-2} = \frac{11}{3}.$$

If we do not require the distribution of T_{12} or T_1, then a simpler procedure will suffice to find μ_{12} and μ_1, as follows. Conditioning on the first step of the chain, and assuming all the expectations exist, we find that

$$\mu_{12} = \frac{1}{3}\mathbf{E}(T_{12}|X_1 = 1) + \frac{2}{3}\mathbf{E}(T_{12}|X_1 = 2).$$

But, by the Markov property,

$$\mathbf{E}(T_{12}|X_1 = 1) = 1 + \mathbf{E}(T_{12}),$$

and obviously $\mathbf{E}(T_{12}|X_1 = 2) = 1$. Hence $\mu_{12} = 1 + \frac{1}{3}\mu_{12}$ as above. Likewise we find $\mu_{21} = 4$, and using conditional expectation again yields $\mu_1 = 1 + \frac{2}{3}\mu_{21} = \frac{11}{3}$ as before. ●

For a rather different type of behaviour consider the following.

(6) **Example** Let X have transition matrix

$$\boldsymbol{P} = \begin{pmatrix} \frac{1}{3} & \frac{1}{3} & \frac{1}{3} \\ \frac{1}{4} & \frac{3}{4} & 0 \\ 0 & 0 & 1 \end{pmatrix}.$$

Since $p_{33} = 1$, state 3 is absorbing, which is to say that upon entering 3 the chain never leaves it subsequently. Hence $T_{12} = r$ occurs when the first $r - 1$ visits to 1 are followed by a step to 2. Thus

$$\mathbf{P}(T_{12} = r) = \left(\frac{1}{3}\right)^r; \quad r \geqslant 1.$$

Hence

$$\mathbf{P}(T_{12} < \infty) = \sum_{r=1}^{\infty} \left(\frac{1}{3}\right)^r = \frac{2}{3}$$

and $\mu_{12} = \infty$. Likewise

$$\mathbf{P}(T_1 = r) = \begin{cases} \frac{1}{3} & ; \quad r = 1 \\ \frac{1}{3}\left(\frac{3}{4}\right)^{r-2} \cdot \frac{1}{4}; & r \geqslant 2 \end{cases}$$

and so $\mathbf{P}(T_1 < \infty) = \frac{2}{3}$, and $\mu_1 = \infty$. ●

These examples demonstrate that the properties of recurrence and first passage times depend very strongly on the nature of the transition matrix $\boldsymbol{P}$. In fact we are going to show that, for any finite regular chain, both μ_k and μ_{ik} are finite (with finite expectation) for all i and k. First we need to clear the ground a little. Since we are only considering finite chains with $|S| = d$, we can without loss of generality set $k = d$. (If you like mnemonics you can think of d as the destination of the chain.) Also, as we are only interested in the progress of the chain until it arrives at d, it is natural to focus attention on the probabilities

(7) $$r_{ik}(n) = \mathbf{P}(X_n = k, n < T_{id}|X_0 = i), \quad i \neq d \neq k.$$

These are the transition probabilities of the chain before entering d, and we denote the array $(r_{ik}(n))$ by $\boldsymbol{R}_n$.

By definition, for one step,

(8) $$r_{ik}(1) = p_{ik} \quad \text{for} \quad i \neq d \neq k.$$

For $n > 1$, the n-step d-avoiding probabilities are given by the following.

(9) **Theorem** For $i \neq d \neq k$

(10) $$r_{ik}(n) = \sum_{j_1 \neq d} \sum_{j_2 \neq d} \cdots \sum_{j_{n-1} \neq d} p_{ij_1} p_{j_1 j_2} \cdots p_{j_{n-1}k},$$

or in matrix form $\boldsymbol{R}_n = \boldsymbol{R}_1^n$.

Proof We use the idea of paths. Every distinct path of the chain which goes from i to k in n steps and does not enter d, is of the form $i, j_1, j_2, \ldots, j_{n-1}, k$ where $j_r \in S\backslash d$ for $1 \leq r \leq n-1$. Such a path has probability $p_{ij_1} \ldots p_{j_{n-1}k}$, and one of them is used, so $r_{ik}(n)$ is just the sum of all these probabilities as given on the right hand side of (10). ■

Corollary For any state i of a regular chain

(11) $$\lim_{n\to\infty} \sum_{k\neq d} r_{ik}(n) = 0,$$

and more strongly

(12) $$\sum_{n=1}^{\infty}\sum_{k\neq d} r_{ik}(n) < \infty.$$

Proof First suppose that the chain is positive so that for every i, $p_{id} > 0$. Hence there exists t such that

(13) $$\sum_{k\neq d} p_{ik} < t < 1.$$

Therefore, using (13) on the last sum in (10),

$$\sum_{k\neq d} r_{ik}(n) \leq \sum_{j_1\neq d} \cdots \sum_{j_{n-1}\neq d} p_{ij_1} \ldots p_{j_{n-2}j_{n-1}} t.$$

Hence

$$\sum_{k\neq d} r_{ik}^{(n)} \leq t^n$$

on using (13) to bound each summation successively. Since $t < 1$, (11) and (12) follow in this case. If the chain is regular but not positive, we first note that because $\sum_k r_{jk} \leq 1$ we have that

$$\sum_{k\neq d} r_{ik}(n+1) = \sum_{j\neq d} r_{ij}(n) \sum_k r_{jk} \leq \sum_{j\neq d} r_{ij}(n).$$

Thus $\sum_k r_{ik}(n)$ is non-increasing in n. Since the chain is regular, there is an m_0 such that $p_{id}(m_0) > 0$ for all i. By the argument of the first part, for some $t_0 < 1$,

(14) $$\sum_{k\neq d} r_{ik}(nm_0) < t_0^n < 1.$$

Hence, since $\sum_k r_{ik}(n)$ is non-decreasing (11) follows. Finally

$$\sum_n \sum_k r_{ik}(n) \leq m_0\left(1 + \sum_{n,k} r_{ik}(m_0 n)\right) \leq m_0\left(\frac{1}{1-t_0^n}\right) < \infty$$

proving (12). ■

With these preliminaries completed we can get on with proving the main claim of the paragraph following Example 6.

(15) Theorem For a regular chain, T_{id} is finite with probability 1, and has finite mean. More precisely, $\mathbf{P}(T_{id} > n) < c\lambda^n$ for some constants $c < \infty$ and $\lambda < 1$.

Proof By the remark preceding (4)

$$\mathbf{P}(T_{id} > n) = \sum_{k \neq d} r_{ik}(n) \to 0$$

as $n \to \infty$ by (11). Therefore T_{id} is finite with probability 1. Also

$$\mathbf{E}(T_{id}) = \sum_{n=0}^{\infty} \mathbf{P}(T_{id} > n) = \sum_{n=0}^{\infty} \sum_{k \neq d} r_{ik}(n) < \infty$$

by (12). The second statement of the theorem follows easily from (14). ■

There is a simple and useful generalization of this result, as follows.

(16) Theorem Let X be a regular Markov chain, and let D be a subset of the state space S. For $i \notin D$, define the first passage time

$$T_{iD} = \min\{n\colon X_n \in D | X_0 = i\}.$$

Then $\mathbf{E}(T_{iD}) < \infty$.

Proof This is an exercise for you. ■

It should be remarked that $\mathbf{E}(T_i) < \infty$ is a trivial consequence of Theorem 15.

As discussed above, first passage times are interesting in themselves for practical reasons, but they are even more interesting because of a crucial theoretical property. Informally it says that given the state of a chain at a first passage time T, the future of the chain is independent of the past. The following example makes this more precise.

(17) Example: Preservation of Markov property at first passage times Let X be a regular Markov chain with transition matrix $\boldsymbol{P}$, and let T be the first passage time of the chain to d. Show that for any $m > 0$, and $x_r \neq d$, we have

(18) $$\mathbf{P}(X_{T+m} = k | X_r = x_r \text{ for } 1 \leq r \leq T, X_T = d) = p_{dk}(m).$$

Solution Let us denote the event $\{X_r = x_r \neq d \text{ for } 1 \leq r < T\}$ by $A(T)$. Then, using conditional probability, the left hand side of (18) may be written as

(19) $$\frac{\mathbf{P}(X_{T+m} = k, A(T), X_T = d)}{\mathbf{P}(A(T), X_T = d)}$$

Now the numerator can be expanded as

$$\sum_{t=1}^{\infty} \mathbf{P}(X_{T+m} = k, A(t), X_t = d, T = t)$$

$$= \sum_{t=1}^{\infty} \mathbf{P}(X_{t+m} = k | A(t), X_t = d) \mathbf{P}(A(t), X_t = d, T = t)$$

$$= p_{dk}(m) \sum_{t=1}^{\infty} \mathbf{P}(A(t), X_t = d, T = t) \quad \text{by the Markov property,}$$

$$= p_{dk}(m)\mathbf{P}(A(T), X_T = d).$$

Finally, substitution into (19) yields (18). ●

It would be difficult to over-emphasize the importance of this result in the theory of Markov chains; it is used repeatedly.

To conclude this section we show that the mass functions of T_{id} and of T_d are related to the transition probabilities $p_{ik}(n)$ by very elegant and useful identities. Let $f_{id}(n) = \mathbf{P}(T_{id} = n)$, $i \neq d$, and $f_{dd}(n) = \mathbf{P}(T_d = n)$. Define the generating functions

$$P_{ik}(z) = \sum_{n=0}^{\infty} p_{ik}(n) z^n$$

and

$$F_{id}(z) = \sum_{n=0}^{\infty} f_{id}(n) z^n$$

with the convention that $p_{ii}(0) = 1$, $p_{ij}(0) = 0$, for $i \neq j$, and $f_{ij}(0) = 0$ for all i and j.

(20) Theorem When $i \neq k$ we have

(21) $$P_{ik}(z) = F_{ik}(z) P_{kk}(z),$$

and otherwise

(22) $$P_{ii}(z) = 1 + F_{ii}(z) P_{ii}(z).$$

Proof The idea of the proof is much the same as that of Example 17. For each k in S let us define the event $A_m = \{X_m = k\}$, and let B_m be the event that the first visit to k after time 0 takes place at time m. That is

$$B_m = \{X_r \neq k \quad \text{for} \quad 1 \leq r < m,\ X_m = k\}.$$

Then following a now familiar route, we write

(23) $$p_{ik}(m) = \mathbf{P}(A_m | X_0 = i) = \sum_{r=1}^{m} \mathbf{P}(A_m \cap B_r | X_0 = i)$$

$$= \sum_{r=1}^{m} \mathbf{P}(A_m | B_r, X_0 = i)\mathbf{P}(B_r | X_0 = i)$$

$$= \sum_{r=1}^{m} \mathbf{P}(A_m | X_r = k)\mathbf{P}(B_r | X_0 = i) \quad \text{by the Markov property,}$$

$$= \sum_{r=1}^{m} p_{kk}(m - r) f_{ik}(r).$$

The right hand side of (23) is a convolution, so multiplying both sides by z^m and summing over all $m \geq 1$ gives $P_{ik}(z) - \delta_{ik} = F_{ik}(z) P_{kk}(z)$ where $\delta_{ik} = \begin{cases} 1 & \text{if } i = k \\ 0 & \text{otherwise} \end{cases}$ as required. ■

(24) **Example: Weather** Successive days are either hot or cold, and they are also either wet or dry. From one day to the next either the temperature changes with probability α, or the precipitation changes with probability $1 - \alpha$. Let $f(n)$ be the probability that it is again hot and dry for the first time on the nth day, given that it was hot and dry on day zero. Show that

(25) $$F(z) = \sum_{n=1}^{\infty} f(n)z^n = z^2 \frac{1 + (1 - 2z^2)(1 - 2\alpha)^2}{2 - z^2 - z^2(1 - 2\alpha)^2}.$$

Solution It is helpful to visualize this Markov chain as a random walk on the vertices of a square, in which steps are taken along a horizontal edge with probability α, or a vertical edge with probability $1 - \alpha$. We identify the four states of the chain with the vertices of the square; the origin is hot and dry. The walk can return to the origin only after an even number $2n$ of steps, of which $2k$ are horizontal and $2n - 2k$ are vertical. Hence $p_0(2n)$, the probability of returning on the $2n$th step (not necessarily for the first time), is

$$\begin{aligned} p_0(2n) &= \sum_{k=0}^{n} \alpha^{2k}(1 - \alpha)^{2n-2k}\binom{2n}{2k} = \frac{1}{2}(\alpha + (1 - \alpha))^{2n} + \frac{1}{2}(\alpha - (1 - \alpha))^{2n} \\ &= \frac{1}{2}((1 - 2\alpha)^{2n} + 1). \end{aligned}$$

Hence

$$P_0(z) = \sum_{0}^{\infty} p_0(2n)z^{2n} = \frac{1}{2}\left(\frac{1}{1 - (1 - 2\alpha)^2 z^2} + \frac{1}{1 - z^2}\right).$$

Hence by (22) we have

$$F(z) = \frac{P_0(z) - 1}{P_0(z)} = z^2 \frac{1 + (1 - 2z^2)(1 - 2\alpha)^2}{2 - z^2(1 + (1 - 2\alpha)^2)}$$

which is (25). ●

If you have read Section 6.7 you will have noticed much in common with the above analysis and the results of that section. This is of course because the visits of a Markov chain to some given state k form a renewal process. We explore this link a little in Example 9.13.

9.4 Stationary distributions

We now consider one of the most important properties of the transition matrix $\boldsymbol{P}$. That is, for any $d \times d$ stochastic matrix $\boldsymbol{P}$, the set of equations

(1) $$x_k = \sum_{1 \leq i \leq d} x_i p_{ik}; \quad 1 \leq k \leq d,$$

always has a solution such that

(2) $$x_i \geq 0$$

and

(3) $$\sum_{i=1}^{d} x_i = 1.$$

Such a solution is thus a probability mass function, and it is commonly denoted by $\boldsymbol{x} = \boldsymbol{\pi} = (\pi_1, \ldots, \pi_d)$. It may not be unique.

(4) Example (a) If

$$\boldsymbol{P} = \begin{pmatrix} \frac{1}{2} & \frac{1}{2} \\ \frac{1}{2} & \frac{1}{2} \end{pmatrix}$$

then clearly $\boldsymbol{\pi} = (\frac{1}{2}, \frac{1}{2})$.

(b) If

$$\boldsymbol{P} = \begin{pmatrix} 0 & 1 \\ 1 & 0 \end{pmatrix}$$

then it is also clear that $\boldsymbol{\pi} = (\frac{1}{2}, \frac{1}{2})$.

(c) If

$$\boldsymbol{P} = \begin{pmatrix} 1 & 0 \\ 0 & 1 \end{pmatrix}$$

then we have $\boldsymbol{\pi} = (\alpha, 1 - \alpha)$ for any $\alpha \in [0, 1]$.

Note that the first chain is regular, the second periodic, and the third has two absorbing states; these chains evolve in very different ways. ●

The mass function $\boldsymbol{\pi}$ is called a *stationary distribution* of the chain for the following reason. Suppose that $\boldsymbol{\pi}$ is the mass function of X_0, then X_1 has mass function

$$\alpha_k(1) = \sum_i \pi_i p_{ik} = \pi_k$$

because $\boldsymbol{\pi}$ is a solution of (1). Hence X_1 has mass function $\boldsymbol{\pi}$, and by a trivial induction so does X_n for all n:

(5) $$\mathbf{P}(X_n = k) = \pi_k; \quad n \geqslant 0.$$

In formal terms (1) says that $\boldsymbol{P}$ has a positive left eigenvector corresponding to the eigenvalue 1. Experience of student calculations leads us to stress that $\boldsymbol{\pi}$ is a *left* eigenvector. If your stationary vector $\boldsymbol{\pi}$ is constant, check that you have not inadvertently found the right eigenvector. (And see (22) below.)

Here is a less trivial example.

(6) Example: Random walk with retaining barriers Let X have state space $\{0, 1, 2, \ldots, d\}$ and transition probabilities

$$p_{i,i+1} = p; \quad 0 \leqslant i \leqslant d - 1$$
$$p_{i,i-1} = 1 - p = q; \quad 1 \leqslant i \leqslant d$$
$$p_{00} = q$$
$$p_{dd} = p.$$

Then a stationary distribution must satisfy

(7)
$$\pi_i = p\pi_{i-1} + q\pi_{i+1}; \quad 1 \leqslant i \leqslant d-1$$
$$p\pi_0 = q\pi_1$$

and

$$q\pi_d = p\pi_{d-1}.$$

Simple substitution shows that if $p \neq q$, then $\pi_i = \pi_0(\frac{p}{q})^i$. Since $\Sigma_i \pi_i = 1$, it now follows that

$$\pi_i = \frac{1-\dfrac{p}{q}}{1-\left(\dfrac{p}{q}\right)^{d+1}}\left(\frac{p}{q}\right)^i.$$ ●

(8) **Example** Let X have transition matrix $\boldsymbol{P}$, and suppose that there exists a stationary distribution $\boldsymbol{\pi}$ satisfying (1). Define the Markov chain Y_n by $Y_n = (X_n, X_{n+1})$; $n \geqslant 0$. Show that Y has stationary distribution

(9)
$$\eta_{ij} = \pi_i p_{ij}; \quad i \in S,\ j \in S.$$

Solution We just have to check that $\boldsymbol{\eta}$ satisfies (1) and (3). Recall from Example 9.1.10 that Y has transition probabilities

$$\mathbf{P}(Y_{n+1} = (k, l) | Y_n = (i, j)) = p_{kl}\delta_{jk},$$

so that

$$\sum_{i,j} \eta_{ij} p_{kl} \delta_{jk} = \sum_{i,j} \pi_i p_{ij} p_{kl} \delta_{jk} = \sum_j \pi_j p_{kl} \delta_{jk} = \pi_k p_{kl} = \eta_{kl}.$$

Furthermore

$$\sum_{i,j} \eta_{ij} = \sum_{i,j} \pi_i p_{ij} = \sum_j \pi_j \quad \text{by (1)}$$
$$= 1 \quad \text{by (3)}.$$

Hence $\boldsymbol{\eta}$ is the stationary distribution of Y. ●

(10) **Example: Non-homogeneous random walk** Let $(S_n;\ n \geqslant 0)$ be a Markov chain with transition matrix given by

$$p_{i,i+1} = \lambda_i,$$
$$p_{i,i-1} = \mu_i,$$
$$p_{i,k} = 0, \quad \text{if } |i-k| \neq 1,$$

where $\lambda_i + \mu_i = 1$. This may be regarded as a random walk, taking positive or negative unit steps on the integers, such that the step probabilities depend on the position of the particle.

Is there a stationary distribution $\boldsymbol{\pi}$?

For simplicity let us suppose that $\mu_0 = 0$ and $S_0 \geqslant 0$, so that the walk is confined to the non-negative integers. Then if $\boldsymbol{\pi}$ exists it satisfies

$$\pi_0 = \mu_1\pi_1$$
$$\pi_1 = \lambda_0\pi_0 + \mu_2\pi_2$$
$$\pi_2 = \lambda_1\pi_1 + \mu_3\pi_3$$

and in general, for $k > 1$,

$$\pi_k = \lambda_{k-1}\pi_{k-1} + \mu_{k+1}\pi_{k+1}.$$

Solving these equations in order of appearance gives

$$\pi_1 = \frac{\lambda_0}{\mu_1}\pi_0;\ \pi_1 = \frac{\lambda_0\lambda_1}{\mu_1\mu_2}\pi_0;\ \pi_3 = \frac{\lambda_0\lambda_1\lambda_2}{\mu_1\mu_2\mu_3}\pi_0;$$

and so on. It is now easy to verify that for $n > 0$,

$$\pi_n = \frac{\lambda_0\lambda_1\ldots\lambda_{n-1}}{\mu_1\mu_2\ldots\mu_n}\pi_0.$$

This is a stationary distribution if

$$1 = \sum_{n=0}^{\infty}\pi_n = \pi_0 + \pi_0\sum_{n=1}^{\infty}\frac{\prod_0^{n-1}\lambda_r}{\prod_1^n\mu_r},$$

and so we deduce that a stationary distribution exists if this sum converges. ●

Having examined some consequences of (1) we now turn to the question of proving it.

The existence of $\boldsymbol{x}$ satisfying (1), (2) and (3) is a very famous result with many algebraic and analytical proofs. Most of these are neither elementary nor probabilistic. We prefer to give a proof which uses the ideas of probability theory, and is elementary.

(11) **Theorem** A regular Markov chain with transition matrix $\boldsymbol{P}$ has a stationary distribution $\boldsymbol{\pi}$.

Proof Let s be an arbitrary state of the chain with recurrence time T_s and mean recurrence time μ_s. For all $k \in S$ let $\rho_k(s)$ be the expected number of visits to k between successive visits to s; with the convention that $\rho_s(s) = 1$. We shall show that

(12) $$\pi_k = \mu_s^{-1}\rho_k(s); \quad 1 \leqslant k \leqslant d$$

is a stationary distribution of the chain.

First, let I_n denote the indicator of the event that the chain visits k at the nth step and has not previously revisited s, given that it started in s. Then the total number of visits to k between visits to s is

$$R_k = \sum_{n=1}^{\infty} I_n; \quad k \neq s,$$

and in accord with our convention above, when $k = s$ we have $R_s = 1$. Now

$$T_s = 1 + \sum_{k \neq s} R_k.$$

It follows that the expected value $\rho_k(s)$ of R_k is finite, and also that

(13) $$\mu_s = \sum_k \rho_k(s).$$

Furthermore

$$\rho_k(s) = \mathbf{E}(R_k) = \sum_{n=1}^{\infty} \mathbf{E}(I_n) = \sum_{n=1}^{\infty} \mathbf{P}(X_n = k, T_s \geqslant n | X_0 = s).$$

Now for $n = 1$,

$$\mathbf{P}(X_1 = k, T_s \geqslant 1 | X_0 = s) = p_{sk}.$$

For $n \geqslant 2$,

$$\begin{aligned}
&\mathbf{P}(X_n = k, T_s \geqslant n | X_0 = s) \\
&= \sum_{j \neq s} \mathbf{P}(X_n = k, X_{n-1} = j, T_s \geqslant n | X_0 = s) \\
&= \sum_{j \neq s} \mathbf{P}(X_n = k | X_{n-1} = j, T_s \geqslant n, X_0 = s) \mathbf{P}(X_{n-1} = j, T_s \geqslant n | X_0 = s) \\
&\qquad \text{by conditional probability} \\
&= \sum_{j \neq s} p_{jk} \mathbf{P}(X_{n-1} = j, T_s \geqslant n - 1 | X_0 = s) \quad \text{by the Markov property.}
\end{aligned}$$

Hence

(14) $$\begin{aligned}
\rho_k(s) &= p_{sk} + \sum_{j \neq s} p_{jk} \sum_{n=2}^{\infty} \mathbf{P}(X_{n-1} = j, T_s \geqslant n - 1 | X_0 = s) \\
&= \rho_s(s) p_{sk} + \sum_{j \neq s} p_{jk} \rho_j(s) = \sum_j \rho_j(s) p_{jk}.
\end{aligned}$$

Dividing throughout by μ_s yields the result (12) as required. ■

In view of the appearance of mean recurrence times in the above proof, it is perhaps not surprising to discover another intimate link between $\boldsymbol{\pi}$ and $\boldsymbol{\mu}$.

(15) Theorem For a regular Markov chain the stationary distribution is unique and satisfies $\pi_k \mu_k = 1$; $k \in S$. Hence

(16) $$\rho_k(s) = \frac{\mu_s}{\mu_k}.$$

Proof Recall that $T_{ik} = \min\{n \geqslant 0: X_n = k | X_0 = i\}$, so that in particular $T_{kk} = 0$, and $T_k = \min\{n \geqslant 1: X_n = k | X_0 = k\}$. Conditioning on the outcome of the first transition of the chain we have, for $i \neq k$

$$\mu_{ik} = \mathbf{E}(\mathbf{E}(T_{ik}|X_1)) = 1 + \sum_j p_{ij}\mu_{jk}. \tag{17}$$

Also,

$$\mu_k = 1 + \sum_j p_{kj}\mu_{jk}. \tag{18}$$

By using the Kronecker delta

$$\delta_{ik} = \begin{cases} 1 & \text{if } i = k \\ 0 & \text{otherwise} \end{cases} \tag{19}$$

these may be combined as one equation valid for all i:

$$\mu_{ik} + \delta_{ik}\mu_k = 1 + \sum_j p_{ij}\mu_{jk}. \tag{20}$$

Now if $\boldsymbol{\pi}$ is a stationary distribution, we multiply (20) by π_i and sum over all i to give

$$\sum_i \pi_i\mu_{ik} + \sum_i \pi_i\delta_{ik}\mu_k = 1 + \sum_i\sum_j \pi_i p_{ij}\mu_{jk} = 1 + \sum_j \pi_j\mu_{jk}$$

on using the fact that $\boldsymbol{\pi} = \boldsymbol{\pi P}$. Hence using (19) in the second sum we have $\pi_k\mu_k = 1$. Since μ_k is uniquely determined and finite, the required results follow. ■

(21) Example: Cube Suppose that a particle performs a random walk on the vertices of a cube, in such a way that when it is at a vertex it is equally likely to move along any one of the three edges which meet there, to a neighbouring vertex. Find the mean recurrence time of each vertex.

Solution The state space can be chosen as $S = \{i: 1 \leqslant i \leqslant 8\}$ and the transition probabilities are

$$p_{ij} = \begin{cases} \frac{1}{3} & \text{if } i \text{ and } j \text{ are joined by an edge} \\ 0 & \text{otherwise.} \end{cases}$$

Hence, $\sum_{i \in S} p_{ij} = 1$, and so the stationary distribution is $\pi_i = \dfrac{1}{8}$; $1 \leqslant i \leqslant 8$. By Theorem 15 $\mu_i = 8$; $1 \leqslant i \leqslant 8$. ●

More generally we note that for any finite regular doubly stochastic Markov chain, all states have the same mean recurrence time. This follows easily from the observation that in the doubly stochastic case we have

$$\frac{1}{d} = \sum_{i \in S} \frac{1}{d}\, p_{ij} = \frac{1}{d}. \tag{22}$$

Hence $\pi_i = d^{-1}$ is a stationary distribution, and $\mu_i = d$.

(23) Example: Library books My local lending library permits me to borrow one book at a time. Each Saturday I go to the library; if I have not finished reading the book I renew it, otherwise I borrow another. It takes me W_r weeks to read to rth book, where $(W_r;\ r \geqslant 1)$ is a sequence of independent random variables, which are identically distributed. Let X_n be the number of times that I have renewed the book which I take out of the library on the nth Saturday. Show that X_n is a Markov chain and find its transition matrix $\boldsymbol{P}$.

Find the stationary distribution of $\boldsymbol{P}$ when W_r is uniformly distributed on $\{1, \ldots, d\}$.

Solution Let W_r have mass function $f(k)$ and distribution function $F(k)$. Let R denote the record of borrowings and renewals up to, but not including, the book I am currently reading, and suppose that $X_n = i$. Either I renew it again, so $X_{n+1} = i + 1$, or I borrow a new one in which case $X_{n+1} = 0$. Because the W_r are independent and identically distributed

(24)
$$
\begin{aligned}
\mathbf{P}(X_{n+1} = i + 1 | X_n = i, R) &= \mathbf{P}(W_1 \geqslant i + 1 | W_1 = i) \\
&= \frac{\mathbf{P}(W_1 \geqslant i + 1)}{\mathbf{P}(W_1 \geqslant i)} \quad \text{by conditional probability} \\
&= \frac{1 - F(i + 1)}{1 - F(i)} = \mathbf{P}(X_{n+1} = i + 1 | X_n = i) \\
&= p_{i,i+1}; \quad i \geqslant 0.
\end{aligned}
$$

Otherwise

(25)
$$
\mathbf{P}(X_{n+1} = 0 | X_n = i, R) = 1 - p_{i,i+1} = \frac{f(i)}{1 - F(i)}.
$$

Hence X is a Markov chain with transition probabilities given by (24) and (25).

If W_r is uniform on $\{1, \ldots, d\}$ then

$$
p_{i,i+1} = \frac{d - i - 1}{d - i}; \quad 0 \leqslant i < d - 1
$$

and $p_{d-1,0} = 1$. Hence any stationary distribution $\boldsymbol{\pi}$ satisfies

(26)
$$
\pi_{i+1} = \pi_i \frac{d - i - 1}{d - i}; \quad 0 \leqslant i < d - 1
$$

and $\pi_0 = \pi_{d-1}$. Iterating (26) gives $\pi_{i+1} = \frac{d-i-1}{d} \pi_0$, and since $\sum_i^d \pi_i = 1$, it follows that

$$
\pi_{i+1} = \frac{2(d - i - 1)}{d(d + 1)}.
$$
●

9.5 The long run

It is natural to speculate about the behaviour of the Markov chain X in the long run, that is as $n \to \infty$. As usual we obtain insight from examples before turning to the general case.

Example 9.2.9 revisited: Survival Recall that

$$p_{ik}(n) = \begin{cases} p^n & \text{if } k = i + n \\ qp^k & \text{if } 0 \leqslant k \leqslant n - 1. \end{cases}$$

Now allowing $n \to \infty$ shows that

$$p_{ik}(n) \to qp^k.$$

Notice that $\sum_k qp^k = 1$, and that the collection $\pi_k = (qp^k;\ k \geqslant 0)$ is a stationary distribution. To see this just check that $\boldsymbol{\pi} = \boldsymbol{\pi} \boldsymbol{P}$ because

$$\pi_0 = q \sum_0^\infty \pi_k = q$$

and $\pi_k = p\pi_{k-1}$. This stationary distribution does not depend on the starting point of the chain. ●

Example (9.2.7) revisited: Simple random walk In this case for all i and k

$$\lim_{n\to\infty} p_{ik}(n) = 0.$$ ●

Example: Gambler's ruin This is a simple random walk which stops when it reaches 0 or K, and is therefore a Markov chain. The probability of ruin starting from i is p_i, and we have shown above that the probability of winning from i is $1 - p_i$. Hence for $0 \leqslant i \leqslant K$, as $n \to \infty$

$$\begin{aligned} p_{i0}(n) &\to p_i \\ p_{iK}(n) &\to 1 - p_i \\ p_{ij}(n) &\to 0 \quad \text{for } 0 < j < K. \end{aligned}$$

The pair $\{p_i, 1 - p_i\}$, is a stationary distribution, but it depends on the initial state of the chain. ●

These examples illustrate the possibilities, and agree with our intuition. Roughly speaking, if a chain can get to absorbing states then it may eventually be absorbed in one of them; on the other hand if it has no stationary distribution then the chance of finding it in any given state vanishes in the long run.

The most interesting case, as you might expect, arises when the chain has a unique stationary distribution, and the principal result for a finite chain is then the following theorem.

(1) **Theorem** Let X be regular with transition probabilities p_{ik}. Then as $n \to \infty$, for any i and k,

$$p_{ik}(n) \to \pi_k > 0.$$

Furthermore

(i) π_k does not depend on i
(ii) $\sum_{k\in S} \pi_k = 1$

and

(iii) $\pi_k = \sum_{i \in S} \pi_i p_{ik}$,

and so $\boldsymbol{\pi}$ is the stationary distribution of X.

Our proof of Theorem 1 will rely on the idea of *coupling*. This technique has many forms and applications; we use a very simple version here. Suppose we run two independent chains X and Y with the same transition matrix, and they first take the same value s at time T, say. Now as we have shown above, the Markov property is preserved at such first passage times, so given $X_T = Y_T = s$, the further progress of X and Y is independent of their activities before T. Hence, on the event $T \leqslant n$ we have

(2) $$\mathbf{P}(X_n = k;\ T \leqslant n) = \mathbf{P}(Y_n = k;\ T \leqslant n),$$

since given $T = t$ both sides are equal to $p_{sk}(n-t)$. The chains are coupled at T.

Now we can tackle the theorem.

Proof of Theorem 1 Let X and Y be independent regular Markov chains with the same state space S and transition matrix p_{ij}. Let X_0 have mass function

$$\mathbf{P}(X_0 = i) = 1,$$

and Y_0 have the stationary distribution of p_{ij} so that

$$\mathbf{P}(Y_0 = i) = \pi_i.$$

Define the Markov chain

$$W = (X, Y),$$

and let T be the first passage time of W to the set $D = \{(x, y)\colon x = y\}$, namely $T = \min\{n\colon X_n = Y_n\}$. Now by (9.2.11), since X and Y are regular, so is W. Hence T is finite with probability 1 (and has finite mean) by Theorem 9.3.15.

Now, bearing in mind our preparatory remarks above, we can say

(3) $$\begin{aligned} |p_{ik}(n) - \pi_k| &= |\mathbf{P}(X_n = k) - \mathbf{P}(Y_n = k)| \\ &= |\mathbf{P}(X_n = k, n \geqslant T) - \mathbf{P}(Y_n = k, n \geqslant T) \\ &\quad + \mathbf{P}(X_n = k, n < T) - \mathbf{P}(Y_n = k, n < T)| \\ &= |\mathbf{P}(X_n = k, n < T) - \mathbf{P}(Y_n = k, n < T)| \quad \text{by (2)} \\ &\leqslant \mathbf{P}(T > n), \end{aligned}$$

where the last inequality follows because $|\mathbf{P}(A \cap B) - \mathbf{P}(A \cap C)| \leqslant \mathbf{P}(A)$ for any events A, B and C. Since $\mathbf{P}(T > n) \to 0$ as $n \to \infty$, we have

$$p_{ik}(n) \to \pi_k$$

as required. The rest of the assertions follow because $\boldsymbol{\pi}$ is the stationary distribution of X. ■

This is a rather useful result; to find the long term behaviour of the chain we just solve $\boldsymbol{\pi} = \boldsymbol{\pi} \mathbf{P}$, which gives the limiting distribution of X. Indeed, we know from the results of Section 9.3 that this distribution is approached rather quickly, since from

Theorem 9.3.15

(4) $$|p_{ij}(n) - \pi_j| < \mathbf{P}(T > n) < c\lambda^n$$

for some constants $c < \infty$ and $\lambda < 1$.

The probabilities $p_{ij}(n)$ are said to approach π_j *geometrically fast*.

(5) Example Let X have state space $\{1, 2\}$ and transition matrix

$$\boldsymbol{P} = \begin{pmatrix} 1-\alpha & \alpha \\ \beta & 1-\beta \end{pmatrix}.$$

From the results of Example 9.2.20 we see that when $0 < \alpha + \beta < 2$, as $n \to \infty$

$$p_{11}(n) \to \beta,\ p_{21}(n) \to \beta$$
$$p_{12}(n) \to \alpha,\ p_{22}(n) \to \alpha.$$

And of course

$$(\beta, \alpha)\begin{pmatrix} 1-\alpha & \alpha \\ \beta & 1-\beta \end{pmatrix} = (\beta, \alpha),$$

so $\left(\dfrac{\beta}{\alpha+\beta}, \dfrac{\alpha}{\alpha+\beta}\right)$ is the stationary distribution as it must be.

When $\alpha + \beta = 0$, the chain is not irreducible, and when $\alpha + \beta = 2$, the chain is not regular (being periodic). ●

(6) Example: Entropy of a Markov source Let $X = (X_n;\ n \geqslant 1)$ be a collection of jointly distributed random variables, and write $\boldsymbol{X}_n = (X_1, \ldots, X_n)$. Recall that in Example 9.2.28 we defined the conditional entropy function

$$H(X_{n+1}|\boldsymbol{X}_n) = -\mathbf{E}[\mathbf{E}(\log f(X_{n+1}|\boldsymbol{X}_n)|\boldsymbol{X}_n)].$$

If $H_X = \lim_{n\to\infty} H(X_{n+1}|\boldsymbol{X}_n)$ exists, then H_X is said to be the *entropy* or *uncertainty* of X.

Now let X be a regular Markov chain with transition matrix $\boldsymbol{P}$. Show that H_X does indeed exist and is given by

(7) $$H_X = -\sum_i \pi_i \sum_k p_{ik} \log p_{ik}$$

where $\boldsymbol{\pi}$ is the stationary distribution of $\boldsymbol{P}$.

Solution In Example 9.2.28 it was shown that for a Markov chain X,

(8) $$\begin{aligned} H(X_{n+1}|\boldsymbol{X}_n) &= H(X_{n+1}|X_n) \\ &= -\sum_i\sum_k \mathbf{P}(X_{n+1} = k|X_n = i) \times \log(\mathbf{P}(X_{n+1} = k|X_n = i))\mathbf{P}(X_n = i) \\ &= -\sum_i\sum_k \alpha_i(n) p_{ik} \log p_{ik}. \end{aligned}$$

Now, by Theorem 1, as $n \to \infty$, $\alpha_i(n) \to \pi_i$, and therefore taking the limit as $n \to \infty$ of the right hand side of (8) gives (7) as required. ●

The basic limit theorem (1) tells us that in the long run the probability of finding the regular chain X in state k converges to π_k, for each $k \in S$. It seems very plausible that, also in the long run, the proportion of time that X spends visiting k should converge to π_k. The following theorem shows that a more precise version of this vague statement is indeed true. It may be thought of as a type of weak law of large numbers for Markov chains.

(9) Theorem Let X be regular with transition matrix $\boldsymbol{P}$ and stationary distribution $\boldsymbol{\pi}$. Let $V_k(n)$ be the number of visits to the state k by X up to time n. Then for any $\epsilon > 0$

(10)
$$\mathbf{P}\left(\left|\frac{1}{n+1}V_k(n) - \pi_k\right| > \epsilon\right) \to 0$$

as $n \to \infty$.

Proof Some groundwork is required before setting about the proof of (10). Let $I_k(n)$ be the indicator of a visit to k at time n, so

$$I_k(n) = \begin{cases} 1 & \text{if } X_n = k \\ 0 & \text{otherwise.} \end{cases}$$

By the basic property of indicators

(11)
$$\mathbf{E}(I_k(n)) = \mathbf{P}(X_n = k) = \alpha_k(n)$$

and for $m \neq r$

(12)
$$\mathbf{E}(I_k(m)I_k(r)) = \alpha_k(s)p_{kk}(t)$$

where $s = \min\{m, r\}$ and $t = |m - r|$. These indicators will be useful because

(13)
$$V_k(n) = \sum_{r=0}^{n} I_k(r).$$

Now we recall that for some constants c and λ with $1 \leqslant c < \infty$ and $0 < \lambda < 1$,

(14)
$$|\alpha_k(n) - \pi_k| < c\lambda^n$$

and

(15)
$$|p_{ik}(n) - \pi_k| < c\lambda^n.$$

At last we are in a position to tackle (10). By Chebyshov's inequality

$$\begin{aligned}
&\mathbf{P}\left(\left|\frac{1}{n+1}V_k(n) - \pi_k\right| > \epsilon\right) \\
&\leqslant \mathbf{E}\left(\left(\frac{V_k(n) - (n+1)\pi_k}{(n+1)\epsilon}\right)^2\right) \\
&= \frac{1}{(n+1)^2\epsilon^2}\mathbf{E}\left(\left(\sum_{r=0}^{n}(I_k(r) - \pi_k)\right)^2\right) \\
&= \frac{1}{(n+1)^2\epsilon^2}\mathbf{E}\left(\sum_{m=0}^{n}\sum_{r=0}^{n}(I_k(m)I_k(r) - \pi_k I_k(m) - \pi_k I_k(r) + \pi_k^2)\right)
\end{aligned}$$

$$= \frac{1}{(n+1)^2\epsilon^2} \sum_{m,r} ((\alpha_k(s) - \pi_k)(p_{kk}(t) - \pi_k)$$
$$+ \pi_k[(\alpha_k(s) - \pi_k) + (p_{kk}(t) - \pi_k) - (\alpha_k(m) - \pi_k) - (\alpha_k(r) - \pi_k)]) \quad \text{by (11) and (12)}$$

$$\leqslant \frac{1}{(n+1)^2\epsilon^2} \sum_{m,r} 2c^2(\lambda^s + \lambda^t) \quad \text{by (14) and (15)}$$
$$\to 0$$

as $n \to \infty$, establishing (10). ■

(16) Corollary For any bounded function $g(x)$, and any $\epsilon > 0$

(17) $$\mathbf{P}\left(\left|\frac{1}{n+1}\sum_{r=0}^{n} g(X_r) - \sum_{k\in S} \pi_k g(k)\right| > \epsilon\right) \to 0$$

as $n \to \infty$.

Proof The key to this lies in the observation that

$$\sum_{r=0}^{n} g(X_r) = \sum_{k\in S} g(k) V_k(n).$$

Hence we can rewrite (17) as

$$\mathbf{P}\left(\left|\sum_{\substack{k\in S\\ g(k)\neq 0}} g(k)\left(\frac{V_k(n)}{n+1} - \pi_k\right)\right| > \epsilon\right) \leqslant \sum_{\substack{k\in S\\ g(k)\neq 0}} \mathbf{P}\left(\left|\frac{V_k(n)}{n+1} - \pi_k\right| > \frac{\epsilon}{dg(k)}\right) \to 0$$

as $n \to \infty$, by Theorem 9 (using the fact that S is finite). ■

We can give an immediate application of these results.

(18) Example: Asymptotic equipartition for a Markov source Let the regular Markov chain X with transition matrix $\boldsymbol{P}$ and stationary distribution $\boldsymbol{\pi}$ represent the output from a Markov information source, as defined in Example 9.1.5. Let $\boldsymbol{X}_n = (X_0, \ldots, X_n)$ have joint mass function $f(x_0, \ldots, x_n) = \mathbf{P}(X_0 = x_0, \ldots, X_n = x_n)$, and recall from Example 9.5.6 that the entropy of X is

(19) $$H_X = -\sum_{i\in S}\sum_{k\in S} \pi_i p_{ik} \log p_{ik}.$$

Show that, for any $\delta > 0$, as $n \to \infty$

(20) $$\mathbf{P}\left(\left|H_X + \frac{1}{n}\log f(X_0, \ldots, X_n)\right| > \delta\right) \to 0.$$

Solution First, from Example 9.4.8 the sequence $Y_n = \{X_n, X_{n+1}\}$; $n \geqslant 0$, is a Markov chain with stationary distribution $(\pi_i p_{ik};\ i \in S,\ k \in S)$. Second, we have

$$-\frac{1}{n}\log f(X_0, \ldots, X_n) = -\frac{1}{n}\log(p_{X_0X_1}p_{X_1X_2}, \ldots, p_{X_{n-1}X_n})$$
$$= -\frac{1}{n}\sum_{r=0}^{n-1}\log p_{X_rX_{r+1}}.$$

Finally, we note that if we set $g(Y_n) = \log p_{X_nX_{n+1}}$, then Corollary 16 applied to the Markov chain Y shows that

$$\mathbf{P}\left(\left|\frac{1}{n}\sum_{r=0}^{n-1}\log p_{X_rX_{r+1}} - \sum_{i,k}\pi_i p_{ik}\log p_{ik}\right| > \delta\right) \to 0$$

and (20) follows immediately, on remembering (19) and (21). ●

Here is a useful application of the asymptotic equipartition example. If d is the size of the alphabet, then the total number of messages of length n which the source can emit is d^n. Let us divide them into disjoint sets T and A, where

(22) $$T = \left\{(x_1, \ldots, x_n): \left|\frac{1}{n}\log f(x_1, \ldots, x_n) + H_X\right| < \delta\right\}$$

and

$$A = \left\{(x_1, \ldots, x_n): \left|\frac{1}{n}\log f(x_1, \ldots, x_n) + H_X\right| \geq \delta\right\}.$$

By Example 18, for any $\epsilon > 0$ and $\delta > 0$, there exists $n_0 < \infty$ such that

(23) $$\mathbf{P}\{(X_1, \ldots, X_{n_0}) \in T\} \geq 1 - \epsilon;$$

since this is arbitrarily near 1, sequences in T are called *typical*. Also by Example 18

$$\mathbf{P}(\{X_1, \ldots, X_{n_0}\} \in A) < \epsilon$$

which is arbitrarily small, so sequences in A are called *atypical*. If you are seeking efficient transmission of messages, it therefore makes sense to concentrate on the typical sequences. It follows that a natural question is, how many typical sequences are there? At this point we recall that by convention the logarithms in Example 18 are taken to base 2. Hence, from (22)

$$2^{-n(H_X+\delta)} < f(x_1, \ldots, x_n) < 2^{-n(H_X-\delta)}.$$

But also, from (23)

$$1 - \epsilon \leq \sum_{x_n \in T} f(x_1, \ldots, x_n) \leq 1.$$

Hence the number $|T|$ of sequences in T satisfies

$$(1-\epsilon)2^{n(H_X-\delta)} \leq |T| \leq 2^{n(H_X+\delta)},$$

which is to say that, roughly speaking, there are about 2^{nH_X} typical messages of length n.

9.6 Markov chains in general

Up to now we have dealt chiefly with finite regular chains, because such chains have useful and elegant properties with elementary proofs. However, as some examples have indicated, many chains are irregular or infinite or both. We therefore give a brief account of some of the important results for more general chains; the proofs are all omitted.

In the above sections it was found that in a finite regular chain, any state d has a recurrence time T_d which is finite with probability 1, and has finite expectation. When X has countably infinite state space, this need no longer be true, as a glance at the unrestricted simple random walk shows immediately. We therefore distinguish these cases.

(1) Definition Let the state d have recurrence time T_d. Then:

(i) if $\mathbf{P}(T_d < \infty) < 1$ then d is said to be *transient*.

(ii) if $\mathbf{P}(T_d < \infty) = 1$, but $\mathbf{E}(T_d) = \infty$, then d is said to be *recurrent null* (or persistent null.) Otherwise d is *recurrent* (or persistent). ▲

These new types of behaviour seem to complicate matters, but the following theorem helps to simplify them again.

(2) Decomposition theorem The state space S can be uniquely partitioned as $S = T \cup C_1 \cup C_2 \cup \dots$ where T is the set of transient states and each C_i is an irreducible closed set of recurrent states.

This means that eventually the chain ends up in some one of the C_i and never leaves it, or it remains forever in the transient states.

Of course if there is only one closed set of recurrent states matters are even simpler, so the following theorem is useful.

(3) Theorem The chain X has a unique stationary distribution $\boldsymbol{\pi}$ if and only if S contains exactly one recurrent non-null irreducible subchain C. For each i in C,

(4) $$\pi_i = \mu_i^{-1}$$

where μ_i is the mean recurrence time of i; and for $i \notin C$, $\pi_i = 0$.

There is also a limit theorem for general Markov chains.

(5) Theorem For any aperiodic state k of a Markov chain

$$p_{kk}(n) \to \frac{1}{\mu_k} \quad \text{as } n \to \infty,$$

where the limit is zero if k is null or transient. If i is any other state of the chain then

$$p_{ik}(n) \to \frac{1}{\mu_k}\mathbf{P}(T_{ik} < \infty).$$

9.7 Markov chains with continuous parameter

We have suggested above that Markov chains can provide a good description of a variety of real systems. However, a moment's thought about real systems is sufficient to see that many of them do not change their state at integer times. Components fail, your telephone rings, meteorites fall, at *any* time.

Spurred by this, it is natural to wish to study collections of random variables of the form

$$X = (X(t);\ t \geq 0).$$

Here $t \in \mathbb{R}$ is often regarded as the time, and then $X(t) \in \mathbb{Z}$ is regarded as the state of the system X at time t. Such a collection is often called a random process. The most obvious thing about X is that it is an uncountable collection of random variables, and it follows that a rigorous account of the behaviour of $X(t)$ is beyond the scope of an elementary text such as this. However, we can discover quite a lot informally in special cases.

First, the remarks in Section 9.1 that motivated our interest in Markov chains apply equally well in continuous time. We therefore make the following definition, analogous to Definition 9.1.1. As usual $X(t) \in S$, where S is a subset of the integers called the state space.

(1) **Definition** The process $X = (X(t);\ t \geq 0)$ taking values in S is a Markov process (or has the Markov property) if

$$\mathbf{P}(X(t) = k \mid X(t_1) = i_1, \ldots, X(t_n) = i_n) = \mathbf{P}(X(t) = k \mid X(t_n) = i_n)$$

for all possible $k, i_1, \ldots, i_n$, and any sequence $0 \leq t_1 < t_2 < \ldots < t_n < t$ of times. We write

$$\mathbf{P}(X(t+s) = k \mid X(s) = i) = p_{ik}(t). \qquad \blacktriangle$$

As in the discrete case $(p_{ik}(t);\ i, k \in S)$ are known as the transition probabilities, and they satisfy the Chapman–Kolmogorov equations as follows.

(2) **Theorem** For any $s > 0$ and $t > 0$ and $i, k \in S$,

$$p_{ik}(s+t) = \sum_{j \in S} p_{ij}(s) p_{jk}(t).$$

Proof By the same arguments as we used in Theorem 9.2.12.

(3)
$$\begin{aligned} p_{ik}(s+t) &= \sum_{j \in S} \mathbf{P}(X(s+t) = k,\ X(s) = j \mid X(0) = i) \\ &= \sum_{j \in S} \mathbf{P}(X(t+s) = k \mid X(s) = j,\ X(0) = i)\mathbf{P}(X(s) = j \mid X(0) = i) \\ &= \sum_j p_{ij}(s) p_{jk}(t). \qquad \blacktriangle \end{aligned}$$

Given this collection of equations it is possible to set about solving them in special cases, without any further ado. In fact we do just that in the next section, but as usual there are a few preliminaries.

First we must ask, do any non-trivial Markov processes exist? (Obviously the trivial process $X(t) = 1$ for all t is a Markov process, but not a very exciting one.) This question is not as stupid as it may appear to you. Recall that we defined

Markov chains by visualizing a counter or particle moving around the vertices of a graph according to some specified distributions, and if necessary we could actually do it. Here, we have started with a collection of probabilities, with no description of how we might actually produce a sequence $X(t)$ having these transition probabilities and joint distributions.

Of course the answer to the above question is, yes they do exist, and you have already met one, namely the Poisson process. This was defined by construction in Definition 8.8.1, and we showed that it had the Markov property in Exercise 8.16.3. Henceforth, where necessary, we assume without proof that the processes we consider exist; in more advanced texts it is shown that they do.

(4) **Example: Poisson process** It has already been shown that if $N = (N(t); t \geqslant 0)$ is a Poisson process with parameter λ, then $N(t) - N(0)$ has a Poisson distribution with parameter λt, which is to say that

$$p_{ik}(t) = \frac{e^{-\lambda t}(\lambda t)^{k-i}}{(k-i)!}.$$

Hence we can calculate

$$\begin{aligned}\sum_{j=i}^{k} p_{ij}(s)p_{jk}(t) &= \sum_{j=i}^{k} \frac{e^{-\lambda s}(\lambda s)^{j-i}}{(j-i)!}\frac{e^{-\lambda t}(\lambda t)^{k-j}}{(k-j)!}\\ &= \frac{e^{-\lambda(t+s)}}{(k-i)!}\sum_{r=0}^{k-i}\binom{k-i}{r}(\lambda s)^r(\lambda t)^{k-i-r}\\ &= \frac{e^{-\lambda(t+s)}(\lambda(s+t))^{k-i}}{(k-i)!} = p_{ik}(s+t).\end{aligned}$$

Thus the transition probabilities of the Poisson process satisfy the Chapman–Kolmogorov equations, as they must of course by Theorem 2. ●

Note that this result of itself does *not* show that N is a Markov process; there are processes which are not Markov, whose transition probabilities nevertheless satisfy (3). The crucial property which makes N a Markov process is the exponential distribution of times between events. This property is in fact characteristic of Markov processes in general; they wait in each successive state for an exponentially distributed time before moving to the next. Naturally it is the lack-of-memory property of the exponential distribution which is basically responsible for this essential role in the theory of Markov processes. However, we can do no more here than state the fact baldly; exploring its ramifications is beyond our scope.

One example will suffice to give some trivial insight into these remarks.

Example: Falling off a log Let $X(t)$ be a Markov chain with two states 0 and 1. Suppose that transitions from 1 to 0 are impossible. Let us consider transitions from 0 to 1. Since $X(t)$ is Markov, the transition probabilities satisfy the Chapman–Kolmogorov equations. Hence, as $p_{10}(t) = 0$, we have

$$p_{00}(s+t) = p_{00}(s)p_{00}(t) + p_{01}(s)p_{10}(t) = p_{00}(s)p_{00}(t).$$

However, as we have remarked previously, the only bounded solutions to the equation $f(x+y) = f(x)f(y)$ are of the form $f(x) = e^{-\lambda x}$. Hence

$$p_{00}(t) = e^{-\lambda t}$$

for some $\lambda \geq 0$. The exponential density is forced upon us by the assumption that $X(t)$ is Markov. ●

9.8 Forward equations; Poisson and birth processes

It is all very well to verify that a previously obtained solution satisfies (9.7.3). A pressing question is, can we solve (9.7.3) without already knowing the answer? We therefore develop a technique for tackling the Chapman–Kolmogorov equations in this section.

First we observe that for the Poisson process, as $t \to 0$

(1) $$p_{k,k+1}(t) = \mathbf{P}(N(t) = 1) = \lambda t e^{-\lambda t} = \lambda t + o(t);^\dagger$$

(2) $$p_{kk}(t) = \mathbf{P}(N(t) = 0) = e^{-\lambda t} = 1 - \lambda t + o(t);$$

for $j < k$

(3) $$p_{kj}(t) = \mathbf{P}(N(t) < 0) = 0;$$

and for $j > k+1$

(4) $$p_{kj}(t) = \mathbf{P}(N(t) > 1) = o(t).$$

Equations (1)–(4) say that:

(5) $N(t)$ is non-decreasing.

(6) The probability of an event in $[s, s+t]$ is proportional to t, for small t, and does not depend on previous events.

(7) The probability of two or more events in $[s, s+t]$, for small t, is $o(t)$.

What we are going to do now is to seek a Markov process $X(t)$ with transition probabilities $p_{ik}(t)$ which satisfy (5), (6) and (7).

Since $p_{ik}(t)$ satisfies (9.7.3), we have for small t

$$\begin{aligned} p_{ik}(s+t) &= \sum_{j=1}^{k} p_{ij}(s) p_{jk}(t) \\ &= p_{ik}(s)(1 - \lambda t + o(t)) + p_{i,k-1}(s)(\lambda t + o(t)) + \sum_{j=i}^{k-2} p_{ij}(s).o(t). \end{aligned}$$

Hence

$$\frac{p_{ik}(s+t) - p_{ik}(s)}{t} = -\lambda p_{ik}(s) + \lambda p_{i,k-1}(s) + o(1)$$

and allowing $t \to 0$ gives

(8) $$\frac{d}{ds} p_{ik}(s) = -\lambda p_{ik}(s) + \lambda p_{i,k-1}(s),$$

valid for all $0 \leq i \leq k$ (remembering that $p_{i,i-1}(s) = 0$).

† We discussed the $o(.)$ notation in Section 7.5.

At $t = 0$, we have the initial condition

(9) $$p_{ii}(0) = 1.$$

The equations (8), as i and k range over all possible values, are known as the *forward equations* for $p_{ik}(t)$, and may be solved in a number of ways.

Theorem The solution of (8) is given by

(10) $$p_{ik}(t) = \frac{e^{-\lambda t}(\lambda t)^{k-i}}{(k-i)!},$$

namely the transition probabilities of the Poisson process.

Proof We give two methods of proof. First, solve (8) with $k = i$, using (9), to find $p_{ii}(t) = e^{-\lambda t}$. Substituting this into (8) with $k = i + 1$ yields

$$p_{i,i+1}(t) = \lambda t e^{-\lambda t}.$$

A simple induction now yields (10). ■

A second method relies on the generating function

$$G(z, t) = \sum_{k=i}^{\infty} p_{ik}(t) z^k = \mathbf{E}(z^{N(t)} | N(0) = i).$$

Multiply (8) by z^k and sum over k, to obtain

(11) $$\frac{\partial G}{\partial t} = \lambda(z-1)G.$$

From (9) we have

(12) $$G(z, 0) = z^i.$$

The solution of (11) that satisfies (12) is

(13) $$G(z, t) = z^i \exp(\lambda t(z-1))$$

and the coefficient of z^k in this expression is just (10). ■

The point of this elaborate re-working is that the simple assumptions (5), (6) and (7) also lead to the Poisson process in a simple and straightforward way. It turns out that a great many useful processes can be analysed by specifying $p_{ik}(t)$ for small t, and all i and k, then obtaining the forward equations, and finally (occasionally) solving them.

(14) Example: The simple birth process A population of individuals grows as follows. Each member of the population in existence at time t may be replaced by two new individuals during $[t, t + h]$ with probability $\lambda h + o(h)$, independently of the other members of the population. Otherwise the given member of the population remains intact during $[t, t + h]$ with probability $1 - \lambda h + o(h)$, also independently of the rest of the population. If the population at time t is $X(t)$, and $X(0) = 1$, show that

(15) $$\mathbf{E}(z^{X(t)}) = \frac{z}{z + (1 - z)e^{\lambda t}}.$$

What is $\mathbf{E}(X(t))$?

Solution Let $p_n(t) = \mathbf{P}(X(t) = n)$ and

$$p_{jk}(t) = \mathbf{P}(X(t) = k | X(0) = j).$$

Suppose that $X(t) = i$. Because each individual is replaced (or not) independently of all the others we have, as $h \to 0$,

$$p_{ii}(h) = (1 - \lambda h + o(h))^i$$
$$p_{i,i+1}(h) = i(\lambda h + o(h))(1 - \lambda h + o(h))^{i-1} = i\lambda h + o(h)$$
$$p_{ik}(h) = o(h); \quad k > i + 1$$
$$p_{ik}(h) = 0; \quad k < i.$$

Following the by now familiar routine, we find

$$p_k(t + h) = (1 - \lambda kh)p_k(t) + \lambda(k - 1)hp_{k-1}(t) + o(h)$$

and so

(16) $$\frac{\partial}{\partial t} p_k(t) = -\lambda k p_k(t) + \lambda(k - 1)p_{k-1}(t).$$

Now we set $G_X(z, t) = \mathbf{E}(z^{X(t)})$, and notice that since probability generating functions are differentiable, at least for $|z| < 1$, we have

(17) $$\frac{\partial G_X}{\partial z} = \sum_{k=1}^{\infty} k p_k(t) z^{k-1}.$$

Now on multiplying (16) by z^k, and summing over k, we notice $\dfrac{\partial G_X}{\partial z}$ appearing on the right hand side. In fact

(18) $$\frac{\partial G_X}{\partial t} = \lambda z(z - 1)\frac{\partial G_X}{\partial z}.$$

Also, since $X(0) = 1$,

(19) $$G_X(z, 0) = z.$$

By inspection, for any differentiable function $h(.)$, the function

(20) $$G(z, t) = h\left(\lambda t + \int^z \frac{1}{v(v - 1)}\, dv\right)$$

satisfies (18). Imposing the boundary condition (19) reveals that

$$z = h\left(\int^z \frac{1}{v(v - 1)}\, dv\right) = h\left(\log\left(\frac{z - 1}{z}\right)\right).$$

Hence the function $h(.)$ is given by $h(y) = (1 - e^y)^{-1}$ and so

$$G_X(z, t) = \frac{1}{1 - \exp\left(\lambda t + \log\left(\frac{z-1}{z}\right)\right)} = \frac{z}{z + (1-z)e^{\lambda t}},$$

as required.

Now to find $\mathbf{E}(X(t))$ we have a choice of methods. Obviously by differentiating (15) with respect to z, and setting $z = 1$ we find

(21)
$$\mathbf{E}(X(t)) = \frac{\partial}{\partial z} G_X(1, t) = e^{\lambda t}.$$

However we could have obtained $\mathbf{E}(X(t))$ without solving (18). If we assume $\mathbf{E}[(X(t))^2]$ exists, then differentiating (18) with respect to z and setting $z = 1$ yields

$$\frac{\partial}{\partial t}\frac{\partial G_X}{\partial z}(1, t) = \lambda \frac{\partial G_X}{\partial z}(1, t).$$

This has solution given by (21). ●

9.9 Forward equations; equilibrium

Guided by our glances at the Poisson and simple birth processes, we can now outline a simple technique for dealing with some elementary Markov chains. The aim is to obtain forward equations, so from the Chapman–Kolmogorov equations we write, for $h > 0$,

(1)
$$\frac{p_{ik}(t+h) - p_{ik}(t)}{h} = \frac{1}{h}\left(\sum_j p_{ij}(t)p_{jk}(h) - p_{ik}(t)\right).$$

We wish to let $h \downarrow 0$. By inspection of (1) this is possible if for some finite numbers $(g_{jk};\ j, k \in S)$ we have as $h \to 0$

(2)
$$p_{jk}(h) = g_{jk}h + o(h),$$

and

(3)
$$p_{kk}(h) = 1 + g_{kk}h + o(h).$$

In this case we obtain the required forward equations by letting $h \downarrow 0$ in (1), to give

(4)
$$\frac{\partial}{\partial t} p_{ik}(t) = \sum_j p_{ij}(t) g_{jk}.$$

The application of this idea is best illustrated by examples. We give perhaps the simplest here, others follow in due course.

(5) Example: Machine A machine can be either up or down. (You can interpret this figuratively (working/not working), or literally (a lift), it makes no difference to the mathematics.) If it is up at time t, then it goes down during $[t, t+h]$ with probability $\alpha h + o(h)$, independently of its past record. Otherwise it stays up with probability $1 - \alpha h + o(h)$ during $[t, t+h]$. Likewise, if it is down at time t, it goes up in $[t, t+h]$ with probability $\beta h + o(h)$, independently of its past, or it stays down with probability $1 - \beta h + o(h)$.

(a) If it is up at $t = 0$, find the probability that it is down at time $t > 0$.
(b) Let $N(t)$ be the number of occasions on which it has gone down during $[0, t]$. Find $\mathbf{E}(N(t))$.
(c) Find the probability generating function $\mathbf{E}(z^{N(t)})$.

Solution (a) Let $X(t)$ be the state of the machine, where $X(t) = 0$ if it is up at t, and $X(t) = 1$ if it is down at t. By the assumptions of the question $X(t)$ is a Markov process, and

$$p_{01}(t+h) = p_{00}(t)\alpha h + p_{01}(t)(1 - \beta h) + o(h)$$
$$p_{00}(t+h) = p_{01}(t)\beta h + p_{00}(t)(1 - \alpha h) + o(h).$$

Hence

(6) $$\frac{d}{dt}p_{01}(t) = -\beta p_{01}(t) + \alpha p_{00}(t)$$

(7) $$\frac{d}{dt}p_{00}(t) = -\alpha p_{00}(t) + \beta p_{01}(t)$$

and, since $X(0) = 0$,

(8) $$p_{00}(0) = 1.$$

Solving (6), (7) and (8) gives the required probability of being down at t

$$p_{01}(t) = \frac{\alpha}{\alpha + \beta}(1 - e^{-(\alpha+\beta)t}).$$

(b) The first thing to realize here is that $N(t)$ is *not* a Markov process. However, if we let $M(t)$ be the number of times the machine has gone up during $[0, t]$, then

$$\boldsymbol{Y}(t) = \{N(t), M(t)\}$$

is a Markov process. By the assumptions of the problem, as $h \to 0$

$$\mathbf{P}(\boldsymbol{Y}(t+h) = (k+1, k) | \boldsymbol{Y}(t) = (k, k)) = \alpha h + o(h)$$
$$\mathbf{P}(\boldsymbol{Y}(t+h) = (k, k) | \boldsymbol{Y}(t) = (k, k)) = 1 - \alpha h + o(h)$$
$$\mathbf{P}(\boldsymbol{Y}(t+h) = (k, k) | \boldsymbol{Y}(t) = (k, k-1)) = \beta h + o(h)$$
$$\mathbf{P}(\boldsymbol{Y}(t+h) = (k, k-1) | \boldsymbol{Y}(t) = (k, k-1)) = 1 - \beta h + o(h).$$

Hence, if $f_{kj}(t) = \mathbf{P}(\boldsymbol{Y}(t) = (k, j))$, the forward equations may be derived routinely as

(9) $$\frac{d}{dt}f_{kk}(t) = -\alpha f_{kk}(t) + \beta f_{k,k-1}(t); \quad k \geqslant 0.$$

(10) $$\frac{d}{dt}f_{k,k-1}(t) = -\beta f_{k,k-1}(t) + \alpha f_{k-1,k-1}(t); \quad k \geqslant 1,$$

where $f_{0,-1}(t) = 0$.

Now consider a Poisson process $Z(t)$ of rate α. By construction of $N(t)$, $\mathbf{P}(N(t) = k) \leqslant \mathbf{P}(Z(t) = k)$ for all k. Hence $\mathbf{E}(N(t))$ exists, since it is less than

$\mathbf{E}(Z(t)) = \alpha t$. In fact

$$\mathbf{P}(N(t) = k) = f_{kk}(t) + f_{k,k-1}(t),$$

and so $\mathbf{E}(N(t)) = m_1(t) + m_2(t)$, where

$$m_1(t) = \sum_{k=1}^{\infty} k f_{kk}(1)$$

and

$$m_2(t) = \sum_{k=1}^{\infty} k f_{k,k-1}(t).$$

Multiplying (9) and (10) by k, and summing gives

$$\frac{dm_1}{dt} = -\alpha m_1 + \beta m_2 \tag{11}$$

and

$$\begin{aligned} \frac{dm_2}{dt} &= -\beta m_2 + \alpha \sum_k (k - 1 + 1) f_{k-1,k-1}(t) \\ &= -\beta m_2 + \alpha m_1 + \alpha \sum_k f_{k-1,k-1}(t) \\ &= -\beta m_2 + \alpha m_1 + \alpha \mathbf{P}(N(t) = M(t)) \\ &= -\beta m_2 + \alpha m_1 + \alpha p_{00}(t), \end{aligned} \tag{12}$$

since the machine is up if $N(t) = M(t)$. Hence, adding (11) and (12), we have

$$\frac{d}{dt}\mathbf{E}(N(t)) = \alpha p_{00}(t),$$

and so, using the result of (a),

$$\mathbf{E}(N(t)) = \int_0^t \alpha\left(\frac{\beta}{\alpha + \beta} + \frac{\alpha e^{-(\alpha+\beta)v}}{\alpha + \beta}\right) dv.$$

It follows that as $t \to \infty$, $t^{-1}\mathbf{E}(N(t)) \to \alpha\beta/(\alpha + \beta)$.

(c) Let

$$x(t, z) = \sum_{k=0}^{\infty} f_{kk}(t) z^k$$

and

$$y(t, z) = \sum_{k=1}^{\infty} f_{k,k-1}(t) z^k.$$

Then $\mathbf{E}(z^{N(t)}) = x(t, z) + y(t, z)$, and multiplying each of (9) and (10) by z^k and summing over k gives

$$\frac{\partial x}{\partial t} = -\alpha x + \beta z y, \quad \text{and} \quad \frac{\partial y}{\partial t} = -\beta y + \alpha z x.$$

This pair of simultaneous differential equations is solved by elementary methods, subject to the initial conditions $x(0, z) = 1$ and $y(0, z) = 0$, to yield

$$x(t, z) = [(\alpha + \lambda_2(z))/(\lambda_2(z) - \lambda_1(z))]e^{\lambda_1(z)t} + [(\alpha + \lambda_1(z))/(\lambda_1(z) - \lambda_2(z))]e^{\lambda_2(z)t}$$

and

$$y(t, z) = (\alpha + \lambda_1)(\alpha + \lambda_2)(e^{\lambda_1(z)t} - e^{\lambda_2(z)t})/(\beta z(\lambda_2 - \lambda_1))$$

where

$$\lambda_1(z) = \frac{1}{2}[-(\alpha + \beta) + ((\alpha - \beta)^2 + 4\alpha\beta z^2)^{\frac{1}{2}}],$$

and

$$\lambda_2(z) = \frac{1}{2}[-(\alpha + \beta) - ((\alpha - \beta)^2 + 4\alpha\beta z^2)^{\frac{1}{2}}].$$ ●

Recalling our results about chains in discrete time, it is natural to wonder whether chains in continuous time have stationary distributions, and whether $p_{ij}(t)$ converges as $t \to \infty$. A detailed answer to these questions is far beyond our scope, but we can make some guarded remarks. Let us start with the simplest.

(13) **Theorem** Let $X(t)$ be a finite Markov process with transition matrix $p_{ij}(t)$. Then $\lim_{t\to\infty} p_{ij}(t)$ exists for all i and j.

If $X(t)$ is irreducible then the limit is independent of i, we write

$$\lim_{t\to\infty} p_{ij}(t) = \pi_j.$$

Furthermore π_j satisfies $\Sigma_j \pi_j = 1$, and

$$\pi_j = \sum_i \pi_i p_{ij}(t); \quad t \geqslant 0,$$

$\boldsymbol{\pi}$ is the stationary distribution of $X(t)$.

(A chain is irreducible if for each i, j there is some finite t such that $p_{ij}(t) > 0$.)

(14) **Example 5 revisited** In this case we have

$$p_{01}(t) = \frac{\alpha}{\alpha + \beta}(1 - e^{-(\alpha+\beta)t}) \to \begin{cases} \dfrac{\alpha}{\alpha + \beta} & \text{if } \alpha + \beta > 0 \\ 0 & \text{otherwise,} \end{cases}$$

with three similar results for p_{00}, p_{10} and p_{11}. If $\alpha = \beta = 0$ then $X(t) = X(0)$ for all t. On the other hand, if $\alpha\beta > 0$ then the chain is irreducible and has stationary distribution $\boldsymbol{\pi} = (\frac{\beta}{\alpha+\beta}, \frac{\alpha}{\alpha+\beta})$. We can check that for all t

$$\begin{aligned} \pi_0 &= \frac{\beta}{\alpha + \beta} \\ &= \frac{\beta}{\alpha + \beta}\left(\frac{\beta}{\alpha + \beta} + \frac{\alpha e^{-(\alpha+\beta)t}}{\alpha + \beta}\right) + \frac{\alpha}{\alpha + \beta}\left(\frac{\beta}{\alpha + \beta} - \frac{\beta e^{-(\alpha+\beta)t}}{\alpha + \beta}\right) \\ &= \pi_0 p_{00}(t) + \pi_1 p_{10}(t). \end{aligned}$$ ●

In practice the state space is often countably infinite, and of course we would like to use the forward equations (4). The following theorem is relevant.

(15) Theorem Let $X(t)$ be an irreducible Markov process with transition matrix $p_{ij}(t)$ satisfying (2) and (3) above. Then $\lim_{t\to\infty} p_{ij}(t)$ exists and is independent of i, for all j.

There are two possibilities.

Either (a)
$$\lim_{t\to\infty} p_{ij}(t) = \pi_j > 0$$

where $\Sigma\pi_i = 1$ and

$$\pi_j = \sum_i \pi_i p_{ij}(t)$$

and

(16)
$$\sum_i \pi_i g_{ij} = 0$$

for all j;

or (b)
$$\lim_{t\to\infty} p_{ij}(t) = 0.$$

We give no proof of this result, but you may notice with some pleasure that there are no tiresome reservations about periodicity.

(17) Example: Queue Let $X(t)$ be the length of queue formed before a single service point at time t. The times between arrivals are exponentially distributed with parameter λ; each individual is served on reaching the head of the queue; each service time is exponentially distributed with parameter μ; interarrival times and service times are all independent of each other.

It follows that as $h\to 0$, when $X(t) > 0$,

$$\mathbf{P}(X(t+h) - X(t) = 1) = \mathbf{P}(\text{one arrival; no service completed in } (t, t+h))$$
$$= \lambda h(1-\mu h) + o(h) = \lambda h + o(h).$$

Likewise

$$\mathbf{P}(X(t+h) - X(t) = -1) = \mu h + o(h)$$

and

$$\mathbf{P}(X(t+h) - X(t) = 0) = 1 - (\lambda+\mu)h + o(h).$$

The process $X(t)$ is Markov, by the properties of the exponential density, and the above statements show that

$$p_{i,i+1}(h) = \lambda h + o(h),$$
$$p_{i,i-1}(h) = \mu h + o(h); \quad i \neq 0$$
$$p_{ii}(h) = 1 - (\lambda+\mu)h + o(h); \quad i \neq 0.$$

When $i = 0$, no service can be completed, so we have $p_{01}(h) = \lambda h + o(h)$ and $p_{00}(h) = 1 - \lambda h + o(h)$. These supply us with all the numbers g_{ij}, and so by (16) to find the stationary distribution we seek a solution to the equations $\boldsymbol{\pi}\mathbf{G} = 0$, that is

$$-\lambda\pi_0 + \mu\pi_1 = 0$$
$$\lambda\pi_{i-1} - (\lambda+\mu)\pi_i + \mu\pi_{i+1} = 0; \quad i \geqslant 1.$$

Solving recursively shows that

$$\pi_i = \left(\frac{\lambda}{\mu}\right)^i \pi_0; \quad i \geqslant 0,$$

and so a stationary distribution exists if $(\frac{\lambda}{\mu}) < 1$, and it is given by

$$\pi_i = \left(1 - \frac{\lambda}{\mu}\right)\left(\frac{\lambda}{\mu}\right)^i. \qquad \bullet$$

WORKED EXAMPLES AND EXERCISES

9.10 Example: Crossing a cube

One vertex O of a unit cube is at the origin $(0, 0, 0)$. The others are at $(0, 0, 1)$, $(0, 1, 0)$ and so on. A particle performs a random walk on the vertices of this cube as follows. Steps are of unit length, and from any vertex it steps in the x direction with probability α, the y direction with probability β or the z direction with probability γ, where $\alpha + \beta + \gamma = 1$.

(a) Let T be the first passage time from O to V. Find $\mathbf{E}(s^T)$, and deduce that

$$\mathbf{E}(T) = 1 + \frac{1}{\alpha} + \frac{1}{\beta} + \frac{1}{\gamma}.$$

(b) Let X be the number of visits that the walk makes to V before the first return to O. Show that $\mathbf{E}(X) = 1$.

Solution The walk visits O whenever an even number of steps has been taken in all three possible directions (x, y and z directions). Thus

$$u(2n) = \mathbf{P}(\text{the walk visits } O \text{ at the } 2n\text{th step}) = \sum_{i+j+k=n} \alpha^{2i}\beta^{2j}\gamma^{2k} \frac{(2n)!}{(2i)!(2j)!(2k)!}.$$

Now write $\beta + \gamma - \alpha = a$, $\alpha + \gamma - \beta = b$, and $\alpha + \beta - \gamma = c$. It is easy to check, by expanding each side, that

$$\begin{aligned} 4u(2n) &= (\alpha + \beta + \gamma)^{2n} + (\beta + \gamma - \alpha)^{2n} + (\alpha + \gamma - \beta)^{2n} + (\alpha + \beta - \gamma)^{2n} \\ &= 1 + a^{2n} + b^{2n} + c^{2n}. \end{aligned}$$

Hence

$$U(s) = \sum_{n=0}^{\infty} s^{2n} u(2n) = \frac{1}{4}\left(\frac{1}{1 - s^2} + \frac{1}{1 - a^2 s^2} + \frac{1}{1 - b^2 s^2} + \frac{1}{1 - c^2 s^2}\right).$$

Similarly, starting from O, the walk visits V whenever an odd number of steps has been taken in all three possible directions. Hence

$$\begin{aligned} u_V(2n + 1) &= \mathbf{P}(\text{the walk visits } V \text{ at the } (2n + 1)\text{th step}) \\ &= \sum_{i+j+k=n-1} \alpha^{2i+1}\beta^{2j+1}\gamma^{2k+1} \frac{(2n + 1)!}{(2i + 1)!(2j + 1)!(2k + 1)!}. \end{aligned}$$

Now it is easy to check as above that $4u_V(2n + 1) = 1 - a^{2n+1} - b^{2n+1} - c^{2n+1}$.

Hence

$$U_V(s) = \sum_{n=1}^{\infty} u_V(2n+1)s^{2n+1} = \frac{s^3}{4}\left(\frac{1}{1-s^2} - \frac{a^3}{1-a^2s^2} - \frac{b^3}{1-b^2s^2} - \frac{c^3}{1-c^2s^2}\right).$$

Now we use (9.3.21) to see that

$$\mathbf{E}(s^T) = \frac{U_V(s)}{U(s)}.$$

Hence, evaluating $\frac{d}{ds}\mathbf{E}(s^T)$ at $s = 1$ yields $\mathbf{E}(T) = 1 + \alpha^{-1} + \beta^{-1} + \gamma^{-1}$ as required.

(b) Let ρ be the probability that the walk returns to O before ever visiting V. By symmetry this is also the probability that a walk started at V never visits O before returning to V. For $X = k$ it is necessary for the walk to reach V before revisiting O, then revisit V on $k - 1$ occasions without visiting O, and finally return to O with no further visit to V. Hence $\mathbf{P}(X = k) = (1-\rho)\rho^{k-1}(1-\rho)$, by the Markov property. Therefore

$$\mathbf{E}(X) = \sum_{k=1}^{\infty} k\rho^{k-1}(1-\rho)^2 = 1.$$

(1) Exercise Suppose that at every step the walk may remain at its current vertex with probability δ, where now $\alpha + \beta + \gamma + \delta = 1$. Find:

(a) the mean recurrence time of O; (b) $\mathbf{E}(X)$; (c) $\mathbf{E}(T)$.

(2) Exercise Let W be the vertex $(1, 1, 0)$, and define $\hat{T}$ to be the number of steps until the walk first visits V or W starting from O. (That is to say, $\hat{T}$ is the first passage time from O to $\{V, W\}$.) Show that $\mathbf{E}(\hat{T}) = (\alpha^{-1} + \beta^{-1})\gamma$.

9.11 Example: Reversible chains

A collection of random variables $(X(n);\ n \in \mathbb{Z})$ is called *reversible* if $(X(n_1), X(n_2), \ldots, X(n_r))$ has the same distribution as $(X(m-n_1), X(m-n_2), \ldots, X(m-n_r))$ for all m and $n_1 < n_2 < \ldots < n_r$. Let X be an irreducible non-null recurrent aperiodic Markov chain.

(a) Prove that the Markov chain X with transition probabilities p_{ij} is reversible if and only if it is stationary and there exist $(\pi_i;\ i \in S)$ such that for all $i, j \in S$

(1) $$\pi_i > 0,$$

(2) $$\sum_{i \in S} \pi_i = 1,$$

(3) $$\pi_i p_{ij} = \pi_j p_{ji}.$$

(b) Prove further that if X is stationary then it is reversible if and only if

(4) $$p_{i_1 i_2} p_{i_2 i_3} \cdots p_{i_r i_1} = p_{i_1 i_r} p_{i_r i_{r-1}} \cdots p_{i_2 i_1}$$

for any finite sequence of states $i_1, i_2, \ldots, i_r$ in S.

Solution (a) The truth of (1), (2) and (3) implies that π_i is the stationary distribution of X, for summing (3) over j yields $\sum_i \pi_i p_{ij} = \pi_j$, as required.

Next we note that (3) implies

(5) $$\pi_i p_{ij}(n) = \pi_j p_{ji}(n).$$

To see this consider any n-step path from i to j, and then using (3) gives

$$\pi_i p_{ii_1} p_{i_1 i_2} \dots p_{i_{n-1}j} = p_{i_1 i} \pi_{i_1} p_{i_1 i_2} \dots p_{i_{n-1}j} = p_{i_1 i} p_{i_2 i_1} \dots p_{j i_{n-1}} \pi_j$$

after repeated appliations of (3). Now summing over all paths from i to j gives (5).

Now applying (5) shows that

$$\mathbf{P}(X_{n_1} = i_1, X_{n_2} = i_2, \dots, X_{n_r} = i_r) = \pi_{i_1} p_{i_1 i_2}(n_2 - n_1) \dots p_{i_{r-1} i_r}(n_r - n_{r-1})$$
$$= \mathbf{P}(X_{m-n_1} = i_1, \dots, X_{m-n_r} = i_r)$$

as required. The converse is obvious, and so we have finished (a).

(b) If the chain is reversible then (3) holds. Hence

$$p_{i_1 i_2} \dots p_{i_r i_1} = \frac{\pi_{i_2}}{\pi_{i_1}} p_{i_2 i_1} \dots \frac{\pi_{i_1}}{\pi_{i_r}} p_{i_1 i_r},$$

which is (4) since all the π_i cancel successively. Conversely, if (4) holds, we may choose two states i_1 and i_r (say) as fixed and equal, so $i_1 = i_r = j$. Then summing (4) over the remaining indices yields

$$p_{ij}(r-1) p_{ji} = p_{ji}(r-1) p_{ij}.$$

Now, as the chain is aperiodic and non-null recurrent, we can let $r \to \infty$ and obtain (3) as required.

Remark This example is important because a remarkable number of Markov chains are reversible in equilibrium (particularly those encountered in examinations). In the exercises you may assume that X is aperiodic and irreducible.

(6) Exercise Let $(X_n; n \in \mathbb{Z})$ be a Markov chain, and let $Y_n = X_{-n}$ be the chain X reversed. Show that for $n_1 < n_2 < \dots < n_r$

$$\mathbf{P}(Y_{n_r} = k \mid Y_{n_1}, \dots, Y_{n_{r-1}}) = \mathbf{P}(Y_{n_r} = k \mid Y_{n_{r-1}}).$$

Show that if X has stationary distribution $\boldsymbol{\pi}$ and transition probabilities $\boldsymbol{P}$, then, in equilibrium, Y is a Markov chain with transition probabilities $q_{ij} = \pi_j \pi_i^{-1} p_{ji}$.

(7) Exercise 6 continued Let X be a Markov chain, with transition probabilities p_{ij}. Show that if there exists a stochastic matrix (q_{ij}) and a mass function (π_i) such that $\pi_i q_{ij} = \pi_j p_{ji}$, then q_{ij} is the transition matrix of X reversed, and π_i is its stationary distribution.

(8) Exercise Let X_n be a Markov chain on the non-negative integers with transition matrix

$$p_{ij} = \begin{cases} \lambda_i & \text{if } j = i+1 > 0 \\ \mu_i & \text{if } j = i-1 \geqslant 0 \\ 0 & \text{otherwise} \end{cases}$$
$$p_{00} = \mu_0.$$

Show that X is reversible in equilibrium. Deduce that X has a stationary distribution if and only if $\sum_{n=1}^{\infty} \prod_{k=1}^{n} \frac{\lambda_{k-1}}{\mu_k} < \infty$.

(9) Exercise Let X_n and Y_n be independent reversible Markov chains with stationary distributions $\boldsymbol{\pi}$ and $\boldsymbol{\nu}$ respectively. Let $Z_n = (X_n, Y_n)$. Show that Z_n is reversible in equilibrium.

9.12 Example: Diffusion models

(a) Two separate containers together contain m distinguishable particless. At integer times, $t = 1, 2, \ldots$ one of the particles is selected at random and transferred to the other container. Let X_n be the number of particles in the first container after the nth transfer.

(i) Show that $(X_n;\ n \geq 0)$ is a Markov chain and write down its transition matrix.

(ii) Find the stationary distribution of X.

(b) The two containers C_1 and C_2 are now separated by a semipermeable membrane. At each time t a particle is selected at random; if it is in C_1 then it is transferred with probability α, if it is in C_2 it is transferred with probability β. Otherwise the particles remain where they are. Show that X has stationary distribution

(1) $$\pi_i = \frac{\alpha^{m-i}\beta^i}{(\alpha+\beta)^m}\binom{m}{i}.$$

Solution (a) Given $X_0, \ldots, X_n = j$, the probability that a particle in C_1 is selected for transfer is j/m, and the probability that a particle in C_2 is selected for transfer is $(m - j)/m$. Hence X is a Markov chain, and

$$p_{j,j+1} = (m-j)/m,$$
$$p_{j,j-1} = j/m,$$
$$p_{jk} = 0 \quad \text{when} \quad |k - j| \neq 1.$$

The chain is of the type considered in Exercise 9.11.8. Therefore the chain is reversible, and the stationary distribution may be obtained by solving $\pi_i p_{ij} = \pi_j p_{ji}$. Thus

$$\pi_{i+1} = \pi_i p_{i,i+1}/p_{i+1,i} = \pi_i(m-i)/(i+1) = \pi_0\binom{m}{i+1} \quad \text{on iterating.}$$

Since $\sum_0^m \pi_i = 1$, it follows that

$$\pi_0^{-1} = \sum_{i=0}^{m}\binom{m}{i} = 2^m$$

and so $\pi_i = \binom{m}{i}2^{-m}$. This is a symmetric binomial distribution.

(b) By the same reasoning as given in (a), this is a Markov chain with

$$p_{i,i+1} = \beta(m-i)/m$$
$$p_{i,i-1} = \alpha i/m$$
$$p_{ii} = 1 - \alpha i/m - \beta(m-i)/m$$
$$p_{ij} = 0 \quad \text{when} \quad |i - j| > 1.$$

Again, this is reversible, so the stationary distribution is given by

(2) $$\pi_{i+1} = \pi_i p_{i,i+1}/p_{i+1,i} = \pi_i\beta(m-i)/(\alpha(i+1))$$

(3) $$= \pi_0\left(\frac{\beta}{\alpha}\right)^{i+1}\binom{m}{i+1} \quad \text{on iterating.}$$

Now requiring $\sum \pi_i = 1$, yields

$$\pi_i = \frac{\alpha^{m-i}\beta^i}{(\alpha+\beta)^m}\binom{m}{i},$$

which is an asymmetric binomial distribution.

Remark This is the Ehrenfest model for diffusion (and heat transfer) (1907).

(4) Exercise Is it true that as $n \to \infty$, $p_{ij}(n) \to \pi_j$ in either case (a) or case (b)?

(5) Exercise (a) Show that $\mathbf{E}(X_n) = \left(1 - \frac{2}{m}\right)^n \left(\mathbf{E}(X_0) - \frac{m}{2}\right) + \frac{m}{2}$, when $\alpha = \beta = 1$.

(b) What is $\mathbf{E}(X_n)$ when $\alpha \neq 1 \neq \beta$?

(6) Exercise: Bernoulli diffusion model Two adjacent containers C_1 and C_2 each contain m particles. Of these $2m$ particles, m are of type A and m are of type B. At $t = 1, 2, \ldots$ one particle is selected at random from each container, and these two particles are each transferred to the other container. Let X_n be the number of type A particles in C_1.

(a) Show that X is a Markov chain and write down its transition matrix.
(b) Find the stationary distribution.
(c) Is the stationary distribution the limiting distribution of X?

(7) Exercise In the above exercise suppose that the containers are separated by a semi-permeable membrane. In this case, if particles of different types are chosen they are exchanged with probability α if the type A particle is in C_1, or with probability β if the type A particle is in C_2. Find the stationary distribution of X.

9.13 Example: The renewal chains

Let $(f_n;\ n \geq 1)$ satisfy $f_n \geq 0$ and

$$\sum_{i=1}^{n} f_i \leq 1. \tag{1}$$

Define a sequence $(u_n;\ n \geq 0)$ by $u_0 = 1$ and

$$u_n = \sum_{r=1}^{n} f_r u_{n-r}; \quad n \geq 1. \tag{2}$$

Such a sequence $(u_n;\ n \geq 0)$ is called a *renewal sequence*.

(a) Show that $(u_n;\ n \geq 0)$ is a renewal sequence if and only if there is a Markov chain U_n such that for some state $s \in S$,

$$u_n = \mathbf{P}(U_n = s | U_0 = s).$$

(b) Let X be a random variable having the probability mass function $(f_n;\ n \geq 1)$. Show that the chain U is recurrent if $\sum_n f_n = 1$, and non-null if $\mathbf{E}(X) < \infty$.

(c) Explain why u_n is called a renewal sequence.

Solution (a) Let u_n and f_n be as defined above. Define the sequence $(F_n;\ n \geq 0)$ by $F_0 = 0$ and

$$F_n = \sum_{r=1}^{n} f_r; \quad n \geq 1.$$

Next let $(U_n;\ n \geqslant 0)$ be a Markov chain taking values in the non-negative integers, with $U_0 = 0$, and having transition probabilities

(3) $$p_{i0} = \frac{f_{i+1}}{1 - F_i}$$

(4) $$p_{i,i+1} = 1 - p_{i0} = \frac{1 - F_{i+1}}{1 - F_i}; \quad i \geqslant 0.$$

Now let us calculate the first return probability $f_{00}(n)$, that is the probability that the chain first returns to 0 at the nth step. At each stage the chain either does so return, or it increases by 1. Hence

(5) $$\begin{aligned} f_{00}(n) &= p_{01}p_{12}\cdots p_{n-2,n-1}p_{n-1,0} \\ &= \frac{1 - F_1}{1 - F_0}\cdot\frac{1 - F_2}{1 - F_1}\cdots\frac{1 - F_{n-1}}{1 - F_{n-2}}\cdot\frac{f_n}{1 - F_{n-1}} = f_n. \end{aligned}$$

It follows by conditional probability, and the Markov property, that

(6) $$\mathbf{P}(U_n = 0) = \sum_{r=1}^{n} f_{00}(r)\mathbf{P}(U_{n-r} = 0) = \sum_{r=1}^{n} f_r\mathbf{P}(U_{n-r} = 0).$$

Since $\mathbf{P}(U_0 = 0) = 1$, it follows by comparison with (2) that $u_n = \mathbf{P}(U_n = 0 | U_0 = 0)$ as required.

(b) First observe that if $F_j = 1$ for some $j < \infty$, then the chain is finite. Obviously in this case $\mathbf{E}(X) < \infty$ and the chain is recurrent and non-null. Otherwise, if $\Sigma f_i = 1$ then $\sum_n f_{00}(n) = 1$ by (5), and hence the chain is recurrent. Since it is irreducible we can settle whether or not it is non-null by seeking a stationary distribution $\boldsymbol{\pi}$. Any such stationary distribution satisfies

$$\begin{aligned} \pi_{i+1} &= \pi_i p_{i,i+1}; \quad i \geqslant 0 \\ &= \pi_i \frac{1 - F_{i+1}}{1 - F_i} = \pi_0(1 - F_{i+1}) \quad \text{on iterating.} \end{aligned}$$

Hence $\boldsymbol{\pi}$ is a stationary distribution satisfying $\sum_i \pi_i = 1$ if $\sum_i (1 - F_i) < \infty$. That is to say U is non-null if $\mathbf{E}(X) < \infty$, and then

(7) $$\pi_i = \frac{1 - F_i}{\mathbf{E}(X)} = \frac{\mathbf{P}(X > i)}{\mathbf{E}(X)}.$$

(c) Recall the recurrent event or renewal processes defined in Section 6.7. Events may occur at integer times and the intervals between successive events are independent and identically distributed random variables $(X_i;\ i \geqslant 1)$ where $f_X(r) = f_r$. An event occurs at $n = 0$. Now the construction of the chain U in(3) and (4) allows us to identify visits of U to zero with the occurrence of an event in this renewal process. And at any time n, the state U_n of the chain is just the time elapsed since the most recent event of the renewal process. Thus U_n is the current life (or age) of the renewal process at time n.

Finally, u_n is just the probability that an event of the renewal process occurs at time n.

(8) Exercise Show that if u_n and v_n are renewal sequences, then $u_n v_n$ is a renewal sequence.

(9) Exercise Show that if u_n is a renewal sequence then $(u_{nd};\ u \geqslant 0)$ is a renewal sequence for any fixed d.

(10) Exercise Let $(X_i;\ i \geqslant 1)$ be the inter-event times of a discrete renewal process, and at any time n let B_n be the time until the next following event of the process. (That is, B_n is the excess life or balance of life.) Show that B_n is a Markov chain, and find the stationary distribution when it exists.

(11) Exercise Write down the transition probabilities of the chain U_n reversed in equilibrium, and also write down the transition probabilities of B_n reversed in equilibrium. Explain your answers.
[Hint: use Exercise 9.11.6.]

(12) Exercise Use Theorem 9.6.5 to show that $\lim_{n\to\infty} u_n = \pi_0$.

(13) Exercise Recall the recurrent event (renewal) processes of Section 6.7. Use (12) to show that as $n \to \infty$, $u_n \to \frac{1}{\mathbf{E}(X_2)}$ and $v_n \to \frac{1}{\mathbf{E}(X_2)}$.

9.14 Example: Persistence

Let i and j be states of the Markov chain X. We write $i \to j$ if there exists $n < \infty$ such that $p_{ij}(n) > 0$. Let V_{ij} be the number of occasions on which X visits j, given that initially $X_0 = i$. Define

$$\eta_{ij} = \mathbf{P}(V_{ij} = \infty),$$

and recall that T_{ij} is the first passage time from i to j.

(a) Show that

$$\eta_{ii} = \begin{cases} 1 & \text{if } i \text{ is persistent} \\ 0 & \text{if } i \text{ is transient.} \end{cases}$$

(b) Show that

$$\eta_{ij} = \begin{cases} \mathbf{P}(T_{ij} < \infty) & \text{if } j \text{ is persistent} \\ 0 & \text{if } j \text{ is transient.} \end{cases}$$

(c) Show that if $i \to j$ and i is persistent then $\eta_{ij} = \eta_{ji} = 1$.

Solution (a) First note that $V_{ii} \geqslant 1$ if and only if $T_i < \infty$, so $\mathbf{P}(V_{ii} \geqslant 1) = \mathbf{P}(T_i < \infty)$. Next

$$\mathbf{P}(V_{ii} \geqslant 2) = \mathbf{P}(V_{ii} \geqslant 2 | V_{ii} \geqslant 1)\mathbf{P}(V_{ii} \geqslant 1) = \mathbf{P}(V_{ii} \geqslant 2 | T_i < \infty)\mathbf{P}(V_{ii} \geqslant 1).$$

However, we have shown in Example 9.3.17 that the Markov property is preserved at T_i when it is finite, and therefore $\mathbf{P}(V_{ii} \geqslant 2 | T_i < \infty) = \mathbf{P}(V_{ii} \geqslant 1)$.

Now an obvious induction shows that

$$\mathbf{P}(V_{ii} \geqslant n) = (\mathbf{P}(V_{ii} \geqslant 1))^n = (\mathbf{P}(T_i < \infty))^n \to \begin{cases} 1 & \text{if } \mathbf{P}\ (T_i < \infty) = 1 \\ 0 & \text{if } \mathbf{P}\ (T_i < \infty) < 1 \end{cases}$$

as $n \to \infty$, as required.

(b) Using the same idea as in (a) we write

$$\begin{aligned} \mathbf{P}(V_{ij} \geqslant m) &= \mathbf{P}(V_{ij} \geqslant m | V_{ij} \geqslant 1)\mathbf{P}(V_{ij} \geqslant 1) \\ &= \mathbf{P}(V_{ij} \geqslant m | T_{ij} < \infty)\mathbf{P}(T_{ij} < \infty) = \mathbf{P}(V_{jj} \geqslant m - 1)\mathbf{P}(T_{ij} < \infty) \end{aligned}$$

because the Markov property is preserved at T_{ij}. Now allowing $m \to \infty$ gives the result, by (a).

(c) Since i is persistent

$$\begin{aligned}1 = \mathbf{P}(V_{ii} = \infty) &= \mathbf{P}(\{V_{ij} = 0\} \cap \{V_{ii} = \infty\}) + \mathbf{P}(\{V_{ij} > 0\} \cap \{V_{ii} = \infty\}) \\ &\leqslant \mathbf{P}(V_{ij} = 0) + \mathbf{P}(\{T_{ij} < \infty\} \cap \{V_{ii} = \infty\}) \\ &= 1 - \mathbf{P}(T_{ij} < \infty) + \mathbf{P}(T_{ij} < \infty)\mathbf{P}(V_{ji} = \infty)\end{aligned}$$

because the Markov property is preserved at T_{ij}.

Hence $\mathbf{P}(V_{ji} = \infty) \geqslant 1$, and therefore $\eta_{ji} = 1$. Hence $\mathbf{P}(T_{ij} < \infty) = 1$ and so $j \to i$. It follows that $\eta_{ij} = 1$.

(1) **Exercise** Show that if i is persistent and $i \to j$, then j is persistent.

(2) **Exercise** Show that $\eta_{ij} = 1$ if and only if $\mathbf{P}(T_{ij} < \infty) = \mathbf{P}(T_j < \infty) = 1$.

(3) **Exercise** Show that if $i \to j$ and $j \to i$ then i and j have the same class and period.

(4) **Exercise** Show that if X is irreducible and persistent then $\mathbf{P}(T_{ij} < \infty) = \mathbf{P}(T_{ji} < \infty) = 1$.

(5) **Exercise** Show that if $i \to j$ but $j \not\to i$, then i is transient.

9.15 Example: First passages and Bernoulli patterns

Let X be a finite Markov chain with n-step transition probabilities $p_{ij}(n)$. As usual T_{ij} is the first passage time from i to j, with mean μ_{ij}; T_j is the recurrence time of j with mean μ_j. We write $F_{ij}(s) = \mathbf{E}(s^{T_{ij}})$, and $P_{ij}(s) = \sum_{n=0}^{\infty} p_{ij}(n)s^n$.

(a) A biased coin is tossed repeatedly, let X_n be the outcome of tosses $n+1$, $n+2$ and $n+3$; for $n \geqslant 0$. (Thus $(X_n;\ n \geqslant 0)$ is a Markov chain with state space S comprising all triples using H and T, namely HHH, HTH, HHT, and so on.) Show that for any i and $j \in S$,

(1)
$$\mu_{ij} = [1 + p_{jj}(1) + p_{jj}(2) - p_{ij}(1) - p_{ij}(2)]\mu_j.$$

Deduce that if $i = HHH$, and $j = THT$, and the coin is fair, then $\mu_{ij} = 10$ and $\mu_{ji} = 14$.

(b) Let D be a given subset of the states of a finite Markov chain $(Y_n;\ n \geqslant 0)$, and let T_{sD} be the first passage time from the state $s \notin D$ into D, with mean $\mu_{sD} = \mathbf{E}(T_{sD})$. Also let ϕ_{sj} be the probability that the chain first enters D at the state j. Show that for $i \in D$

(2)
$$\mu_{si} = \mu_{sD} + \sum_{j \in D} \phi_{sj}\mu_{ji}.$$

(c) Hence show that in an unlimited sequence of tosses of a fair coin, the probability that the consecutive sequence THT occurs before HHH is $\frac{7}{12}$.

Solution (a) From Theorem 6.2.13 we have that

(3)
$$\mu_{ij} = \lim_{s \uparrow 1} \frac{1 - F_{ij}(s)}{1 - s} = \lim_{s \uparrow 1} \frac{P_{jj}(s) - P_{ij}(s)}{(1 - s)P_{jj}(s)} \quad \text{by Theorem 9.3.20.}$$

Now because tosses of the coin are independent we have

(4)
$$p_{ij}(n) = p_{jj}(n) \quad \text{for} \quad n \geqslant 3.$$

Also, using Theorems 6.2.13 and 9.3.20 again gives

(5) $$\lim_{s \uparrow 1} (1-s)P_{jj}(s) = \mu_j^{-1},$$

and now (3), (4) and (5) give (1).

If the chance of a head is p, then $p_{ij}(1) = p_{ij}(2) = p_{jj}(1) = 0$, and $p_{jj}(2) = p(1-p)$. Hence

$$\mu_{ij} = [1 + p(1-p)]\mu_j.$$

When the coin is fair $p = \frac{1}{2}$ and $\mu_j = 8$, so $\mu_{ij} = 10$. Likewise $p_{ji}(1) = p_{ji}(2) = 0$, and $p_{ii}(1) = p$, $p_{ii}(2) = p^2$. Hence when the coin is fair

$$\mu_{ji} = \left(1 + \frac{1}{2} + \frac{1}{4}\right) 8 = 14.$$

(b) Let D_j denote the event that the chain first enters D at j. Then

(6) $$\mu_{si} = \mathbf{E}(T_{si}) = \mathbf{E}(T_{si} - T_{sD} + T_{sD}) = \mathbf{E}(T_{si} - T_{sD}) + \mu_{sD}$$

$$= \sum_{j \in D} \mathbf{E}(T_{si} - T_{sj} | D_j)\phi_{sj} + \mu_{sD}.$$

However, given D_j, the chain continues its journey to i independently of the past, that is to say

(7) $$\mathbf{E}(T_{si} - T_{sj} | D_j) = \mu_{ji}.$$

Substituting (7) into (6) gives (2).

(c) Now let the outcomes of successive tosses be $(S_n;\ n \geqslant 1)$ and define the Markov chain $(Y_n;\ n \geqslant 0)$ by

$$Y_0 = \phi = s$$
$$Y_1 = S_1$$
$$Y_2 = S_1 S_2$$
$$Y_n = S_{n-2}S_{n-1}S_n; \quad n \geqslant 3.$$

Thus on the third step the chain enters the closed irreducible subset of sequences of length 3. Setting $HHH \equiv 1$, $THT \equiv 2$, so $D = \{HHH, THT\} = \{1, 2\}$, we have from (2) that

$$\phi_{s1} = \frac{\mu_{s2} + \mu_{21} - \mu_{s1}}{\mu_{12} + \mu_{21}}.$$

Now we showed in Example 6.14 that $\mu_{s1} = 14$ and $\mu_{s2} = 10$. Also in (b) we established that $\mu_{12} = 10$ and $\mu_{21} = 14$. Hence

$$\phi_{s1} = \frac{10 + 14 - 14}{10 + 14} = \frac{5}{12}$$

and $\phi_{s2} = \frac{7}{12}$ is the probability that THT occurs before HHH.

(8) Exercise Show that for a fair coin the expected number of tosses to obtain HHH after HTH is 12, and the expected number required to obtain HTH after HHH is 8.

(9) Exercise Show that the probability that HHH is observed before HTH is $\frac{3}{10}$.

(10) Exercise Show that the probability that TTH is observed before HHH is $\frac{7}{10}$.

(11) Exercise A fairground showman offers to play the following game. On payment of an entry fee of £1, a customer names a possible outcome of a sequence of 3 coin tosses; the showman then names another possible outcome, and a fair coin is tossed repeatedly until one of the named sequences is obtained in three successive throws. The player who named that sequence wins.

(i) Show that the probability that THH beats HHH is $\frac{7}{8}$.
(ii) Show that the probability that TTH beats THH is $\frac{2}{3}$.
(iii) Show that the probability that TTH beats THT is $\frac{7}{8}$.
(iv) Show that the probability that HTT beats TTH is $\frac{2}{3}$.
(v) If the showman wishes to make on average 30 p per game, what prize money should he offer:

(a) if customers choose sequences at random?
(b) if customers make the best possible choice?

Remark This game was named Penney–Ante by W Penney in 1969.

9.16 Example: Poisson processes

Let $X(t)$ be a Markov process taking values in the non-negative integers; suppose that $X(t)$ is non-decreasing, with $X(0) = 0$. Suppose that as $h \to 0$ the transition rates satisfy

$$\mathbf{P}(X(t+h) = i+1 | X(t) = i) = \lambda(t)h + o(h)$$
$$\mathbf{P}(X(t+h) = i | X(t) = i) = 1 - \lambda(t)h + o(h),$$

so that $X(t)$ changes its value by jumps of size one. Denote the times at which $X(t)$ jumps by $T_1, T_2, T_3, \ldots$

(a) Show that for fixed t, $X(t)$ has a Poisson mass function with parameter $\Lambda(t) = \int_0^t \lambda(u)\,du$. Hence find the density of T_1, the time of the first jump.
(b) Find the joint density of T_1 and T_2, and hence find the conditional density of T_1 given T_2.

Remark $X(t)$ is called a non-homogeneous Poisson process with intensity (or rate) function $\lambda(t)$.

Solution (a) Let $p_n(t) = \mathbf{P}(X(t) = n)$. Then by conditional probability

$$p_n(t+h) = p_n(t)(1 - \lambda(t)h) + p_{n-1}(t)\lambda(t)h + o(h).$$

Hence we obtain the forward equations in the usual manner as

(1) $$\frac{dp_n(t)}{dt} = -\lambda(t)p_n(t) + \lambda(t)p_{n-1}(t); \quad n \geqslant 0,$$

where $p_{-1}(t) = 0$. Setting $G(z, t) = \sum_0^\infty z^n p_n(t)$ we find using (1) that

(2) $$\frac{\partial G}{\partial t} = +\lambda(t)(z-1)G.$$

Since $X(0) = 0$, we have $G(z, 0) = 1$, and so, by inspection, (2) has solution

(3) $$G(z, t) = \exp\left((z-1)\int_0^t \lambda(u)du\right).$$

This of course is the p.g.f. of the Poisson distribution with parameter $\Lambda(t)$, as required.

Now we note that

$$\mathbf{P}(T_1 > t) = \mathbf{P}(X(t) = 0) = G(0, t) = \exp(-\Lambda(t)).$$

(b) From (2) and (3) we can now see that for $w > t$,

$$\mathbf{E}(z^{X(w)}|X(t)) = z^{X(t)}\exp\left\{(z-1)\int_t^w \lambda(u)du\right\},$$

and

$$\mathbf{E}\{y^{X(t)}z^{X(w)-X(t)}\} = \mathbf{E}(y^{X(t)})\mathbf{E}(z^{X(w)-X(t)}).$$

It follows that this non-homogeneous Poisson process also has independent increments. Now

$$\begin{aligned}\mathbf{P}(T_1 > t, T_2 > w) &= \mathbf{P}(X(w) \in \{0, 1\}, X(t) = 0)\\ &= \mathbf{P}(X(t) = 0)\mathbf{P}(X(w) - X(t) \le 1)\\ &\qquad\text{by the independence of increments}\\ &= e^{-\Lambda(t)}\,[e^{-(\Lambda(w)-\Lambda(t))}(1 + \Lambda(w) - \Lambda(t))].\end{aligned}$$

Hence, differentiating, T_1 and T_2 have joint density

$$f(t, w) = \lambda(t)\lambda(w)e^{-\Lambda(w)};\quad 0 < t < w < \infty.$$

Integrating with respect to t shows that the density of T_2 is $f_{T_2}(w) = \lambda(w)\Lambda(w)e^{-\Lambda(w)}$, and so the conditional density is

$$f_{T_1|T_2}(t|w) = \frac{\lambda(t)}{\Lambda(w)};\quad 0 < t < w.$$

(4) **Exercise** Show that $\mathbf{P}(T_1 < \infty) = 1$ if and only if $\lim_{t\to\infty}\Lambda(t) = \infty$.

(5) **Exercise** If $\lambda(t) = \lambda e^{-\lambda t}$ for $\lambda > 0$, show that $\lim_{t\to\infty}\mathbf{P}(X(t) = k) = 1/ek!$

(6) **Exercise: Compound Poisson process** Let $(Y_n;\ n \ge 1)$ be independent and identically distributed and independent of $X(t)$. Let $Z(t) = \sum_{n=1}^{X(t)} Y_n$.

Show that $\mathbf{E}(e^{\theta Z(t)}) = \exp\left(\int_0^t \lambda(u)du(M(\theta) - 1)\right)$ where $M(\theta) = \mathbf{E}(e^{\theta Y_1})$.

(7) **Exercise: Doubly stochastic Poisson process** Suppose that $X(t)$ is a non-homogeneous Poisson process with random intensity $\lambda(t)$; that is to say, for any realization of the process $X(t)$, $\lambda(t)$ is a realization of a random process $Y(t)$, where $\mathbf{E}(\exp[\theta\int_0^t Y(t)dt]) = M(\theta)$. Show that $X(t)$ has probability generating function $M(z - 1)$. Find the mean and variance of $X(t)$ in this case.

9.17 Example: Decay

Let $(T_n = \sum_{i=1}^n X_i;\ n \ge 1)$ be the partial sums of the independent exponential random variables $(X_i;\ i \ge 1)$ having parameter λ. A certain class of particles has the property that when freshly produced their (independent) lifetimes are exponential with parameter μ. At the ends of their lives they disappear. At time T_n a number Y_n

of fresh particles is released into a chamber; the random variables $(Y_n;\ n \geq 0)$ are independent and identically distributed with p.g.f. $G_Y(z)$. At time t the number of particles in the chamber is $N(t)$. Show that

(1) $$\mathbf{E}(z^{N(t)}) = \exp\left(\lambda\int_0^t [G_Y((z-1)e^{-\mu v}+1)-1]dv\right).$$

Solution By construction, the batches arrive at the instants of a Poisson process. Hence, given that k batches have arrived at time t, their arrival times are independently and uniformly distributed over $(0, t)$. For any particle in a batch of size Y which arrived at time U, the chance of survival to t is $e^{-\mu(t-U)}$ independently of all the others. Hence, given $U = u$, the p.g.f. of the number S of survivors of Y at t is

$$\mathbf{E}((ze^{-\mu(t-u)}+1-e^{-\mu(t-u)})^Y) = G_Y(ze^{-\mu(t-u)}+1-e^{-\mu(t-u)}).$$

Hence

$$\mathbf{E}(z^S) = \mathbf{E}(\mathbf{E}(z^S|U)) = \frac{1}{t}\int_0^t G_Y((z-1)e^{-\mu v}+1)dv.$$

Finally, recalling that the total number of particles at t are the survivors of a Poisson number of such batches we obtain (1).

(2) Exercise What is $\mathbf{E}(z^{N(t)})$ when $G_Y(z) = z$? In this case find $\lim_{t\to\infty}\mathbf{P}(N(t) = k)$.

(3) Exercise In the case when $G_Y(z) = z$, show that $N(t)$ is a Markov process such that as $h \to 0$,

$$p_{i,i+1}(h) = \lambda h + o(h)$$
$$p_{i,i-1}(h) = i\mu h + o(h)$$

and

$$p_{ii}(h) = 1 - \lambda h - i\mu h + o(h).$$

Hence obtain your answer to Exercise 2 by using the forward equations.

(4) Exercise In the case when $G_Y(z) = z$, let t_n be the time when $N(t)$ makes its nth jump. Let $Z_n = N(t_n)$ be the imbedded Markov chain which records the successive different values of $N(t)$. Find the stationary distribution of Z_n.

9.18 Example: Disasters

A population evolves as follows. Immigrants arrive according to a Poisson process of rate ν. On arrival each immigrant immediately starts a simple birth process with parameter λ independently of all other immigrants. Disasters occur independently of the population according to a Poisson process of rate δ; when a disaster occurs all individuals then in existence are annihilated. A disaster occurs at $t = 0$.

Let $X(t)$ denote the number of individuals in existence at time $t \geq 0$.

(a) Show that $\lim_{t\to\infty}\mathbf{E}(X(t))$ is finite if and only if $\delta > \lambda$.
(b) Find an expression for $\mathbf{E}(s^{X(t)})$.

Solution Since $X(t)$ is a Markov process, we could proceed by writing down forward equations. However it is neater to use the properties of the Poisson process directly as follows. We start by assembling some facts established in earlier sections.

At time t, let $C(t)$ be the time that has elapsed since the most recent disaster. From Example 8.16 we recall that

(1) $$\mathbf{P}(C(t) > x) = e^{-\delta x}; \quad 0 \leqslant x \leqslant t.$$

Now note that arrivals are a Poisson process independent of disasters, so given that $C(t) = x$, the number of subsequent arrivals up to time t is a Poisson random variable N, with parameter νx.

Next we recall from Theorem 8.8.6 that conditional on $N = k$, these k arrivals are independently and uniformly distributed over the interval $(t - x, t)$, at times $t - Y_1$, $t - Y_2, \ldots, t - Y_k$, say (where the Y_i are uniform on $(0, x)$).

Finally we remember that given $Y_1 = y$, the expected number of descendants at time t from this arrival at $t - y$ is $e^{\lambda y}$; this is from (9.8.21). Now we remove the conditions one by one. First, the expected number of descendants of an arrival at $t - Y_1$ is

(2) $$\frac{1}{x}\int_0^x e^{\lambda y}\,dy = \frac{1}{\lambda x}(e^{\lambda x} - 1)$$

since Y_1 is uniform on $(0, x)$. Second, the expected number of descendants at t of the N arrivals during $(t - x, t)$ is, using (2),

(3) $$\mathbf{E}(N)\frac{1}{\lambda x}(e^{\lambda x} - 1) = \frac{\nu}{\lambda}(e^{\lambda x} - 1).$$

Finally, using (1) and (3), we have

(4) $$\mathbf{E}(X(t)) = \int_0^t \delta e^{-\delta x}\frac{\nu}{\lambda}(e^{\lambda x} - 1)dx + \frac{\nu}{\lambda}e^{-\delta t}(e^{\lambda t} - 1).$$

You can now see (if you want to) that, in more formal terms, what we have done is to say

$$\mathbf{E}(X) = \mathbf{E}(\mathbf{E}(\mathbf{E}[\mathbf{E}(X|C, N, Y_1, \ldots, Y_N)|C, N]|C)),$$

and then to successively evaluate the conditional expectations from the inside out.

So from (4), if $\lambda \geqslant \delta$, $\mathbf{E}(X(t)) \to \infty$ as $t \to \infty$, whereas if $\lambda < \delta$, $\mathbf{E}(X(t)) \to \frac{\nu}{\delta - \lambda}$.

An expression for $\mathbf{E}(s^{X(t)})$ is found by following exactly the same sequence of successive conditional expectations. Thus, given that $C(t) = x$, $N(x) = k$, and $Y_1 = y_1$, this arrival initiates a simple birth process whose size at time y_1 has generating function

$$\frac{se^{-\lambda y_1}}{1 - s + se^{-\lambda y_1}} \quad \text{by Example 9.8.14.}$$

Hence, since Y_1 is uniformly distributed on $[0, x]$, the generating function of the number of descendants at time x of one arrival in $[0, x]$ is

(5) $$\frac{1}{x}\int_0^x \frac{se^{-\lambda y}}{1 - s + se^{-\lambda y}}\,dy = -\frac{1}{\lambda x}\log(1 - s + se^{-\lambda x}).$$

By independence, the generating function of the sum of k such independent arrivals is

$$(\log(1 - s + se^{-\lambda x})^{-\frac{1}{\lambda x}})^k.$$

Next we recall that $N(x)$ is Poisson with parameter νx, so that using conditional expectation again, the generating function of the descendants at t of the arrivals in $[t-x, t]$ is

(6) $$\exp(\nu x(\log(1-s+se^{-\lambda x})^{-\frac{1}{\lambda x}}-1)) = \frac{e^{-\nu x}}{(1-s+se^{-\lambda x})^{\frac{\nu}{\lambda}}}.$$

Now we recall from Example 8.16 that the current life (or age) of a Poisson process has density

$$f_{C(t)}(x) = \delta e^{-\delta x}; \quad 0 \leqslant x \leqslant t$$

with $\mathbf{P}(C(t)=t) = e^{-\delta t}$. Hence, finally

(7) $$\mathbf{E}(s^{X(t)}) = \int_0^t \frac{\delta e^{-\delta x} e^{-\nu x}}{(1-s+se^{-\lambda x})^{\frac{\nu}{\lambda}}}\,dx + e^{-\delta t}\frac{e^{-\nu t}}{(1-s+se^{-\lambda t})^{\frac{\nu}{\lambda}}}$$

(8) Exercise Attempt to obtain (7) by writing down the forward equations for $\mathbf{P}(X(t)=n)$; $n \geqslant 0$, and solving them.

(9) Exercise Suppose that each immigrant gives rise to a simple birth and death process with parameters λ and μ. Show that $\lim_{t\to\infty}\mathbf{E}(X(t)) < \infty$ if and only if $\delta > \lambda - \mu$. [See 9.20].

(10) Exercise Suppose that an ordinary immigration–death process with parameters ν and μ is subject to disasters. Show that the population size $X(t)$ has a stationary distribution with mean $\frac{\nu}{\delta+\mu}$. [Set $\lambda = 0$ in Exercise 9.]

9.19 Example: The general birth process

Let $(Y_n;\ n \geqslant 1)$ be a collection of independent exponentially distributed random variables such that Y_n has parameter λ_{n-1}. Let

$$T_n = \sum_{r=1}^{n} Y_r$$

and

$$N(t) = \max\{n: T_n \leqslant t\}.$$

The process $N(t)$ is a general birth process. Show that if

(1) $$\sum_0^\infty \lambda_r^{-1} < \infty$$

then for $t > 0$

(2) $$\mathbf{P}(N(t) < \infty) < 1.$$

Also show that $\mathbf{E}(N(t)|N(t) < \infty)$ is finite or infinite depending on whether $\sum_0^\infty r\lambda_r^{-1}$ converges or diverges.

Solution First recall the often-used identity

(3) $$\mathbf{P}(T_n \leqslant t) = \mathbf{P}(N(t) \geqslant n).$$

Let T_n have density $f_n(t)$ and moment generating function $M_n(\theta)$, and define

(4) $$T = \lim_{n\to\infty} T_n = \sup\{t: N(t) < \infty\}.$$

Since the Y_n are independent and exponentially distributed it follows that

(5) $$M_n(\theta) = \prod_{r=0}^{n-1} (1 + \theta\lambda_r^{-1})^{-1}.$$

If (1) holds, then as $n \to \infty$ the infinite product converges (uniformly) to a non-zero limit $M(\theta)$. By the continuity theorem this is the moment generating function of the density $f_T(t)$ of T. Hence, by (3)

$$\mathbf{P}(N(t) < \infty) = \mathbf{P}(T > t) = \int_t^\infty f_T(u)du < 1,$$

for $t > 0$.

If (1) does not hold, then the product in (5) diverges to zero as $n \to \infty$, for $\theta \neq 0$, and $f_T(t) = 0$.

Furthermore, from (3)

(6) $$p_n(t) = \mathbf{P}(N(t) = n) = \mathbf{P}(T_n \leq t) - \mathbf{P}(T_{n+1} \leq t) = \int_0^t f_n(u)du - \int_0^t f_{n+1}(u)du.$$

Now using (6), for $\theta < 0$ we have

(7) $$\int_0^\infty e^{\theta t} p_n(t)dt = \frac{1}{\theta}(M_n(\theta) - M_{n+1}(\theta)) \quad \text{on integrating by parts}$$
$$= \frac{1}{\lambda_n} M_{n+1}(\theta) \quad \text{using (5).}$$

Since $M_n(\theta)$ converges uniformly to $M(\theta)$ we can use the inversion theorem on each side of (7) to find that as $n \to \infty$

(8) $$\lambda_n p_n(t) \to f_T(t).$$

Now

$$\mathbf{E}(N(t)|N(t) < \infty) = \frac{\sum_{n=0}^\infty n p_n(t)}{\sum_{n=0}^\infty p_n(t)}$$

which converges or diverges with $\Sigma n p_n(t)$. Using (8) it follows that $\mathbf{E}(N(t)|N(t) < \infty) < \infty$ if and only if $\sum_{n=0}^\infty n\lambda_n^{-1} < \infty$.

(9) Exercise Write down the forward equations for $p_n(t)$, and deduce (7) directly from these.

(10) Exercise Deduce from (7) that $p_n(t) = \frac{1}{\lambda_n}\sum_{i=0}^n a_i\lambda_i e^{-\lambda_i t}$ where

$$a_i = \prod_{\substack{j=0 \\ j\neq i}}^n \frac{\lambda_j}{\lambda_j - \lambda_i}.$$

(11) Exercise Show that if $\lambda_n = n(\log n)^\gamma$; $\gamma > 1$ then for any $\beta > 0$, $\mathbf{E}([N(t)]^\beta|N(t) < \infty) = \infty$.

9.20 Example: The birth–death process

Let the Markov process $X(t)$ represent the number of individuals in a population at time t. During any interval $(t, t + h)$, any individual alive at t may die with probability $\mu h + o(h)$, or split into two individuals with probability $\lambda h + o(h)$. All individuals act independently in these activities. Write down the forward equations for $p_n(t) = \mathbf{P}(X(t) = n)$ and show that if $X(0) = I$ then

(1)
$$\mathbf{E}(s^{X(t)}) = \begin{cases} \left(\dfrac{s + \lambda t(1-s)}{1 + \lambda t(1-s)}\right)^I & \text{if } \lambda = \mu \\ \left(\dfrac{\mu \exp(t(\lambda - \mu)) + \theta(s)}{\lambda \exp(t(\lambda - \mu)) + \theta(s)}\right)^I & \text{if } \lambda \neq \mu \end{cases}$$

where $\theta(s) = (\lambda s - \mu)/(1 - s)$.

Solution Since individuals act independently, the probability of no change during $(t, t+h)$ when $X(t) = k$ is

$$p_{kk}(h) = (1 - \lambda h - \mu h + o(h))^k = 1 - (\mu + \lambda)kh + o(h).$$

Similarly the probability of just one split and no deaths among k individuals during $(t, t+h)$ is

$$p_{k,k+1}(h) = k\lambda h(1 - \lambda h - \mu h + o(h))^{k-1} = k\lambda h + o(h),$$

and likewise the chance of just one death is $p_{k,k-1}(h) = k\mu h + o(h)$. Other transitions have probabilities which are all $o(h)$ as $h \to 0$, and so by conditional probability

$$p_k(t+h) = h\lambda(k-1)p_{k-1}(t) + h\mu(k+1)p_{k+1}(t) + (1 - (\lambda + \mu)kh)p_k(t) + o(h).$$

The forward equations now follow as usual, giving

$$\frac{dp_k(t)}{dt} = \lambda(k-1)p_{k-1}(t) + \mu(k+1)p_{k+1}(t) - (\lambda + \mu)kp_k(t),$$

with the convention that $p_{-1}(t) = 0$. Defining $G(s, t) = \mathbf{E}(s^{X(t)})$, and differentiating G with respect to s, shows that

(2)
$$\frac{\partial G}{\partial t} = \lambda s^2 \frac{\partial G}{\partial s} + \mu \frac{\partial G}{\partial s} - (\lambda + \mu)s \frac{\partial G}{\partial s} = (\lambda s - \mu)(s - 1)\frac{\partial G}{\partial s}.$$

Since $X(0) = I$, we have $G(s, 0) = s^I$, and it is straightforward but dull to verify that (1) satisfies (2) and the initial condition $X(0) = I$.

(3) Exercise Let η be the probability that the population ever falls to zero. Show that

$$\eta = \begin{cases} 1 & \text{if } \mu \geq \lambda \\ \left(\dfrac{\mu}{\lambda}\right)^I & \text{if } \lambda > \mu. \end{cases}$$

(4) Exercise Let T be the time until $X(t)$ first takes the value zero. Show that if $X(0) = 1$,

$$\mathbf{E}(T \mid T < \infty) = \begin{cases} \dfrac{1}{\lambda}\log\dfrac{\mu}{\mu - \lambda}; & \lambda < \mu \\ \dfrac{1}{\mu}\log\dfrac{\lambda}{\lambda - \mu}; & \lambda > \mu. \end{cases}$$

(5) Exercise Let $X(0) = 1$ and define $z(t) = \mathbf{P}(X(t) = 0)$. Show that $z(t)$ satisfies

$$\frac{dz}{dt} = \mu - (\lambda + \mu)z(t) + \lambda(z(t))^2.$$

Hence find $z(t)$. What is $\mathbf{P}(X(t) = 0 | X(s) = 0)$ for $0 < t < s$?

(6) Exercise Suppose that $X(0) = 1$ and $\lambda < \mu$. Show that

$$\lim_{t\to\infty} \mathbf{P}(X(t) = k | X(t) > 0) = \left(\frac{\lambda}{\mu}\right)^k \left(1 - \frac{\lambda}{\mu}\right).$$

(7) Exercise Suppose that new individuals join the population at the instants of a Poisson process with parameter ν (independently of the birth and death process). Write down the forward equations for the process. Deduce that if $\lambda < \mu$ the stationary distribution is

$$\pi_k = \frac{1}{k!}\left(1 - \frac{\lambda}{\mu}\right)^{\frac{\nu}{\lambda}} \left(\frac{\lambda}{\mu}\right)^k \left(\frac{\nu}{\lambda} + k - 1\right)\left(\frac{\nu}{\lambda} + k - 2\right) \dots \frac{\nu}{\lambda}.$$

What is the mean of this distribution?

PROBLEMS

1 Let $(X_n;\ n \geqslant 1)$ be a collection of independent identically distributed non-negative random variables. Define:

(i) $S_n = \sum_{i=1}^n X_i$
(ii) $M_n = \max\{X_1, X_2, \dots, X_n\}$
(iii) $L_n = \min\{X_1, X_2, \dots, X_n\}$
(iv) $K_n = X_n + X_{n-1}$.

(a) Which of the sequences X, S, M, L, K are Markov chains?
(b) For those that are, find the transition probabilities.

2 Classify the chains in Problem 1; that is to say, show whether the states are persistent or not, null or not, periodic or not.

3 Can a reversible chain be periodic?

4 Let $(X_n;\ n \geqslant 1)$ and $(Y_n;\ n \geqslant 1)$ be independent irreducible Markov chains, and set $Z_n = (X_n, Y_n);\ n \geqslant 1$.

(a) Is Z_n irreducible?
(b) If X and Y are reversible and also aperiodic, show that Z is reversible.

5 Let X be a Markov chain. Show that the sequence $(X_i;\ i \geqslant 0)$ conditional on $X_m = r$ still has the Markov property.

6 Show that Definition 9.1.1 is equivalent to each of (9.1.6), (9.1.7) and (9.1.8) as asserted.

7 Let Y_n be the number of heads shown in n tosses of a coin. Let $Z_n = Y_n$ modulo 10. Show that $(Z_n;\ n \geqslant 0)$ is a Markov chain; find its transition probabilities and stationary distribution.

8 Let $(S_n;\ n \geqslant 0)$ be a simple random walk with $S_0 = 0$; show that $Y_n = |S_n|$ is a Markov chain.

9 Let $(X_n;\ n \geqslant 1)$ be a Markov chain. Show that if $g(X_n)$ is any function of X_n, then $\mathbf{E}(\mathbf{E}(g(X_{n+m})|X_n)|X_r) = \mathbf{E}(g(X_{n+m})|X_r)$, for $r \leqslant n$.

10 Let $(u_n;\ n \geqslant 0)$ be a sequence defined by $u_0 = 1$ and $u_n = \sum_{k=1}^n f_k u_{n-k}$, where $f_k > 0$ and $\sum_{k=1}^\infty f_k < 1$.

(a) Show that v_n defined by $u_n = \rho^n u_n;\ n \geqslant 0$, is a renewal sequence as defined in Example 9.13 if $\sum_{n=1}^\infty \rho^n f_n = 1$.
(b) Show that as $n \to \infty$, for some constant c, $\rho^n u_n \to c$.

11 **Murphy's law** Let $(X_n;\ n \geqslant 1)$ be an irreducible aperiodic persistent chain. Let $s = (s_1, \ldots, s_m)$ be any finite sequence of states of the chain such that $p_{s_1 s_2} p_{s_2 s_3} \ldots p_{s_{m-1} s_m} > 0$. Show that with probability 1 the sequence s occurs in finite time. Explain the implications.

12 Let $(X_n;\ n \geqslant 0)$ be a Markov chain. Show that for any constant d the sequence $(X_{nd};\ n \geqslant 0)$ is a Markov chain.

13 Let A be a subset of the states of a regular chain X. Let $T_1 < T_2 < T_3 < \ldots$ be the successive times at which the chain visits A. Show that $(X_{T_r};\ r \geqslant 1)$ is a Markov chain.

14 Let $(X_n;\ n \geqslant 0)$ and $(Y_n;\ n \geqslant 0)$ be Markov chains with the same state space S, and distinct transition matrices $\overset{X}{p}_{ij}$ and $\overset{Y}{p}_{ij}$. Let $(W_n;\ n \geqslant 0)$ be a process defined on S with transition probabilities

$$q_{ij}(n) = \frac{1}{2}(\overset{X}{p}_{ij}(n) + \overset{Y}{p}_{ij}(n)).$$

Show that $q_{ij} \geqslant 0$ and $\Sigma_j q_{ij} = 1$, but that $(W_n;\ n \geqslant 0)$ is not a Markov chain in general.

15 **Truncation** Let $(X_n;\ n \geqslant 0)$ be an irreducible Markov chain with state space S, transition probabilities p_{ij}, and stationary distribution $(\pi_i;\ i \in S)$. Let A be some subset of S, and suppose that a new chain Y is formed by banning transitions out of A. That is to say Y has transition probabilities q_{ij} where for $i \in A$, $q_{ij} = p_{ij}$ for $j \in A$ and $j \neq i$, and $q_{ii} = p_{ii} + \sum_{j \in A^c} p_{ij}$. Show that if X is reversible in equilibrium, then so is Y, and write down the stationary distribution of Y.

16 Let X_n and Y_n be independent simple random walks. Let Z_n be (X_n, Y_n) truncated as in Problem 15 to the region $x \geqslant 0$, $y \geqslant 0$, $x + y \leqslant a$. Find the stationary distribution of Z_n.

17 Let $(X_n;\ n \geqslant 0)$ be a Markov chain with state space S. For each $n \geqslant 0$ independently, X_n is replaced by $s \in S$ with probability p. Is the new sequence a Markov chain?

18 At each time $n = 0, 1, 2, \ldots$ a number Y_n of particles is injected into a chamber, where $(Y_n;\ n \geqslant 0)$ are independent Poisson random variables with parameter λ. The lifetimes of particles are independent and geometric with parameter p. Let X_n be the number of particles in the chamber at time n. Show that X_n is a Markov chain; find its transition probabilities and the stationary distribution.

19 Let $(f_k;\ k \geqslant 0)$ be a probability mass function. Let the irreducible Markov chain X have transition probabilities

$$p_{jk} = f_{k-j+1} \quad \text{if} \quad k - j + 1 \geqslant 0,\ j \geqslant 1,$$

and $p_{0k} = p_{1k}$.

Show that X is recurrent and non-null if $\sum_{k=1}^{\infty} k f_k < 1$.

20 Let $(f_k;\ k \geqslant 0)$ be a probability mass function. Suppose the Markov chain X has transition probabilities

$$p_{jk} = \begin{cases} f_{j-k+1} & \text{for } k > 0,\ j - k + 1 \geqslant 0 \\ \displaystyle\sum_{i=j+1}^{\infty} f_i & \text{for } k = 0 \\ 0 & \text{otherwise.} \end{cases}$$

Show that X is recurrent and non-null if $\sum_{k=1}^{\infty} k f_k > 1$.

21 **Lumping** Let X have state space S, and suppose that $S = \bigcup_k A_k$, where $A_i \cap A_j = \phi$ for $i \neq j$. Let $(Y_n;\ n \geqslant 0)$ be a process which takes the value y_k whenever the chain X lies in A_k. Show that Y is also a Markov chain if $p_{i_1 j} = p_{i_2 j}$ for any i_1 and i_2 in the same set A_k.

22 **Markov times** Let X be a Markov chain. Let T be a positive random variable such that

$\mathbf{P}(T = t|X_0, \ldots, X_t)$ is either zero or one. T is called a *Markov time*. Show that the Markov property is preserved at T.

23 Let S_n be the random walk such that

$$\mathbf{P}(S_{n+1} - S_n = 2) = p,$$

and

$$\mathbf{P}(S_{n+1} - S_n = -1) = q, \quad \text{where} \quad p + q = 1.$$

If the origin is a retaining barrier show that equilibrium is possible with $S_n \geq 0$ if $p < \frac{1}{3}$, and that in this case the stationary distribution has p.g.f.

$$\pi(s) = \frac{(1 - 3p)(s - 1)}{s - q - ps^3}.$$

24 Let $X(t)$ be the two-state chain in continuous time, $t \in \mathbb{R}$, $X(t) \in \{0, 1\}$, having stationary distribution $\{\pi_0, \pi_1\}$.

(a) Show that as $\tau \to \infty$

$$\mathbf{P}(X(0) = 1|X(-\tau), X(\tau)) \to \pi_1.$$

(b) Find $\operatorname{cov}(X(s), X(s + t)); \quad t > 0$.

(c) What is $\lim_{s\to\infty} \operatorname{cov}(X(s), X(s + t))$?

25 Let $N(t)$ and $M(t)$ be independent Poisson processes with parameters λ and μ respectively.

(a) Is $N(t) + M(t)$ a Poisson process?

(b) Is either of $\min\{N(t), M(t)\}$ or $\max\{N(t), M(t)\}$ a Poisson process?

26 Let $N(t)$ be a non-homogeneous Poisson process with rate $\lambda(t)$. Find $\operatorname{cov}(N(s), N(s + t))$; $t > 0$.

27 Mosquitoes land on your neck at the instants of a Poisson process with parameter $\lambda(t)$ and each bites you with probability p independently of the decisions of the others. Show that bites form a Poisson process with parameter $p\lambda(t)$.

28 Let $X(t)$ be a Markov chain with transition probabilities $p_{ij}(t)$ and stationary distribution $\boldsymbol{\pi}$. Let $(T_n;\ n \geq 0)$ be the jump times of a Poisson process independent of $X(t)$. Show that the sequence $Y_n = X(T_n)$ is a Markov chain with the same stationary distribution as $X(t)$.

29 Find the mean and variance of the size $X(t)$ of the population in the birth–death process of Example 9.20.

30 **A non-homogeneous chain** Let $X(t)$ be a Markov chain with $X(0) = I$, and such that as $h \to 0$

$$\mathbf{P}(X(t + h) = k + 1|X(t) = k) = \frac{1 + \mu k}{1 + \mu t} h + o(h)$$

and

$$\mathbf{P}(X(t + h) = k|X(t) = k) = 1 - \frac{1 + \mu k}{1 + \mu t} h + o(h).$$

Show that $G = \mathbf{E}(s^{X(t)})$ satisfies

$$\frac{\partial G}{\partial t} = \frac{s - 1}{1 + \mu t}\left(G + \mu s \frac{\partial G}{\partial s}\right).$$

Hence find $\mathbf{E}(X)$ and $\operatorname{var}(X)$.

31 **Truncation again** Let $(X_n;\ n \geq 0)$ be an irreducible Markov chain, with state space S, stationary distribution $(\pi_i;\ i \in S)$, and transition probabilities p_{ij}. Let A be some subset of S, and suppose that $(Z_n;\, n \geq 0)$ is a Markov chain with state space A and transition probabilities

$$q_{ij} = \frac{p_{ij}}{p_{iA}} \quad \text{for } i, j \in A,$$

where $p_{iA} = \sum_{j \in A} p_{ij}$.

If X is reversible, show that Z is reversible with stationary distribution given by

$$\nu_i = \pi_i p_{iA} \Big/ \sum_{i \in A} \pi_i p_{iA}.$$

32 'Motto' is a coin-tossing game at the start of which each player chooses a sequence of three letters, each of which is either H or T (his 'motto'). A fair coin is then tossed repeatedly, and the results recorded as a sequence of H's and T's (H for 'heads', T for 'tails'). The winner is the first player whose motto occurs as three consecutive letters in this sequence. Four players A, B, C, D choose as their mottoes, respectively, HHT, THH, TTH and HTT. Show that if only A and B take part in a game then B has probability $\frac{3}{4}$ of winning. With what probability does C win if he plays a game with B as the only opponent? If all four players take part simultaneously, what are the respective probabilities of each player winning?
[You may assume that if a fair coin is tossed repeatedly then with probability 1 any motto will occur eventually.]

33 Let $N(t)$ be a non-homogeneous Poisson process. Show that, conditional on $N(t) = k$, the times $T_1, \ldots, T_k$ of the events have conditional joint density

$$k! \prod_{i=1}^{k} \left\{ \frac{\lambda(t_i)}{\Lambda(t)} \right\}, \quad 0 \leq t_1 \leq \ldots \leq t_k \leq t.$$

Appendix: Solutions and hints for selected exercises and problems

No experienced mathematician feels well acquainted with a subject until he has tackled some problems; through attempting and failing, we extend the boundaries of our knowledge and experience. This observation applies to students also. *It would be a big mistake to treat the remarks of this section as a solution sheet*. Many of the hints and comments will be useful only to those who have spent a half hour, say, on the problem already. The remarks vary in style and content between small hints and detailed solutions; some problems receive no comments at all (indicating, perhaps, that they are either very easy or good challenges).

CHAPTER 1

Exercises

1.7.1 $\mathbf{P}(\{1,1\} \cup \{1,2\} \cup \{2,1\} \cup \{6,6\}) = \frac{4}{36} = \frac{1}{9}$.

1.7.2 (a) $\frac{1}{2}$; (b) $\frac{1}{2}$; (c) $\frac{1}{4}$.

1.7.3 $\frac{11}{18}$

1.7.4 $\frac{5}{12}$

1.7.5 $\frac{1}{3} - \frac{1}{18}r$

1.8.1 $\Omega = \{(i, j): 1 \leq i < j \leq 2n\}$ and so $|\Omega| = n(2n-1)$. Likewise $|\{HH\}| = \frac{1}{2}n(n-1)$. Hence $\mathbf{P}(HH) = (n-1)/(2(2n-1))$.

1.8.2 (a) $\frac{1}{2}$; (b) $\frac{1}{2}$.

1.8.3 $\frac{1}{2}$

1.8.4 $\frac{1}{2}$

1.9.1 (a) zero; (b) $\frac{2}{3}$.

1.9.2 Let C_j be the event that the jth cup and saucer match. Then $\mathbf{P}(C_j) = \frac{1}{4}$; $\mathbf{P}(C_i \cap C_j) = \frac{1}{12}$, $i \neq j$, $\mathbf{P}(C_i \cap C_j \cap C_k) = \frac{1}{24}$, $i \neq j \neq k \neq i$. Hence by (1.4.5), (1.4.8), (1.6.1),

$$\mathbf{P}\left(\bigcap_{i=1}^{4} C_i^c\right) = 1 - 4 \cdot \frac{1}{4} + 6 \cdot \frac{1}{12} - 4 \cdot \frac{1}{24} + \frac{1}{24} = \frac{3}{8}.$$

1.10.1 $\mathbf{P}(A) = \frac{36}{91}$; $\mathbf{P}(B) = \frac{30}{91}$; $\mathbf{P}(C) = \frac{25}{91}$.

1.10.2 Let all the players continue rolling, even after first rolling a 6, and let H_r be the event that all three roll a 6 in the rth round. Since $E \subseteq \left(\bigcup_{r=1}^{\infty} H_r\right)^c$ we have

$$\mathbf{P}(E) \leqslant 1 - \mathbf{P}\left(\bigcup_{r=1}^{\infty} H_r\right) = 1 - \sum_{r=1}^{\infty} \mathbf{P}(H_r) = 1 - \sum_{r=1}^{\infty} (6^3 - 1)^{r-1}/6^{3r} = 0.$$

1.10.3 Ignoring Chryseis, $\mathbf{P}(A) = \sum_{n=0}^{\infty} 5^{2n}/6^{2n+1} = \frac{6}{11}$.

1.10.4 As above, let all players continue rolling irrespective of 6's achieved. A gets his first 6 on the $(3r + 1)$th roll in 5^r ways, B and C both have at least one already in $(6^r - 5^r)^2$ ways. Hence $\mathbf{P}(A \text{ last}) = \sum_{r=1}^{\infty} 5^r(6^r - 5^r)^2/6^{3r+1} = \frac{305}{1001}$.

1.11.2 $\mathbf{P}(B_1) = \frac{5}{8}$; $\mathbf{P}(C) = \frac{1}{4}$.

1.11.3 (a) $\mathbf{P}(B_2) = \frac{3}{8}$; $\mathbf{P}(B_3) = \frac{1}{8}$. (b) $\mathbf{P}(B_2) = \frac{1}{8}$; $\mathbf{P}(B_3) = \frac{1}{8}$. (c) $\mathbf{P}(B_2) = \frac{1}{8}$; $\mathbf{P}(B_3) = \frac{1}{8}$.

1.11.4 (a) 0; (b) $\frac{1}{4}$; (c) $\frac{1}{2}$.

1.12.1 $p_2 = \frac{n_7}{36} + \frac{n_{11}}{36} + \frac{1}{(36)^2}(n_4^2 + n_5^2 + n_6^2 + n_8^2 + n_9^2 + n_{10}^2) = \frac{97}{324}$.

1.12.2 $p_3 = p_2 + \frac{2}{(36)^3}(27n_4^2 + 26n_5^2 + 25n_6^2) = \frac{97}{324} + \frac{1}{6} \cdot \frac{107}{324}$.

1.12.3 (a) $\frac{436}{990}$; (b) $\frac{526}{990}$.

1.12.4 Let p_j be the chance of winning when the first die shows j. Then $p_1 = \frac{361}{990}$, $p_3 = p_4 = \frac{502}{990}$, $p_5 = \frac{601}{990}$, so you would fix the first die at 5 if you could.

1.13.3 $\sum_{r=0}^{\infty} \left(\frac{1}{2}\right)^{2r} \frac{1}{2} = \frac{2}{3}$.

1.13.4 Let s be a sequence of length $m + n$ in which the first m terms have x heads and $m - x$ tails, and the next n have x heads and $n - x$ tails. Now change the first x heads to tails and the first $m - x$ tails to heads giving a sequence t of length $m + n$ with n heads. This map is 1–1, so the number of t-sequences equals the number of s-sequences, giving the result.

Problems

1 $\frac{1}{13}$

2 $\frac{1}{13}$

3 (a) $\frac{8}{15}\left(> \frac{1}{2}\right)$; (b) $\frac{112}{225}\left(< \frac{1}{2}\right)$.

4 $\frac{1}{2}$

5 2^n for some integer $n \geqslant 1$.

6 (a) $\frac{25}{216}$; (b) $\frac{125}{216}$; (c) 1; (d) 17 because $\left(\frac{5}{6}\right)^{17} < 0.05 < \left(\frac{5}{6}\right)^{16}$.

7 (a) $(A \cap B) \cup (B \cap C) \cup (C \cap A)$; (b) $(A^c \cap B \cap C) \cup (A \cap B^c \cap C) \cup (A \cap B \cap C^c)$ (c) $(A \cap B \cap C)^c$

8 The loaded die is equivalent to a fair ten-sided die with five faces numbered 6.

(a) $\left(\frac{5}{10}\right)^4 = 81\left(\frac{1}{6}\right)^4$ so the factor is 81; (b) $p_{23} = 4p_{24}$.

9 (a) $\frac{5}{6}$; (b) $\frac{1}{4}$; (c) $\frac{3}{16}$; (d) $\frac{1}{8}$.

10 $\left(\frac{35}{36}\right)^{24} \simeq 0.16$

11 $\frac{19}{36}$

12 (a) Use induction; (b) $\frac{1}{9}$. [*Hint: the cups can be arranged in 90 distinct ways.*]

13 $\frac{2}{3}$

14 $\frac{1}{3}$ in each case.

15 (a) $1 - \left(\frac{5}{6}\right)^4$; (b) $\left(\frac{3}{6}\right)^4 - \left(\frac{2}{6}\right)^4$.

16 (a) $\left(\frac{1}{2}\right)^n - \left(\frac{1}{3}\right)^n \downarrow 0$; (b) $1 - \left(\frac{5}{6}\right)^n \uparrow 1$.

17 (a) $\frac{1}{n}\left[\frac{n}{3}\right]$; $\frac{1}{n}\left[\frac{n}{4}\right]$; $\frac{1}{n}\left[\frac{n}{12}\right]$; $\frac{1}{n}\left(\left[\frac{n}{3}\right] + \left[\frac{n}{4}\right] - \left[\frac{n}{12}\right]\right)$; (b) $\frac{1}{3}$; $\frac{1}{4}$; $\frac{1}{12}$; $\frac{1}{2}$.
(c) You would get the same answers.

19 (c) Use induction.

21 In every case $p = \frac{x(x-1)}{(x+y)(x+y-1)}$ (a) $x = 3$, $y = 1$ and $x = 15$, $y = 6$;
(b) $x = 6$, $y = 10$; (c) when $r = 6$, $x = 2$ and $y = 7$.

22 Always in each case except (c) which holds when $B \subseteq C \subseteq A$, and (d) which holds when $A \cap C = \phi$.

23 1988 was a leap year, so $p = 1 - \frac{(366)!}{(366-m)!(366)^m}$

CHAPTER 2

Exercises

2.5.6 $1 - r^2$

2.5.7 $\mathbf{P}(V|A_n \cup B_n) = \mathbf{P}(A_n)/\mathbf{P}(A_n \cup B_n) = \frac{r^{n-1}p}{r^{n-1}p + r^{n-1}q} = \frac{p}{p+q}$.

[Note the lack of any dependence on n.]

2.6.1 Use induction.
2.6.2 $(c + d)/(b + c + d)$
2.6.3 $(c + nd)/(b + c + nd) \to 1$ as $n \to \infty$
2.6.4 Use (2.6.1).
2.6.5 It is also the probability of getting m cyan balls and n blue balls in any given fixed order.
2.7.4 $(1 - p)^n$; $np(1 - p)^{n-1}$
2.7.5 (a) $1 - \pi_n$; (b) $\sum_{k=1}^{n-1} p_k \pi_{k-1} s_{n-k}$; (c) $1 - (1 - p)^n$; $\frac{1}{2}n(n - 1)p^2(1 - p)^{n-2}$.
2.7.7 $\prod_{k=1}^{\infty}(1 - p_k) > 0$ if and only if $\sum_{k=1}^{\infty} p_k < \infty$.
2.8.1 $\alpha(1 - \alpha) + \alpha^2(1 - \gamma)/(1 - (1 - \alpha)(1 - \gamma))$
2.8.2 $(2 - \alpha + \alpha\gamma(1 - \gamma))^{-1}$
2.8.3 Biggles
2.9.1 (a) $\mathbf{P}(E) = 0.108$; $\mathbf{P}(A^c|D) = \frac{45}{86}$.
(b) $\mathbf{P}(E) = 0.059$; $\mathbf{P}(A^c|D) = \frac{5}{451}$.
2.9.2 $\frac{p(\mu(1-\pi)+v(1-\mu))}{p(\mu(1-\pi)+v(1-\mu))+(1-p)(1-\pi)}$ (a) $\mathbf{P}(A|D^c) = \frac{1}{46} < \frac{45}{86}$;
(b) $\mathbf{P}(A|D^c) = \frac{6}{61} > \frac{5}{451}$
2.9.3 $\mathbf{P}(L) = (1 - \rho)\mathbf{P}(E) + \rho(1 - \mathbf{P}(E))$; $\mathbf{P}(A^c|M) = \frac{(\pi(1 - \rho) + \rho(1 - \pi))(1 - p)}{\rho\mathbf{P}(D^c) + (1 - \rho)\mathbf{P}(D)}$
$\mathbf{P}(A|M^c) = \frac{p[(1 - \rho)(\mu(1 - \pi) + v(1 - \mu)) + \rho(\mu\pi + (1 - \mu)(1 - v))]}{\rho\mathbf{P}(D) + (1 - \rho)\mathbf{P}(D^c)}$
2.10.4 $1 - k/K$
2.10.6 $\frac{(1 - \rho)(1 - \rho^K)}{(1 - \rho^{k+1})(1 - \rho^{K-k})}$ where $\rho = \frac{1 - p}{p}$.
2.11.1 When $\lambda = \mu$.
2.11.2 $\mathbf{P}(A_2|A_1) - \mathbf{P}(A_2) = \frac{(\mu - \lambda)^2}{2(\lambda + \mu)} = 0$ when $\lambda = \mu$.
2.11.3 (a) $(\mu^3 + \lambda^3)/\mu^2 + \lambda^2)$ (b) $\frac{\mu^n + \lambda^n}{\mu^{n-1} + \lambda^{n-1}} \to \max\{\lambda, \mu\}$ as $n \to \infty$
2.11.4 (a) $\frac{\mu}{\lambda + \mu}$; (b) $\frac{\lambda}{\lambda + \mu}$
2.11.5 $\frac{\mu^n}{\mu^n + \lambda^n} \to 1$ as $n \to \infty$ if $\mu > \lambda$.
2.12.3 (a) $\frac{2p_1}{3p_1 + p_2(1 - p_1)}$; (b) $\frac{3p_1}{3p_1 + p_2(1 - p_1)}$.
2.12.4 Yes, if $p_1 < \frac{1}{2}$ and $p_2 = \frac{p_1}{1 - p_1}$.
2.13.1 (a) $a = 1, b = 1, c = 2$; (b) $a = b = c$.
2.13.1 $\frac{16}{41}$
2.13.2 $\frac{20}{41}$

Problems

1 (a) 0.12; (b) 0.61; (c) 0.4758; (d) $\frac{7}{13}$.

3 (i) $\frac{1}{6}$; $\frac{1}{3}$; $\frac{1}{2}$ respectively; (ii) $\frac{7}{9}$; (iii) $\frac{1}{14}$; $\frac{2}{7}$; $\frac{9}{14}$ respectively; (iv) $\frac{6}{7}$.

4 (b) No.

5 (a) 0.36; (b) 0.06; (c) 0.7; (d) $\frac{18}{41}$

6 (a) zero; (b) one.

7 (a) $\frac{3}{5}$; (b) $\frac{2}{3}$; (c) $\frac{5}{6}$; (d) $\frac{4}{5}$.

8 (a) $\frac{1}{2}$; (b)(i) $\frac{1}{2}$; (ii) $\frac{1}{4}$; (c) $\frac{1}{36}$; (d) $\frac{1}{42}$.

9 (b) $\frac{3}{4}$; (c) 0.7

11 $\frac{83}{102}$

12 (a) $(1-p^2)^2$; (b) $1-p+p(1-p^2)^2$; (c) $\dfrac{p(1-p^2)^2}{1-p+p(1-p^2)^2}$

13 (a) $\dfrac{r}{b+r}$; (b) $\frac{1}{2}$; (c) zero if $r < b$, $\frac{1}{2}$ if $r = b$, 1 if $r > b$.

14 $1 - x \leqslant \exp(-x)$.

15 Let $\mathbf{P}(A) = \alpha$ and $\mathbf{P}(B) = \beta$. If the claim is false then $(1-\alpha)(1-\beta) < \frac{4}{9}$ and $\alpha\beta < \frac{4}{9}$ and $\alpha(1-\beta) + \beta(1-\alpha) < \frac{4}{9}$. The intersection of these three regions in the α–β plane is empty, so the claim is true.

16 $\mathbf{P}(E|A = \text{tail}) = (1-(1-p)^{s-1})\mathbf{P}(E|A = \text{head})$

$$\mathbf{P}(E) = \frac{p^{r-1}(1-(1-p)^s)}{1-(1-p^{r-1})(1-(1-p)^{s-1})}$$

17 He must answer k satisfying

$$\left(\frac{u(1-b)}{b(1-u)}\right)^k > \frac{p}{q}\left(\frac{1-b}{1-u}\right)^n.$$

Therefore the examiners must set questions such that $(u/b)^n > p/q$, or the student can never convince them that he is not bluffing.

18 (i) 0.7; (ii) $\frac{2}{9}$; (iii) Use (2.10.2).

19 The second set of rules.

20 $\frac{16}{31}$

21 6^{-2}

22 (a) $\frac{1}{2}\left(1-\left(\frac{2}{3}\right)^n\right)$; (b) $\dfrac{1}{36(\lambda-\mu)}\left(\dfrac{\lambda^n}{1-\lambda} - \dfrac{\mu^n}{1-\mu}\right)$ where $\lambda = \frac{5}{12} + \frac{\sqrt{5}}{4}$ and $\mu = \frac{5}{12} - \frac{\sqrt{5}}{4}$; (c) $\frac{1}{2}$.

23 $\mathbf{P}(\text{ever hit}) \leqslant \frac{1}{4}\sum\left(\frac{2}{3}\right)^{n-1} = \frac{3}{4}$.

24 (i) $\frac{7}{8}$; (ii) $\frac{83}{128}$.

25 (i) $p_A = p_B = \dfrac{1 - \frac{1}{2}(\theta + \theta^2)}{2 - \theta + \theta^2}$; (ii) $\theta = \frac{1}{10}(\sqrt{41} - 1)$; (iii) $\dfrac{2(1-\theta)}{2 - \theta - \theta^2}$.

26 (a) $\frac{197}{450}$; (b) $\frac{77}{225}$; (c) $\frac{25}{148}$.

27 (a) $5^{n-1}.6^{-n}$; (b) $\frac{5}{11}$; (c) $\frac{1}{6}$.

28 (a) $2\left(\dfrac{p\left(\frac{q}{p}\right)^{n-1} - q}{\left(\frac{q}{p}\right)^{n-1} - 1}\right)$; (b) $(p-q)\left(\dfrac{1 + \left(\frac{q}{p}\right)^n}{1 - \left(\frac{q}{p}\right)^n}\right)$ (c) $\dfrac{q-p}{\left(\frac{q}{p}\right)^n - 1} + \dfrac{q\left(\frac{p}{q}\right)^{n-1} - p}{\left(\frac{p}{q}\right)^{n-1} - 1}$.

29 (i) $p^k_m = \frac{1}{2}p^k_{m-1} + \frac{1}{2}p^k_{m+1}$; $p^k_0 = 0$, $p^k_n = p^{k-1}_{n-1}$ (ii) $p^0_m = 1 - \frac{m}{n}$; $\sum_{k=0}^{\infty} p^k_m = 1 - \frac{m}{n} + \frac{m}{n} = 1$.

30 (a) $\sum_{k=1}^{6} \frac{1}{6}\left(1 + \left(\frac{q}{p}\right)^{60/k}\right)^{-1}$ (b) Choose $x = 6$.

31 (a) $\dfrac{1 - p_1}{3 - p_1}$; (b) $\dfrac{(1-p_1)^2}{2 + (1-p_1)^2}$; (c) $\dfrac{1 - p_1}{3 - p_1 - p_2 - p_3}$.

32 (i) $\frac{1}{16}$; (ii) 5 metres and one step; (iii) 5 metres.

33 (a) $\dfrac{1}{(n!)^2}$ (b)$p_t = \begin{cases} \dfrac{t-1}{jk} & \text{for } 2 \leqslant t \leqslant j \wedge k \\ \dfrac{(j \wedge k) - 1}{jk} & \text{for } j \wedge k \leqslant t \leqslant j \vee k \\ \dfrac{j + k - t - 1}{jk} & \text{for } j \vee k \leqslant t \leqslant j + k - 2 \end{cases}$

(c) $j = k = n!$

34 (i) 0.3; 0.1545; (ii) For men 0.5; 0.6; for women 0.1; 0.11.

35 (a) $\frac{11}{23}$; (b) $\frac{12}{23}$.

36 No

39 $p_n = (1-p)\left(\dfrac{1 - (q-p)^{n-1}}{1 - (q-p)^n}\right) \to 1 - p$.

44 **P**(win with 1–6 flat) = 0.534;
P(win with 5–2 flat) = 0.5336.

CHAPTER 3

Exercises

3.8.1 (a) $\dfrac{(n+r)!}{r!}$; (b) $\binom{n+r}{r}$;
(c) $p_{r+1}(n)$.

3.8.2 (i) 462, assuming oranges are indistinguishable.
(ii) 7 = $\underline{x^5}|$ (the coefficient of x^5) in $\prod_{k=1}^{7}(1 - x^k)^{-1}$.

3.8.3 4

3.9.1 (a) $\dfrac{6}{n(n-1)}$; (b) $\dfrac{6(n-3)}{n(n-1)}$; (c) $\dfrac{5!(n-5)(n-6)}{2n(n-1)(n-2)(n-3)}$

3.10.7 $\dfrac{1}{n^r}\sum \dfrac{n!}{k!(r-2k)!(n-r+k)!}$ where the sum is over all k such that $\max\{0, r-n\} \leqslant 2k \leqslant r \leqslant 2n$.

3.10.8 $\dfrac{1}{n^r}\sum_{k=0}^{n}\binom{n}{k}(-)^k M_k$ where $r \geqslant 2n$ and $M_k = (n-k)^{r-k}(n-k+1)^k$.

3.11.1 $n^{-r}\binom{n}{r}$

3.11.3 $p(m, n) - p(m-1, n)$, where $p(m, n)$ is given in Example 3.10.

3.12.3 Rotational symmetry.

3.12.4 As in the example; k disjoint pairs of seats can be chosen in M_k ways, k given pairs of twins can occupy the pairs of seats in $2^k k!$ ways, the rest occupy their seats in $(2n-2k)!$ ways, so $\mathbf{P}$ (no pair adjacent) $= \sum_{k=0}^{n}(-)^k\binom{n}{k}\dfrac{M_k 2^k k!(2n-2k)!}{(2n)!}$

$$= \sum_{k=0}^{n}\frac{(-)^k}{k!}\frac{n!(2n-k-1)!}{(n-k)!(2n-1)!}2^k \to e^{-1} \text{ as } n \to \infty.$$

3.12.5 e^{-2}.

3.14.1 $b\binom{a}{k}\bigg/\binom{n}{k+1}$

3.14.2 Same as 3.14.1

3.14.3 $\left(1+\binom{n-k-1}{a-k+1}\right)\bigg/\binom{n}{a}$

3.14.4 $\left(b\binom{a}{k}+a\binom{b}{k}\right)\bigg/\binom{n}{k+1}$

3.15.2 (a) This depends on the order in which you catch the species, so the answer is a horrible sum. (b) $\binom{n}{x}\dfrac{b^x c^{n-x}}{(b+c)^n}$. Avoid sampling with partial replacement!

3.15.4 $\binom{n}{x}p^x(1-p)^{n-x}$ in all three cases.

3.16.5 $K(20, 3) = \binom{22}{2} = 231$.

3.16.6 $K(17, 3) = \binom{19}{2} = 171$.

3.16.7 Recall Example 3.11.

3.17.5 $p(n, r+s)\bigg/\sum_{j=r}^{n} p(n, j)$. Let $n \to \infty$.

3.17.7 (a) $\dfrac{1}{n!r!}\sum_{k=0}^{n-r}(-)^k(n-r-k)!/k!$; (b) $\sum_{k=0}^{n}(-)^k\dfrac{(n-k)!}{n!k!}$; (c) $(p(n, 0))^2$.

3.17.8 (a) zero; (b) zero; (c) e^{-2}.

Problems

1 $\dfrac{49}{153}$

2 (a) $\dfrac{6a!b!c!}{(a+b+c)!}$; (b) $\dfrac{1}{a!b!c!}$; (c) $\dfrac{6}{(a+b+c)!}$.

3 (a) $\dfrac{4!48!}{(12!)^4}\Big/\dfrac{52!}{(13!)^4}$; (b) $\dfrac{16}{\binom{52}{13}} - \dfrac{72}{\binom{52}{13}\binom{39}{13}} + \dfrac{72}{\binom{52}{13}\binom{39}{13}\binom{26}{13}}$;

(c) $\left(4\binom{39}{13} - 6\binom{26}{13} + 4\right)\Big/\binom{52}{13}$; (d) $4\binom{48}{9}\Big/\binom{52}{13}$.

4 (a) $13\binom{4}{2}\binom{12}{3}4^3\Big/\binom{52}{5}$; (b) $44\binom{4}{2}^2\binom{13}{2}\Big/\binom{52}{5}$; (c) $4^5 10\Big/\binom{52}{5}$;

(d) $4\binom{13}{5}\Big/\binom{52}{5}$; (e) $156\binom{4}{2}\binom{4}{3}\Big/\binom{52}{5}$.

6 $\dfrac{32491}{1 + 10^5}$

7 (i) 10; (ii) S where $S = \max\{n: n^2 \leqslant M\}$.

8 (a) $\binom{m+r}{m}$ is an integer. (b) You have $(k - 1)!$ colours, and k balls of each colour. How many arrangements are there?

9 (a) $5\left(\frac{5}{6}\right)\left(\frac{1}{6}\right)^4$ if the question means exactly 4 aces; or $\frac{26}{6}\left(\frac{1}{6}\right)^4$ if it means at least 4 aces.

(b) 6 times (a). (c) $1 - 3\left(\frac{5}{6}\right)^5 + 3\left(\frac{2}{3}\right)^5 - \left(\frac{1}{2}\right)^5$.

10 (a) $18\Big/\binom{64}{8}$; (b) $8!\Big/\binom{64}{8}$.

11 (a) $1 - \binom{3n}{r}\Big/\binom{4n}{r}$; (b) $\binom{n}{2}\binom{3n}{r-2}\Big/\binom{4n}{r}$;

(c) $\left(\binom{4n}{r} - 4\binom{3n}{r} + 6\binom{2n}{r} - 4\right)\Big/\binom{4n}{r}$.

12 (a) 15; (b) 2; (c) 36.

14 $P_k = \binom{b+w-k-1}{w-1}\Big/\binom{b+w}{w} = \dfrac{w}{b+w}\binom{b}{k}\Big/\binom{b+w-1}{k}$.

15 (b)(i) $\binom{N+m-1}{m-1}$; (ii) $\binom{N-1}{m-1}$.

16 Follows from $-\dfrac{1}{2} + \left(Kn + \dfrac{1}{4}\right)^{\frac{1}{2}} < N < \dfrac{1}{2} + \left(Kn + \dfrac{1}{4}\right)^{\frac{1}{2}}$.

17 (a) k^n; (b) $k!$ if $k \leqslant n$; (c) $\binom{n+k-1}{n}$.

19 See Example 3.10.

20 $\displaystyle\sum_{m=0}^{n} \frac{(-)^m}{m!}\left(\frac{n!}{(n-m)!}\right)^2 \frac{(2n-2m)!}{2n!}2^m \sim \sum_{0}^{\infty} \frac{(-)^m}{m!}\frac{1}{2^m}$.

23 (a) $\dfrac{(n-4)(n-3)}{(n-2)(n-1)}$ whether Arthur sits at random or not.

(b) Number the knights at the first sitting, and then use Problem 21.

25 (a) $\displaystyle\sum_{k=0}^{n}(-)^k\binom{n}{k}\frac{(2n-k)!}{(2n)!} \to e^{-\frac{1}{2}}$.

(b) Problem 21 again. The limit is e^{-1}.

26 $\dfrac{1}{2}$; (ii) $7\Big/\binom{7}{3} = \dfrac{1}{5}$; (iii) $\dfrac{1}{3}\binom{7}{2}2^{-7} + \left(\dfrac{4}{5}\right)\binom{7}{3}2^{-7} + \dfrac{1}{2} = \dfrac{99}{128}$; (iv) $\dfrac{92}{99}$.

27 $\binom{2n}{n}(pq)^n = (4pq)^n\left(1-\frac{1}{2}\right)\left(1-\frac{1}{4}\right)\dots\left(1-\frac{1}{2n}\right) \leqslant (4pq)^n \exp\left(-\frac{1}{2}\sum_1^n k^{-1}\right)$
$\leqslant (4pq)^n \exp\left(-\frac{1}{2}\log n\right) \to 0.$

36 $\frac{1}{4}+\frac{3}{4}\left(\frac{-1}{3}\right)^n.$

37 The number of ways of choosing k non-overlapping triples (three adjacent) is $\binom{3n-2k}{k}$.
Now use inclusion–exclusion.

39 Use induction.

CHAPTER 4

Exercises

4.7.2 (a) 992; (b) $32k!(31)^{k-1}$.
4.7.3 They give the same chance because trials are independent.
4.7.4 $\mathbf{P}(T > j + k | T > j) = (31/32)^k = \mathbf{P}(T > k)$.
4.8.4 Either form a difference equation or rearrange the sum.
4.8.6 (a) $[(n+1)p-1]$; (b) $\left[\frac{k}{p}-1\right]$.
4.8.7 (a) $\frac{1}{2}(1+(2p-1)^n)$; (b) $\frac{1}{2}(1-(2p-1)^n)$;
(c) $\sum_{m=k}^{n} \frac{m!}{(m-k)!}\binom{n}{m}p^m(1-p)^{n-m} = \frac{p^k n!}{(n-k)!}$; $\quad 0 \leqslant k \leqslant n.$
4.9.1 $(1-e^{-\lambda}-\lambda e^{-\lambda})/(1-e^{-\lambda})$
4.9.2 $\lambda(1-e^{-\lambda})^{-1}$
4.9.4 (a) $[\lambda-1]$; (b) k.
4.9.5 (a) $\exp(\lambda(e-1))$; (b) $e^{-\lambda}$; (c) λ; (d) λ^k.
4.10.3 Choose t to minimize

$$L(t) = a\sum_{x\leqslant t}(t-x)f(x) + b\mathbf{P}(X > t).$$

(a) $\hat{t} = \left[\frac{\log a - \log(a+bp)}{\log q}\right]+1$; (b) $\hat{t} = \left[\left(1+\frac{b}{a}\right)^{\frac{1}{2}}\right]$; (c) $\hat{t} = \left[\frac{b}{a}-n\right]$ if this lies in $[-n, n]$, otherwise $\hat{t} = n$ if $b/a > 2n$. (What if a or b can be negative?).
4.10.4 (a) Any median of X; (b) $\mathbf{E}(X)$.
4.10.5 Minimize $L(m) = b\sum_{k=0}^{m}(m-k)p_k(m) + c\sum_{k=m+1}^{m+n}(k-m)p_k(m)$
where $p_k(m) = p^k(1-p)^{m+n-k}\binom{m+n}{k}$. For no overbooking you need $L(0) < L(m)$ for all $m > 0$; solutions are approximate or numerical.
4.11.2 $k(a+b-k)$.
4.11.3 Let p_k be the probability that B wins if A's initial fortune is k. Then $\mathbf{P}(A|B) = \mathbf{P}(B|A)\mathbf{P}(A)/\mathbf{P}(B) = pp_{k+1}/p_k$ and $\mathbf{P}(A^c|B) = (1-p)p_{k-1}/p_k$.
Hence $pp_{k+1}\mathbf{E}(X_{k+1}|B) - p_k\mathbf{E}(X_k|B) + (1-p)p_{k-1}\mathbf{E}(X_{k-1}|B) = -p_k$.
When $p = \frac{1}{2}$ we have $p_k = (a+b-k)/(a+b)$, giving

$$\mathbf{E}(X_k|B) = \frac{1}{3}((a+b)^2-(a+b-k)^2), \quad 0 \leqslant k < a+b.$$

4.11.4 $\sum_{k=0}^{a+b} m_k \binom{a+b}{k} 2^{-(a+b)} = \begin{cases} \dfrac{a+b}{q-p}\left\{\dfrac{1}{2} - \dfrac{1-(2p)^{-(a+b)}}{1-(q/p)^{a+b}}\right\}; & p \neq q \\ \dfrac{1}{4}(a+b)(a+b-1); & p = \dfrac{1}{2}. \end{cases}$

4.11.5 (i) $p \neq q$; $\dfrac{k-(a+b)}{q-q} + \dfrac{p}{(q-p)^2}\left(\left(\dfrac{q}{p}\right)^{a+b} - \left(\dfrac{q}{p}\right)^k\right)$.

(ii) $p = \dfrac{1}{2}$; $(a+b)(a+b-1) - k(k-1)$.

4.12.1 (b) $\dfrac{M+1}{j(j+1)M}$; (c) $\dfrac{1-\exp(-\lambda j)}{1-\exp\left(-\frac{1}{2}\lambda M(M+1)\right)}$.

4.12.2 $\mathbf{E}(X_A) = 2$.

4.12.3 (b) $\sum_{n=1}^{\infty} \dfrac{m+1}{m+n} = \infty$; (c) $e^{-\frac{1}{2}m(m+1)} \sum_{n=1}^{\infty} e^{-\frac{1}{2}\lambda(m+n)(m+n+1)} < \infty$.

4.12.4 (i) Median is ∞; (ii) $\mathbf{E}(X_A | X_A < \infty) = \infty$.

4.13.1 $\hat{m} = \max\left\{0, \left[\left[\dfrac{N(b+d)-c}{b+d+c}\right]\right]\right\}$.

4.13.2 $\hat{m} = [m]$ where m is the positive root of $(m+1)q^m = c/(b+c)$.

4.14.1 You win \$1 with probability 1, but your winning bet has infinite expected value.

4.14.2 $\dfrac{p(1-(2q)^{L+1})}{(1-2q)(1-q^{L+1})}$ if $q \neq p$. This $\to \infty$ if $p < \dfrac{1}{2}$ or $\to p/(2p-1)$ if $p > \dfrac{1}{2}$. If $p = \dfrac{1}{2}$ then expectation is $\dfrac{Lp}{1-q^{L+1}} \to \infty$ as $L \to \infty$.

4.15.2 With an obvious notation $m_{us} = \dfrac{2}{3}\cdot\dfrac{1}{4}\cdot\left(\dfrac{3}{4}\right)^{s-1} = 2.3^{s-2}4^{-s}$; $m_{ds} = \dfrac{1}{3}\cdot\dfrac{3}{4}\cdot\left(\dfrac{1}{4}\right)^{s-1} = 4^{-s}$.

Hence the rth search downstairs comes after the sth search upstairs if $\dfrac{2}{9} > 3^{-s}4^{s-r} > \dfrac{1}{6}$.

The order is *duuuuduuuuudu* . . .

4.15.3 Place $m_{rs} = (1-d_{r1})(1-d_{r2}) \ldots (1-d_{rs})p_r$ in non-increasing order.

4.16.4 $q^2(1+p)/(1-pq) = (1-p^2)^2/(1+p^3)$

4.16.5 $(1-q^2)^2/(1+q^3)$

4.16.6 $\mathbf{E}(X|B^c) = \dfrac{2}{1-pq} + \dfrac{q}{1+q}$

4.16.7 $\mathbf{P}(A_1|B) = \dfrac{p}{1+p}$; $\mathbf{P}(A_2|B) = \dfrac{qp}{1+p}$.

4.16.8 Every number in $[2, 3)$ is a median.

4.16.9 $\mathbf{P}(B) = p^2(1-q^3)/(1-(1-p^2)(1-q^2))$
$\mathbf{E}(X) = (1+pq)(1-2pq)/(1-pq(1-p)(1+q))$

4.16.10 With new rules $\mathbf{P}(B) = \dfrac{q^2}{p^2+q^2}$ which is smaller than old $\mathbf{P}(B)$ if $p > \dfrac{1}{2}$.
Brianchon is making a mistake. $\mathbf{E}(X) = 4$ with new rules.

4.17.2 Since the answer 'yes' is false with probability $(1-p)/(2p)$, individuals should be much more likely to tell the truth. Then $\dfrac{2Y_n}{n} - 1$ should not be too far from p in the long run.

4.17.3 $\mathbf{P}\left(\bigcup_n^{\infty} A_k(\epsilon)\right) \leqslant \sum_n^{\infty} \mathbf{P}(A_k(\epsilon))$.

4.17.4 Use 4.17.3

4.17.5 Use 4.17.1

4.17.6 Use Markov's inequality and Chebyshov's inequality

4.17.8 $a(m, k) = \dfrac{m(m-1)\dots(m-k+1)}{(m+1)\dots(m+k)}$. Hence for large enough m with k fixed

$$\left|\log a(m, k) + \frac{1}{m} + \frac{2}{m} + \dots + \frac{k-1}{m} + \frac{1}{m} \dots + \frac{k}{m}\right| \leqslant \frac{2\sum_{1}^{k} r^2}{m^2} \to 0 \text{ as } m \to \infty.$$

Hence $[a(m, k)]^m e^{k^2} \to 1$. The inequalities follow from

$$\binom{2m}{m} 4^{-m} < \frac{2m(2m-2)\dots 2}{(2m+1)\dots 3.1} = \frac{1}{2m+1}\left(\binom{2m}{m}4^{-m}\right)^{-1}$$

and

$$2\binom{2m}{m} 4^{-m} > \frac{(2m-2)\dots 4.2}{(2m-1)\dots 3.1} = \frac{1}{2m}\left(\binom{2m}{m}4^{-m}\right)^{-1}.$$

4.18.6 For the left inequality, prove and use the fact that for any collection of probabilities $p_1, \dots, p_r$, we have $-\sum_i p_i \log\left(\sum_i p_i\right) < -\sum_i p_i \log p_i$. Equality holds when $g(.)$ is a one–one map. For the right hand inequality note that

$$f_i = \exp(-cg(x_i)) \Big/ \sum_i \exp(-cg(x_i))$$

is a mass function and use 4.18.1. Equality holds if $f_i = f_X(x_i)$ for all i.

Problems

1 (b) $f(0) = \frac{1}{140}$, $f(1) = \frac{18}{140}$, $f(2) = \frac{66}{140}$; $f(3) = \frac{55}{140}$, $\mathbf{E}(X) = \frac{9}{4}$.

3 $\frac{35}{12}$

4 If X is uniform on $\{1, 2, \dots, n\}$ then $\text{var}(X) = \frac{1}{12}(n^2 - 1)$.

5 (a)(i) $(e^2-1)^{-1}$; (ii) $p^{-1} - 1$; (iii) $(\log(1-p)^{-1})^{-1}$; (iv) $6\pi^{-2}$; (v) 1.
(b)(i) $2e^2(e^2-1)^{-1}$; (ii) $(1-p)^{-1}$; (iii) $p((1-p)\log(1-p)^{-1})^{-1}$; (iv) ∞; (v) ∞.

6 Yes, in all cases.

7 $c = \dfrac{4(M+1)(M+2)}{M(M+3)} \to 4$; $\mathbf{E}(X) = \dfrac{2(M+1)}{M+3} \to 2$.

8 Condition on the appearance of the first tail to get $\mathbf{P}(A_n) = \frac{1}{2}\mathbf{P}(A_{n-1}) + \frac{1}{4}\mathbf{P}(A_{n-2}) + \frac{1}{8}\mathbf{P}(A_{n-3})$, $n > 3$. Hence $\mathbf{P}(A_n) = A\alpha^n + B\beta^n + C\gamma^n$ where α, β, γ are roots of $8x^3 - 4x^2 - 2x - 1 = 0$, and A, B, C are chosen to ensure that $\mathbf{P}(A_1) = \mathbf{P}(A_2) = 0$ and $\mathbf{P}(A_3) = \frac{1}{8}$. Similar conditioning gives

$$\mathbf{E}(T) = \frac{1}{2}(1 + \mathbf{E}(T)) + \frac{1}{4}(2 + \mathbf{E}(T)) + \frac{1}{8}(3 + \mathbf{E}(T)) + \frac{3}{8}.$$

Hence $\mathbf{E}(T) = 14$. To find $\mathbf{E}(U)$ consider the event that a sequence of n tosses including no HTH is followed by HTH. Hence either $U = n + 1$ or $U = n + 3$, and so

$$\mathbf{P}(U > n)\frac{1}{8} = \mathbf{P}(U = n + 1)\frac{1}{4} + \mathbf{P}(U = n + 3).$$

Summing over n gives $\frac{1}{8}\mathbf{E}(U) = \frac{1}{4} + 1$; $\mathbf{E}(U) = 10$.

9 $\mathbf{E}(X) = \infty$.

14 (i) $\dfrac{1}{b - a + 1}$ for $a \leqslant k \leqslant b$; (ii) $\dfrac{m - n - k}{m - n}$.

15 e.g. $f(-2) = \frac{1}{2}$, $f(1) = f(3) = \frac{1}{4}$.

16 (i) $\sum_{-\infty}^{\infty} (f(2n) - f(2n + 1))$; (ii) zero.

17 $$F_Y(y) = \begin{cases} F_X\left(\dfrac{y - b}{a}\right); & a > 0 \\ \mathbf{P}\left(X \geqslant \dfrac{y - b}{a}\right); & a < 0 \\ \begin{Bmatrix} 0 & y < b \\ 1 & y \geqslant b \end{Bmatrix}; & a = 0. \end{cases}$$

19 (a)When $\lambda = 0$; (b) $p^{-1} - \log(1 - p)^{-1} - 1$.

20 (a) If the type is uncommon and the population is large. (c) X is roughly Poisson with parameter 10 so $\mathbf{P}$ (this is the only one) $\simeq 10(e^{10} - 1)^{-1}$ which is very small.

21 $f_R(r + 1) = \left(\dfrac{n - r}{n}\right)^2$, $f_R(r - 1) = \left(\dfrac{r}{n}\right)^2$; $\mathbf{E}(R) \to \frac{1}{2} n$.

22 $f_X(1) = \frac{5}{18}$; $f_X(3) = \frac{2}{7}$; $\mathbf{E}(X) = 2 - \frac{5}{18} + \frac{2}{7}$.

24 $\mathbf{P}(X \leqslant n) = \frac{1}{2}(m + n)(m + n + 1)m^{-2}$ for $-m \leqslant n \leqslant 0$. For $0 \leqslant n \leqslant m$, $\mathbf{P}(X \leqslant n)$ $= 1 - \frac{1}{2}(m - n)(m - n - 1)m^{-2}$.

25 Mean μp.

26 (i) $\dfrac{6!}{(2!)^3}$; (ii) $\frac{6}{5} + \frac{16}{15} + \frac{5}{3} = \frac{59}{15}$.

27 (a) $p^k q^{n-k}\dbinom{n - 1}{k - 1}$; (b) $(rp)^k(1 - rp)^{n-k}\dbinom{n - 1}{k - 1}$;

(c) $(1 - r)^k p^k (1 - p)^{n-k}\dbinom{n}{k}(1 - pr)^{-n}$; $\mathbf{E}(X) = \dfrac{n(1 - r)p}{1 - rp}$.

28 (i) Choose A if $2000(1 - p) < 1000p$ i.e. if $p > \frac{2}{3}$.

(ii) Choose A if $2000\left(1 - \dfrac{9p}{4 + 5p}\right) < 1000\left(\dfrac{9p}{4 + 5p}\right)$; but $p > \frac{8}{17}$, so choose B.

29 (a) p^n; (b) $(1 - p)p^{n-1}$; (c) $(1 - p)^{-1}$.

31 $M(n)/n \to C = \dfrac{1}{1 + p}$ as $n \to \infty$.

32 (a) $1 - (1 - p)^7$; (b) $p(1 - p)^{x-1}/(1 - (1 - p)^7)$; (c) $\sum_1^7 x f_X(x) = \dfrac{1}{p} - \dfrac{7(1 - p)^7}{1 - (1 - p)^7}$;

(d) $\dfrac{1}{p}(1 - (1 - p)^4)$.

36 (b) $10^2(1 - p)^{10} + 10^3 p(1 - p)^9$; (c) $[10^2(1 - p)^{10} + 990p(1 - p)^{19}]b + (1 - r)10^2 c$.

37 (c) $\frac{2}{3}(n + 1)$.

38 (a) Y is $\mathrm{B}\left(n, \frac{1}{36}\right)$; $\mathbf{E}(Y) = n/36$. (b) $X - r$ is $\mathrm{B}\left(n - r, \frac{6}{7}\right)$; $\mathbf{E}(X - r) = (n - r)\frac{6}{7}$.

39 (a) $\left(\frac{p^2}{1+p} + \frac{q^2}{1+q}\right)^{-1}$; $\frac{p^2}{1+p}\left(\frac{p^2}{1+p} + \frac{q^2}{1+q}\right)^{-1}$.

(b) $\frac{1}{1-r}\left(\frac{2(1-r)^2 + pq}{(1-r)^2 - pq}\right)$; $\frac{p^2}{p+q}\left(\frac{p^2 + 2q}{p^2 + q^2 + pq}\right)$.

The second set of rules.

41 $\mathrm{B}\left(n, \frac{1}{32}\right)$, use the Poisson approximation.

43
$$\begin{aligned} n - \mathbf{E}(H) = \mathbf{E}(n - H) = \sum_{k=0}^{n}(n-k)p_k &= \sum_{0}^{n-1}(2n-k)\binom{2n-k-1}{n}2^{k-2n} \\ &= \sum_{1}^{n}(2n-k+1)\binom{2n-k}{n}2^{k-1-2n} \\ &= \frac{1}{2}\mathbf{E}(2n+1-H) - \frac{2n+1}{2^{2n+1}}\binom{2n}{n}. \end{aligned}$$

So $\mathbf{E}(H) = \frac{2n+1}{2^{2n}}\binom{2n}{n} - 1$.

44 $\sum_{k} q(1-q)^{k-1}a_k$

45 With an obvious notation, $m = 1 + \frac{1}{2}m_1 + \frac{1}{3}m_2 + \frac{1}{6}m_3$, also $m_1 = 1 + \frac{1}{2}m_1 + \frac{1}{3}m_{12}$ $+ \frac{1}{6}m_{13}$ (and two similar equations), also $m_{12} = 1 + \frac{5}{6}m_{12}$ (and two similar equations). Solve to get m.

CHAPTER 5

Exercises

5.10.3 $\rho = -\left(\frac{qr}{(1-q)(1-r)}\right)^{\frac{1}{2}}$

5.10.4 $\binom{m}{x}\left(\frac{p}{p+q}\right)^{x}\left(\frac{q}{p+q}\right)^{m-x}$

5.10.5 $\sum_{k=0}^{9}\frac{18!(pq)^{9-k}r^{2k}}{((9-k)!)^2(2k)!}$

5.10.6 $\frac{p}{p+r}(n-Y)$

5.10.7 $n - Y - Z$

5.11.1 S is $\mathrm{B}(m, p^2)$ with mean mp^2, variance $mp^2(1-p^2)$.

5.11.5 A is the sum of two binomial random variables $\mathrm{B}(m, \phi)$ and $\mathrm{B}(m, \mu)$. Hence

$$\begin{aligned} f_A(a) &= \sum_{k=0}^{a}\phi^k(1-\phi)^{m-k}\binom{m}{k}\mu^{a-k}(1-\mu)^{m-a+k}\binom{m}{a-k} \\ &= \mathbf{E}(A) = m(\phi + \mu). \end{aligned}$$

5.11.6 $\mathbf{E}(A|S) = 2S + (m-S)\phi + (m-S)\mu$

5.12.3 $\mathbf{E}(R_n) = 2 + \frac{2}{3}\frac{2^n - 1}{2^{n-1}+1} + \frac{1}{3}\frac{2^{n-1}-1}{2^n+1} \to \frac{7}{2}$

5.12.4 $p_r = \frac{r}{r+b}$

5.12.5 $\frac{m}{n}$

5.13.4 When $X_i = c$ where c is constant.

5.13.5 We assume that 'at random' means an individual is selected at random from n independent families $X_1, \ldots, X_n$. Define

$$I(X_i \geqslant k) = \begin{cases} 1 & \text{if } X_i \geqslant k \\ 0 & \text{if } X_i < k. \end{cases}$$

Then

$$f_R(k) = \mathbf{E}\left(\frac{\sum_{i=1}^n I(X_i \geqslant k)}{\sum_{i=1}^n X_i}\right) = n\mathbf{E}\left(\frac{I(X_1 \geqslant k)}{\sum_{i=1}^n X_i}\right)$$

and

$$\mathbf{E}(R) = \mathbf{E}\left(\frac{\sum_k kI(X_1 \geqslant k)}{\sum_1^n X_i/n}\right).$$

5.13.6 Let X be uniform on $\{x_1, \ldots, x_n\}$ and Y uniform on $\{y_1, \ldots, y_n\}$.

5.14.1 $S + (n-S)\frac{p - p\gamma}{1 - \gamma p}$

5.14.2 $N\gamma$

5.14.3 $\rho(N, S) = \left(\frac{\gamma(1-p)}{1-\gamma p}\right)^{\frac{1}{2}}$

5.14.4 (i) $\mathbf{P}(T = k, S = j, N = i) = \binom{j}{k}\tau^k(1-\tau)^{j-k}\binom{i}{j}\gamma^j(1-\gamma)^{i-j}\binom{n}{i}p^i(1-p)^{n-i}$ for $k \leqslant j \leqslant i \leqslant n$.

(ii) $\mathbf{P}(N = i|T = k) = \binom{n-k}{n-i}\left(\frac{1-p}{1-p\gamma\tau}\right)^{n-i}\left(\frac{p - p\gamma\tau}{1-p\gamma\tau}\right)^{i-k}$ which is binomial.

5.14.5 $\binom{n-s}{i-s}\left(\frac{p(1-\gamma)}{1-p\gamma}\right)^{i-s}\left(\frac{1-p}{1-p\gamma}\right)^{n-i}\binom{s}{k}\tau^k(1-\tau)^{s-k}; \quad k \leqslant s \leqslant i.$

5.14.6 Zero

5.16.2 No

5.16.3 $\mathbf{E}(Z) = \frac{1}{2}$; $\operatorname{var}(Z) = \frac{5}{12}$.

5.16.4 $\frac{1}{6}n$; $\frac{5}{36}n$.

5.17.7 Recall the ballot theorem.

5.18.3 $na(b-a)(b-n)/(b^2(b-1))$.

5.18.4 $\mathbf{E}(Z_m) = \frac{b}{a} + \frac{b-1}{a-1} + \ldots + \frac{b-m+1}{a-m+1}$.

5.18.5 $\mathbf{P}(X = k|X + Y = j) = \binom{n}{k}\binom{n}{j-k}\Big/\binom{2n}{j}$.

5.19.5 $\mathbf{E}(R_r) = \infty$

5.19.7 For a walk starting at zero the expected number of visits to zero *including* the first is $\frac{1}{|p-q|}$. Hence, for $p < q$ and $r > 0$, $\mathbf{E}(V) = \left(\frac{p}{q}\right)^r\frac{1}{|p-q|}$;

for $p < q$ and $r < 0$, $\mathbf{E}(V) = \dfrac{1}{|p-q|}$. Likewise if $p > q$

$$\mathbf{E}(V) = \begin{cases} \left(\dfrac{p}{q}\right)^r \dfrac{1}{|p-q|} & r < 0 \\ \dfrac{1}{|p-q|} & r > 0. \end{cases}$$

5.19.8 $\mathbf{E}(V) = \sum_n \mathbf{P}(S_X = S_Y = 0) = \sum_n \dfrac{1}{4^{2n}} \sum_{k=0}^{n} \dfrac{(2n)!}{(k!)^2((n-k)!)^2} = \sum_n \dfrac{1}{4^{2n}} \binom{2n}{n}^2$

$$= \sum_n \frac{(2n-1)^2(2n-3)^2 \ldots 1^2}{(2n)^2(2n-2)^2 \ldots 2^2} \geqslant \sum_n \frac{1}{2n} = \infty.$$

5.20.5 $\mathbf{P}(X \geqslant Y) = \int F_Y(x) f_X(x)dx \geqslant \int F_X(x) f_X(x)dx = \dfrac{1}{2}$.

5.20.7 (i) Let I_j be independent Bernoilli random variables with parameter p. Then $\sum_1^m I_j \leqslant \sum_1^n I_j$ for $m \leqslant n$.

(ii) Let I_j be independent Bernoulli with parameter p_2 and K_j independent Bernoulli with parameter $\dfrac{p_1}{p_2}$. Then $I_j K_j$ is Bernoulli with parameter p_1 and $\sum_1^n I_j K_j \leqslant \sum_1^n I_j$.

Problems

1 $f(2,0) = f(0,2) = \dfrac{1}{36}$, $f(1,0) = f(0,1) = \dfrac{8}{36}$, $f(0,0) = \dfrac{16}{36}$, $f(1,1) = \dfrac{2}{36}$; $\operatorname{cov}(X, Y) = -\dfrac{1}{18}$; $\rho(X, Y) = -\dfrac{1}{5}$.

2 (a) zero; (b) $\dfrac{1}{16}$; (c) $\dfrac{1}{2}$; (d) 1.

3 e.g. $X = \pm 1$ with probability $\dfrac{1}{2}$ each, $Y = |X|$.

7 $\operatorname{cov}(U, V) = ac + bd + (ad + bc)$; $\rho(X, Y) = 0$ for many choices of a, b, c, d.

8 (i) $\mathbf{P}(\text{correct}) = 2\left(p - \dfrac{1}{2}\right)^2 + \dfrac{1}{2}$; (ii) $\mathbf{P}(\text{correct}) = p^3 + 3p(1-p)^2$.

9 (b) $\mathbf{P}(U = m, V = n) = p^{m+1}q^n + q^{m+1}p^n$; $\operatorname{cov}(U, V) = (4pq - 1)/(pq)$; $\rho(U, V) = -|p - q|$.

10 You need results like $\sum_{i=1}^n i^2 = \dfrac{1}{3} n(n^2 - 1) + \dfrac{1}{2} n(n+1)$ and $\sum_{\substack{i \neq j \\ 1 \leqslant i, j \leqslant n}} ij = \left(\sum_{i=1}^n i\right)^2 - \sum_{i=1}^n i^2$.

Then $\operatorname{cov}(X, Y) = -\dfrac{n+1}{12}$; $\rho(X, Y) = -\dfrac{1}{n-1} \to 0$.

11 (b) $a - 4a^2$; (c) $\mathbf{E}(X|Y = 0) = \dfrac{a}{1 - 2a}$; $\mathbf{E}(X|Y = 1) = \dfrac{1}{2}$; (d) $a = \dfrac{1}{4}$.

12 $\operatorname{cov}(U, V) = \left(\dfrac{5}{6}\right)^2 \left(\dfrac{7}{6}\right)^2$.

13 (a) $\mathbf{E}(|XY|) = \mathbf{E}(|X||Y|) = \mathbf{E}(|X|)\mathbf{E}(|Y|) < \infty$; (b) $\mathbf{E}(X^2)\mathbf{E}(Y^2) < \infty$.

14 (a) yes when $\theta = 3 - 2\sqrt{2}$; (b) no; (c) yes when $\theta = \dfrac{1}{2}(\sqrt{5} - 1)$;

(d) yes when $\theta = \frac{1}{2}(3 - 2\sqrt{2})$ and independence holds; (e) yes when $\frac{\alpha\beta}{1-\beta} = 1$;
(f) yes when $\alpha = 6\pi^{-2}$.

15 (a) $f_X(i) = \begin{cases} \dfrac{1+\theta}{1-\theta}\theta^{|i|}, & i \neq 0 \\ \dfrac{2\theta}{1-\theta}, & i = 0 \end{cases}$

(c) $f_X(i) = \dfrac{\theta^{2i+3}}{1-\theta}, \quad i \geq 0$ (d) $f_X(i) = \dfrac{\theta^{i+1}}{1-\theta}, \quad i \geq 0$

(e) $f_X(i) = \alpha\left(\dfrac{i\beta}{c - i\beta} - \dfrac{(i-1)\beta}{c-(i-1)\beta}\right), 1 \leq i \leq c$ (f) $f_Y(j) = \alpha j^{-2}, 1 \leq j$.

17 (c) $\mathbf{P}(X_1 = x_1) = \binom{a_1}{x_1}\binom{\sum_2^k a_i}{\sum_2^k x_i}\Big/\binom{\sum_1^k a_i}{\sum_1^k x_i}$.

18 (a) $\dfrac{1-p^c}{1-p}$;

(b) $\mathbf{P}(\min\{X, Y\} > n) = p_1^n p_2^n$ so $\mathbf{E}(Z) = \dfrac{1}{1 - p_1 p_2}$.

20 (a) $\dfrac{p\beta}{1-q\beta}$; (b) $\dfrac{\alpha p}{\beta q}\dfrac{\beta^m q^n}{1-\beta^m q^n}$.

26 Let (x, y, z) take any of the 8 values $(\pm 1, \pm 1, \pm 1)$. Then

$$\begin{aligned} 1 - xy &= |1 - xy| = |(1-xy)||(-xz)| \quad \text{since } |(-xz)| = 1, \\ &= |(1-xy)(-xz)| = |yz - xz| \geq \pm(yz - xz). \end{aligned}$$

Now use Corollary 5.3.2 to get the result.

27 $f_Y(j) = \dfrac{1}{2n+1}$; $f_X(i) = \dfrac{1}{2m+1}$.

28 $f(0, 1) = f(1, 2) = f(2, 0) = \frac{1}{3}$. Then $f_X(i) = f_Y(i) = \frac{1}{3}$, and $\mathbf{P}(X < Y) = \frac{2}{3}$,

$\mathbf{P}(Y < X) = \frac{1}{3}$.

29 (a) $\mathbf{E}(U_r) = \dfrac{U^2}{U+V} + \dfrac{UV}{U+V}\left(1 - \dfrac{U+V}{UV}\right)^r \to \dfrac{U^2}{U+V}$.

(b) Let T be the number of tosses to the first head,

$$\mathbf{E}(U_{T-1}) = \frac{U^2}{U+V} + \frac{pU^2V^2}{(U+V)(UVp + U + V - p(U+V))}.$$

When $U = V = \dfrac{1}{p}$ this is $\dfrac{U}{2}\left(1 + \dfrac{U}{3U-2}\right) \sim \dfrac{2}{3}U$.

31 The total $T = \sum_1^k S_i$ where S_i is the score on the ith ball.

(a) $\mathbf{E}(T) = \frac{1}{2}k(n+1)$, $\mathbf{E}(S_iS_j) = \dfrac{2}{n(n-1)}\sum_{i>j} ij = \dfrac{1}{12}(3n+2)(n+1)$.

Hence $\operatorname{var}(T) = \dfrac{1}{12}(n+1)k(n-k)$. If M is the maximum,

$$\mathbf{P}(M = m) = \binom{m-1}{k-1}\Big/\binom{n}{k} \quad \text{without replacement;}$$

$$\mathbf{P}(M = m) = \left(\frac{m}{n}\right)^k - \left(\frac{m-1}{n}\right)^k \quad \text{with replacement.}$$

33 (a) $\frac{1}{2}$; (b) $\frac{3\sqrt{5}-1}{6}$; (c) $\frac{5}{6}$.

34 Use the argument of Theorem 5.6.7; $pr^3 - r + q = 0$; $r = (-p + (p^2 + 4pq)^{\frac{1}{2}})/(2p)$.

35 $\binom{n}{r}\frac{3^{n-r}}{4^n}$; $\sum_{n=r}^{\infty}\binom{n}{r}\frac{3^{n-r}}{4^n}\frac{e^{-8}8^n}{n!} = \frac{e^{-2}2^r}{r!}$; mean = variance = 12.

38 (a) $\mathbf{P}(M \geqslant r) = \left(\frac{p}{q}\right)^r$; $\mathbf{P}(M = r) = \left(\frac{p}{q}\right)^r - \left(\frac{p}{q}\right)^{r+1}$; $\mathbf{E}(M) = \frac{p}{q-p}$.

(b) $\mathbf{P}(M = r|S_0 = -k) = \alpha\beta^k\left(1 - \frac{p}{q}\right)\left(\frac{p}{q}\right)^{r+k}$; $\mathbf{P}(M = r) = \frac{\alpha(q-p)}{q-\beta p}\left(\frac{p}{q}\right)^r$;

$\mathbf{P}(S_0 = -k|M = r) = \left(1 - \frac{\beta p}{q}\right)\left(\frac{\beta p}{q}\right)^k$; $\quad k \geqslant 0$.

40 No

CHAPTER 6

Exercises

6.9.7 $\mathbf{E}(s^{T_{a0}}) + \mathbf{E}(s^{T_{aK}})$

6.9.8 $(\lambda_2(s))^a$

6.9.9 (i) $\mathbf{P}(T_{a0} < \infty) = \begin{cases} 1 & p \leqslant q \\ \frac{q}{p} & p \geqslant q \end{cases}$

(ii) $\mathbf{E}(T_{a0}|T_{a0} < \infty) = \frac{a}{|p-q|}$.

6.9.10 (i) $\mathbf{E}(s^T) = ps\mathbf{E}(s^{T_{10}}) + qs\mathbf{E}(s^{T_{01}})$;

(ii) $\mathbf{E}(T|T < \infty) = 1 + \frac{1}{|p-q|}$.

6.10.1 (0, 2, 2, 4) and (2, 3, 3, 4).

6.10.2 (b) $f(x)$ and $g(x)$ have non-negative coefficients and $\frac{f(x)}{f(1)}\frac{g(x)}{g(1)} = \left(\frac{x(1-x^{12})}{12(1-x)}\right)^2$.

6.10.4 Yes, trivially.

6.10.5 No.

6.11.10 $n\sigma^2$

6.11.13 r is the chance of extinction derived in Example 6.3.16.

6.12.7 Use induction.

6.12.9 (a) By Jensen's inequality (4.6.14) we have

$$\begin{aligned}\mathbf{E}(X^2) &= \mathbf{E}(X^2|X > 0)\mathbf{P}(X > 0) \geqslant (\mathbf{E}(X|X > 0))^2\mathbf{P}(X > 0) \\ &= \mathbf{E}(X|X > 0)\mathbf{E}(X).\end{aligned}$$

(b) Hence $\mathbf{E}(Z_n\rho^{-n}|Z_n > 0) \leqslant \mathbf{E}(Z_n^2\rho^{-2n})$

6.12.10 Let $\mathbf{E}(s^{Z_n^*}) = G_n^*(s)$. Then

$$\mathbf{E}(s^{Z_n^*}t^{Z_{n+m}^*}) = G_{m-1}^*(t)G_n^*(sG_m(t)) \to sG_{m-1}^*(t)G_m(t)(\rho - 1)/(\rho sG_m(t) - 1).$$

6.13.4 Set $z = y + 1$ in (3) and equate coefficients.

6.14.10 For HHH, $\mathbf{E}(X) = 2 + 4 + 8 = 14$; for HTH, $\mathbf{E}(X) = 2 + 8 = 10$; for HHT, $\mathbf{E}(X) = 8$; for THH, $\mathbf{E}(X) = 8$. The others all follow by symmetry from these.

6.14.16 $\frac{1}{p} + \frac{1}{p^2q} + \frac{1}{p^3q^2}$ (= 42 in the fair case).

6.15.6 Arguing directly

$$\mathbf{E}(T) = \mathbf{E}(X_1 I_a) + \mathbf{E}((X_1 + T')I_a^c) = \mathbf{E}(X_1 I_a) + \mathbf{E}(X_1 I_a^c) + \mathbf{E}(T)\mathbf{E}(I_a^c).$$

Hence

$$\mathbf{E}(T) = \frac{\mathbf{E}(X_1)}{1 - \mathbf{E}(I_a^c)} = \frac{\mathbf{E}(X_1)}{\mathbf{P}(X_1 \leqslant a)} = \frac{1}{q(1 - p^a)}$$

if X_1 is geometric.

6.15.7 $\mathbf{E}(T) = b + \mathbf{E}(X_1 I_b^c)/\mathbf{P}(X_1 > b) = (1 - p^b)/(qp^b)$ if X_1 is geometric.

6.15.8 This is (1) with $a = 0$ and $b = r$.

6.15.9 This is (8) with X geometric.

6.15.10 $\mathbf{P}(L_n < r) = \mathbf{P}(W > n)$ where W is as defined in (9). Hence

$$1 + \sum_n s^n \pi_{n,r} = \sum_n s^n \mathbf{P}(W > n) = \frac{1 - \mathbf{E}(s^W)}{1 - s} = \frac{1 - p^r s^r}{1 - s + qp^r s^{r+1}}.$$

Problems

1 $\sum \mathbf{P}(X < k)s^k = \sum \mathbf{P}(X \leqslant k)s^{k+1}$

2 (a) $G = \dfrac{1}{n}\dfrac{n - s^{n+1}}{1 - s}$ (b) $G = \dfrac{1}{2n+1}\dfrac{s^{-n} - s^{n+1}}{1 - s}$; $s \neq 0$

(c) $G = 1 - (1 - s^{-1})\log(1 - s)$; $|s| \leqslant 1$

(d) $G = 1 - \frac{1}{2}(1 - s)\log(1 - s^{-1}) - \frac{1}{2}(1 - s^{-1})\log(1 - s)$; $|s| = 1$

(e) $G = \dfrac{1 - c}{1 + c}\left\{1 + \dfrac{cs}{1 - cs} + \dfrac{cs^{-1}}{1 - cs^{-1}}\right\}$; $|s| = 1$.

3 (a) a p.g.f. wherever $G_X(s)$ exists; (b) not a p.g.f. (c) a p.g.f. for $|s| < p^{-1}$; (d) a p.g.f. for all s; (e) a p.g.f. for $|s| \leqslant 1$; (f) a p.g.f. if $\alpha \log(1 + \beta) = 1$, $\beta < 0$, for $|\beta s| < 0$.

5 Let N have p.m.f. $f_N(k) = 2^{-k}$, $k \geqslant 1$, and $(X_i;\, i \geqslant 1)$ be independent and identically distributed with p.g.f. G, then $Y = \sum_1^N X_i$.

6 If it were possible then $1 - s^{11} = (1 - s)R_1(s)R_2(s)$ where R_1 and R_2 are polynomials (with real positive coefficients) of degree five. Since the imaginary roots of unity form conjugate pairs this is impossible.

7 (b) Yes, make it a flat with $f(2) = 0$.

9 (b) $G_Y''(1) + G_Y'(1) - (G_Y'(1))^2 = \mathrm{var}(N)(\mathbf{E}(X))^2 + \mathbf{E}(N)\,\mathrm{var}(X)$.

10 (c) $\frac{1}{3}(G_X(1) + G_X(\omega) + G_X(\omega^2))$ where ω is a complex cube root of unity.

11 $G = G_X$ if and only if $G = \dfrac{(1 + \mu)^{-1}}{1 - s(1 + \mu)^{-1}}$, so X is geometric.

13 Use conditioning. So $\mathbf{E}(s^N) = \dfrac{s}{2}\mathbf{E}(s^N) + \dfrac{s^2}{4}\mathbf{E}(s^N) + \dfrac{s^2}{4}$;

$\mathbf{E}(N) = \frac{1}{2}(1 + \mathbf{E}(N)) + \frac{1}{4}(2 + \mathbf{E}(N)) + \frac{1}{2}$; $\mathbf{E}(N) = 6$.

14 (a) $\frac{1}{2}\lambda + 1$. (b) $\mathbf{P}(R = r) = \sum \mathbf{P}(R = r \mid X = x)\mathbf{P}(X = x) = \displaystyle\sum_{x=r}^{\infty} \frac{cp^x}{x(x+1)}$.

Hence $(1 - s)G_R(s) = cd - cs\left(1 - \left(1 - \dfrac{1}{ps}\right)\log(1 - ps)\right)$ where $d = \left(\dfrac{1}{p}\log(1 - p)\right)$.

15 By the independence $\mathrm{var}(H - T) = \mathrm{var}(H) + \mathrm{var}(T) = \lambda = \mathrm{var}(N)$.

16 With the notation of Problem 15,

$$\mathbf{E}(s^H t^T) = \mathbf{E}(s^H t^{N-H}) = \mathbf{E}((ps+qt)^N) = G_N(ps+qt).$$

If H and T are independent then

$$G_N(ps+qt) = G_H(s)G_T(t) = G_N(ps+q)G_N(p+qt).$$

Write $s = x+1$, $t = y+1$, $G_N(v) = f(v-1)$ to get $f(px+qy) = f(px)f(qy)$. The only continuous solutions of this are $f(z) = e^{\lambda z}$, so $G_N(s) = e^{\lambda(s-1)}$.

17 $G_{X_n}(s) = \left(\dfrac{p}{1-qs}\right)^n = \left(\dfrac{1-\dfrac{\lambda}{n}}{1-\dfrac{\lambda s}{n}}\right)^n \to e^{\lambda(s-1)}.$

18 (b) $\mathbf{E}(s^N) = s + (s-1)(\exp(se^{-a}) - 1)$; $\mathbf{E}(N) = e^{e^{-a}}$.

19 Do not differentiate $G(s)$!

20 (i) $G_X(s) = \dfrac{s}{3-2s}$. (ii) $G_Y(s) = s\cdot\dfrac{(m-1)s}{m-s}\cdot\dfrac{(m-2)s}{m-2s}\cdots\dfrac{s}{m-(m-1)s}$.

Do not differentiate this to find the mean!

21 Use L'Hopital's rule.

22 $G_n(s) = \dfrac{s}{2-G_{n-1}(s)}$; $G_n(1) = 1$; $G_n'(1) = n$.

24 Differentiate.

25 $\left(\dfrac{a}{1+a-s}\right)^\lambda$

26 $\mathbf{E}(s^{X+Y}t^{X-Y}) = \exp(\lambda(st-1) + \mu(st^{-1}-1))$. Hence for $X-Y$, $\kappa_r = \lambda + (-)^r\mu$ and for $X+Y$, $\mu^{(k)} = (\lambda+\mu)^k$.

27 $(ps+q)^n(p+qs^{-1})^m s^m = (ps+q)^{m+n}$.

28 (a) $\dfrac{\alpha p}{1-(1-p+(1-\alpha)s)t}$; (b) $\dfrac{p(1-\alpha)}{1-\alpha p}$.

29 For all $0 \leqslant r \leqslant n$ we have

$$\sum_{k=0}^{n}\binom{n-r}{k}x^k = (1+x)^{n-r} = \left(1-\frac{x}{1+x}\right)^r(1+x)^n = \sum_{k=0}^{n}\binom{r}{k}\left(\frac{-x}{1+x}\right)^k(1+x)^n.$$

Since the sums are polynomials in r of degree at most n, it follows that they must be identically equal. Hence setting $r = -n-1$ and $x = 1$ gives

$$\sum_{k=0}^{n}\binom{n+k}{k}2^{-k+n} = \sum_{k=0}^{n}\binom{2n+1}{k} = \frac{1}{2}\sum_{k=0}^{2n+1}\binom{2n+1}{k} = 2^{2n}.$$

Hence $\sum_{k=0}^{n} a_k = \dfrac{1}{2^{n+1}}\sum_{k=0}^{n}\binom{n+k}{k}2^{-k} = \dfrac{1}{2}$.

Recognising the p.g.f. of the negative binomial distribution with parameter $\frac{1}{2}$, this says that in a sequence of coin tosses the chance of getting up to n tails before $n+1$ heads equals the chance of getting $n+1$ or more tails before $n+1$ heads equals $\frac{1}{2}$. Now remember the ant of Example 3.7.1.

30 Let $S_n = X_n + Y_n$. Then S_n is a simple random walk with $p = \alpha_1 + \alpha_2$ and $q = \beta_1 + \beta_2 = 1-p$.

31 S_n is symmetric so (a) $\mathbf{E}(T) = \infty$ and (b) $\mathbf{E}(s^{T_1})|_{s=1} = 1$.

(c) Let $U_n = X_n - Y_n$ and $V_n = X_n + Y_n$, so

$$\mathbf{E}(s^{U_1}T^{V_1}) = \frac{1}{4}(st + st^{-1} + ts^{-1} + s^{-1}t^{-1})$$

$$= \frac{1}{2}(s + s^{-1})\frac{1}{2}(t + t^{-1}).$$

Hence U_n and V_n are independent simple random walks and

$$\mathbf{E}(s^{X_T - Y_T}) = \mathbf{E}(\mathbf{E}(s^{V_T}|T))$$
$$= \mathbf{E}\left(\left(\frac{s + s^{-1}}{2}\right)^T\right)$$
$$= \left(F_1\left(\frac{s + s^{-1}}{2}\right)\right)^m \quad \text{where } F_1(s) = \frac{1 - (1 - s^2)^{\frac{1}{2}}}{s}.$$

40 $\mathbf{E}(s^{X_m}|X_m > 0) = (\mathbf{E}(s^{X_m}) - \mathbf{P}(X_m = 0))/\mathbf{P}(X_m > 0) = \left(\left(\frac{p}{1 - qs}\right)^m - p^m\right)/(1 - p^m)$

$= p^m(e^{-m\log(1-qs)} - 1)/(1 - e^{m\log p}) = p^m(m\log(1 - qs) + O(m^2))/(m\log p + O(m^2))$

$\to \dfrac{\log(1 - qs)}{\log(1 - q)}$ as $m \to 0$. ($O(.)$ is defined on p. 241.)

41 We know $U(s) = \sum u_{2k}s^{2k} = (1 - s^2)^{-\frac{1}{2}}$. Let $r_{2n} = \sum_0^n u_{2k}$. Then $\sum s^{2k} r_{2k} = \dfrac{U(s)}{1 - s^2}$

$= (1 - s^2)^{-3/2} = \dfrac{1}{s}\dfrac{d}{ds}\left(\dfrac{1}{(1 - s^2)^{\frac{1}{2}}}\right) = \dfrac{1}{s}\sum \dfrac{d}{ds}\binom{2k}{k}4^{-k}s^{2k}$

$= \sum 2k\binom{2k}{k}4^{-k}s^{2k-2} = \sum(2k + 2)\binom{2k + 2}{k + 1}4^{-k-1}s^{2k}.$

The result follows.

CHAPTER 7

Exercises

7.10.4 $f(x) = \frac{3}{4} + \frac{1}{4}\cdot 12\left(x - \frac{1}{2}\right)^2$, so you toss a coin twice and set

$$X = \begin{cases} \left(\dfrac{U}{4} - \dfrac{1}{8}\right)^{\frac{1}{3}} + \dfrac{1}{2} & \text{if you get two heads} \\ U & \text{otherwise.} \end{cases}$$

7.10.5 $\mathbf{P}(Y \leqslant y) = \mathbf{P}\left(-\log U \leqslant \left(\frac{y}{\gamma}\right)^\beta\right) = 1 - \exp\left(-\left(\frac{y}{\gamma}\right)^\beta\right)$

7.10.7 (a) $(\pi(1 + x^2))^{-1}$; $-\infty < x < \infty$ (b) $2(\pi(1 + x^2))^{-1}$; $0 \leqslant x < \infty$.

7.11.7 First note that the coefficient of x^n in H_n is 1, so $D^n H_n = n!$ Now integrating by parts

$$\int_{-\infty}^{\infty} H_n H_m \phi = [(-)^{m-1} H_n H_{m-1}\phi]_{-\infty}^{\infty} + \int_{-\infty}^{\infty} DH_n(-)^{m-1}D^{m-1}\phi.$$

The first term is zero, and repeated integration by parts gives zero if $m > n$, or $\int_{-\infty}^{\infty} \phi D^n H_n = n!$ if $m = n$.

7.11.8 By Taylor's theorem $\phi(x)\sum \dfrac{t^n H_n(x)}{n!} = \sum t^n(-)^n D^n\phi(x)/n! = \phi(x - t)$.

Hence $\sum \dfrac{t^n H_n(x)}{n!} = e^{-\frac{1}{2}(x-t)^2 + \frac{1}{2}x^2} = e^{-\frac{1}{2}t^2 + xt}$.

7.11.9 Set $\phi = -\phi'/x$ in the integral in (6), and integrate by parts again.

7.11.10 $\mathbf{E}(X - t|X > t) = \int_t^\infty \frac{1 - F(x)}{1 - F(t)} dx = e^{(\lambda t)^2} \int_t^\infty e^{-(\lambda x)^2} dx = e^{(\lambda t)^2} \pi^{\frac{1}{2}} \lambda^{-1} (1 - \Phi(\lambda + \sqrt{2}))$
and the inequality follows using (3).

7.12.2 Still $\frac{1}{3}$.

7.13.5 The policy is essentially the same with the one difference that $\hat{x} = \hat{y}$.

7.13.6 The new expected cost function λ^* is related to λ by $\lambda^*(x) = \lambda(x) + m\mathbf{P}(Z > x) = \lambda(x) + m(1 - F(x))$. Then $\frac{\partial \mu}{\partial y} = 0$ yields
$0 = c - h - (h + p + m\lambda) \exp(-\lambda(\hat{y} - a))$. Thence $\lambda^*(\hat{x}) + c\hat{x} = k + c\hat{y} + \lambda^*(\hat{y})$.

7.14.6 Let $g(s) = \log s - (s - 2)(s + 1)$. At $s = 1$ we have $g(1) = 2 > 0$; at $s = e^4$, we have $g(e^4) = 4 - (e^4 - 2)(e^4 + 1) = (3 - e^4)(2 + e^4) < 0$. There is thus at least one root. However $\log s$ lies below its tangent, and $s^2 - s - 2$ lies above, so there can be no more than one root in this interval.

7.14.7 Use $\log s \leqslant s - 1$.

7.15.2 $r(t) = \lambda$. Your part has no memory.

7.15.3 (a) Use Bayes' theorem. (b) $\pi \to 1$ if $\lambda > \mu$; $\pi \to 0$ if $\lambda < \mu$; $\pi = p$ if $\lambda = \mu$.

7.15.4 $\frac{d^2}{d\theta^2} \log M(\theta) = \left(\frac{\mathbf{E}(X^2 e^{\theta X})\mathbf{E}(e^{\theta X}) - (\mathbf{E}(X e^{\theta X}))^2}{(M(\theta))^2} \right)$.

But by Cauchy–Schwarz $(\mathbf{E}(X e^{\theta X}))^2 = [\mathbf{E}(X e^{\theta X/2} e^{\theta X/2})]^2 \leqslant \mathbf{E}(X^2 e^{\theta X})\mathbf{E}(e^{\theta X})$.

7.15.5 $\mathbf{P}(T > t) = \mathbf{E}(\exp(-\Lambda t)) = M_\Lambda(-t)$. But $\frac{dr(t)}{dt} = -\frac{d^2}{dt^2}\mathbf{P}(T > t) = -\frac{d^2}{dt^2} M_\Lambda(-t) < 0$, by (4). Hence T is DFR.

7.15.6 As above $r'(t) = \mathbf{E}(f'_{T_\Lambda})(1 - \mathbf{E}(F_{T_\Lambda})) + (\mathbf{E}(f_{T_\Lambda}))^2$. Now since F_{T_λ} is DFR we have $(1 - F_{T_\lambda}) f'_{T_\lambda} + f^2_{T_\lambda} \leqslant 0$, and hence $(\mathbf{E}(f_{T_\Lambda}))^2 \leqslant (\mathbf{E}(-f'_{T_\Lambda}(1 - F_{T_\Lambda}))^{\frac{1}{2}})^2 \leqslant \mathbf{E}(-f'_{T_\Lambda})\mathbf{E}(1 - F_{T_\Lambda})$ by Cauchy–Schwarz. Hence $r'(t) \leqslant 0$.

7.16.4 (i) $\frac{d}{dt} \frac{1}{t} \int_0^t r(v)dv = \frac{r(t)}{t} - \frac{1}{t^2} \int_0^t r(v)dv = \frac{1}{t^2} \int_0^t [r(t) - r(v)]dv > 0$ if $r'(v) > 0$ for all v. Hence IFR $\Rightarrow$ IFRA. (ii) Use (7.8.6) and Theorem 7.8.7.

7.16.5 (i) $\mathbf{E}(T - t|A_t) = \int_0^\infty \frac{\mathbf{P}(T > t + s)}{\mathbf{P}(T > t)} ds \leqslant \int_0^\infty \mathbf{P}(T > s)ds$ if NBU, by Definition 7.7.8 (iii). Hence NBU $\Rightarrow$ NBUE.

7.16.6 $\mathbf{P}(T > t) = \int_t^\infty \lambda^2 x e^{-\lambda x} dx = (1 + \lambda t)e^{-\lambda t}$. Hence $H(t) = -\log(1 + \lambda t) + \lambda t$,
$r(t) = \lambda - \frac{1}{1 + \lambda t}$, and $r'(t) = \frac{1}{(1 + \lambda t)^2} > 0$.

7.17.3 Uniform on $\left(0, \frac{1}{2}\right)$.

7.17.4 $\mathbf{E}\{X \wedge (1 - X)\} = \frac{1}{4}$; $\mathbf{E}\{X \vee (1 - X)\} = \frac{3}{4}$.

7.17.5 $\mathbf{E}(\sin \Theta) = 2\int_0^{\frac{1}{2}} \frac{x}{(x^2 + (1 - x)^2)^{\frac{1}{2}}} dx = \frac{1}{\sqrt{2}}(1 + \log(1 + \sqrt{2})) - 1$.

Likewise $\mathbf{E}(\cos \Theta) = \frac{1}{\sqrt{2}}(-1 + \log(1 + \sqrt{2})) + 1$, so

$$\frac{\mathbf{E}(\sin \Theta)}{\mathbf{E}(\cos \Theta)} = \frac{\log(1 + \sqrt{2}) + 1 - \sqrt{2}}{\log(1 + \sqrt{2}) - (1 - \sqrt{2})} \simeq 0.36.$$

7.17.7 $\mathbf{E}(\cot\Theta) = \int_0^{\frac{1}{2}} \frac{1}{B(a,b)} x^{a-2}(1-x)^b\,dx + \int_{\frac{1}{2}}^1 \frac{1}{B(a,b)} x^a(1-x)^{b-2}\,dx.$

7.18.6 Remember (or prove) that $\Gamma(n) = (n-1)!$ when n is an integer.

7.18.7 By the reflection principle, the number of paths which visit b on the way from $(0, 0)$ to $(2n, 0)$ is the same as the number of paths from $(0, 0)$ to $(2n, 2b)$, namely $\binom{2n}{n-b}$. Hence the probability required is $\binom{2n}{n-b}\Big/\binom{2n}{n} = \frac{(n!)^2}{(n-b)!(n+b)!}$

$$\simeq \frac{n^{2n+1}}{(n-b)^{n-b+\frac{1}{2}}(n+b)^{n+b+\frac{1}{2}}} \text{ using Stirling's formula}$$

$$= \left(1-\frac{b}{n}\right)^{-n+b-\frac{1}{2}}\left(1+\frac{b}{n}\right)^{-n-b+\frac{1}{2}} \to e^{-y^2}.$$

7.18.8 (a) Use Stirling's formula. (b) Take logs. (c) Use the Riemann integral.

7.18.9 The number of paths from $(0, 0)$ to $(2n, 2j)$ is $\binom{2n}{n-j}$; the number from $(0, 0)$ to $(2r, 0)$ is $\binom{2r}{r}$; and the number from $(2r, 0)$ to $(2n, 2j)$ which do not visit 0 is $\frac{j}{n-r}\binom{2n-2r}{n-r+j}$; (recall the reflection principle, or the hitting time theorem).

Hence the required probability is $\frac{j}{n-r}\binom{2n-2r}{n-r+j}\binom{2r}{r}\Big/\binom{2n}{n-j} = f_r.$

Now use Stirling's formula and take logs in the usual way.

Problems

1 x lies between α and β, $c(\alpha,\beta)^{-1} = \left|\int_\alpha^\beta (x-\alpha)(\beta-x)\,dx\right|.$

Q **2** $\mathbf{P}(X = x) = \lim_{n\to\infty}\left(F(x) - F\left(x - \frac{1}{n}\right)\right).$

3 $\operatorname{var}(X) = \frac{B(a+2,b)}{B(a,b)} - \left(\frac{B(a+1,b)}{B(a,b)}\right)^2.$

4 $1 = c\int_0^{\pi/2} (\sin x)^\alpha(\cos x)^\beta\,dx = \frac{c}{2}B\left(\frac{\alpha-1}{2}, \frac{\beta-1}{2}\right).$

5 $\frac{2}{\pi}\sin^{-1} x$

6 $\exp(-\exp(-x))$

7 $\mathbf{E}(Y) < \infty$ for $\lambda > a$

9 $\left(1 - \Phi\left(x + \frac{a}{x}\right)\right)\Big/(1-\Phi(x)) = \frac{\phi(x+\frac{a}{x})r(x+\frac{a}{x})}{\phi(x)r(x)} \to e^{-a},$

using the properties of Mills' ratio $r(x)$, and $\phi(x)$.

11 (i) $f(x) = (b - 4\pi^2 ml_0 x^{-2})/(b-a)$ for $2\pi\left(\frac{ml_0}{b}\right)^{\frac{1}{2}} \leqslant x \leqslant 2\pi\left(\frac{ml_0}{a}\right)^{\frac{1}{2}}.$

(ii) $\mathbf{E}(X) = (\pi(ml_0)^{\frac{1}{2}})/(\sqrt{a} + \sqrt{b}).$

12 Choose x_l such that $F(x_l) < 2^{-(n+1)}$ and x_u such that $1 - F(x_u) < 2^{-(n+1)}$. Then set $S_n(X) = x_l + r\epsilon$ for $x_l + r\epsilon < X \leqslant x_l + (r+1)\epsilon$, for all r in $0 \leqslant r \leqslant \left[\frac{x_u - x_l}{\epsilon}\right].$

14 $\mathbf{P}(X > x) = (1 - 2\sqrt{3}x)^2$, so $f_X(x) = 4\sqrt{3}(1 - 2\sqrt{3}x).$

15 $c = \mathbf{E}(X)$

16 $\exp\left(-\frac{2}{3}\lambda t^{\frac{3}{2}}\right), \quad t > 0$

17 $F'(x) = g(x)$ and $F(1) = 1$

18 $\kappa_1 = \mu$; $\kappa_2 = \sigma^2$; $\kappa_r = 0$, $r \geqslant 3$

19 $\log M_X(t) = -\log\left(1 - \frac{t}{\lambda}\right) = \sum_{r=1}^{\infty} \frac{t^r}{r\lambda^r}$, so $\kappa_r = \frac{(r-1)!}{\lambda^r}$

20 Set $X = \begin{cases} U & \text{if } 0 \leqslant V \leqslant \frac{3}{5} \\ U^{\frac{1}{2}} & \text{if } \frac{3}{5} < V \leqslant \frac{9}{10} \\ U^{\frac{1}{3}} & \text{if } \frac{9}{10} < V \leqslant 1 \end{cases}$

21 (a) Set $u = v/a$ (b) Set $u = b/v$ after differentiating. (c) Integrate $\frac{\partial I}{\partial b} = -2aI$ with the boundary condition $I(a, 0) = \pi^{\frac{1}{2}}/(2a)$.

22 (a) Set $x = v^2$ to get $M_X(t) = 2\alpha I((\gamma - t)^{\frac{1}{2}}, \beta^{\frac{1}{2}}) = \alpha\left(\frac{\pi}{\gamma - t}\right)^{\frac{1}{2}} \exp(-2(\beta(\gamma - t))^{\frac{1}{2}})$, so $\alpha = \left(\frac{\gamma}{\pi}\right)^{\frac{1}{2}} \exp(-2(\beta\gamma)^{\frac{1}{2}})$.) (b) $\exp\left(-2\left(-\frac{t}{2}\right)^{\frac{1}{2}}\right)$ for $t \leqslant 0$.

23 Use Problem 22.

24 $\mathbf{E}(e^{tX^2}) = \int (2\pi)^{-\frac{1}{2}} \exp\left((2t - 1)\frac{x^2}{2}\right) dx = (1 - 2t)^{-\frac{1}{2}}$. Hence X^2 has the $\chi^2(1)$ density.

25 $\frac{\mu}{\mu + t} + \frac{\lambda}{\lambda - t}$ for $-\mu < t < \lambda$.

26 $\mathbf{P}(X > k) = \mathbf{P}\left(\frac{\log U}{\log(1 - p)} \geqslant k\right) = \mathbf{P}(U \leqslant q^k) = q^k$.

27 $X = \begin{cases} U & \text{with probability } \frac{24}{25} \\ \left(\frac{1}{4}U - \frac{1}{8}\right)^{\frac{1}{3}} + \frac{1}{2} & \text{with probability } \frac{1}{25}. \end{cases}$

28 Use Example 7.9.8.

29 $\int_0^{\infty} f(\theta, x) g(\theta) d\theta = \frac{v}{(v + x)^2}$. This is non-negative and $\int_0^{\infty} \frac{v}{(v + x)^2} dx = 1$.

30 $\mathbf{E}(\exp(t(X_n - n)n^{-\frac{1}{2}})) = e^{-tn^{\frac{1}{2}}} \exp(n(e^{-tn^{-\frac{1}{2}}} - 1)) = \exp(-tn^{\frac{1}{2}} + tn^{\frac{1}{2}} + \frac{1}{2}t^2 + O(n^{-\frac{1}{2}})) \to e^{\frac{1}{2}t^2}$;

now use the continuity theorem.

31 $(U + 1)^{-d^{-1}}$

32 X_α has mean and variance α, and m.g.f. $\frac{1}{(1 - t)^\alpha}$. Hence

$$M_\alpha(t) = \mathbf{E}\left(\exp\left(\frac{t}{\sqrt{\alpha}}(X_\alpha - \alpha)\right)\right) = e^{-t/\sqrt{\alpha}}\left(1 - \frac{t}{\sqrt{\alpha}}\right)^{-\alpha}.$$

Now for fixed t, and $\left|\frac{t}{\sqrt{\alpha}}\right| < 1$,

$$\log M_\alpha(t) = -\frac{t}{\sqrt{\alpha}} - \alpha \log\left(1 - \frac{t}{\sqrt{\alpha}}\right) = -\frac{t}{\sqrt{\alpha}} + \alpha \sum_{r=1}^{\infty} \frac{1}{r}\left(\frac{t}{\sqrt{\alpha}}\right)^r \to \frac{t^2}{2} \text{ as } \alpha \to \infty.$$

Using the continuity theorem gives the result.

33 Use Chebyshov's inequality.

CHAPTER 8

Exercises

8.10.3 By definition $f_{Y|X}(y|x) = f(x, y)/f_X(x) = \dfrac{1}{(2\pi(1-\rho^2))^{\frac{1}{2}}} \exp\left(-\dfrac{(y-\rho x)^2}{2(1-\rho^2)}\right)$.

This is $N(\rho x, 1-\rho^2)$; therefore $\mathbf{E}(e^{tY}|X) = \exp\left(\rho X t + \frac{1}{2}\left(1-\rho^2\right)t^2\right)$.

8.10.4 By conditional expectation

$$\mathbf{E}(e^{sX+tY}) = \mathbf{E}(\mathbf{E}(e^{sX+tY}|X)) = \mathbf{E}(e^{(s+\rho t)X})e^{\frac{1}{2}(1-\rho^2)t^2} = \exp\left(\frac{1}{2}\left(s+\rho t\right)^2 + \frac{1}{2}(1-\rho^2)t^2\right)$$

as above.

8.10.5 $\mathbf{E}(e^{sW+tZ}) = \mathbf{E}\left(\exp \sum(\alpha_i s + \beta_i t)X_i\right) = \exp\left(\frac{1}{2}\sum(\alpha_i s + \beta_i t)^2\right)$. This factorizes as required for the independence if and only if $\sum \alpha_i \beta_i = 0$ (or in geometrical terms, $\boldsymbol{\alpha}.\boldsymbol{\beta} = 0$).

8.10.6 $\mathbf{E}(e^{t(aX+bY)}) = \exp\left(\frac{1}{2}\, t^2(a^2 + 2\rho ab + b^2)\right)$. Hence $aX + bY$ is $N(0, a^2 + 2\rho ab + b^2)$.

8.11.4 A triangle is feasible if $U < V + W$ and $V < U + W$ and $W < U + V$. In terms of X and Y this gives (when $X < Y$) the constraints $X < \frac{1}{2}$, $Y - X < \frac{1}{2}$ and $Y > \frac{1}{2}$. A similar possibility arises when $X > Y$. Now a sketch of these two regions shows that they form two triangles with combined area $\frac{1}{4}$, and this is the required probability, since (X, Y) is uniform on the unit square.

8.11.6 By symmetry it is the same as $X_{(1)}$, namely $n(1-x)^{n-1}$; $0 \leqslant x \leqslant 1$.

8.11.7 By symmetry this is the same as the joint density of $X_{(1)}$ and $1 - X_{(n)}$. Now $\mathbf{P}(X_{(1)} > x, 1 - X_{(n)} > y) = (1 - x - y)^n$, so $f = n(n-1)(1-x-y)^{n-2}$.

8.11.8 Given neither point is on the diameter, then the density of the angle they make at the midpoint of the diameter is given by (1) with $a = \pi$. Hence the expected area in this case is $\int_0^\pi \frac{2(\pi - x)}{\pi^2} \frac{1}{2} \sin x dx = \frac{1}{\pi}$.

Given one on the diameter and one not, they are jointly uniform on $(0, \pi) \times (-1, 1)$, so the expected area is

$$2\int_0^\pi \frac{1}{2\pi} \int_0^1 \frac{y \sin x}{2}\, dx dy = \frac{1}{2\pi}.$$

Hence the expected area is $\dfrac{1}{\pi}\left(\dfrac{\pi}{\pi+2}\right)^2 + \dfrac{1}{2\pi}\dfrac{4\pi}{(\pi+2)^2} = \dfrac{1}{2+\pi}$.

8.12.4 The easy method uses (1) and (2) to see that $\mathbf{P}(A^c \cap B^c) = 1 - \mathbf{P}(A \cap B) - \mathbf{P}(A \cap B^c) - \mathbf{P}(A^c \cap B)$ which gives the answer. The other method observes that Q

can be divided into 5 regions in which (given $C = (x, y)$), $\mathbf{P}(A^c \cap B^c)$ takes the values

$$0, 1 - \frac{2}{\pi}\cos^{-1}\left(\frac{y}{l}\right), 1 - \frac{2}{\pi}\cos^{-1}\left(\frac{x}{l}\right), 1 - \frac{2}{\pi}\cos^{-1}\left(\frac{x}{l}\right) - \frac{2}{\pi}\cos^{-1}\left(\frac{y}{l}\right),$$

respectively. Identify the regions and do the integrals.

8.12.5 The easy method allows $b \to \infty$ in (2). You should also do it via an integral.

8.12.6 Draw a picture with no B-lines to see that the probability of an intersection is

$$\frac{1}{\pi a}\int_0^{\pi/2} \min\{l\cos\theta, a\}\,d\theta.$$

8.12.7 $\dfrac{(a-l)(b-l)}{ab}$ for $l \leqslant a \wedge b$. Evens if $2l^2 - 2(a+b)l + ab = 0$ which implies that the coin has diameter $a + b - (a^2+b^2)^{\frac{1}{2}}$.

8.13.1 $(2n+3)^{-1}$

8.13.2 $\pi n/(n+1)$

8.13.3 $\pi(n-1)/(n+1)$

8.14.4 Integrate (2) remembering that (1) is a density and $\Gamma(\alpha+\beta) = \int_0^\infty \lambda^{\alpha+\beta}u^{\alpha+\beta-1}e^{-\lambda u}du.$

8.14.5 Let X and Y be independent with density $\dfrac{1}{\Gamma\left(\frac{1}{2}\right)}x^{-\frac{1}{2}}e^{-x}$. Then $U = X + Y$ has density

$$\left(\Gamma\left(\frac{1}{2}\right)\right)^{-2} e^{-u}\int_0^u v^{-\frac{1}{2}}(u-v)^{-\frac{1}{2}}dv = \pi e^{-u}\left(\Gamma\left(\frac{1}{2}\right)\right)^{-2}.$$

The result follows.

8.14.6 Set $(1+x^2)^{-1} = v$ and use Exercise 5.

8.15.4 $\mathbf{P}(N = n) = \left(1 - \dfrac{1}{a}\right)^n \dfrac{1}{a}$ with $\mathbf{E}(N) = a - 1$. Hence a should be as small as (1) permits.

8.15.5 We choose a so that ae^{-x} is as small as possible (since e^{-X} is uniform). Hence

$$a = \sup_x\left\{e^{x-\frac{1}{2}x^2}\left(\frac{2}{\pi}\right)^{\frac{1}{2}}\right\} = \sup_x\left\{\left(\frac{2}{\pi}\right)^{\frac{1}{2}}e^{-\frac{1}{2}(x-1)^2+\frac{1}{2}}\right\} = \left(\frac{2e}{\pi}\right)^{\frac{1}{2}}.$$

Thus we get a variable with density $f_S(x)$ if we set $X = -\log U_1$, whenever $e^{\frac{1}{2}}U_1U_2 < \exp(-(\log U_1)^2/2)$.

8.15.6 Now $-\log U_i$ is exponential with parameter 1, so

$$\mathbf{P}(X \leqslant x | Y > \tfrac{1}{2}(X-1)^2) \propto \int_0^x\int_{\frac{1}{2}(v-1)^2}^\infty e^{-y}e^{-v}dydv \propto \int_0^x e^{-\frac{1}{2}v^2}dv.$$

Hence $f_{X|A} = \left(\dfrac{2}{\pi}\right)^{\frac{1}{2}}e^{-x^2/2}$.

8.16.3 Use independence of increments.

8.16.4 Conditional on $N(24) = k$, the k calls are independent and uniform over $(0, 24)$. Given $X = x$ and $Y = y$, a call at time U finds you in the shower if $x < U < x + y$, with probability $y/24$. Hence $\mathbf{P}$(a call at U finds you in the shower $|N = k) = \mathbf{E}(Y)/24 = p$ (say). Hence the number Z of calls that finds you in the shower given $N = k$ is binomial with parameters k and p. Hence $\mathbf{E}(s^Z|N = k) = (ps + 1 - p)^k$; hence $\mathbf{E}(s^Z) = \mathbf{E}((ps+1-p)^N) = \exp(24\lambda(ps+1-p)) = \exp(\lambda\mathbf{E}(Y)(s-1))$.

8.16.5 Argue as in (4). Given $N(t) = n$ then R_1 and R_2 have a trinomial mass function with p.g.f. $\mathbf{E}(x^{R_1}y^{R_2}|N = n) = (p_1x + p_2y + 1 - p_1 - p_2)^n$. Hence $\mathbf{E}(x^{R_1}y^{R_2})$ factorizes into two Poisson p.g.f.s.

8.16.6 $\lambda\min\{s, t\}$

8.17.5 X_{R-1} is always the smallest of $X_1, \ldots, X_R$. So $\mathbf{P}(X_{R-1} \geqslant x) = \sum_{r=2}^{\infty} \frac{r-1}{r!} \times$ $(1-F(x))^r = 1 - Fe^{1-F}$. Hence $\mathbf{P}(X_{R-1} \leqslant x) = Fe^{1-F}$, with density $f(x) = (1-F(x))\exp(1-F(x))$.

8.17.6 $T > n$ if and only if $X_1 = X_{(n)}$; by symmetry therefore $\mathbf{P}(T > n) = \frac{1}{n}$. Hence

$$\mathbf{P}(T = n) = \frac{1}{n-1} - \frac{1}{n} = \frac{1}{n(n-1)}, \; n \geqslant 2. \text{ When } T = n, X_T = X_{(n)}, \text{ so}$$

$$\mathbf{P}(X_T \leqslant x) = \sum_{n=2}^{\infty} \frac{(F(x))^n}{(n-1)n} \quad \text{as required.}$$

8.17.7 If X_1 represents your loss at some hazard, and $(X_r;\, r \geqslant 1)$ represents the losses of your successors, then the expected time until someone does worse than you is infinite. The argument is symmetrical of course, but we don't feel so strongly about our good luck.

Problems

1 f is a density if $a + 1 \geqslant 0$, $b + 1 \geqslant 0$, $a + b = -\frac{1}{2}$. Independence is impossible.

2 $\text{cov}(X, Y) = (1 - ab)/144$.

3 (a) $c = 1$; (b) $2e^{-1}$; (c) $\frac{1}{2}$.

4 $f_Z(z) = \int_0^z g(u)\, du$.

5 (a) $c = (2\pi)^{-1}$; (b) $\int_0^y (a + y^2)^{-\frac{3}{2}} dy = \frac{y}{a(a+y^2)^{\frac{1}{2}}}$, so X has a Cauchy density.

7 $f(x, y) = 4y^3 x(1-x)$ for $0 < x < 1$, $0 < y < 2$. Hence you can simulate X with density $6x(1-x)$ by forming $U^{\frac{1}{2}}/(U^{\frac{1}{2}} + V^{\frac{1}{2}})$ and accepting it as a value of X if $U^{\frac{1}{2}} + V^{\frac{1}{2}} \leqslant 1$.

9 $6x(1-x)$

10 $\mathbf{E}(e^{sU+tV}) = \mathbf{E}(e^{(s+t)X+(s-t)Y}) = \exp\left(\mu_X(s+t) + \frac{1}{2}\sigma_X^2(s+t)^2 + \mu^Y(s-t) + \frac{1}{2}\sigma_Y^2(s-t)^2\right)$ which factorizes if $\sigma_X = \sigma_Y$.

11 Given $Z < 1$, the point $(2U - 1, 2V - 1)$ has the uniform density over the unit disc, namely $\frac{r}{\pi}$ in polar coordinates. Thus $Z = R^2$ and $0 < \Theta < 2\pi$. Make the transformation $X = (2\log R^{-2})^{\frac{1}{2}}\cos\Theta$, $Y = (2\log R^{-2})^{\frac{1}{2}}\sin\Theta$, with inverse

$$r^2 = \exp\left(-\frac{1}{2}(x^2 + y^2)\right), \; \theta = \tan^{-1}\frac{y}{x} \text{ and } J = \frac{1}{2}\exp\left(-\frac{1}{4}(x^2+y^2)\right).$$

The result follows.

12 (i) zero for $a \leqslant 1$; (ii) $\frac{a\mu}{\lambda + a\mu}$ for $a \geqslant 1$.

13 $ye^{-xy}/(1 - e^{-y^2})$

14 Let A be at the top of the melon; let angle AOB be θ, where O is the centre of the melon. Then the probability that all three remain is, when $\frac{\pi}{2} < \theta < \pi$, $(\pi - \theta)\frac{\sin\theta}{4\pi}$.

Hence $\mathbf{P}(\text{all three remain}) = \int_{\pi/2}^{\pi} (\pi - \theta)\frac{\sin\theta}{4\pi}\, d\theta = \frac{1}{4\pi}$.

Likewise $\mathbf{P}(\text{any one of the three remains}) = 3\mathbf{P}(\text{given one remains})$

$$= 3\int_{\pi/2}^{\pi} \theta\frac{\sin\theta}{4\pi}\, d\theta = \frac{3(\pi - 1)}{4\pi}.$$

16 $\mathbf{P}(U = X) = \mathbf{P}(Y > X) = \dfrac{\lambda}{\lambda + \mu}$.

17 Use induction

18 (a) $\frac{1}{2}(X + Y)$ by symmetry.

(b) $\mathbf{E}(X|X + Y = V) = \dfrac{\sigma^2 + \rho\sigma\tau}{\sigma^2 + 2\rho\sigma\tau + \tau^2}V$; $\mathbf{E}(Y|X + Y = V) = \dfrac{\tau^2 + \rho\sigma\tau}{\sigma^2 + 2\rho\sigma\tau + \tau^2}V$

19 $\mathbf{P}(T \geqslant j + 1) = \mathbf{P}\left(\sum_{i=1}^{j} X_i < 1\right) = p_j$ (say). Trivially $p_1 = \dfrac{1}{1!}$, $p_2 = \dfrac{1}{2!}$. Now p_j is the volume of the 'j-dimensional pyramid' with apex O and corners $(1, 0, \ldots, 0)$, $(0, 1, \ldots, 0)$, etc. Since $p_j = \int_0^1 x^{j-1} p_{j-1} dx$ the result follows by induction. Finally

$\mathbf{E}(T) = \sum \mathbf{P}(T \geqslant j + 1) = e$.

20 A triangle is impossible if $X_1 > X_2 + X_3$. This has the same probability as $1 - X_1 > X_2 + X_3$, namely $\dfrac{1}{3!}$ by Problem 19. Two more similar constraints give

$\mathbf{P}(\text{triangle}) = 1 - 3.\dfrac{1}{3!} = \dfrac{1}{2}$.

21 $\mathbf{P}(n(1 - M_n) > x) = \left(\dfrac{1 - x}{n}\right)^n \to e^{-x}$.

22
$$\begin{aligned}
\mathbf{E}\left(\exp\left(t\sum_1^N X_i\right)\right) &= \mathbf{E}\left(\mathbf{E}\left(\exp\left(t\sum_1^N X_i\right)\Big|N\right)\right) \\
&= \mathbf{E}((\mathbf{E}e^{tX_1})^N) \\
&= (1 - p)\left(\frac{\mu}{\mu - t}\right)\Big/\left(1 - \frac{p\mu}{\mu - t}\right) \\
&= \frac{(1 - p)\mu}{\mu(1 - p) - t}.
\end{aligned}$$

So Y is exponential with parameter μp.

23
$$\begin{aligned}
\mathbf{E}(C(t)X_1) &= \mathbf{E}(X_1\mathbf{E}(C(t)|X_1)) \\
&= \int_t^\infty ut\lambda e^{-\lambda u} du + \int_0^t u\left(\frac{1}{\lambda} - \frac{1}{\lambda}e^{-\lambda(t-u)}\right)\lambda e^{-\lambda u} du \\
&= t^2 e^{-\lambda t} + \frac{te^{-\lambda t}}{\lambda} + \frac{1}{\lambda^2} - \frac{te^{-\lambda t}}{\lambda} - \frac{e^{-\lambda t}}{\lambda^2} - \frac{1}{2}t^2 e^{-\lambda t}.
\end{aligned}$$

So $\operatorname{cov}(C(t), X_1) = \frac{1}{2}t^2 e^{-\lambda t}$.

24 Let the condition be A. Now we notice that

$$\mathbf{P}(X \leqslant x, A) \propto \int^x x^{\alpha-1} e^{-\frac{\alpha-1}{\alpha}} e^{-\frac{x}{\alpha}} dx = x^{\alpha-1} e^{-x}.$$

Hence the result is true by the rejection method Example 8.15, provided that $\left(\dfrac{ex}{\alpha}\right)^{\alpha-1} e^{-\frac{\alpha-1}{\alpha}x} \leqslant 1$ for $x \geqslant 0$. Since $\alpha - 1 \geqslant 0$, this is equivalent to $\log\left(\dfrac{x}{\alpha}\right) \leqslant \dfrac{x}{\alpha} - 1$, so the result holds. This clearly provides a method for simulating gamma random variables, given a supply of uniform random variables.

26 (a) Recalling Theorem 8.4.6 on quotients gives the required density as

$\displaystyle\int_{-\infty}^{\infty} \frac{1}{|u|}\frac{u^2}{w^2}\frac{1}{2\pi}\exp\left(-\frac{1}{2}\left(u^2 + \frac{u^2}{w^2}\right)\right)du = 2\cdot\frac{1}{2\pi w^2}\cdot\frac{1}{(1 + 1/w^2)}$ as required.

(b) $\mathbf{E}(e^{tX_1X_2}) = \mathbf{E}(\mathbf{E}(e^{tX_1X_2}|X_1)) = \mathbf{E}(e^{\frac{1}{2}t^2X_1^2}) = \int_{-\infty}^{\infty}\frac{1}{\sqrt{2\pi}}e^{-\frac{1}{2}x^2(1-t^2)}dx = (1-t^2)^{-\frac{1}{2}}$.

Hence $\mathbf{E}e^{t(X_1X_2+X_3X_4)} = \frac{1}{1-t^2} = \frac{1}{2}\left[\frac{1}{1-t}+\frac{1}{1+t}\right] = \frac{1}{2}(\mathbf{E}(e^{tY})+\mathbf{E}(e^{-tY}))$

where Y is exponential with parameter 1. The result follows.

28 Let $V = \max\{U_1, \dots, U_Y\}$. First we notice that by conditioning on Y

$$F_V = \mathbf{P}(\max\{U_1,\dots,U_Y\} \le v) = \sum_{y=1}^{\infty}\frac{v^y}{(e-1)y!} = \frac{e^v-1}{e-1}.$$

Now let us find the m.g.f. of Z,

$$\mathbf{E}(e^{tZ}) = \mathbf{E}(e^{tX})\mathbf{E}(e^{-tV}) = \frac{(e-1)e^{t-1}}{1-e^{t-1}}\int_0^1 e^{-tv}\frac{e^v}{e-1}dv = \frac{1}{1-t}.$$

Hence Z is exponential with parameter 1. Alternatively you can find F_V directly.

29 $\frac{X}{Z}$ has a beta density.

30 Consider their joint m.g.f. $\mathbf{E}(e^{sX+tZ}) = \mathbf{E}(e^{(s+t\rho)X+t(1-\rho^2)^{\frac{1}{2}}Z}) = e^{\frac{1}{2}(s^2+2\rho st+t^2)}$ as required.

35 $f(x) = e^{-x}$. Calculate:

$$\mathbf{E}(e^{tU(X_1+X_2)}) = \mathbf{E}\left(\mathbf{E}(e^{tU(X_1+X_2)}|U)\mathbf{E}\left(\frac{1}{(1-Ut)^2}\right)\right) = \int_0^1\frac{1}{(1-ut)^2}du = \frac{1}{1-t} = \mathbf{E}(e^{tY}).$$

CHAPTER 9

Exercises

9.10.1 (a) Column sums are one, as well as row sums. Hence $\pi_i = \frac{1}{8}$ satisfies $\boldsymbol{\pi} = \boldsymbol{\pi}\boldsymbol{P}$, and so $\mu_0 = 8 = \mu_V$. (b) $\mathbf{E}(X) = 1$ by the same argument. (c) $\mathbf{E}(T)$ is different. We use the following device. T is the sum of the M steps at which the walk moves to a different vertex and the steps at which it does not move. The number N of non-moving steps before leaving O has the same expectation (and distribution) as the number at every other vertex on the way from O to V, so $\mathbf{E}(T) = \mathbf{E}(M)\mathbf{E}(N)$. By the example, $\mathbf{E}(M) = 1 + (\alpha+\beta+\gamma)(\alpha^{-1}+\beta^{-1}+\gamma^{-1})$, and it is easy to find that $\mathbf{E}(N) = \frac{\delta}{1-\delta}$. Hence $\mathbf{E}(T) = \delta((\alpha+\beta+\gamma)^{-1}+\alpha^{-1}+\beta^{-1}+\gamma^{-1})$.

9.10.2 Consider a random walk on a unit square which takes x-steps with probability p and y-steps with probability q. Then if T is the first passage time from (0, 0) to (1, 1), arguments similar to those of the example show that

$$\mathbf{E}(s^T) = \frac{U_V(s)}{U(s)}$$

where
$$U(s) = \frac{1}{2}\left(\frac{1}{1-s^2}+\frac{1}{1-(p-q)^2s^2}\right)$$

and
$$U_V(s) = \frac{s^2}{2}\left(\frac{1}{1-s^2}-\frac{(p-q)^2}{1-(p-q)^2s^2}\right)$$

which yields $\mathbf{E}(T)$ after some plod. More simply by conditional expectation we have $\mathbf{E}(T) = 1 + p(1+p\mathbf{E}(T)) + q(1+q\mathbf{E}(T))$ which yields $\mathbf{E}(T) = p^{-1}+q^{-1}$. If the walk can wait at vertices with probability r, then by the same device as used in (1), we find $\mathbf{E}(T) = \frac{r}{p}+\frac{r}{q}$. Now we recognise that the question is equivalent to this problem with $p = \alpha$, $q = \beta$, $r = \gamma$.

9.11.6 $LHS = \mathbf{P}(Y_{n_r} = k, Y_{n_{r-1}} = k_1, \ldots, Y_{n_1} = k_{r-1})/\mathbf{P}(Y_{n_{r-1}} = k_1, \ldots, Y_{n_1} = k_{r-1})$

$$= \frac{\mathbf{P}(X_{-n_r} = k, \ldots, X_{-n_1} = k_{r-1})}{\mathbf{P}(X_{-n_{r-1}} = k_1, \ldots, X_{-n_1} = k_{r-1})}$$

$$= \frac{\mathbf{P}(X_{-n_r} = k | X_{-n_{r-1}})\mathbf{P}(X_{-n_1} = k_{r-1}, \ldots | X_{-n_{r-1}})}{\mathbf{P}(X_{-n_1} = k_{r-1}, \ldots | X_{-n_{r-1}})}$$

$$= \mathbf{P}(Y_{n_r} = k | Y_{n_{r-1}}),$$

where we used (9.1.8) at the crucial step. In equilibrium

$$q_{ij} = \mathbf{P}(Y_2 = j, Y_1 = i)/\mathbf{P}(Y_1 = i)$$
$$= \mathbf{P}(X_{-1} = i | X_{-2} = j)\mathbf{P}(X_{-2} = j)/\mathbf{P}(Y_1 = i) = p_{ji}\pi_j\pi_i^{-1}.$$

9.12.4 No for (a) because of periodicity. Yes for (b).

9.12.5 (a) $\mathbf{E}(X_{n+1}|X_n) = 1 - \dfrac{2}{m}X_n + 1$

(b) $\mathbf{E}(X_n) = \left(1 - \dfrac{\alpha + \beta}{m}\right)^n \left(\mathbf{E}(X_0) - \dfrac{\beta m}{\alpha + \beta}\right) + \dfrac{\beta m}{\alpha + \beta}.$

9.13.8 Since v_n is a renewal sequence there is a Markov chain V_n such that $v_n = \mathbf{P}(V_n = 0 | V_0 = 0)$. Let U_n and V_n be independent. Then $((U_n, V_n); n \geqslant 0)$ is a Markov chain and $u_n v_n = \mathbf{P}((U_n, V_n) = (0, 0) | (U_0, V_0) = (0, 0))$, thus $(u_n v_n; n \geqslant 0)$ is a renewal sequence.

9.13.9 Consider the chain $(U_{nd}; n \geqslant 0)$.

9.13.10 If $B_n > 0$ then $B_n - 1 = B_{n+1}$, and if $B_n = 0$ then B_{n+1} is the time to the next event, less the elapsed unit of time. Hence B is a Markov chain with

$$p_{i,\, i-1} = 1; \quad i > 0$$

and

$$p_{0j} = f_X(j + 1) = \mathbf{P}(X = j + 1).$$

Hence for a stationary distribution $\boldsymbol{\pi}$ with $\pi(s) = \sum_i s^i \pi_i$,

$$\pi_j = \pi_{j+1} + \pi_0 f_X(j + 1),$$

whence

$$\pi(s) = \pi_0 \frac{G_X(s) - 1}{s - 1},$$

and so if $\sum_i \pi_i = 1$,

$$\pi(s) = \frac{1}{\mathbf{E}(X)} \frac{1 - G_X(s)}{1 - s} \quad \text{if } \mathbf{E}(X) < \infty.$$

Hence

$$\pi_i = \frac{\mathbf{P}(X > i)}{\mathbf{E}(X)}.$$

9.13.11 The transition probabilites of U reversed are

$$q_{i,\, i-1} = \frac{1 - F_i}{1 - F_{i-1}} \cdot \frac{\pi_{i-1}}{\pi_i} = 1; \quad i > 0$$

and those of B reversed are

$$q_{i,\, i+1} = 1 . \frac{\pi_{i+1}}{\pi_i} = \frac{1 - F_{i+1}}{1 - F_i}; \quad i \geqslant 0.$$

Hence U reversed is B, and B reversed is U.

9.14.1 Using (b) and (c) shows that j is persistent.

9.14.2 Follows from (b).

9.14.3 By assumption, $p_{ij}(n) > 0$ and $p_{ji}(m) > 0$ for some finite n and m. Hence $p_{jj}(m+r+n) \geqslant p_{ji}(m)p_{ii}(r)p_{ij}(n)$.

Now sum over r to get $\sum p_{jj} = \infty$ if $\sum p_{ii}(r) = \infty$. So if i is persistent so is j. Interchange the roles of i and j.

If j has period t, let $r = 0$ to find that when $p_{jj}(m+n) \geqslant 0$, $m+n$ is a multiple of t. Hence the right hand side is non-zero only when r is a multiple of t, so i has period t.

9.15.8 With $HTH = 1$, $HHH = 2$, we have

$$p_{12}(1) = 0,\ p_{12}(2) = p^2 = \tfrac{1}{4} = p^{22}(2) = p_{21}(2) = p_{11}(2)$$
$$p_{22}(1) = 0 = p_{21}(1) = p_{11}(1);\ p_{22}(1) = p = \tfrac{1}{2}.$$

Hence

$$\mu_{12} = (1 + \tfrac{1}{2} + \tfrac{1}{4} - \tfrac{1}{4})8 = 12$$

and

$$\mu_{21} = (1 + p^2 - p^2)8 = 8.$$

9.15.9 We set $HHH = 1$, $HTH = 2$, then

$$\phi_{s1} = \frac{10 + 12 - 14}{8 + 12} = \frac{3}{10}.$$

9.15.10 Here $TTH = 1$, $HHH = 2$. Calculate $\mu_{s1} = 8$. Now

$$p_{11}(1) = p_{11}(2) = p_{21}(1) = p_{21}(2) = p_{12}(1) = 0,$$
$$p_{12}(2) = \frac{1}{4},\ p_{22}(1) = \frac{1}{2},\ p_{22}(2) = \frac{1}{4}.$$

Hence

$$\phi_{s1} = \frac{14 + 8 - 8}{12 + 8} = \frac{7}{10}.$$

9.16.4 $\mathbf{P}(T_1 < \infty) = \mathbf{P}(X(t) > 0) = 1 - \exp(-\Lambda(t))$. Now let $t \to \infty$.

9.16.5 $\mathbf{E}(s^{X(t)}) = \exp\left((z-1)\int_0^t \lambda e^{-\lambda u}\,du\right) \to e^{z-1} = e^{-1}\sum z^k (k!)^{-1}$, and we use the continuity theorem for p.g.f.s.

9.16.6 By conditional expectation

$$\mathbf{E}(e^{\theta Z}) = \mathbf{E}(\mathbf{E}(e^{\theta Z}|X)) = \mathbf{E}((\mathbf{E}e^{\theta Y})^X).$$

You can also get this using forward equations with a lot more work.

9.16.7 $\mathbf{E}(z^X) = \mathbf{E}(\mathbf{E}(z^X|Y)) = \mathbf{E}\left(\exp\left[(z-1)\int_0^t Y(u)\,du\right]\right)$

$$G'_X(z)|_{z=1} = M'(0) = \int_0^t \mathbf{E}(Y(u))\,du.$$
$$G''_X(1) + G'_X(1) - (G'_X(1))^2 = M''(0) - (M'(0))^2 + M'(0)$$
$$= \operatorname{var}\int_0^t Y(u)\,du + \int_0^t \mathbf{E}(Y(u))\,du.$$

9.17.2 $\exp\left(\frac{\lambda}{\mu}(z-1)(1-e^{-\mu t})\right) \to \exp\left(\frac{\lambda}{\mu}(z-1)\right)$

9.17.3 The forward equations are:

$$p'_n(t) = \lambda p_{n-1}(t) + \mu(n+1)p_{n+1}(t) - (\lambda + \mu n)p_n(t).$$

Set $G = \sum s^n p_n$ to get $\dfrac{\partial G}{\partial t} = (s-1)\left(\lambda G - \mu\dfrac{\partial G}{\partial s}\right)$.

9.17.4 $\pi_k = \dfrac{1}{2(k!)}\left(1 + \dfrac{k\mu}{\lambda}\right)\left(\dfrac{\lambda}{\mu}\right)^k \exp\left(-\dfrac{\lambda}{\mu}\right).$

9.18.10
$$\mathbf{E}(s^{X(t)}) = e^{-\delta t} + \int_0^t \delta e^{-\delta x} \exp\left(\frac{\nu}{\mu}(s-1)(1-e^{-\mu x})\right)dx$$

where ν is the immigration rate, μ the death rate and δ the disaster rate. Hence

$$\lim_{t\to\infty} \mathbf{E}(s^{X(t)}) = \int_0^1 \frac{\delta}{\mu} \exp\left(\frac{\nu}{\mu}(s-1)(1-y)\right) y^{(\delta/\mu)-1} dy$$

where we have set $e^{-\mu x} = y$ in the intergrand. Differentiating with respect to s and setting $s = 1$ gives the stationary mean $\nu/(\delta + \mu)$.

9.19.9 The forward equations are

$$\frac{dp_n(t)}{dt} = \lambda_{n-1}p_{n-1}(t) - \lambda_n p_n(t).$$

From (3) we have $f_{n+1}(t) = \sum_{n+1}^{\infty} p'_k(t) = \lambda_n p_n(t)$, so $M_{n+1}(\theta) = \lambda_n \int_0^\infty p_n(t)e^{\theta t}dt$ as required.

9.19.10 Use (5) and partial fractions.

9.20.3 $\eta = \lim_{t\to\infty} \mathbf{P}(X(t) = 0) = \lim_{t\to\infty} G(0, t)$

9.20.4 $\mathbf{P}(T > t) = 1 - G(0, t)$. If $\lambda < \mu$ then extinction is certain, so

$$\mathbf{E}(T) = \int \mathbf{P}(T > t)dt = \int_0^\infty \frac{(\mu-\lambda)\exp((\lambda-\mu)t)}{\mu - \lambda\exp((\lambda-\mu)t)}dt = \frac{1}{\lambda}\log\frac{\mu}{\mu-\lambda}.$$

On the other hand, if $\lambda > \mu$, then $\mathbf{P}(T < \infty) = \eta = \dfrac{\mu}{\lambda}$, so

$$\mathbf{E}(T|T<\infty) = \int\left(1 - \frac{\mathbf{P}(T \leqslant t)}{\eta}\right)dt = \int_0^\infty 1 - \frac{G(0,t)}{\eta}dt$$
$$= \int_0^\infty \frac{(\lambda-\mu)\exp((\mu-\lambda)t)}{\lambda - \mu\exp((\mu-\lambda)t)}dt = \frac{1}{\mu}\log\frac{\lambda}{\lambda-\mu}.$$

9.20.5 Condition on events in $(0, h)$, (and use the fact that if the first individual splits the replacements act independently), to get

$$\eta(t+h) = 1.\mu h + \eta(t)(1 - \lambda h - \mu h) + (\eta(t))^2\lambda h + o(h).$$

Rewrite the equation as $\dfrac{d\eta}{(1-\eta)(\mu-\lambda\eta)} = dt$ and integrate to get

$\eta(t) = G(0, t)$, as expected. $\mathbf{P}(X(t) = 0|X(s) = 0) = \dfrac{\eta(t)}{\eta(s)}$.

9.20.6 $\mathbf{E}(s^{X(t)}|X(t) > 0) = \dfrac{G(s,t) - G(0,t)}{1 - G(0,t)} \to \dfrac{(\mu-\lambda)s}{\mu - \lambda s}.$

9.20.7
$$p_{i,i+1}(h) = (\nu + \lambda i)h,\ p_{i,i-1}(h) = \mu i h$$
$$p_{ii}(h) = 1 - (\nu + \lambda i)h - \mu i h + o(h).$$

Hence

$$\frac{dp_n}{dt} = \mu(n+1)p_{n+1} - (\nu + (\lambda+\mu)n)p_n + (\nu + \lambda(n-1))p_{n-1}.$$

Setting $\dfrac{dp_n}{dt} = 0$, you can check that the given π_k satisfies the resulting equations.

Problems

1 (α) X_n is, with $p_{ij} = \mathbf{P}(X_1 = j)$. ($\beta$) S_n is, with $p_{ij} = \mathbf{P}(X_1 = j - i)$. ($\gamma$) M_n is, with $p_{ij} = \mathbf{P}(X_1 = j)$; $j > i$, $p_{ii} = \mathbf{P}(X_1 \leqslant i)$. ($\delta$) L_n is, with $p_{ij} = \mathbf{P}(X_1 = j)$, $j < i$, $p_{ii} = \mathbf{P}(X_1 \geqslant i)$. ($\epsilon$) K_n is not a Markov chain.

2 X is persistent; S is transient unless $\mathbf{P}(X_1 = 0) = 1$; M is absorbing if X_1 is bounded, transient otherwise; L is absorbing.

3 Only if the period is 2.

4 (a) Not necessarily if X and Y are periodic.

5 Check that $\mathbf{P}(X_{k+1} = j | X_0, X_1, \ldots, X_k, X_n) = \mathbf{P}(X_{k+1} = j | X_k, X_n)$ by expanding the conditional probabilities.

6 These all work by expanding the appropriate conditional probabilities and rearranging them.

7 It is Markov because coins are independent. Then $p_{i,\,i+1} = p$; $0 \leqslant i \leqslant 8$, $p_{90} = p$, $p_{ii} = q$; $\pi_i = 10^{-1}$.

10 $\rho^n u_n = \sum_1^n \rho^{n-k} u_{n-k} \rho^k f_k$, so v_n satisfies $v_n = \sum_1^n f_k^* v_{n-k}$ with $f_k^* = \rho^k f_k$. Hence it is a renewal sequence provided $\sum \rho^n f_n = 1$. It follows that there is a Markov chain such that $v_n = \mathbf{P}(X_n = s) \to \pi_s$ as $n \to \infty$. Hence $\rho^n u_n \to \pi_s = c$.

11 $Z_n = (X_n, X_{n+1}, \ldots, X_{n+m-1})$ is also a persistent chain.

13 The Markov property is preserved at each first passage time.

14 In the obvious notation we require $Q^2 = \frac{1}{4}(P_X + P_Y)^2 = Q(2)$; multiplying out $(P_X + P_Y)^2$ shows that this requires $(P_X - P_Y)^2 = 0$. Hence W is Markov if $P_X = P_Y$.

17 No. Pick $j \neq s \neq i$, then $\mathbf{P}(X_{n+1} = j | X_n = s, X_{n-1} = i) = p_{ij}(2) \neq \mathbf{P}(X_{n+1} = j | X_n = s)$.

18 The lack-of-memory of the geometric distribution means that X is a Markov chain. If $X_n = i$ then X_{n+1} is the survivors S_n of X_n plus the new arrivals. The probability that a geometric lifetime survives one step is $p = 1 - q$, so S_n is binomial with parameters X_n and p. Hence $p_{ij} = \mathbf{P}(S_n + Y_n = j)$. In equilibrium X_{n+1} and X_n have the same distribution, so

$$\begin{aligned}\mathbf{E}(s^{X_{n+1}}) = G(s) &= \mathbf{E}(s^{S_n + Y_n})\\ &= \mathbf{E}(s^{Y_n})\mathbf{E}(\mathbf{E}(s^{S_n}|X_n)) = \mathbf{E}(s^{Y_n})\mathbf{E}((ps + 1 - p)^{X_n})\\ &= e^{\lambda(s-1)}G(ps + 1 - p) = e^{\lambda/q(s-1)}\end{aligned}$$

after a simple induction. So the stationary distribution is Poisson with parameter λ/q, just *after* the fresh particles. The stationary distribution just *before* the fresh particles is Poisson with parameter $\lambda p/q$.

19 We seek a stationary distribution which must satisfy

$$\pi_j = \pi_0 f_j + \sum_{i=1}^{j+1} \pi_i f_{j-i+1} \text{ for } j \geqslant 0.$$

The sum on the right is nearly a convolution, so we introduce $\pi(s) = \sum \pi_i s^i$, and $G(s) = \sum f_i s^i$ to get

$$\pi(s) = \pi_0 G(s) + \frac{1}{s}(\pi(s) - \pi_0)G(s).$$

Hence

$$\pi(s) = \frac{\pi_0(s-1)G(s)}{s - G(s)},$$

which has $\pi(1) = 1$ if $\pi_0 > 0$ and $G'(1) < 1$.

20 Seek a stationary distribution which satisfies

$$\pi_k = \sum_{j=k-1}^{\infty} \pi_j f_{j-k+1} \quad \text{for } k \geq 1, \text{ with } \pi_0 = \sum_{j=0}^{\infty} \pi_j \sum_{i=j+1}^{\infty} f_i.$$

Let $\sum \theta^r f_r = G(\theta)$. In an optimistic spirit we seek a solution of the form $\pi_j = (1-\theta)\theta^j$, giving

$$\theta^k = \sum_{j=k-1}^{\infty} \theta^j f_{j-k+1} = \sum_{r=0}^{\infty} \theta^{k+r-1} f_r = \theta^{k-1} G(\theta),$$

and

$$1 = \sum_{j=0}^{\infty} \theta^j \sum_{i=j+1}^{\theta} f_i = \frac{1-G(\theta)}{1-\theta}.$$

These both reduce to $G(\theta) = \theta$. Hence π_j exists if $G(\theta) = \theta$ has a root less than 1. By convexity it does if $G'(1) > 1$, as required.

22 Recalling Example 9.3.17 we define the event $A(t)$ as we did there; that is, as the event that the chain follows a path consistent with $T = t$. Then the condition of the problem can be rewritten as

$$\mathbf{P}(T = t, A(t), X_t = j) = \text{zero or } \mathbf{P}(X_t = j, A(t)).$$

Now the proof follows exactly the same route as Example 9.3.17.

23 $\boldsymbol{\pi}$ must satisfy $\pi_j = p\pi_{j-2} + q\pi_{j+1}$ with $\pi_0 = q\pi_0 + q\pi_1$ and $\pi_1 = q\pi_2$. Hence $s\pi(s) = ps^3\pi(s) + q(\pi(s) - \pi_0) + qs\pi_0$. This gives $\pi(s) = \dfrac{q(s-1)\pi_0}{s - q - ps^3}$, now insist that $\pi_j \geq 0$ and $\pi(1) = 1$.

24 (a) $\mathbf{P}(X(0) = 1 | X(-\tau) = i, X(\tau) = j)$

$$= \mathbf{P}(X(-\tau) = i, X(0) = 1, X(\tau) = j)/\mathbf{P}(X(-\tau) = i, X(\tau) = j)$$

$$= \frac{\mathbf{P}(X(-\tau) = i) p_{i1}(\tau) p_{1j}(\tau)}{\mathbf{P}(X(-\tau) = i) p_{ij}(2\tau)} \to \frac{\pi_1 \pi_j}{\pi_j}.$$

(b) $\mathbf{P}(X(s+t) = X(s) = 1) - \mathbf{P}(X(s) = 1)\mathbf{P}(X(s+t) = 1)$ (c) $\to \pi_1(p_{11}(t) - \pi_1)$, as $s \to \infty$.

25 (a) Yes, with parameter $\lambda + \mu$; (b) No.

26 $\Lambda(s)$

28 Let $A(t)$ be any event defined by $X(s)$ for $s \leq t$. Then by the Markov property

$$\mathbf{P}(Y_{n+1} = k | Y_n = j, T_n = t, A(t)) = \mathbf{P}(Y_{n+1} = k | Y_n = j, T_n = t).$$

Hence

$$\mathbf{P}(Y_{n+1} = j | Y_n = i, A(T_n)) = \int_0^{\infty} \mathbf{P}(Y_{n+1} = j | Y_n = i, T_n = t) f_{T_n}(t)\,dt$$

$$= \mathbf{P}(Y_{n+1} = j | Y_n = i), \quad \text{so } Y \text{ is Markov}$$

$$= \int_0^{\infty} p_{ij}(t)\lambda e^{-\lambda t}\,dt = q_{ij}, \quad \text{say}.$$

Then $\sum \pi_i q_{ij} = \int \sum \pi_i p_{ij}(t) \lambda e^{-\lambda t}\,dt = \int \pi_j \lambda e^{-\lambda t}\,dt = \pi_j$, as required.

29 $\mathbf{E}(X(t)) = Ie^{(\lambda-\mu)t}$ $\quad \text{var}(X(t)) = \begin{cases} 2I\lambda t; & \lambda = \mu \\ I\dfrac{\lambda+\mu}{\lambda-\mu} e^{(\lambda-\mu)t}(e^{(\lambda-\mu)t} - 1); & \lambda \neq \mu. \end{cases}$

Further reading

There are many very attractive books on probability. To compile a list as short as this has required regrettably ruthless selection from their number.

Intermediate Probability

If you wish to read further at an intermediate level, then high on your list should be the classic text:

Feller, W. (1968) *An introduction to probability theory and its applications*, Vol. I (3rd edn.) John Wiley, New York.

Other books at this level or a little above include:

Grimmett, G.R. and Stirzaker, D.R. (1992) *Probability and random processes* (2nd edn.) Clarendon Press, Oxford.
(1992) *Probability and random processes; problems and solutions*, Clarendon Press, Oxford.
Ross, S.M. (1989) *Introduction to probability models* (4th edn.) Academic Press, Orlando.

Combinatorics

A classic text on combinatorics for probabilists is:

Whitworth, W.A. (1901) *Choice and Chance* (5th edn) reprinted 1948 by Hafner, New York.

Recent introductions include:

Anderson, I. (1974) *A first course in combinatorial mathematics*, Clarendon Press, Oxford.
Hall, M. (1967) *Combinatorial theory*, Blaisdell, Waltham, Mass.
Slomson, A. (1991) *An introduction to combinatorics*, Chapman and Hall, London.

Advanced probability

To advance in probability requires the student to plunge into measure theory. Excellent texts at this level include:

Billingsley, P. (1985) *Probability and measure* (2nd edn.) John Wiley, New York.
Dudley, R.M. (1989) *Real analysis and probability*, Wadsworth, Belmont, Calif.
Durrett, R. (1991) *Probability: theory and examples*, Wadsworth, Belmont. Calif.
Feller, W. (1971) *An introduction to probability theory and its applications*, Vol II (2nd edn.) John Wiley, New York.
Laha, R.G. and Rohatgi, V.K. (1979) *Probability theory*, John Wiley, New York.
Shiryayev, A.N. (1984) *Probability*, Springer, New York.
Williams, D (1991) *Probability with martingales*, Cambridge University Press.

Markov chains and other random processes

Most of the above books contain much material on Markov chains and other random processes at their own levels. However, mention should be made of the classic text:

Doob, J.L. (1953) *Stochastic processes*, John Wiley, New York.

History

Finally, if you wish to find out more about the origins of probability read:

Hald, A. (1990) *A history of probability and statistics and their applications before 1750*, John Wiley, New York.

Index

Abbreviations used in this index: m.g.f. moment generating function; p.g.f. probability generating function; r.v. random variable; r.w. random walk